TETRAHEDRON ORGANIC CHEMISTRY SERIES
*Series Editors:* J E Baldwin, FRS & R M Williams

VOLUME 20

# Palladium in Heterocyclic Chemistry

## A Guide for the Synthetic Chemist

**Related Pergamon Titles of Interest**

**BOOKS**

*Tetrahedron Organic Chemistry Series:*
CARRUTHERS: Cycloaddition Reactions in Organic Synthesis
CLARIDGE: High-Resolution NMR Techniques in Organic Chemistry
FINET: Ligand Coupling Reactions with Heteroatomic Compounds
GAWLEY & AUBÉ: Principles of Asymmetric Synthesis
HASSNER & STUMER: Organic Syntheses Based on Name Reactions and Unnamed Reactions
McKILLOP: Advanced Problems in Organic Reaction Mechanisms
OBRECHT & VILLALGORDO: Solid-Supported Combinatorial and Parallel Synthesis of Small-Molecular-Weight Compound Libraries
PERLMUTTER: Conjugate Addition Reactions in Organic Synthesis
SESSLER & WEGHORN: Expanded, Contracted & Isomeric Porphyrins
TANG & LEVY: Chemistry of *C*-Glycosides
WONG & WHITESIDES: Enzymes in Synthetic Organic Chemistry

**JOURNALS**

BIOORGANIC & MEDICINAL CHEMISTRY
BIOORGANIC & MEDICINAL CHEMISTRY LETTERS
TETRAHEDRON
TETRAHEDRON: ASYMMETRY
TETRAHEDRON LETTERS

*Full details of all Elsevier Science publications are available on www.elsevier.com or from your nearest Elsevier Science office*

# Palladium in Heterocyclic Chemistry

## A Guide for the Synthetic Chemist

JIE JACK LI

*Pfizer Global Research and Development,*
*Ann Arbor, Michigan, USA*

and

GORDON W. GRIBBLE

*Dartmouth College, Hanover, New Hampshire, USA*

PERGAMON

An imprint of Elsevier Science

Amsterdam - Lausanne - New York - Oxford - Shannon - Singapore - Tokyo

ELSEVIER SCIENCE Ltd
The Boulevard, Langford Lane
Kidlington, Oxford OX5 1GB, UK



First edition 2000
Second impression 2002

Library of Congress Cataloging-in-Publication Data

Li, Jie Jack.
Palladium in heterocyclic chemistry : a guide for the synthetic chemist / Jie Jack Li and Gordon W. Gribble.
p. cm. -- (Tetrahedron organic chemistry series ; v. 20)
Includes bibliographical references and index.
ISBN 0-08-043705-2 -- ISBN 0-08-043704-4
1. Heterocyclic compounds--Synthesis. 2. Organopalladium compounds. 3. Palladium catalysts. I. Gribble, Gordon W. II. Title. III. Series.

QD400.5.S95 L52 2000
547'.590459--dc21

00-060616

ISBN: 0 08 043705 2 (hardbound)
ISBN: 0 08 043704 4 (paperback)

♾ The paper used in this publication meets the requirements of ANSI/NISO Z39.48-1992 (Permanence of Paper).
Printed in The Netherlands.

*To Sherry and Louise*

# *Table of Contents*

# Foreword

After an inordinately long induction period, organopalladium chemistry has finally been embraced by synthetic organic chemists. Currently, it is being utilized across the spectrum of organic synthesis, from applications to complex natural product syntheses to the synthesis of polymers. A substantial portion of organopalladium methodology has been developed in the context of heterocyclic chemistry and applications to heterocyclic syntheses abound.

In this new book, Jack Li and Gordon Gribble have compiled an impressive array of richly referenced examples of the use of palladium in heterocyclic chemistry. The book is organized by class of heterocycle (pyrroles, indoles, pyridines, etc.) and each chapter contains the syntheses of heterocyclic precursors as well as details of uses of palladium to both synthesize and functionalize these heterocyclic systems. This book will appeal to anybody involved in heterocyclic chemistry, and will provide an easy entry into the field for those unfamiliar with the area.

Louis S. Hegedus
Department of Chemistry
Colorado State University
Fort Collins, CO
USA, 2000

# Preface

Palladium chemistry, despite its immaturity, has rapidly become an indispensable tool for synthetic organic chemists. Today, palladium-catalyzed coupling is the method of choice for the synthesis of a wide range of biaryls and heterobiaryls. The number of applications of palladium chemistry to the syntheses of heterocycles has grown exponentially.

Then, is there a need for a monograph dedicated solely to the palladium chemistry in heterocycles? The answer is a resounding "yes!":

1. Palladium chemistry of heterocycles has its "idiosyncrasies" stemming from their different structural properties from the corresponding carbocyclic aryl compounds. Even activated chloroheterocycles are sufficiently reactive to undergo Pd-catalyzed reactions. As a consequence of $\alpha$ and $\gamma$ activation of heteroaryl halides, Pd-catalyzed chemistry may take place regioselectively at the activated positions, a phenomenon rarely seen in carbocyclic aryl halides. In addition, another salient peculiarity in palladium chemistry of heterocycles is the so-called "heteroaryl Heck reaction". For instance, while intermolecular palladium-catalyzed arylations of carbocyclic arenes are rare, palladium-catalyzed arylations of azoles and many other heterocycles readily take place. Therefore, the principal aim of this book is to highlight important palladium-mediated reactions of heterocycles with emphasis on the unique characteristics of individual heterocycles.
2. A myriad of heterocycles are biologically active and therefore of paramount importance to medicinal and agricultural chemists. Many heterocycle-containing natural products (they are highlighted in boxes throughout the text) have elicited great interest from both academic and industrial research groups. Recognizing the similarities between the palladium chemistry of arenes and heteroarenes, a critical survey of the accomplishments in heterocyclic chemistry will keep readers abreast of such a fast-growing field. We also hope this book will spur more interest and inspire ideas in such an extremely useful area.

We have compiled important preparations of heteroaryl halides, boranes and stannanes for each heterocycle. The large body of data regarding palladium-mediated polymerization of heterocycles in material chemistry is not focused here; neither is coordination chemistry involving palladium and heterocycles.

We are much indebted to Susan E. Hagan, Douglas S. Johnson, Michael Palucki, W. Howard Roark, Roderick J. Sorenson, Peter L. Toogood, Sharon Ward, and Kim Werner for proofreading the manuscript. We are grateful to Professor Louis S. Hegedus of Colorado State University who read the entire manuscript and offered many invaluable comments and suggestions. Professor Rick L. Danheiser of Massachusetts Institute of Technology also read part of the manuscript and provided very insightful suggestions. Any remaining errors and omissions are, of course, entirely our own.

Last, but not least, we wish to thank Wendy O. Berryman of Dartmouth College for typing part of the manuscript and Sharon Ward of Elsevier Science for editorial assistance throughout the whole project.

Jack Li and Gordon Gribble, June 2000

## Abbreviations

| | |
|---|---|
| AIBN | 2,2'-azobisisobutyronitrile |
| AMPA | amino-3-hydroxy-5-methyl-4-isoxazolepropionic acid |
| 9-BBN | 9-borabicyclo[3.3.1]nonane |
| BHT | 2,6-di-*tert*-butyl-4-methylphenol |
| BINAP | 2,2'-bis(diphenylphosphino)-1,1'-binaphthyl |
| Boc | *tert*-butyloxycarbonyl |
| *t*-Bu | *tert*-butyl |
| cAMP | adenosine cyclic 3',5'-phosphate |
| Cbz | benzyloxycarbonyl |
| CCK-A | cholecystokinin-A |
| CNS | central nerve system |
| *m*-CPBA | *m*-chloroperoxybenzoic acid |
| CuTC | copper thiophene-2-carboxylate |
| DABCO | 1,4-diazabicyclo[2.2.2]octane |
| dba | dibenzylideneacetone |
| DCE | dichloroethane |
| Δ | solvent heated under reflux |
| DDQ | 2,3-dichloro-5,6-dicyano-1,4-benzoquinone |
| DIBAL | diisobutylaluminum hydride |
| DMA | *N,N*-dimethylacetamide |
| DME | 1,2-dimethoxyethane |
| DMF | dimethylformamide |
| DMI | 1,3-dimethyl-2-imidazolidinone |
| DMPU | *N,N*-dimethylpropyleneurea (1,3-dimethyl-3,4,5,6-tetrahydro-2-(1*H*)-pyrimidinone) |
| DMSO | dimethylsulfoxide |
| DMT | dimethoxytrityl |
| DNA | deoxyribonucleic acid |
| dppb | 1,4-bis(diphenylphosphino)butane |
| dppe | 1,2-bis(diphenylphosphino)ethane |
| dppf | 1,1'-bis(diphenylphosphino)ferrocene |
| dppp | 1,3-bis(diphenylphosphino)propane |
| ECE | endothelin conversion enzyme |
| EDC | 1-(3-dimethylaminoprpyl)3-ethylcarbodiimide hydrochloride |

| | |
|---|---|
| GABA | γ-aminobutyric acid |
| HMG-CoA | hydroxymethylglutaryl coenzyme A |
| HMPA | hexamethylphosphoric triamide |
| HT | hydroxytryptamine (serotonin) |
| LDA | lithium diisopropylamide |
| LHMDS | lithium hexamethyldisilazane |
| LTMP | lithium 2,2,6,6-tetramethylpiperidine |
| NADH | reduced nicotinamide adenine dinucleotide |
| NBH | 1,3-dibromo-5,5-dimethylhydantoin |
| NBS | *N*-bromosuccinimide |
| NCS | *N*-chlorosuccinimide |
| NIS | *N*-iodosuccinimide |
| NK | neurokinin |
| NMP | 1-methyl-2-pyrrolidinone |
| PMB | *para*-methoxybenzyl |
| PPA | polyphosphoric acid |
| RaNi | Raney Nickel |
| Red-Al® | sodium bis(2-methoxyethoxy)aluminum hydride |
| RNA | ribonucleic acid |
| SEM | 2-(trimethylsilyl)ethoxymethyl |
| $S_NAr$ | nucleophilic substitution on an aromatic ring |
| TBAF | tetrabutylammonium fluoride |
| TBDMS | *tert*-butyldimethylsilyl |
| TBS | *tert*-butyldimethylsilyl |
| Tf | trifluoromethanesulfonyl (triflyl) |
| TFA | trifluoroacetic acid |
| TFP | tri-*o*-furylphosphine |
| THF | tetrahydrofuran |
| TIPS | triisopropylsilyl |
| TMEDA | *N,N,N',N'*-tetramethylethylenediamine |
| TMG | tetramethyl guanidine |
| TMP | tetramethylpiperidine |
| TMS | trimethylsilyl |
| Tol-BINAP | 2,2'-bis(di-*p*-tolylphosphino)-1,1'-binaphthyl |
| TTF | tetrathiafulvalene |

# CHAPTER 1

## Introduction

A number of books have been published describing different facets of the fast growing field of palladium chemistry and its applications to organic synthesis [1–5]. Also found in the literature are several review articles and book chapters summarizing the development of palladium chemistry involving heterocycles [6–11]. In this chapter, we will highlight some important name reactions involving *catalytic* palladium (throughout this monograph, all palladium involved is catalytic unless specified otherwise) and their mechanisms in the context of heterocyclic chemistry.

Palladium chemistry involving heterocycles has its unique characteristics stemming from the heterocycles' inherently different structural and electronic properties in comparison to the corresponding carbocyclic aryl compounds. One example illustrating the striking difference in reactivity between a heteroarene and a carbocyclic arene is the "heteroaryl Heck reaction" (*vide infra*, see Section 1.4). We define a "heteroaryl Heck reaction" as an intermolecular or an intramolecular Heck reaction occurring onto a heteroaryl recipient. Intermolecular Heck reactions of carbocyclic arenes as the recipients are rare [12a–d], whereas heterocycles including thiophenes, furans, thiazoles, oxazoles, imidazoles, pyrroles and indoles, *etc.* are excellent substrates. For instance, the heteroaryl Heck reaction of 2-chloro-3,6-diethylpyrazine (**1**) and benzoxazole occurred at the C(2) position of benzoxazole to elaborate pyrazinylbenzoxazole **2** [12e].

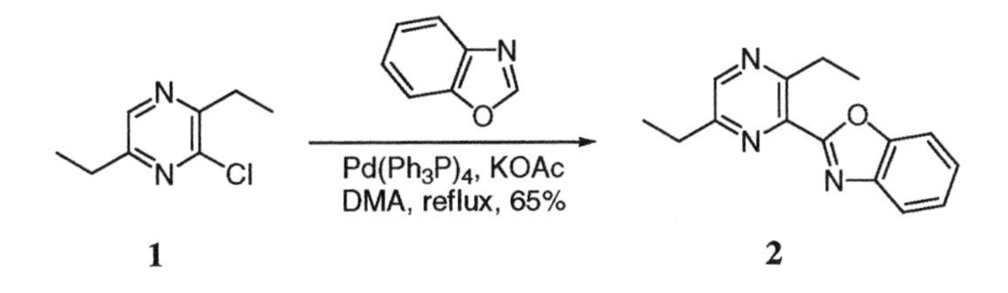


While intermolecular Heck reaction of a carbocyclic arene as the recipient is reluctant to occur, intramolecular Heck reaction of carbocyclic arenes has been well-precedented as illustrated by the following two examples [13].

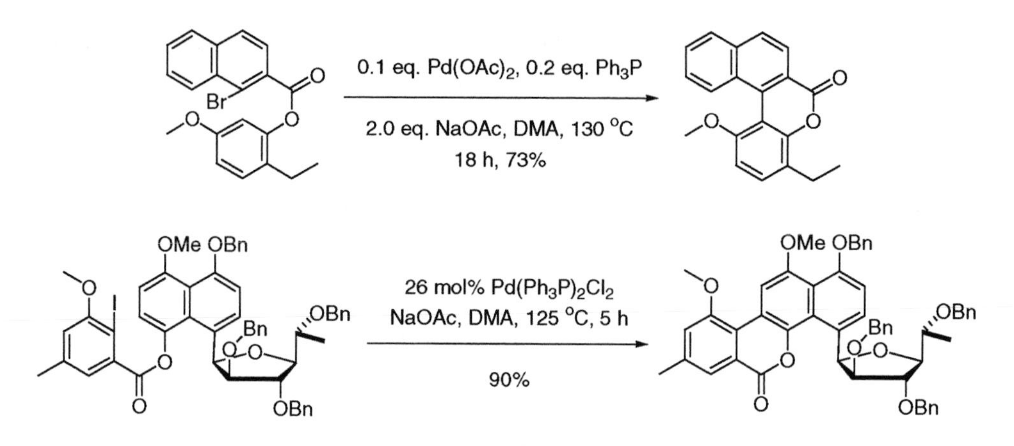


The second salient feature of heterocycles is the marked activation at positions α- and γ- to the heteroatom. For N-containing heterocycles, the presence of the N-atom polarizes the aromatic ring, thereby activating the α and γ positions, making them more prone to nucleophilic attack. The order of $S_NAr$ displacement of heteroaryl halides with $EtO^-$ is [14]:

chloropyrimidine > chloroquinoline > chloropyridine >> chlorobenzene

$7\times10^5$ $3\times10^2$ 1

There is certain similarity in the order of reactivities between $S_NAr$ displacement reactions and oxidative additions in palladium chemistry. Therefore, the ease with which the oxidative addition occurs for these heteroaryl chlorides has a comparable trend. Even α- and γ-chloroheterocycles are sufficiently activated for Pd-catalyzed reactions, whereas chlorobenzene requires sterically hindered, electron-rich phosphine ligands.

The α- and γ-position activation has a remarkable impact on the regiochemical outcome for the Pd-catalyzed reaction of heterocycles. For example, the palladium-catalyzed reaction of 2,5-dibromopyridine takes place regioselectively at the C(2) position, although the halogen-metal exchange reaction takes place predominantly at C(5). Pd-catalyzed reactions of chloropyrimidines take place at C(4) and C(6) more readily than at C(2).

## 1.1 Oxidative coupling/cyclization

The oxidative coupling/cyclization process occurs *via* stoichiometric carbo-palladation using a Pd(II) species, typically $Pd(OAc)_2$. In an early example, submission of diphenylamines **3** to the palladium(II)-promoted oxidative intramolecular cyclization conditions yielded carbazoles **4** [15–

18]. The role of acetic acid in such oxidative cyclization processes is to protonate the acetate ligand, making Pd(II) more electrophilic, thereby promoting the initial electrophilic palladation of the aromatic ring.

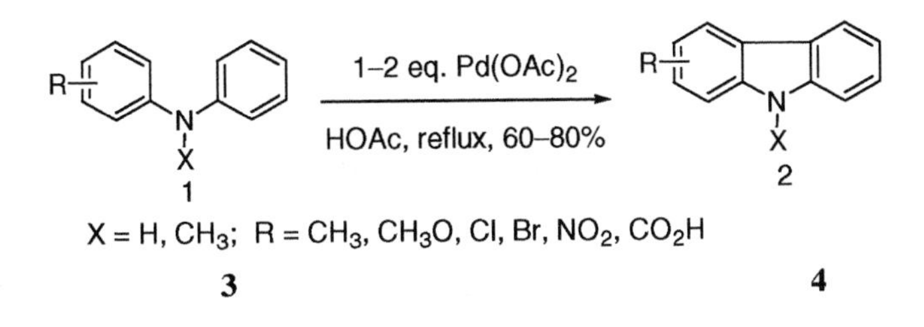


Presumably, the oxidative cyclization of **3** commences with direct palladation at the α position, forming σ-arylpalladium(II) complex **5** *in a fashion analogous to a typical electrophilic aromatic substitution* (this statement will be useful in predicting the regiochemistry of oxidative additions). Subsequently, in a manner akin to an intramolecular Heck reaction, intermediate **5** undergoes an intramolecular insertion onto the other benzene ring, furnishing **6**. β-Hydride elimination of **6** then results in carbazole **4**.

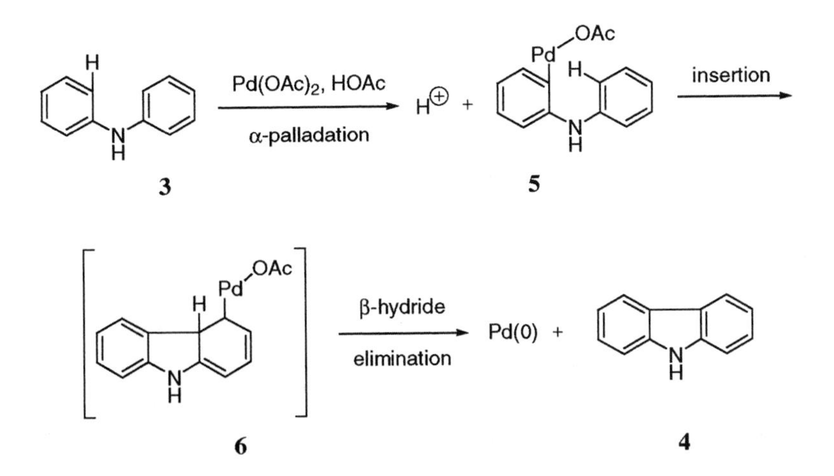


Oxidative addition consumes one equivalent of expensive $Pd(OAc)_2$ in most cases. However, progress has been made towards the catalytic oxidative addition pathway. Knölker's group described one of the first oxidative cyclizations using *catalytic* $Pd(OAc)_2$ in the synthesis of indoles [19]. They reoxidized Pd(0) to Pd(II) with cupric acetate similar to the Wacker reaction, making the reaction catalytic with respect to palladium [20].

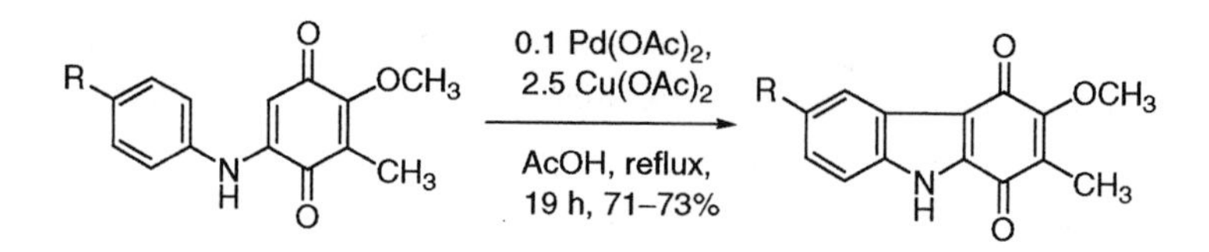


## 1.2 Cross-coupling reactions with organometallic reagents

Palladium-catalyzed cross-coupling reactions of organohalides (or triflates) with organometallic reagents follow a general mechanistic cycle. The 14-electron Pd(0) [the active catalyst PdL is 12 electron when P(*o*-Tol)$_3$ is used as the ligand] catalyst **10** is sometimes reduced from a Pd(II) species **7** by an organometallic reagent $R_1M$ (**8**). The transmetalation product **9** from **7** and **8** undergoes a reductive elimination, giving rise to Pd(0) species **10**, along with the homocoupling product $R_1$—$R_1$. This is one of the reasons why the organometallic coupling partners are often used in a slight excess relative to the electrophilic partners. When the Pd(0) catalyst **10** is

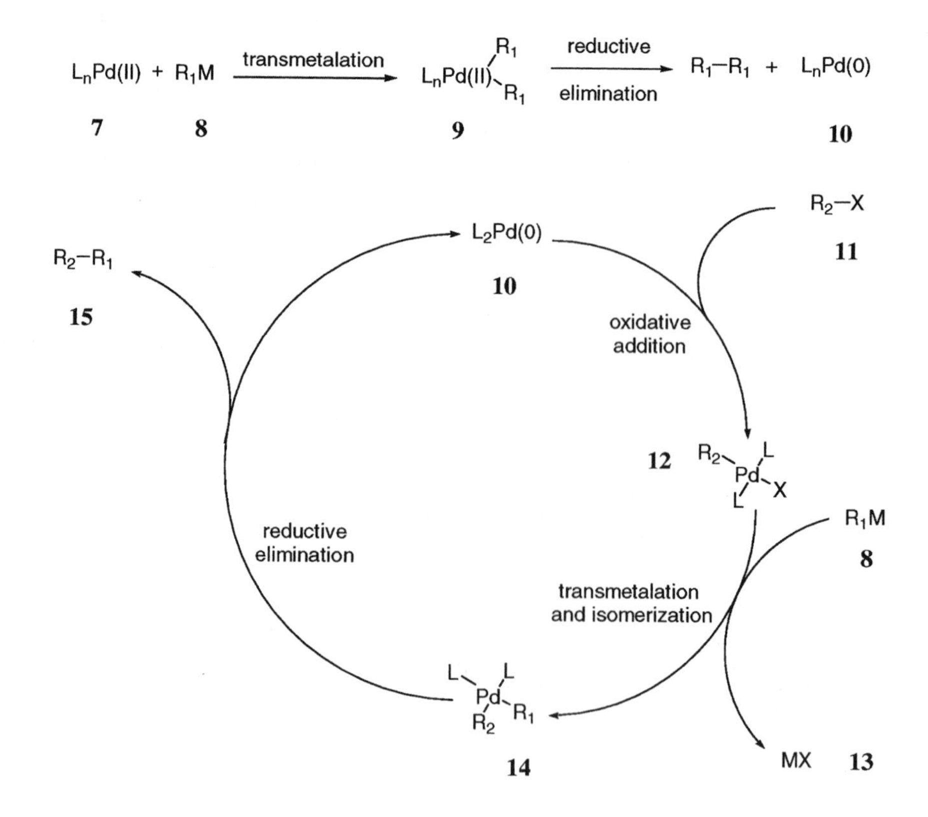

generated, the catalytic cycle goes through a three-step sequence. (a) Electrophile $R_2$—X (**11**) undergoes an *oxidative addition* to Pd(0) to afford a 16-electron Pd(II) intermediate **12**; (b) subsequently, **12** undergoes a *transmetalation* step with the organometallic reagent $R_1M$ (**8**) to produce intermediate **14**. When there is more than one group attached to metal M, the order of transmetalation for different substituents is:

alkynyl > vinyl > aryl > allyl ~ benzyl >> alkyl

The transmetalation step, often rate-limiting, is the step to which attention should be directed if the reaction goes awry; (c) finally, with appropriate *syn* geometry, intermediate **14** undergoes a facile *reductive elimination* step to produce the coupling adduct $R_2$—$R_1$ (**15**), regenerating palladium(0) catalyst **10** to close the catalytic cycle.

### 1.2.1 The Negishi coupling

The Negishi reaction is the palladium-catalyzed cross-coupling between organozinc reagents and organohalides (or triflates) [21–23]. It is compatible with many functional groups including ketones, esters, amines and nitriles. Organozinc reagents are usually generated and used *in situ* by transmetalation of Grignard or organolithium reagents with $ZnCl_2$. In addition, some halides may oxidatively add to Zn(0) to give the corresponding organozinc reagents. In one case, Knochel's group has developed a facile method to prepare organozinc reagents *via* direct oxidative addition of organohalides to Zn(0) dust. This approach is more advantageous than the usual transmetalation using organolithium or Grignard reagents because of better tolerance of functional groups [24, 25]. In another case, Evans *et al.* prepared organozinc reagent **17** from quantitative metalation of iodide **16** using an activated Zn/Cu couple in the presence of both an ester and an amine. Treatment of 2-iodoimidazole **18** with 3 equivalents of **17** in the presence of $PdCl_2(Ph_3P)_2$ then afforded adduct **19** [26].

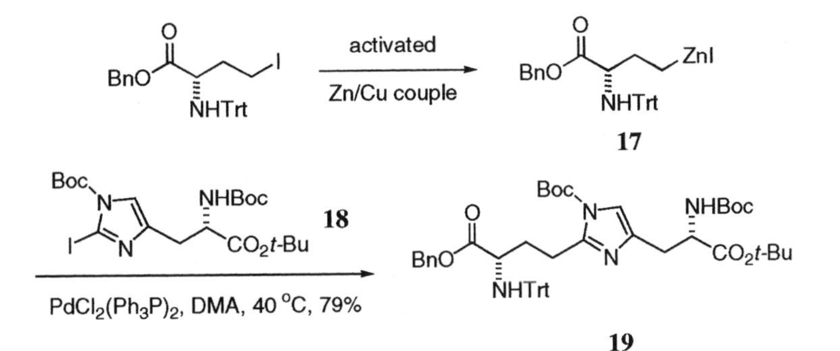

The successful Negishi coupling between *alkylzinc reagent* **17** and 2-iodoimidazole **18** is a good example to correct a misconception that the Negishi reaction does not work for alkyl groups without proximal unsaturation. Although β-hydride elimination of the palladium Pd(II) intermediate is a competing pathway, there are many precedents in which the couplings of alkylzinc reagents were successful because the reductive elimination took place before the β-hydride elimination. Usually, the cross-coupling of alkylmetallic reagents succeeds when reductive elimination > β-hydride elimination > oxidative addition > transmetalation.

In addition, with respect to the Suzuki coupling, various alkylboron reagents have also been successfully coupled with electrophile partners.

### 1.2.2 The Suzuki coupling

The Suzuki reaction represents the palladium-catalyzed cross-coupling reaction between organoboron reagents and organohalides (or triflates) [27, 28]. In comparison to the abundance of heteroarylstannanes, heteroarylboron reagents are not as prevalent. There are three major reasons why one should consider the Suzuki coupling when designing a Pd-catalyzed reaction in heteroaryl synthesis. First, a growing number of heteroarylboron reagents are now known. Second, judiciously designing the coupling partners will enable the use of a heteroaryl halide to couple with a known organoboron reagent for the synthesis of certain molecules. Third, there is no toxicity issue involved in organoboron reagents. Therefore, for a large-scale setting, a Suzuki coupling is a more attractive choice than a Stille coupling.

Since the C—B bond is almost completely covalent, transmetalation of an organoboron reagent to transfer the organic group will not occur without coordination of a negatively charged base or a $F^-$ to the boron atom. As a consequence, the Suzuki reaction normally needs to be carried out in a basic solution, which poses some limitations to base-sensitive substrates.

As each of the ensuing chapters will entail, pyrrolyl-, indolyl-, pyridinyl-, pyrimidinyl-, thienyl- and furylboron reagents have been documented, although those from thiazoles, oxazoles, imidazoles and pyrazines are yet to be seen. The most popular methods for synthesizing heteroarylboron reagents include: a. halogen-metal exchange of a heteroaryl halide followed by treatment with a borate; b. direct metalation of a heteroarene followed by quenching with a borate. Here, we will use the coupling of a pyrimidineboronic acid with a 2-bromothiophene as an example to showcase the utility of Suzuki reaction in heteroaryl synthesis.

Simple 5-pyrimidineboronic acid is not trivial to make because the requisite lithiopyrimidine would add to the azomethine bond. The tendency towards these side reactions is less severe in case of 2,4-di-*tert*-butoxy-5-bromopyrimidine (**20**) and the halogen-metal exchange can be conducted at –75 °C [29]. The nucleophilic attack towards the azomethine bond is retarded due to the steric hindrance. Therefore, the halogen-metal exchange of 2,4-di-*t*-butoxy-5-bromopyrimidine (**20**) followed by quenching with *n*-butylborate, basic hydrolysis and acidification provided 2,4-di-*t*-butoxy-5-pyrimidineboronic acid (**21**). The Suzuki coupling of boronic acid **21** and a variety of heteroaryl halides was conducted to prepare 5-substituted heteroarylpyrimidines such as **22**, which were then hydrolyzed to 5-substituted uracils as potential antiviral agents [30, 31].

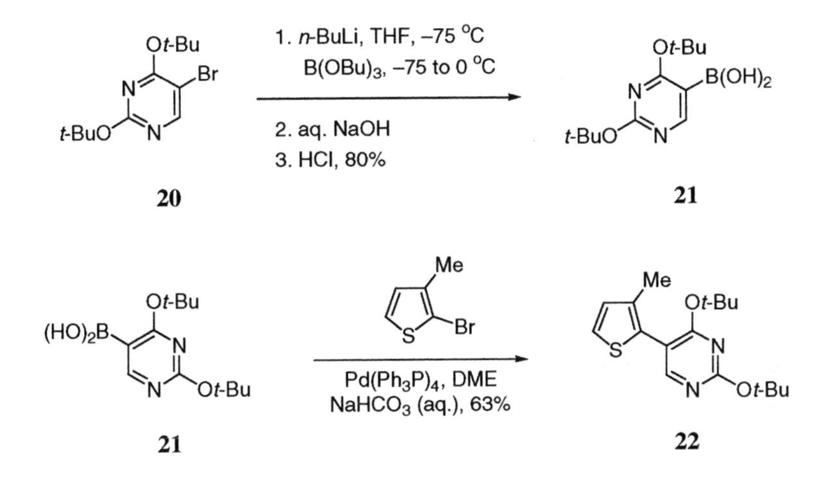


Recognizing the distinct difference in reactivity for each site of *N*-protected 2,4,5-triiodoimidazole **23**, Ohta's group successfully arylated **23** regioselectively [32, 33]. In the total synthesis of nortopsentin C, they coupled **23** with one equivalent of 3-indolylboronic acid **24** to elaborate imidazolylindole **25**. The Suzuki reaction occurred regioselectively at C(2) of the imidazole ring.

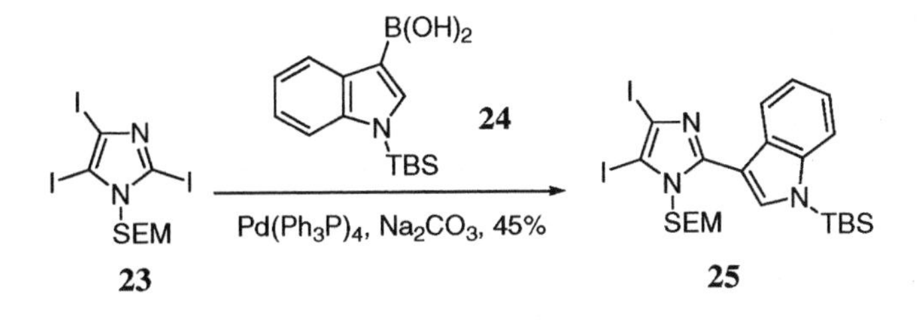


Recently, the groups of Fu and Buchwald have coupled aryl chlorides with arylboronic acids [34, 35]. The methodology may be amenable to large-scale synthesis because organic chlorides are less expensive and more readily available than other organic halides. Under conventional Suzuki conditions, chlorobenzene is virtually inert because of its reluctance to oxidatively add to Pd(0). However, in the presence of sterically hindered, electron-rich phosphine ligands [e.g., $P(t\text{-}Bu)_3$ or tricyclohexylphosphine], enhanced reactivity is acquired presumably because the oxidative addition of an aryl chloride is more facile with a more electron-rich palladium complex. For example, *p*-chloroaniline was coupled with phenylboronic acid in the presence of catalytic $Pd_2(dba)_3$ and $P(t\text{-}Bu)_3$ to afford *p*-aminobiphenyl [34].

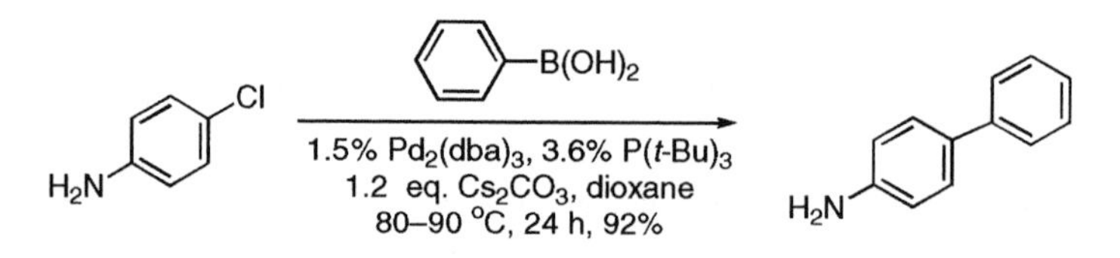


The implications of these discoveries in heterocyclic chemistry reside on the Suzuki coupling of non-activated heteroaryl chlorides. Extensions of these methods to the Suzuki couplings using inexpensive heteroaryl chlorides will be an important development.

As in the Negishi reaction, various *alkylboron reagents* have also been successfully coupled with electrophile partners. Suzuki *et al.* coupled 1-bromo-1-phenylthioethene with 9-[2-(3-cyclohexenyl)ethyl]-9-BBN (**27**), prepared by a simple addition of 9-borabicyclo[3.3.1]nonane (9-BBN) to 4-vinyl-1-hexene (**26**), to furnish 4-(3-cyclohexenyl)-2-phenylthio-1-butene (**28**) in good yield [36].

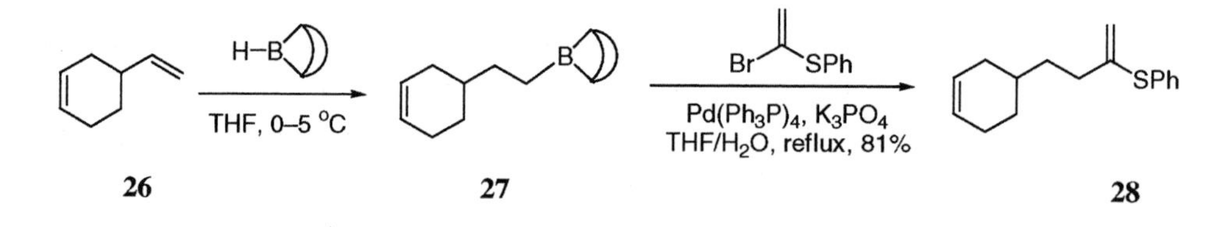

### 1.2.3 The Stille coupling

The Stille reaction is the palladium-catalyzed cross-coupling between an organostannane and an electrophile to form a new C—C single bond [37–41]. It is regarded as the most versatile method in Pd-catalyzed cross-coupling reactions with organometallic reagents for two reasons. First, the organostannanes are readily prepared, purified and stored. Second, the Stille conditions tolerate a wide variety of functional groups. In contrast to the Suzuki, Kumada and the Heck reactions that are run under basic conditions, the Stille reactions can be run under neutral conditions. The pitfall of the Stille reaction is the toxicity of stannanes, making it not suitable for large-scale synthesis.

For the Stille coupling of organic triflates in THF, LiCl often facilitates the reaction. Presumably, the catalytic cycle is locked unless $Cl^-$ replaces the triflate ligand at the palladium intermediate generated from the oxidative addition. However, LiCl is not needed when the Stille reaction is carried out in polar solvents like NMP.

Among many co-catalysts promoting the Stille coupling, Cu(I) is most prevalent. Liebeskind and Farina postulated that Cu(I) performs dual roles [42]. In ethereal solvents, Cu(I) acts as a ligand scavenger to facilitate formation of the coordinatively unsaturated Pd(II) intermediate needed to effect transmetalation. On the other hand, Cu(I) reacts with organostannane to form a more reactive organocopper reagent in highly polar solvents such as NMP in the presence of soft ligands ($AsPh_3$).

Heteroarylstannanes are more prevalent than their heteroarylboron counterparts. As the following chapters will entail, all of the heteroarylstannanes are known with the one exception of quinoxalinylstannanes at the C(2) and C(3) positions. The exception could be simply a result of lack of necessity — because the same adduct can be obtained simply by using quinoxalinyl halides with other stannanes.

In comparison to the Stille couplings of simple aryl precursors, those using heteroaryl substrates have their own characteristics. First of all, many heterocycles have activated positions such as the $\alpha$ and $\gamma$ positions of pyridines. As a consequence, many heteroaryl chlorides are suitable precursors for Stille couplings. Secondly, also due to activation of different positions on a heteroarene, regioselective coupling is often possible. Such a phenomenon is not commonly seen in the Stille couplings of simple carbocyclic aryl substrates. A good example is 2,3-dibromofuran **29**. It is known that nucleophilic substitutions on 2,3-dibromofurans bearing an acceptor substituent at C(5) occur preferentially at C(2). As the trend of oxidative addition is similar to that of nucleophilic substitution, the first Stille coupling of **29** with an allylstannane occurred predominantly at C(2), giving rise to **30** [43]. Under more forcing conditions, the second Stille coupling then took place to afford 2,3,5-trisubstituted furan **31**.

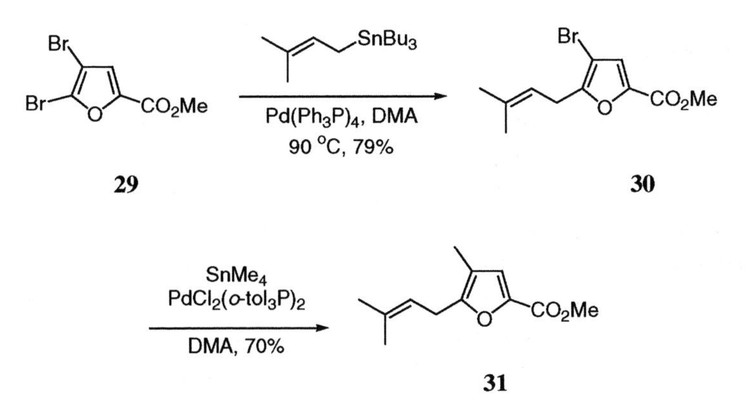

### 1.2.4 The Stille–Kelly coupling

The Stille–Kelly reaction is the Pd-catalyzed intramolecular aryl dihalide cyclization using ditin [44]. In 1990, *en route* to the total synthesis of pradimicinone, Kelly *et al.* synthesized tricyclic **34** by treating dibromide **32** with hexamethylditin in the presence of $Pd(Ph_3P)_4$. The intermediacy of monostannane **33** was confirmed by three experiments. (1), $Pd(Ph_3P)_4$ alone does not promote the cyclization of **32** to **34**; (2), under the agency of $Pd(Ph_3P)_4$ catalysis, independently prepared **33** was converted to **34** in the absence of $Me_3Sn$—$SnMe_3$; (3), workup of the $Me_3Sn$—$SnMe_3$-mediated cyclization in the middle of the reaction reveals the presence of the monostannane **33**. When the dibromide was replaced with diiodide or ditriflate, the cyclization worked as well. The Stille–Kelly reaction offers a means for predetermining the regiochemical outcome of the cyclization of unsymmetrical stilbenes, such control is not always available in photosynthesis or oxidative cyclization.

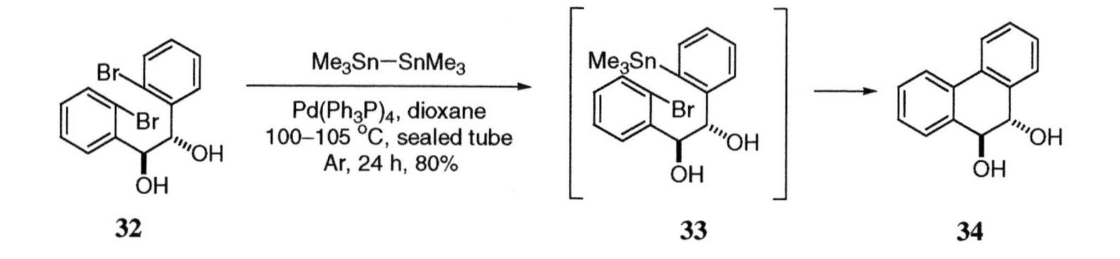


In Grigg's approach to hippadine (**37**), he established the connection between the two phenyl rings *via* the Stille–Kelly reaction [45]. When diiodide **35** was submitted to the Pd(0)/ditin catalyst system, the intramolecular cyclization was realized to establish the C—C bond in lactam **36**. Oxidation of the indoline moiety in **36** using 2,3-dichloro-5,6-dicyano-1,4-benzoquinone (DDQ) then delivered hippadine (**37**). Analogously, the intramolecular Stille coupling of dibromide **38** led directly to hippadine (**37**) [46].

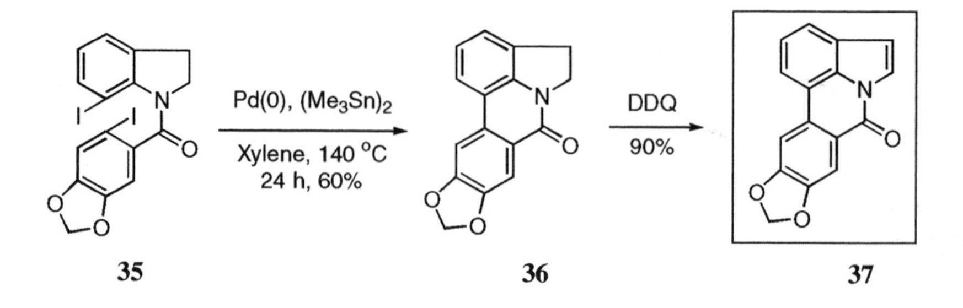

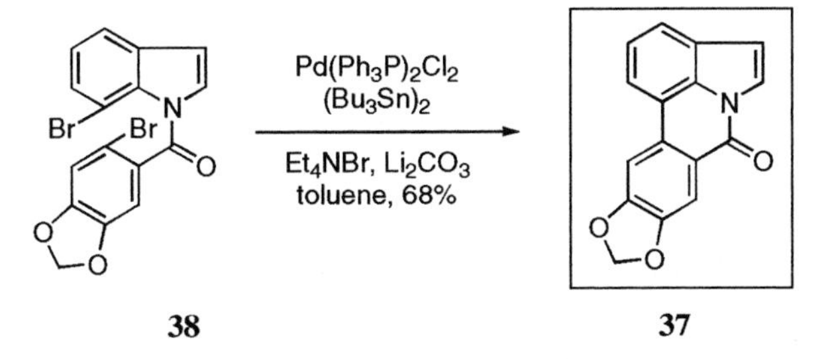


Applications of the aforementioned methodology are also found in the total synthesis of plagiochin D to link a 16-membered biaryl system [47], as well as to the intramolecular cyclization of di-benzyl halides [45, 48]. Additional examples include dithienothiophene (**40**) from dithienyl bromide **39** [49] and carbazole **42** from diarylbromide **41** [50].

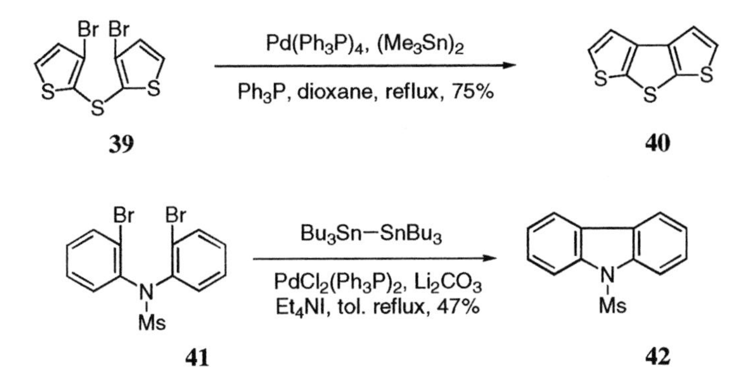


### 1.2.5 The Kumada coupling

The Kumada coupling represents the palladium-catalyzed cross-coupling of a Grignard reagent with an electrophile such as an alkenyl-, aryl- and heteroaryl halide or triflate [51]. Advantage of this reaction is that numerous Grignard reagents are commercially available. Those that are not commercially available may be readily prepared from the corresponding halides. Another advantage is that the reaction can often be run at room temperature or lower. Drawback of this method is the intolerance of many functional groups by the Grignard reagents. In the synthesis of thienylbenzoic acid, interestingly, the carboxylic acid moiety survived under the coupling conditions [52].

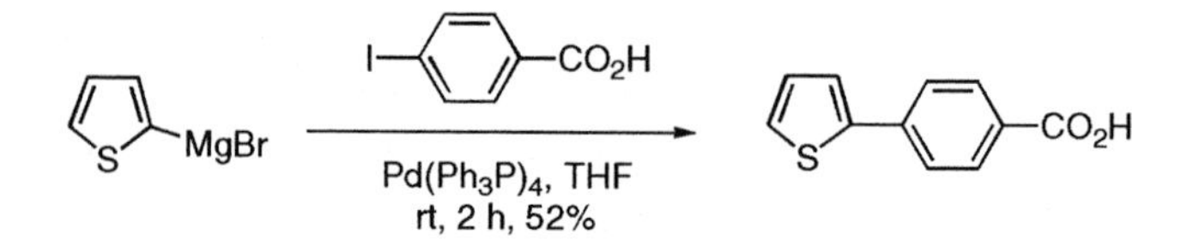


Similar to other Pd-catalyzed cross-couplings of heteroaryl halides, regioselective Kumada coupling may be achieved as well. For example, 2,5-dibromopyridine was mono-heteroarylated regioselectively at C(2) with one equivalent of indolyl Grignard affording indolylpyridine **43**. Subsequent Kumada coupling of **43** with the second Grignard, thienylmagnesium bromide, under more rigorous conditions gave heterotriaryl **44** [53].

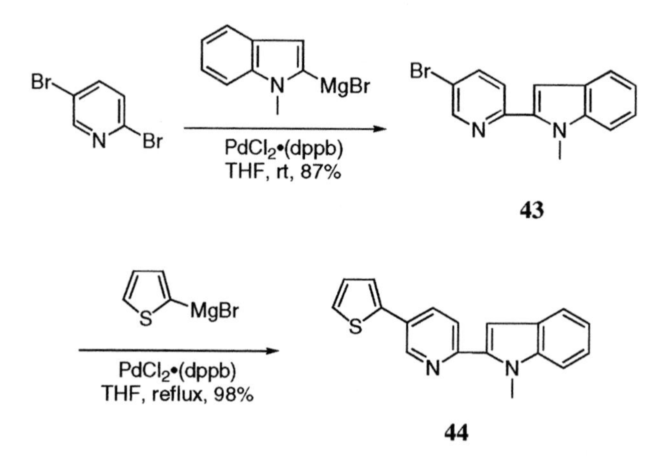


## 1.2.6 The Hiyama coupling

In comparison to the transmetalation of organometallic reagents including RMg, RZn and RSn, transmetalation of an organosilicon reagent does not occur under normal palladium-catalyzed cross-coupling conditions because the C—Si bond is much less polarized. However, a C—Si bond can be activated by a nucleophile such as $F^-$ or $HO^-$ through formation of a pentacoordinated silicate, which weakens the C—Si bond by enhancing the polarization. As a result, the transmetalation becomes more facile and the cross-coupling proceeds readily. One of the advantages of the Hiyama coupling is that organosilicon reagents are innocuous [54, 55]. Another advantage is better tolerance of functional groups in comparison to other strong nucleophilic organometallic reagents. The combination of these two characteristics makes the Hiyama coupling an attractive alternative to other Pd-catalyzed cross-couplings. Therefore, this reaction has much promise for future exploration and exploitation.

The Hiyama couplings of heterocycles are still being developed to their full potential. There is no report yet on Pd-catalyzed cross-coupling using a heteroarylsilicon reagent. Nevertheless, several heteroaryl halides have been cross-coupled with arylsilicon reagents. For instance, in the presence of catalytic $\pi^3$-allylpalladium chloride dimer and two equivalents of KF, the cross-coupling of ethyl(2-thienyl)difluorosilane (**45**) and methyl 3-iodo-2-thiophenecarboxylate led to bis-thiophene **46** under relatively forcing conditions [56].

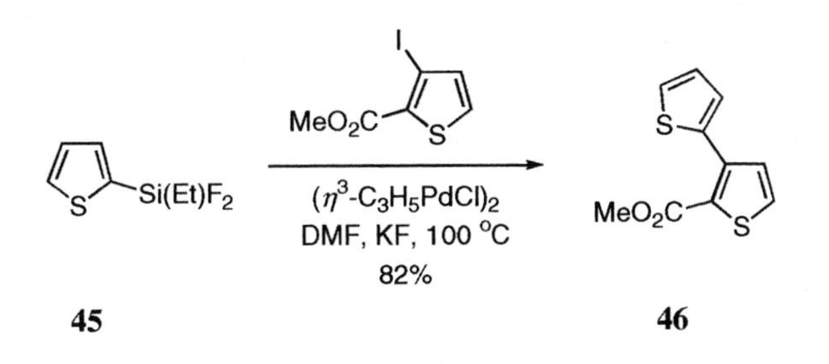


The Hiyama coupling of aryltrimethoxysilane **47** and 3-bromopyridine assembled arylpyridine **48** with the aid of TBAF [57]. In contrast to chlorosilanes, which are susceptible to hydrolysis, aryltrialkoxysilanes are not.

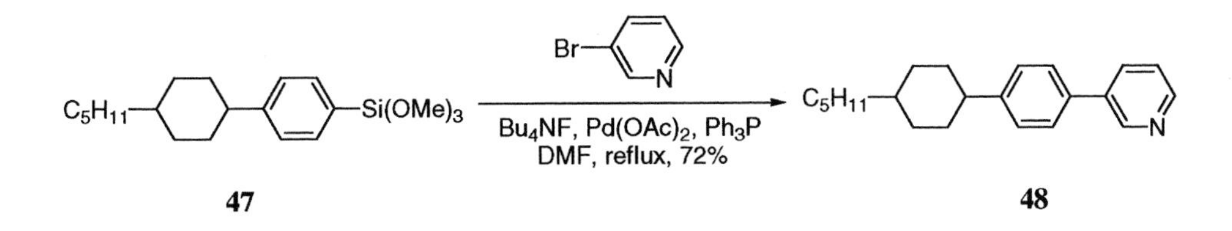


## 1.3 The Sonogashira reaction

The Stephens–Castro reaction is the cross-coupling reaction of copper(I) arylacetylenes with iodoalkenes [58]. Its scope is sometimes limited by the vigorous reaction conditions and by the difficulty in preparing cuprous acetylides. The Sonogashira reaction [59] is the palladium-catalyzed version of the Stephens–Castro reaction. By adding catalytic bis(triphenylphosphine)-palladium dichloride and CuI as the co-catalyst, Sonogashira *et al.* successfully cross-coupled terminal alkynes with aryl- and vinyl halides in the presence of an aliphatic amine under mild conditions. The formation of diphenylacetylene exemplifies the original Sonogashira reaction conditions:

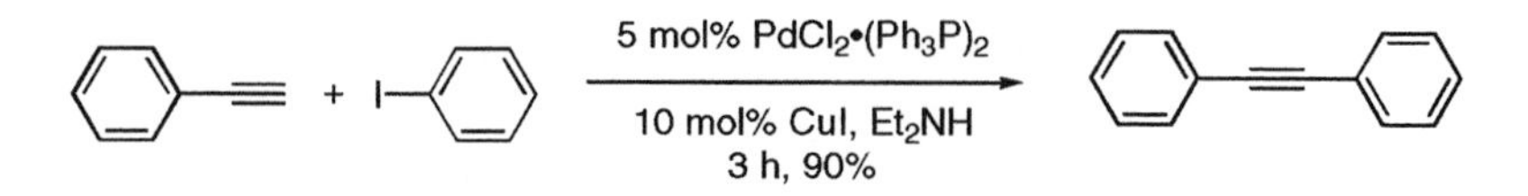


It is speculated that an alkynylcopper species, which undergoes the transmetalation process more readily, is generated during the reaction with the aid of an amine. The aliphatic amine also serves as a reducing agent to generate Pd(0). For recent reviews on the Sonogashira reaction, see references [60] and [61].

The Sonogashira reaction has enjoyed tremendous success in the synthesis of almost all types of heteroarylacetylenes because of the extremely mild reaction conditions and great tolerance of nearly all types of functional groups. A representative reaction with conditions akin to Sonogashira's original conditions is shown below [62]:

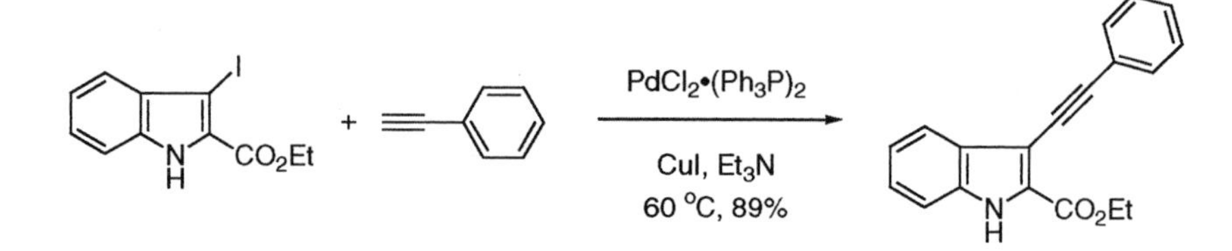


Quite different from the original Sonogashira conditions, the following cross-coupling was achieved under the Jeffery's "ligand-free" conditions in the absence of cuprous iodide [63] [64].

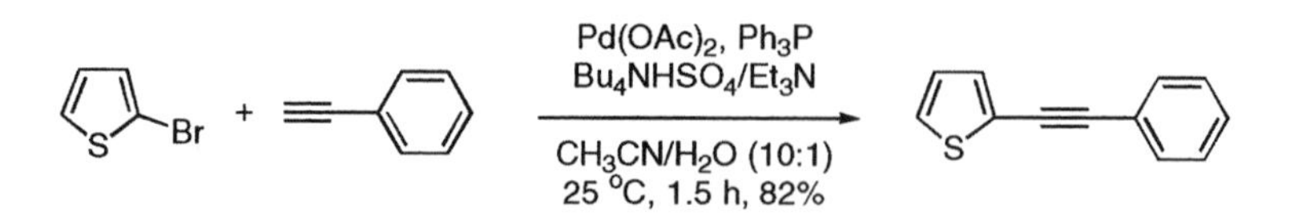


## 1.4 The Heck, intramolecular Heck and heteroaryl Heck reactions

The Heck reaction, first disclosed by the Mori and Heck groups in the early 1970s [65, 66], is the Pd-catalyzed coupling reaction of organohalides (or triflates) with olefins. Nowadays, it has become an indispensable tool for organic chemists. Inevitably, many applications to heterocyclic chemistry have been pursued and successfully executed. In one case, Ohta *et al.* reacted 2-chloro-3,6-dimethylpyrazine (**49**) with styrene to furnish (*E*)-2,5-dimethyl-3-styrylpyrazine (**50**) [67]. Here, only the *E* isomer was observed. The outcome will become apparent during the ensuing discussions on the mechanism.

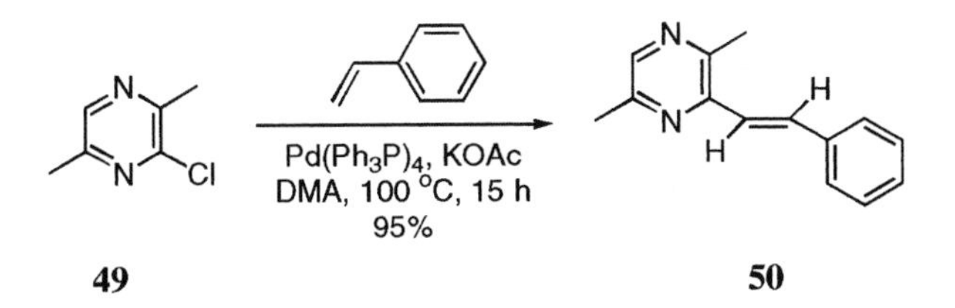


Presumably, 2-chloropyrazine **49** oxidatively adds to the coordinatively unsaturated Pd(0) at the first step, giving rise to σ-pyrazinylpalladium(II) complex **51**. All the ligands here are omitted for clarity. Subsequently, complex **51** undergoes a *cis*-insertion to styrene to afford another Pd(II) complex **52**. Steric effects govern the regiochemistry in this example although electronic effects can be dominant in some other cases. Therefore, the addition occurs at the sterically less hindered, unsubstituted side of the styrene. Since the β-hydride elimination only takes place in a *syn*-orientation, **52** must rotate to conformational isomer **53** before β-hydride elimination occurs. Because the last two operations are thermodynamically-controlled and reversible, the thermodynamically more stable olefin **50** is generated predominantly with concomitant release of hydridopalladium(II) chloride (**54**). With the aid of a base, reductive elimination of HCl then takes place, regenerating Pd(0) to close the catalytic cycle.

While the transmetalation step is often the rate-determining step for Pd-catalyzed reactions with organometallics, the oxidative addition step is often the rate-determining step in the Heck reactions, although olefin insertion can be rate-limiting in some cases — this is why the Heck reactions of tri- and tetra-substituted olefins sometimes proceed slower than those of di-substituted and terminal olefins.

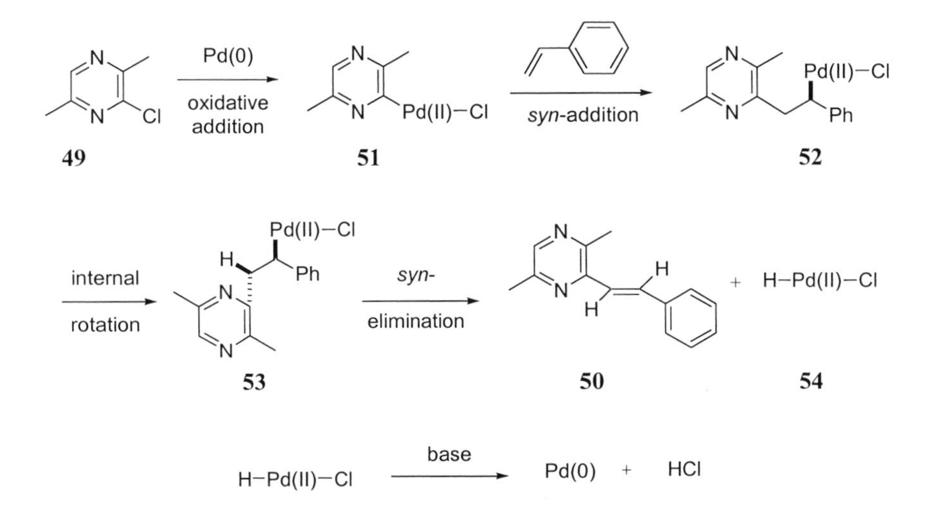

The intramolecular version of the Heck reaction has been extremely fruitful, enabling elegant synthesis of many complex molecules [68, 69]. The Mori-Ban indole synthesis (Section 1.10, *vide infra*) is a good example of such method. In addition, Rawal *et al.* carried out an intramolecular Heck cyclization of pentacyclic lactam **55** with a pendant vinyl iodide moiety [70]. Employing Jeffery's "ligand-free" conditions, **55** was converted to a hexacyclic *strychnan* alkaloid precursor **56** with complete stereochemical control.

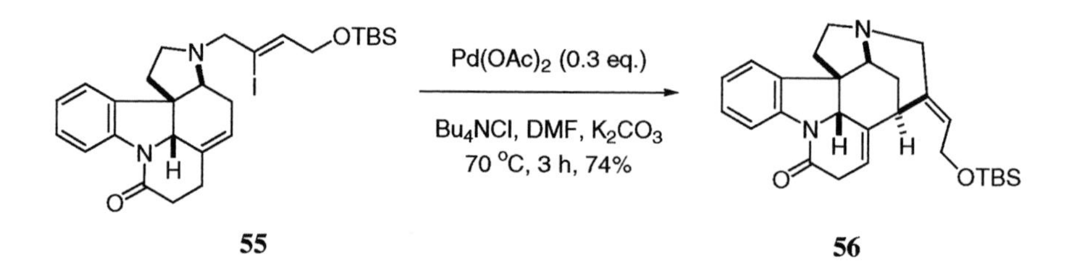


Overman's group [71, 72] enlisted an intramolecular Heck reaction to form a quaternary center in their efforts toward (±)-gelsemine. When the cyclization precursor **70** was submitted to the "ligandless" conditions [$Pd_2(dba)_3$, $Et_3N$] in the weakly coordinating solvent toluene, the quaternary center was formed as a 9:1 ratio of diastereomers (**72**:**71** = 89:11). Addition of a silver salt in polar solvent THF completely reversed the sense of asymmetric induction in the cyclization reaction (**72**:**71** = 3:97).

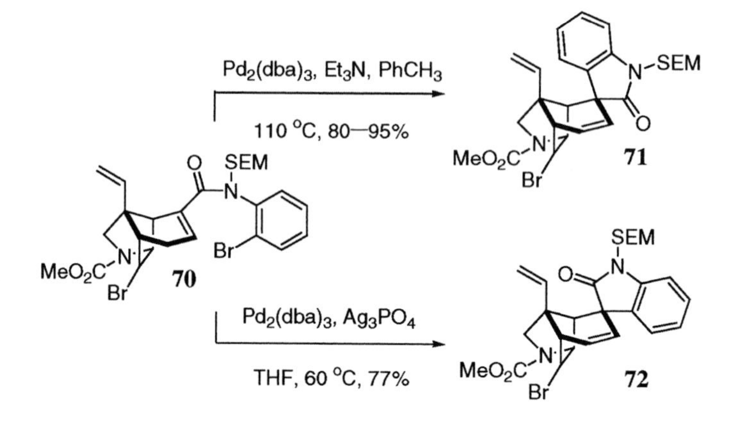


One active field of research involving the Heck reaction is *asymmetric Heck reactions* (AHR). The objective is to achieve enantiomerically-enriched Heck products from racemic substrates using a *catalytic* amount of chiral ligands, making the process more practical and economical. Although intermolecular Heck reactions that occurred onto carbocyclic arenes are rare, they readily take place onto many heterocycles including thiophenes, furans, thiazoles, oxazoles, imidazoles, pyrroles and indoles, *etc*. To distinguish this unique characteristic of heterocycles, we define a "heteroaryl Heck reaction" as an intermolecular or an intramolecular Heck reaction that occurs onto a heteroaryl recipient. Such heteroaryl Heck reactions may be exemplified by the Pd-catalyzed reaction between iodobenzene and thiazole to assemble 5-phenylthiazole [73]. The addition occurs regioselectively at the electron-rich position C(5) of the thiazole ring.

$Pd(Ph_3P)_4$, $Ph_3P$, CuI

$Cs_2CO_3$, DMF, 140 °C, 11 h, 68%

**73**

In one possible mechanism, oxidative addition of iodobenzene to Pd(0) gives Pd(II) intermediate **74**, which subsequently inserts into thiazole regioselectively at the C(5) position to form the σ-adduct of arylpalladium(II) **75**. The order of reactivity is similar to the electrophilic substitution, which is known to be C(5) > C(4) > C(2) [74]. Treatment of the insertion adduct **75** with a base regains the aromaticity after deprotonation, giving rise to **73** along with Pd(0) for the next catalytic cycle.

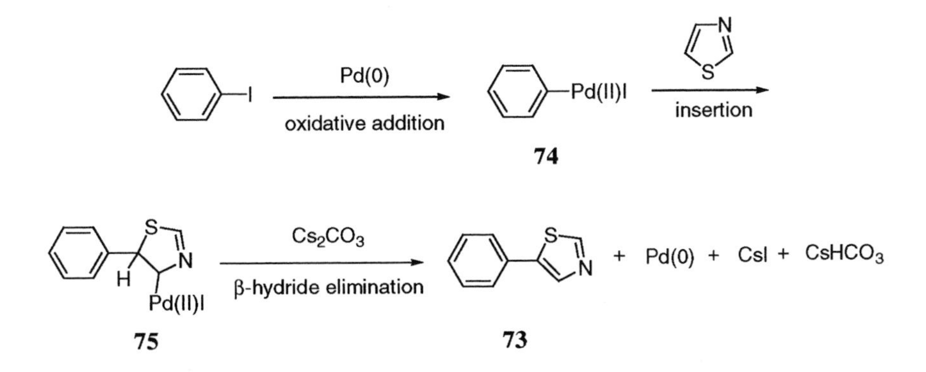


An intramolecular heteroaryl Heck was the pivotal step in the synthesis of 5-butyl-1-methyl-1*H*-imidazo[4,5-*c*]quinolin-4(5*H*)-one (**76**), a potent antiasthmatic agent [75]. The optimum yield was obtained under Jeffery's "ligand-free" conditions,

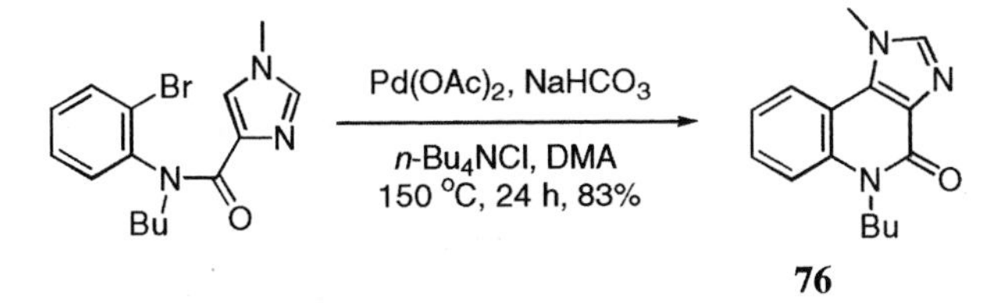

Normally, the oxidative addition of an aryl chloride to Pd(0) is reluctant to take place. But such a process is greatly accelerated in the presence of sterically hindered, electron-rich phosphine ligands [e.g., $P(t\text{-}Bu)_3$ or tricyclohexylphosphine]. In late 1990s, Reetz [76] and Fu [77] successfully conducted intermolecular Heck reactions using arylchlorides as substrates, as exemplified by the conversion of *p*-chloroanisole to adduct **77** [77]. The applications of this discovery will surely be reflected on future Heck reactions of non-activated heteroaryl chlorides.

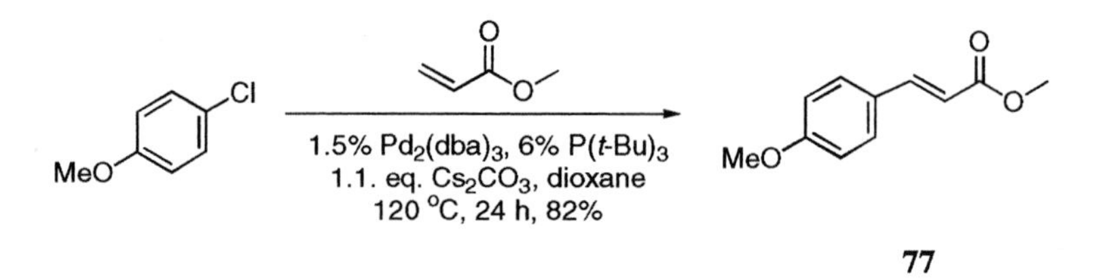


## 1.5 The carbonylation reactions

Pd-catalyzed carbonylation of heteroaryl halides provides a quick entry to heteroaryl carbonyl compounds such as heteroaryl aldehydes, carboxylic acids, ketones, esters, amides, α-keto esters and α-keto amides. In addition, Pd-catalyzed alkoxycarbonylation and aminocarbonylation are compatible with many functional groups, and therefore, are more advantageous than conventional methods for preparing esters and amides [78].

Alkoxycarbonylation of 2,3-dichloro-5-(methoxymethyl)pyridine (**78**) took place regioselectively at C(2) to give ester **79** [79]. Aminocarbonylation of 2,5-dibromo-3-methylpyridine also proceeded preferentially at C(2) to give amide **80** despite the steric hindrance of the 3-methyl group [80].

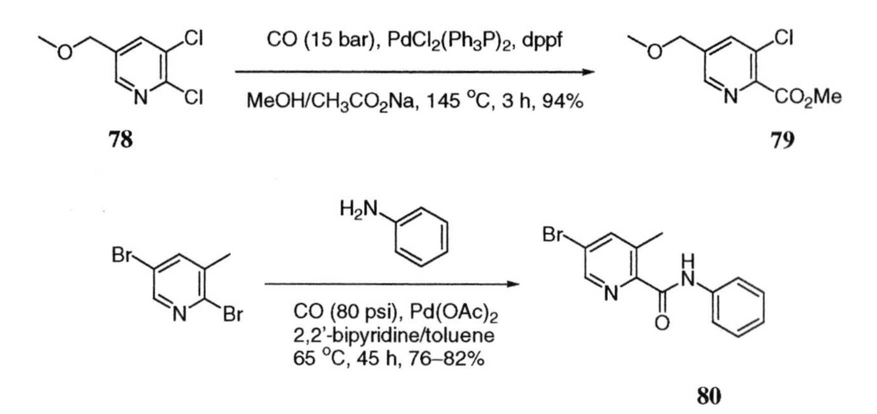

## 1.6 The Pd-catalyzed C—P bond formation

The C—P bond formation of $sp^3$ hybridized C—P bonds is readily achievable using the Michaelis-Arbuzov reaction. Such a method is not applicable to form heteroaryl C$sp^2$—P bonds, but Pd-catalyzed reactions provide suitable approaches to such compounds.

The first report of Pd-catalyzed C—P bond formation was by Hirao *et al.* [81–83]. They synthesized diethyl pyridylphosphonate *via* phosphonation of 3-bromopyridine with diethyl phosphite and a catalytic amount of $Pd(Ph_3P)_4$ [82, 83]. The mechanism that Hirao proposed involves oxidative addition of 3-bromopyridine to Pd(0), giving rise to intermediate **81**. The ligand exchange of dialkyl phosphite on the pyridylpalladium complex **81** leads to **82** which undergoes a reductive elimination to afford diethyl pyridylphosphonate along with Pd(II) hydride which reacts with triethylamine to regenerate Pd(0) species with the deposition of $Et_3N$•HBr. Thus obtained Pd(0) species is now available for another catalytic cycle.

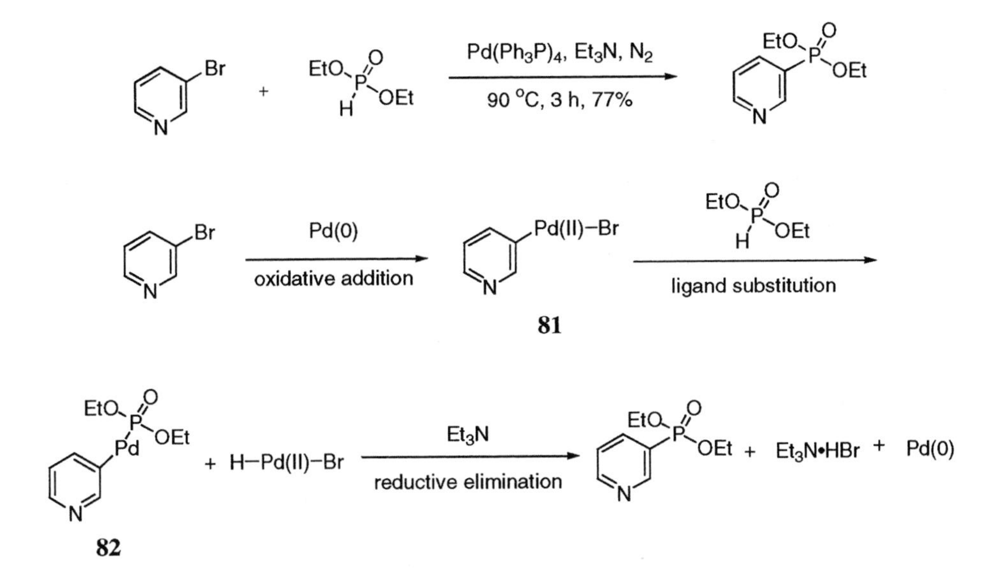


Xu *et al.* have further expanded the scope of such Pd-catalyzed C$sp^2$—P bond formation reactions [84, 85]. In one instance, they coupled 2-bromothiophene with *n*-butyl phenylphosphite to form *n*-butyl arylphosphinate **83**. They also prepared alkyl arylphenylphosphine oxides, functionalized alkyl arylphenylphosphinates, alkenyl arylphenylphosphinates, alkenylbenzyl-phosphophine oxide, as well as chiral, nonracemic isopropyl arylmethylphosphinates in the same fashion [86]. Intramolecular Pd-catalyzed C$sp^2$—P bond formation has also been reported [86].

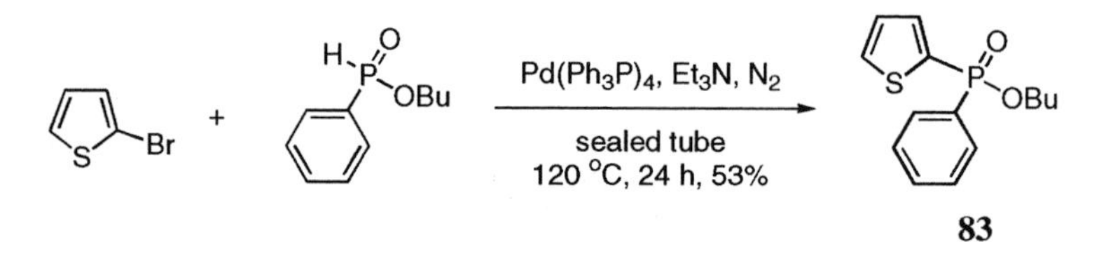


The mechanism for the Pd-catalyzed C$sp^2$—P bond formation proposed by Xu *et al.* is virtually the same as Hirao's with a slight variation. Oxidative addition of 2-bromothiophene to Pd(0) results in Pd(II) intermediate **84**, which then undergoes a ligand exchange to give intermediate **85** with the aid of triethylamine. Triethylamine here serves as a base to neutralize HBr so that the reaction is driven forward. Finally, reductive elimination of **85** furnishes unsymmetrical alkyl arylphosphinate **83**, regenerating Pd(0).

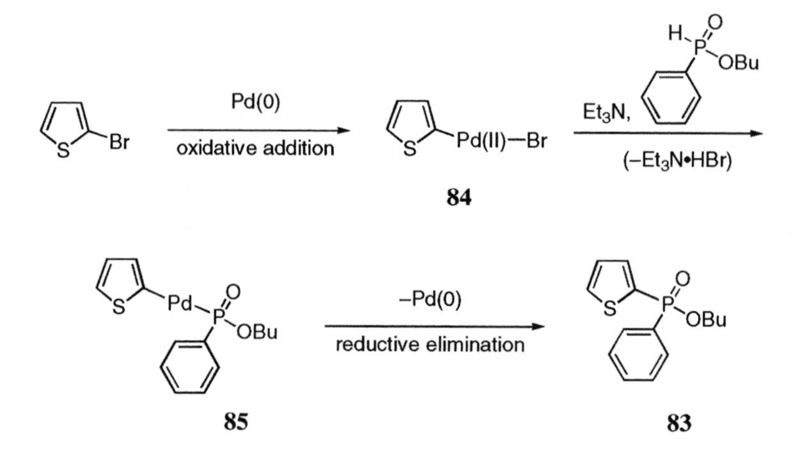


In addition, aryl triflates have proven to be viable substrates for the Pd-catalyzed C$sp^2$—P bond formation reactions [87–90]. Intriguingly, phosphorylation can be achieved from the Pd-catalyzed coupling of alkenyl triflate with not only dialkylphosphites, but also with *hypophosphorous acid* [88]. Thus, phosphinic acid **87** was obtained when triflate **86** was treated with hypophosphorus acid in the presence of $Pd(Ph_3P)_4$. Due to the abundance of alkenyl triflates and milder reaction conditions, alkenyl triflates have certain advantages over the corresponding alkenyl halides as substrates for Pd-catalyzed phosphorylations to make alkenyl phosphonates or phosphinates.

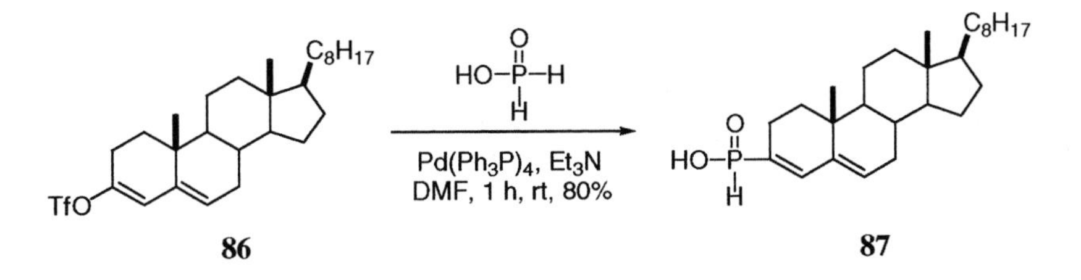


## 1.7 The Buchwald–Hartwig C—N bond and C—O bond formation reactions

### 1.7.1 Pd-catalyzed C—N bond formation

The first Pd-catalyzed C—N bond formation was described by the Migita group in 1983. They cross-coupled bromobenzene with *N,N*-diethylaminotributyltin to prepare *N,N*-diethylaminobenzene [91]. The Migita approach, while employing a catalytic amount of palladium, suffers from the use of toxic and unstable aminostannanes. In 1984, a Pd-mediated C—N bond formation was carried out *via* the intramolecular amination of an aryl halide in Boger's synthesis of lavendamycin [92]. Not only is Boger's method substrate-dependent, but also requires using 1.5 equivalents of expensive $Pd(Ph_3P)_4$, possibly because of a competitive oxidative-addition of liberated HBr to the palladium reagent.

Br + $Et_2N{-}SnBu_3$ —[ $PdCl_2(o\text{-}tol_3P)_2$ ; tol., 100 °C, 3 h, 81% ]→ $NEt_2$ + $Br{-}SnBu_3$

In 1995, Buchwald and Hartwig independently discovered the direct Pd-catalyzed C—N bond formation of aryl halides with amines in the presence of stoichiometric amount of base [93, 94]. This field is becoming rapidly mature and many reviews covering the scope and limitations of this amination have been published since 1995 [95–102]. In the context of heteroaryl synthesis, one example is given to showcase the utility and mechanism of this reaction. Applications to individual heterocycles may be found in their respective chapters.

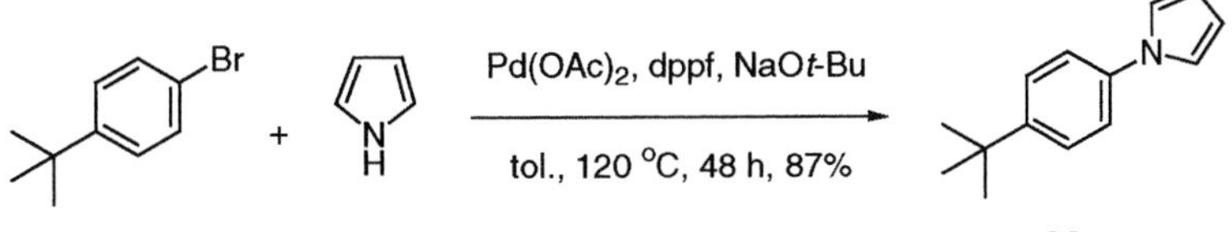

The Pd-catalyzed amination of *p-tert*-butylphenyl bromide with pyrrole in the presence of $Pd(OAc)_2$, dppf and one equivalent of NaO*t*-Bu led to the *N*-arylation product **88**. A simplified version of the mechanism commences with the oxidative addition of *p-tert*-butylphenyl bromide to Pd(0), giving rise to the palladium complex **89**. Ligand exchange with pyrrole followed by deprotonation by the base (NaO*t*-Bu) results in amido complex **90**. Reductive elimination of **90** then gives the amination product **88** with concomitant regeneration of Pd(0) catalyst. If the amine had a β-hydride in amido complex **90**, a β-hydride elimination would be a competing pathway, although reductive elimination is faster than β-hydride elimination in most cases.

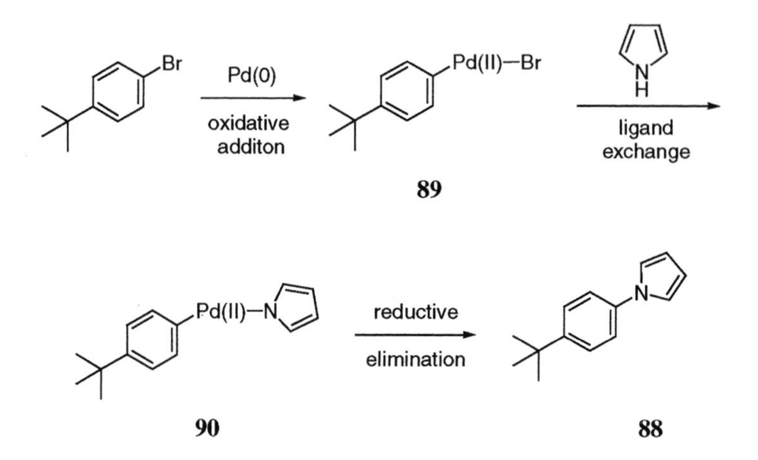


The electrophiles in such reactions can be either aryl halides or triflates, possessing electron-rich, neutral or electron-poor ring systems, whereas amines can range from aliphatic to aromatic and primary to tertiary amines. The Pd-catalyzed C—N bond formation works both inter- and intramolecularly.

### 1.7.2 Pd-catalyzed C—O bond formation

In comparison to the N- and S-counterparts, alkoxides possess lower nucleophilicity. Therefore, the reductive elimination process to form the C—O bond is much slower than those to form C—N and C—S bonds [103]. Palucki, Wolfe and Buchwald developed the first intramolecular Pd-catalyzed synthesis of cyclic aryl ethers from *o*-haloaryl-substituted alcohols [104]. For example, 3-(2-bromophenyl)-2-methyl-2-butanol (**91**) was converted to 2,2-dimethylchroman (**92**) under the agency of catalytic $Pd(OAc)_2$ in the presence (*S*)-(–)-2,2'-bis(di-*p*-tolylphosphino)-1,1'-binaphthyl (Tol-BINAP) as the ligand and $K_2CO_3$ as the base. The method worked well for the tertiary alcohols, moderately well for cyclic secondary alcohols, but not for acyclic secondary alcohols.

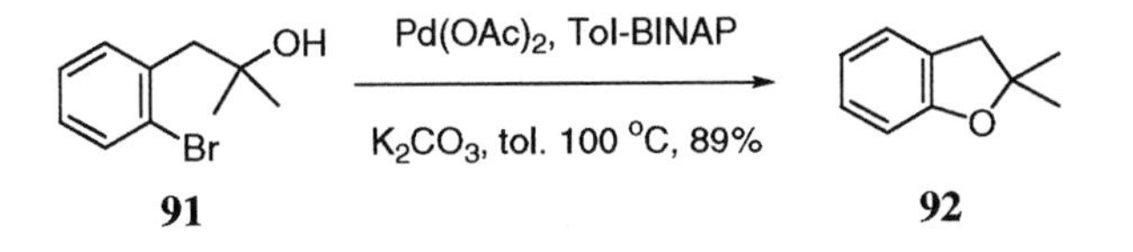


The Pd-catalyzed intermolecular C—O bond formation has also been achieved [105–108]. Novel electron-rich bulky phosphine ligands utilized by Buchwald *et al.* greatly facilitated the Pd-catalyzed diaryl ether formation [109]. When 2-(di-*tert*-butylphosphino)biphenyl (**95**) was used as the ligand, the reaction of triflate **93** and phenol **94** elaborated diaryl ether **96** in the presence of $Pd(OAc)_2$ and $K_3PO_4$. The methodology also worked for electron-poor, neutral and electron-rich aryl halides.

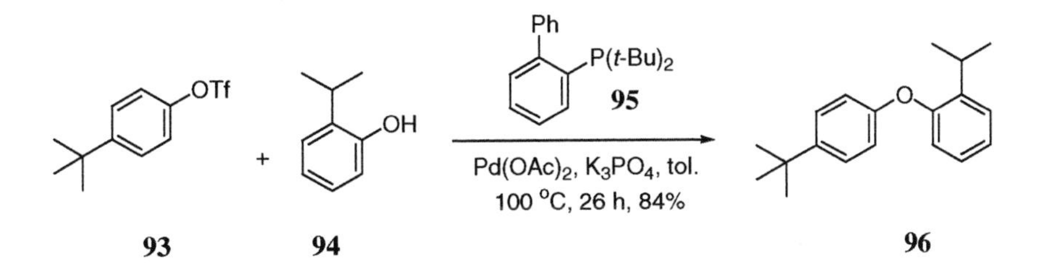


The mechanism for Pd-catalyzed C—O bond formation is similar to that of C—N bond formation. Application of this method to heterocyclic chemistry is yet to be seen, partially because the $S_NAr$ displacements of many heteroaryl halides with alkoxides are facile without the aid of palladium.

## 1.8 The Tsuji–Trost reaction

The Tsuji–Trost reaction is the palladium-catalyzed allylation of nucleophiles [110–113]. In an application to the formation of an *N*-glycosidic bond, the reaction of 2,3-unsaturated hexopyranoside **97** and imidazole afforded *N*-glycopyranoside **99** regiospecifically at the anomeric center with retention of configuration [114]. Therefore, the oxidative addition of allylic substrate **97** to Pd(0) forms the π-allyl complex **98** with inversion of configuration, then nucleophilic attack by imidazole proceeds with a second inversion of configuration to give **99**.

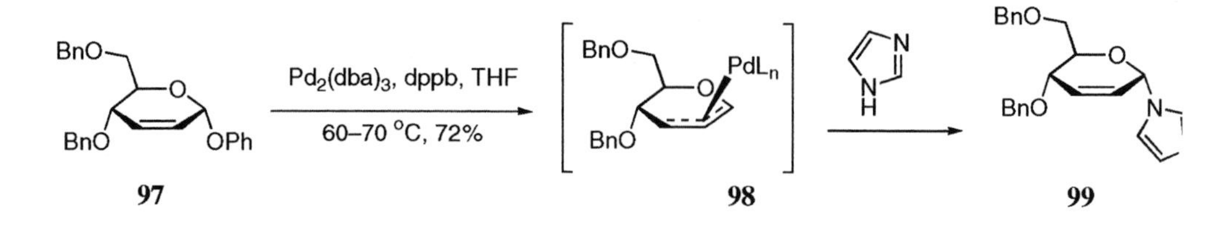


## 1.9 Jeffery's "ligand-free" conditions

In 1984, Jeffery discovered that under "ligand-free" conditions, Pd-catalyzed vinylation of organic halides proceeds at or near room temperature, whereas normal Heck reactions require relatively higher temperature [63, 115–120]. The so-called Jeffery's "ligand-free" conditions have found many applications in various palladium-catalyzed reactions. The transformation of **100** into **101** is a good example [121]. Although classic Heck conditions of *N*-allyl-*N*-benzyl(3-bromoquinoxalin-2-yl)amine (**100**) produced the desired intramolecular Heck product, 1-benzyl-3-methylpyrrolo[2,3-*b*]quinoxaline (**101**), the reaction was slow and low yielding. The low rate and yield may be attributed to the poisoning of the palladium catalyst *via* complexation to the aminoquinoxalines. The enhanced reactivity and yield under Jeffery's "ligand-free" conditions are presumably due to the coordination and thereby solvation of the palladium intermediates by bromide ions present in the reaction mixture, preventing the precipitation of Pd(0). Once the "locked" palladium catalyst is released from the substrates, the catalytic cycle continues smoothly.

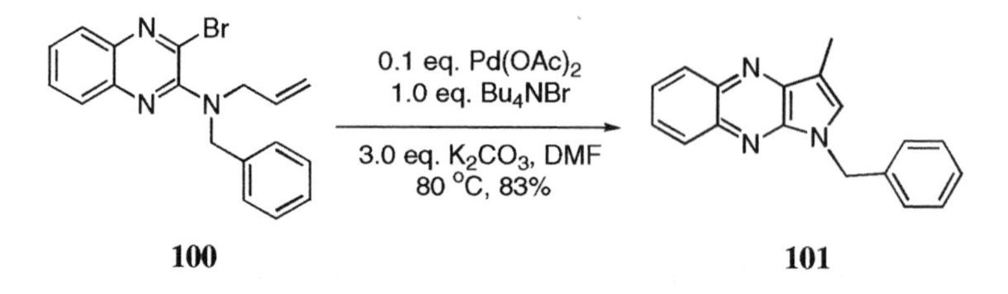


## 1.10 Mori–Ban, Hegedus and Larock indole syntheses

Among early reported Pd-catalyzed reactions, the Mori-Ban indole synthesis has proven to be very useful for pyrrole annulation. In 1977, based on their success of nickel-catalyzed indole synthesis from 2-chloro-*N*-allylaniline, the group led by Mori and Ban disclosed Pd-catalyzed intramolecular reactions of aryl halides with pendant olefins [122]. Compound **102**, easily prepared from 2-bromo-*N*-acetylaniline and methyl bromocrotonate, was adopted as the cyclization precursor. Treatment of **102** with $Pd(OAc)_2$ (2 mol%), $Ph_3P$ (4 mol%) and $NaHCO_3$

in DMF provided indole **103**. Although yields from the initial report were moderate, they have been greatly improved over the last two decades.

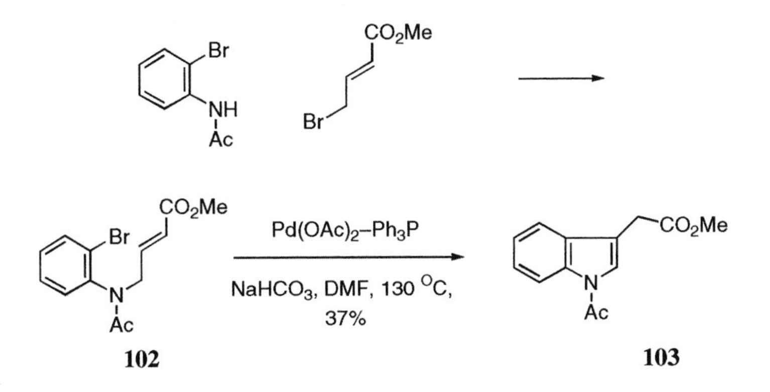


The mechanism for the Mori-Ban indole formation is representative of many Pd-catalyzed pyrrole annulation processes [123]. Reduction of $Pd(OAc)_2$ by $PPh_3$ generates Pd(0) species accompanied by triphenylphosphine oxide and acetic anhydride.

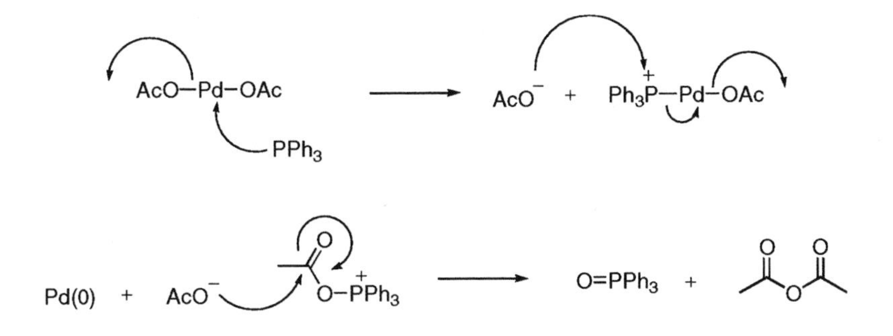


With Pd(0) generated *in situ*, the oxidative addition of aryl bromide **102** to Pd(0) proceeds to form Pd(II) intermediate **104**. Migratory insertion of **104** then occurs to furnish the cyclized indoline intermediate **105**. Subsequent reductive elimination of **105** takes place in a *cis* fashion, giving rise to *exo*-cyclic olefin **107**, which then tautomerizes spontaneously to the thermodynamically more stable indole **103**. The reductive elimination by-product as a palladium hydride species **106** reacts with base, regenerating Pd(0) to close the catalytic cycle.

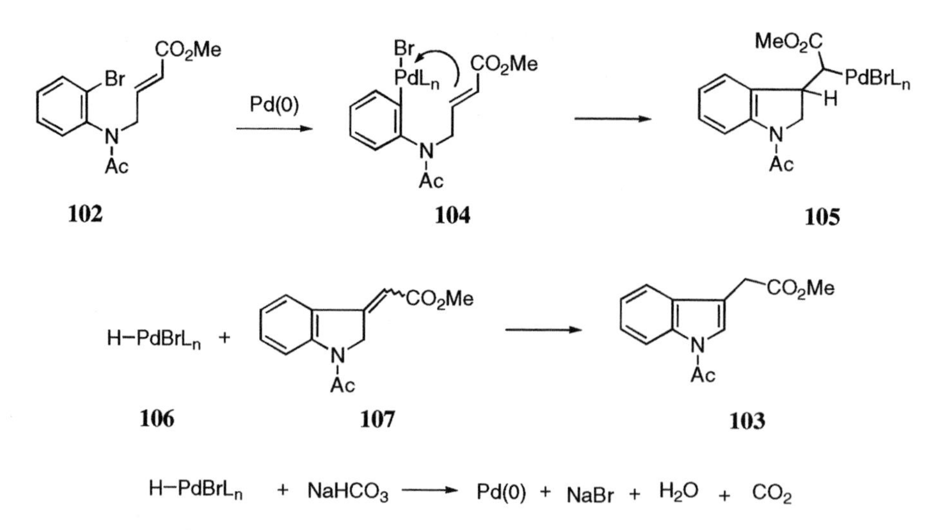


While the Mori-Ban indole synthesis is catalyzed by a Pd(0) species, the Hegedus indole synthesis is catalyzed by a Pd(II) complex [124–126]. In addition, the Mori-Ban indole synthesis is accomplished *via* a Pd-catalyzed vinylation, whereas the Hegedus indole synthesis established the pyrrole ring *via* a Pd-catalyzed amination (a Wacker-type reaction). In 1976, Hegedus *et al.* described an indole synthesis using a Pd-assisted intramolecular amination of olefins, which tolerated a range of functionalities. For instance, the requisite *o*-allylaniline **108** was prepared in high yield by the reaction of 5-methoxycarbonyl-2-bromoaniline with π-allylnickel bromide. Addition of **108** to a suspension of stoichiometric $PdCl_2(CH_3CN)_2$ in THF produced a yellow precipitate (the putative intermediate **110**), which upon treatment with $Et_3N$ gave rise to indole **109** and deposited metallic palladium.

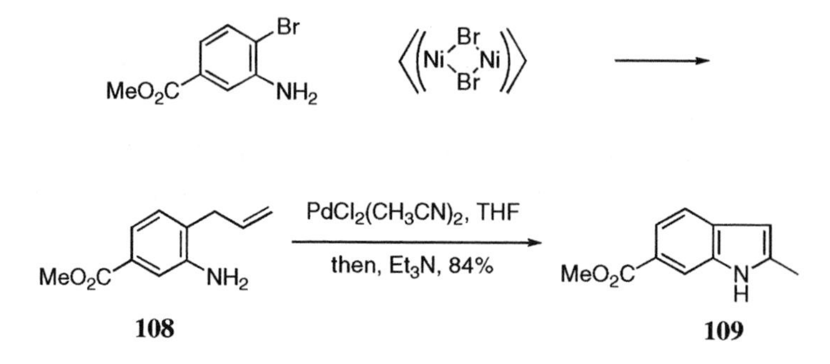


Hegedus proposed the probable course of the cyclization reaction, which follows a Wacker-type reaction mechanism. Coordination of the olefin to Pd(II) results in precipitate **110**, which upon treatment with $Et_3N$ undergoes intramolecular amination to afford intermediate **111**. As expected, the nitrogen atom attack occurs in a *5-exo-trig* fashion to afford **112**. Hydride elimination of **112** gives rise to exocyclic olefin **113**, which spontaneously rearranges to indole **109**. A stoichiometric amount of Pd(II) is needed in such a process because Pd(0) is not converted to useful Pd(II) species. However, addition of oxidants such as benzoquinone obviates the use of a full equivalent of expensive palladium. Hence, the transformation only requires catalytic amount of palladium because Pd(0) is now reoxidized to Pd(II).

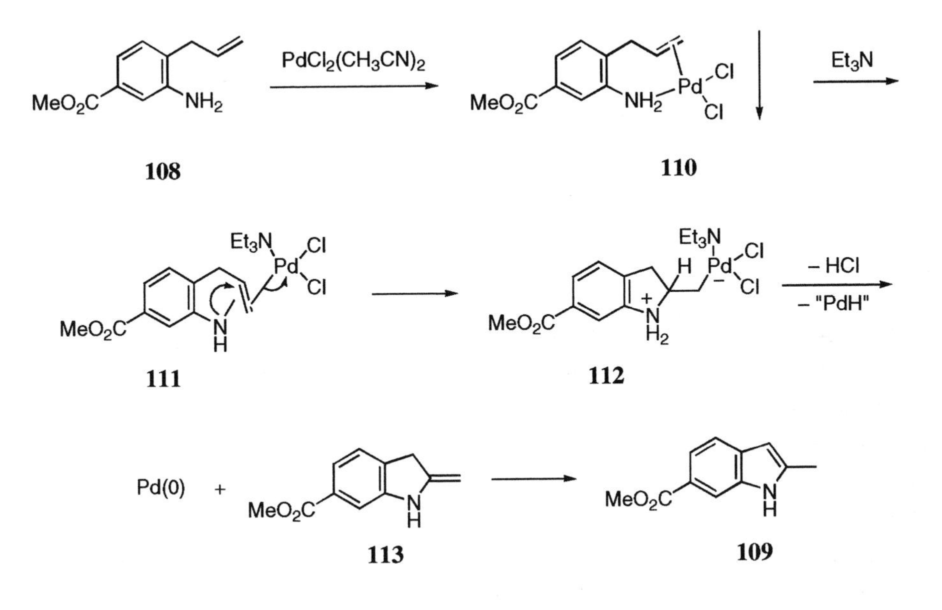


In 1991 [127–129], Larock *et al.* described an indole synthesis *via* a Pd-catalyzed heteroannulation of internal alkynes using *o*-iodoaniline and its derivatives, as exemplified by the Pd-catalyzed reaction between *o*-iodoaniline and propargyl alcohol **114** to give indole **115**. The pronounced directive effect of neighboring alcohol groups on the regiochemistry of alkyne insertion appears to be the result of coordination of the alcohol to the palladium during the insertion step (see intermediate **118** in the mechanism scheme below). Generally, the annulation of unsymmetrical alkynes has proven to be highly regioselective, providing one regioisomer in most cases. The more sterically bulky group ends up near the nitrogen atom in the indole product.

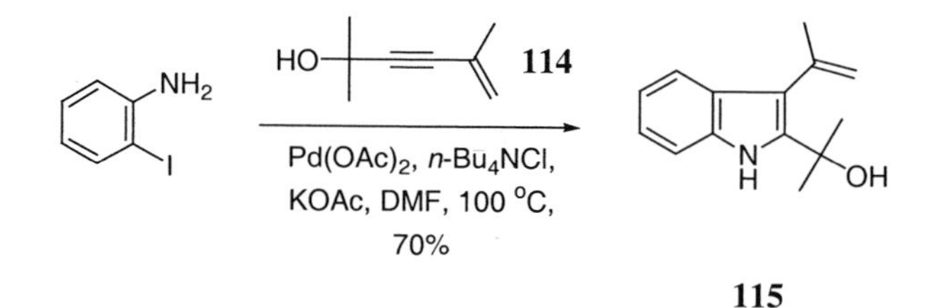


The mechanism of the Larock indole synthesis is postulated as following: reduction of $Pd(OAc)_2$ with $Me_2NH$ contaminant in DMF generates Pd(0), which is subsequently "solvated" by chloride to afford a species denoted as $L_2PdCl^-$. Oxidative addition of *o*-iodoaniline to Pd(0) gives rise to the palladium intermediate **116**, which then coordinates with the internal alkyne, affording complex **117**. Regioselective *syn*-insertion of the alkyne into the arylpalladium bond furnishes vinylic palladium intermediate **118**, which forms a six-membered palladacyclic intermediate **119** from the nitrogen displacement of the iodide. Finally, reductive elimination of **119** provides indole **115**, meanwhile regenerating Pd(0).

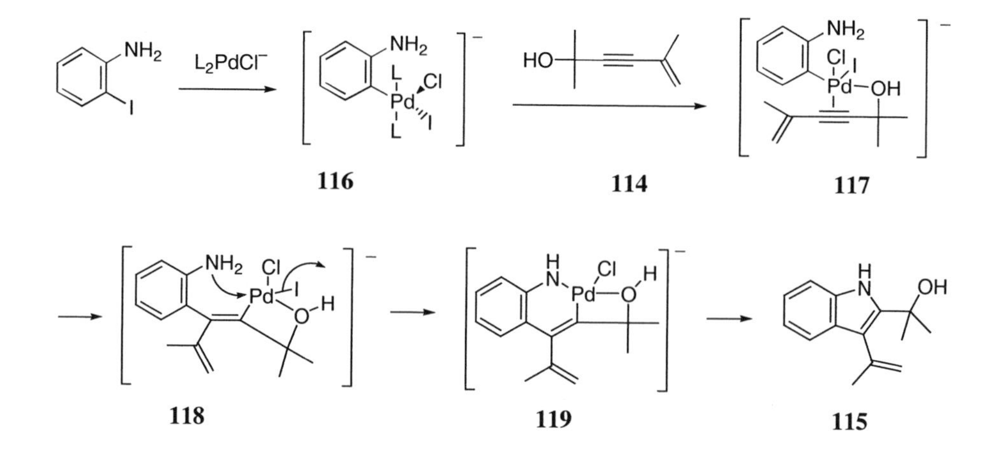


## 1.11 References


1. *Metal-catalyzed Cross-coupling Reactions*; Diederich, F.; Stang, P. J. Eds. Wiley-VCH: Weinhein, Germany, **1998**.
2. (a) Tsuji, J. *Perspectives in Organopalladium Chemistry for the 21st Century s* Publisher: Elsevier, **1999**; (b) Tsuji, J. *Palladium Reagents and Catalysts: Innovations in Organic Synthesis* Publisher: (Wiley, Chichester, UK), **1995**.

3. Farina, V.; Krishnamurthy, V.; Scott, W. J. *The Stille Reaction* Publisher: Wiley, New York, N. Y., **1998**.
4. Hegedus, L. S. *Transition Metals in the Synthesis of Complex Organic Molecules* 2nd Ed., Publisher: (University Science Books, Mill Valley, USA), **1999**.
5. Malleron, J.-L.; Fiaud, J.-C.; Legros, J.-Y. *Handbook of Palladium-catalyzed Organic Reactions* Academic Press, **1997**.
6. Kalinin, V. N. *Synthesis* **1992**, 413–32.
7. (a) Li, J. J. in *Alkaloids, Chemical and Biological Perspectives, Vol. 14*, Pelletier, S. W. Ed., Pergamon: Amsterdam, Netherlands, **1999**, 437–503. (b) Undheim, K.; Benneche, T. in *Adv. Heterocyclic Chem.* **1995**, *62*, 305–418.
8. Stanforth, S. P. *Tetrahedron* **1998**, *54*, 263–303.
9. Sakamoto, T.; Kondo, Y.; Yamanaka, H. *Heterocycles* **1988**, *27*, 2225–49.
10. Godard, A.; Marsais, F.; Plé, N.; Trecourt, F. Turck, A.; Quéguiner, G. *Heterocycles* **1995**, *40*, 1055–91.
11. Undheim, K.; Benneche, T. *Heterocycles* **1990**, *30*, 1155–93.
12. Intermolecular arylation of benzanilides, (a) Kametani, Y.; Satoh, T.; Miura, M.; Nomura, M. *Tetrahedron Lett.* **2000**, *41*, 2655–8. Intermolecular arylation of phenolates, (b) Satoh, T.; Kawamura, Y.; Miura, Y.; Nomura, M. *Angew. Chem., Int. Ed. Engl.* **1936**, *27*, 1740–2. (c) Satoh, T.; Inoh, J.-I.; Kawamura, Y.; Miura, Y.; Nomura, M. *Bull. Chem. Soc. Jpn.* **1998**, *71*, 2239–46. (d) Kawamura, Y.; Satoh, T.; Miura, M.; Nomura, M. *Chem. Lett.* **1999**, 961–2. Heteroaryl Heck reaction, (e) Aoyagi, Y.; Inoue, A.; Koizumi, I.; Hashimoto, R.; Tokunaga, K.; Gohma, K.; Komatsu, J.; Sekine, K.; Miyafuji, A.; Konoh, J. Honma, R. Akita, Y.; Ohta, A. *Heterocycles* **1992**, *33*, 257–72.
13. (a) Bringmann, G.; Heubes, M.; Breuning, M.; Göbel, L.; Ochse, M.; Schöner, B.; Schupp, O. *J. Org. Chem.* **2000**, *65,* 722–8. (b) Deshpande, P. P.; Martin, O. R. *Tetrahedron Lett.* **1990**, *31*, 6313–6. (c) Hosoya, T.; Takashira, E.; Matsumoto, T.; Suzuki, K. *J. Am. Chem. Soc.* **1994**, *116,* 1004–15.
14. Chapman, N. B.; Russell-Hill, D. Q. *J. Chem. Soc.* **1956**, 1563–72.
15. Åkermark, B.; Eberson, L.; Jonsson, E.; Pettersson, E. *J. Org. Chem.* **1975**, *40,* 1365–7.
16. For an early review, see, Hegedus, L. S. *Angew. Chem., Int. Ed. Engl.* **1988**, *27*, 1113–26.
17. Knölker, H.-J.; Fröhner, W. *J. Chem. Soc., Perkin Trans. 1* **1998**, 173–6.
18. Knölker, H.-J.; O'Sullivan, H. *Tetrahedron* **1994**, *50*, 10893–908.
19. For reoxidation of Pd(0) using *tert*-butyl hydroperoxide, see, Åkermark, B.; Oslob, J. D.; Heuschert, U. *Tetrahedron Lett.* **1995**, *36*, 1325–6.
20. Knölker, H.-J.; Reddy, K. R.; Wagner, A. *Tetrahedron Lett.* **1998**, *39*, 8267–70.
21. Negishi, E.-i.; Baba, S. *J. Chem. Soc., Chem. Commun.* **1976**, 596–7.
22. Negishi, E.-i. *Acc. Chem. Res.* **1982**, *15,* 340–8.

23. Erdik, E. *Tetrahedron* **1992**, *48,* 9577–648.
24. Prasad, A. S.; B.; Stevenson, T. M.; Citineni, J. R.; Zyzam, V.; Knochel, P. *Tetrahedron* **1997**, *53*, 7237–54.
25. Knochel, P.; Singer, R. D. *Chem. Rev.* **1993**, *93*, 2117–88.
26. Evans, D. A.; Bach, T. *Angew. Chem., Int. Ed. Engl,* **1993**, *32,* 1326–7.
27. Miyaura, N.; Suzuki, A. *Chem. Rev.* **1995**, *95*, 2457–83.
28 Suzuki, A. in *Metal-catalyzed Cross-coupling Reactions*; Diederich, F.; Stang, P. J. Eds. Wiley-VCH: Weinhein, Germany, **1998**, chapter 2, 49–97.
29 Gronowitz, S.; Hörnfeldt, A.-B.; Musil, T. *Chem. Scr.* **1986**, *26*, 305–9.
30. Peters, D.; Hörnfeldt, A.-B.; Gronowitz, S. *J. Heterocycl. Chem.* **1990**, *27*, 2165–73.
31. Wellmar, U.; Hörnfeldt, A.-B.; Gronowitz, S. *J. Heterocycl. Chem.* **1995**, *32*, 1159–63.
32. Kawasaki, I.; Yamashita, M.; Ohta, S. *Chem. Pharm. Bull.* **1996**, *44*, 1831–9. and references cited therein.
33. Kawasaki, I.; Katsuma, H.; Nakayama, Y.; Yamashita, M.; Ohta, S. *Heterocycles* **1998**, *48,* 748–50.
34. Littke, A. F.; Fu, G. C. *Angew. Chem., Int. Ed.* **1998**, *37*, 3387–8.
35. Wolfe, J. P.; Buchwald, S. L. *Angew. Chem., Int. Ed.* **1999**, *38*, 2413–6.
36. Ishiyama, T.; Nishijima, K.; Miyaura, N.; Suzuki, A. *J. Am. Chem. Soc.* **1993**, *115*, 7219–25.
37. Milstein, D.; Stille, J. K. *J. Am. Chem. Soc.* **1978**, *100*, 3636–8.
38. Milstein, D.; Stille, J. K. *J. Am. Chem. Soc.* **1979**, *101*, 4992–8.
39. Stille, J. K. *Angew. Chem., Int. Ed. Engl.* **1986**, *25*, 508–24.
40. Farina, V.; Krishnamurphy, V.; Scott, W. J. *Organic Reactions* **1997**, *50*, 1–652.
41. For an excellent review on the intramolecular Stille reaction, see, Duncton, M. A. J.; Pattenden, G. *J. Chem. Soc., Perkin Trans. 1* **1999**, 1235–46.
42. Liebeskind, L. S.; Fengl, R. W. *J. Org. Chem.* **1990**, *55*, 5359–64.
43. Bach, T.; Krüger, L. *Synlett* **1998**, 1185–6.
44. Kelly, T. R.; Li, Q.; Bhushan, V. *Tetrahedron Lett.* **1990**, *31*, 161–4.
45. Grigg, R.; Teasdale, A.; Sridharan, V. *Tetrahedron Lett.* **1991**, *32*, 3859–62.
46. Sakamoto, T.; Yasuhara, A.; Kondo, Y.; Yamanaka, H. *Heterocycles* **1993**, *36*, 2597–600.
47. Fukuyama, Y.; Yaso, H.; Nakamura, K.; Kodama, M. *Tetrahedron Lett.* **1999**, *40*, 105–8.
48. Seiders, T. J.; Baldridge, K. K.; Elliott, E. L.; Grube, G. H.; Siegel, J. S. *J. Am. Chem. Soc.* **1999**, *121*, 7439–40.
49. Iyoda, M.; Miura, M.i; Sasaki, S.; Kabir, S. M. H.; Kuwatani, Y.; Yoshida, M. *Tetrahedron Lett.* **1997**, *38*, 4581–2.
50. Iwaki, T.; Yasuhara, A.; Sakamoto, T. *J. Chem. Soc., Perkin Trans. 1* **1999**, 1505–10.

51. Tamao, K.; Sumitani, K.; Kiso, Y.; Zembayashi, M.; Fujioka, A.; Kodma, S.-i.; Nakajima, I.; Minato, A.; Kumada, M. *Bull. Chem. Soc. Jpn.* **1976**, *49*, 1958–69.
52. Amatore, C.; Jutand, A.; Negri, S.; Fauvarque, J. F. *J. Organomet. Chem.* **1990**, *390*, 389–98.
53. Minato, A.; Suzuki, K.; Tamao, K.; Kumada, M. *J. Chem. Soc., Chem. Commun.* **1984**, 511–3.
54. Hiyama, T.; Hatanaka, Y. *Pure Appl. Chem.* **1994**, *66*, 1471–8.
55. Hiyama, T. in *Metal-Catalyzed Cross-Coupling Reactions* **1998**, Eds: Diederich, F.; Stang, P.J. Publisher: Wiley–VCH Verlag GmbH, Weinheim, Germany, chapter 10, 421–53.
56. Hatanaka, Y.; Fukushima, S.; Hiyama, T. *Heterocycles* **1990**, *30*, 303–6.
57. Shibata, K.; Miyazawa, K.; Goto, Y. *Chem. Commun.* **1997**, 1309–10.
58. Stephens, R. D.; Castro, C. E. *J. Org. Chem.* **1963**, *28*, 3313–5.
59. Sonogashira K.; Tohda, Y.; Hagihara, N. *Tetrahedron Lett.* **1975**, 4467–70.
60. Rossi, R. Carpita, A.; Belina, F. *Org. Prep. Proc. Int.* **1995**, *27*, 129–60.
61. Campbell, I. B. in *Organocopper Reagents*, **1994**, 217–35. Ed.: Taylor, R. J. K. Publisher: IRL Press, Oxford, UK.
62. Sakamoto, T.; Nagano, T.; Kondo, Y.; Yamanaka, H. *Chem. Pharm. Bull.* **1988**, *36*, 2248–52.
63. Jeffery, T. *Synthesis*, **1987**, 70–1.
64. Nguefack, J.-F.; Bolitt, V.; Sinou, D. *Tetrahedron Lett.* **1996**, *37*, 5527–30.
65. Mizoroki, T.; Mori, K.; Ozaki, A. *Bull. Chem. Soc. Jpn.* **1971**, *44*, 581.
66. Heck, R. F.; Nolley, J. P., Jr. *J. Org. Chem.* **1972**, *37*, 2320–2.
67. Akita, Y.; Inoue, A.; Mori, Y.; Ohta, A. *Heterocycles* **1986**, *24*, 2093–7.
68. Bräse, S.; de Meijere, A. in *Metal-catalyzed Cross-coupling Reactions*; Diederich, F.; Stang, P. J. Eds. Wiley-VCH: Weinhein, Germany, **1998**, chapter 3, 99–166.
69. Link, J. T.; Overman, L. E. in *Metal-catalyzed Cross-coupling Reactions*; Diederich, F.; Stang, P. J. Eds. Wiley-VCH: Weinhein, Germany, **1998**, chapter 6, 231–69.
70. Rawal V. H.; Iwasa, H. *J. Org. Chem.* **1994**, *59*, 2685–6.
71. Earley, W. G.; Oh, T.; Overman, L. E. *Tetrahedron Lett.* **1988**, *29*, 3785–8.
72. Madin, A.; Overman, L. E. *Tetrahedron Lett.* **1992**, *33*, 4859–62.
73. Pivsa-Art, S.; Satoh, T.; Kawamura, Y.; Miura, M.; Nomura, M. *Bull. Chem. Soc. Jpn.* **1998**, *71*, 467–73.
74. Potts, K. T. *Comprehensive Heterocyclic Chemistry* Pergamon Press, Oxford, **1984**, Vols 5 and 6.
75. Kuroda, T.; Suzuki, F. *Tetrahedron Lett.* **1991**, *32*, 6915–8.
76. Reetz , M. T.; Lohmer, G.; Schwickardi, R. *Angew. Chem., Int. Ed.* **1998**, *37*, 481–3.

77. Littke, A. F.; Fu, G. C. *J. Org. Chem.* **1999**, *64*, 10–1.
78. Tsuji, J. *Palladium Reagents and Catalysts: Innovations in Organic Synthesis* Publisher: (Wiley, Chichester, UK), **1995**, 340–5.
79. Bessard, Y.; Roduit, J. P. *Tetrahedron* **1999**, *55*, 393–404.
80. Wu, G. G.; Wong, Y.; Poirier, M. *Org. Lett.* **1999**, *1*, 745–7.
81. Hirao, T.; Masunaga, T.; Ohshiro, Y.; Agawa, T. *Tetrahedron Lett.* **1980**, *21,* 3595–8.
82. Hirao, T.; Masunaga, T.; Ohshiro, Y.; Agawa, T. *Synthesis* **1981**, 56–7.
83. Hirao, T.; Masunaga, T.; Yamada, Y.; Ohshiro, Y.; Agawa, T. *Bull. Chem. Soc. Jpn.* **1982**, *55,* 909–13.
84. Xu, Y.; Li, Z.; Xia, J.; Guo, H.; Huang, Y. *Synthesis* **1983**, 377–8.
85. Xu, Y.; Zhang, J. *Synthesis* **1984**, 778–10.
86. Xu, Y.; Wei, H.; Zhang, J.; Huang, G. *Tetrahedron Lett.* **1989**, *30,* 949–52. And references cited therein.
87. Petrakis, K. S.; Nagabhushan, T. L. *J. Am. Chem. Soc.* **1987**, *109*, 2831–3.
88. Holt, D. A.; Erb, J. M. *Tetrahedron Lett.* **1989**, *30,* 5393–6.
89. Kurz, L.; Lee, G.; Morgan, D., Jr.; Waldyke, M. J.; Ward, T. *Tetrahedron Lett.* **1990**, *31,* 6321–4.
90. Uozumi, Y.; Tanahashi, A.; Lee, S.Y.; Hayashi, T. *J. Org. Chem.* **1993**, *58*, 1945–8.
91. Kosugi, M.; Kameyama, M.; Migita, T. *Chem. Lett.* **1983**, 927–8.
92. Boger, D. L.; Panek, J. S. *Tetrahedron Lett.* **1984**, *25*, 3175–8.
93. Paul, F.; Patt, J.; Hartwig, J. F. *J. Am. Chem. Soc.* **1994**, *116*, 5969–70.
94. Guram, A. S.; Buchwald, S. L. *J. Am. Chem. Soc.* **1994**, *116*, 7901–2.
95. Hartwig, J. F. *Synlett* **1997**, 329–40.
96. Hartwig, J. F. *Acc. Chem. Res.* **1998**, *31*, 852–60.
97. Hartwig, J. F. *Angew. Chem., Int. Ed.* **1998**, *37*, 2090–3.
98. Hartwig, J. F. *Angew. Chem., Int. Ed.* **1998**, *37*, 2046–67.
99. Wolfe, J. P.; Wagaw, S.; Marcoux, J.-F.; Buchwald, S. L. *Acc. Chem. Res.* **1998**, *31*, 805–18.
100. Frost, C. G.; Mendonça, P. *J. Chem. Soc., Perkin Trans. 1* **1998**, 2615–24.
101. Yang, B. H.; Buchwald, S. L. *J. Organomet. Chem.* **1999**, *576*, 125–46.
102. Mann, G.; Hartwig, J. F.; Driver, M. S.; Fernandez-Rivas, C. *J. Am. Chem. Soc.* **1998**, *120*, 827–8.
103. Hartwig, J. F. *Angew. Chem., Int. Ed.* **1998**, *37*, 2046–67.
104. Palucki, M.; Wolfe, J. P.; Buchwald, S. L. *J. Am. Chem. Soc.* **1996**, *118*, 10333–4.
105. Palucki, M.; Buchwald, S. L. *J. Am. Chem. Soc.* **1997**, *119*, 11108–9.
106. Mann, G.; Hartwig, J. F. *J. Org. Chem.* **1997**, *62*, 5413–8.
107. Mann, G.; Hartwig, J. F. *Tetrahedron Lett.* **1997**, *38*, 8005–8.

108. Mann, G.; Incarvito, C.; Rheingold, A. L.; Hartwig, J. F. *J. Am. Chem. Soc.* **1999**, *121*, 3224–5.
109. Aranyos, A.; Old, D. W.; Kiyomori, A.; Wolfe, J. P.; Sadighi, J. P.; Buchwald, S. L. *J. Am. Chem. Soc.* **1999**, *121*, 4369–78.
110. Tsuji, J.; Takahashi, H.; Morikawa, M. *Tetrahedron Lett.* **1965**, 4387–10.
111. Tsuji, J. *Acc. Chem. Res.* **1969**, *2*, 144–52.
112. S.A. Godleski, in "*Comprehensive Organic Synthesis*" (Trost, B. M.; and Fleming, I.; eds.), vol. 4. Chapter 3.3. Pergamon, Oxford, **1991**.
113. M. Moreno–Mañas, R. Pleixats, in "*Advances in Heterocyclic Chemistry*" (A.R. Katritzky, ed.), **1996**, *66*, 73–129, Academic Press.
114. Bolitt, V.; Chaguir, B.; Sinou, D. *Tetrahedron Lett.* **1992**, *33*, 2481–4.
115. Jeffery, T. *J. Chem. Soc., Chem. Commun.* **1984**, 1287–9.
116. Jeffery, T. *Tetrahedron Lett.* **1999**, *40,* 1673–6.
117. Jeffery, T. *Tetrahedron Lett.* **1994**, *35,* 4103–6.
118. Jeffery, T. *Adv. Met.-Org. Chem.* **1996**, *5*, 153–260.
119. Jeffery, T. *Tetrahedron* **1996**, *52,* 10113–30.
120. Jeffery, T. *Tetrahedron Lett.* **1998**, *39,* 5751–4.
121. Li, J. J. *J. Org. Chem.* **1999**, *64*, 8425–7.
122. (a) Mori, M.; Chiba, K.; Ban, Y. *Tetrahedron Lett.* **1977**, *12*, 1037–40; (b) Ban, Y.; Wakamatsu, T.; Mori, M. *Heterocycles* **1977**, *6,* 1711–5.
123. (a) Amatore C.; Carre, E.; Jutand, A.; M'Barki, M. A.; Meyer, G. *Organometallics* **1995**, *14*, 5605–14; (b) Amatore C.; Carre, E.; M'Barki, M. A. *Organometallics* **1995**, *14*, 1818–26; (c) Amatore C.; Jutand, A.; M'Barki, M. A. *Organometallics* **1992**, *11*, 3009–13; (d) Amatore C.; Azzabi, M; Jutand, A. *J. Am. Chem. Soc.* **1991**, *113,* 8375–84.
124. Hegedus, L. S.; Allen, G. F.; Waterman, E. L. *J. Am. Chem. Soc.* **1976**, *98,* 2674–6.
125. Hegedus, L. S.; Allen, G. F.; Bozell, J. J.; Waterman, E. L. *J. Am. Chem. Soc.* **1978**, *100,* 5800–7.
126. Hegedus, L. S. *Angew. Chem., Int. Ed. Engl.* **1988**, *27,* 1113–26.
127. Larock, R. C.; Yum, E. K. *J. Am. Chem. Soc.* **1991**, *113,* 6689–90.
128. Larock, R. C.; Yum, E. K.; Refvik, M. D. *J. Org. Chem.* **1998**, *63,* 7652–62.
129. Larock, R. C. *J. Organomet. Chem.* **1999**, *576,* 111–24.

# CHAPTER 2

## Pyrroles

The pyrrole ring is widely distributed in nature. It occurs in both terrestrial and marine plants and animals [1–3]. Examples of simple pyrroles include the *Pseudomonas* metabolite pyrrolnitrin, a recently discovered seabird hexahalogenated bipyrrole [4], and an ant trail pheromone. An illustration of the abundant complex natural pyrroles is konbu'acidin A, a sponge metabolite that inhibits cyclin-dependent kinase 4. The enormous reactivity of pyrrole in electrophilic substitution reactions explains the occurrence of more than 100 naturally occurring halogenated pyrroles [2, 3].

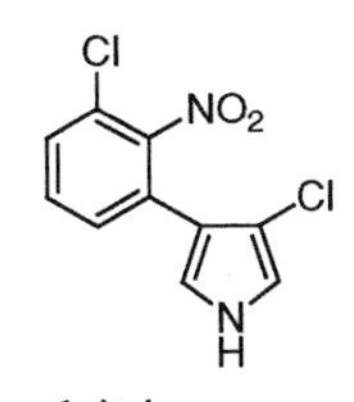


pyrrolnitrin

seabird compound

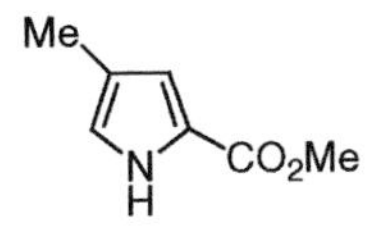


ant trail pheromone

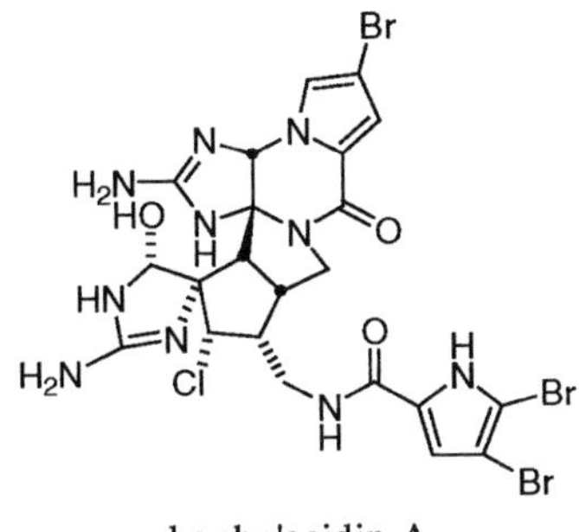


konbu'acidin A

Following the discovery of the unique electronic properties of polypyrrole, numerous polymers of pyrrole have been crafted. A copolymer of pyrrole and pyrrole-3-carboxylic acid is used in a glucose biosensor, and a copolymer of pyrrole and *N*-methylpyrrole operates as a redox switching device. Self-doping, low-band gap, and photorefractive pyrrole polymers have been synthesized, and some examples are illustrated [1, 5].

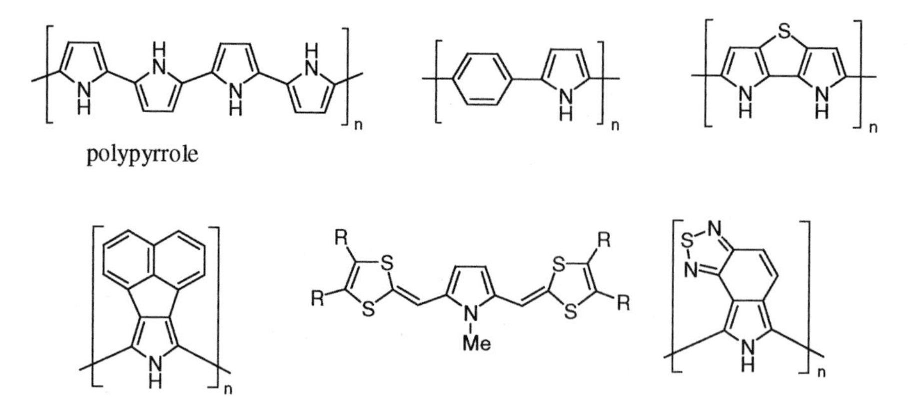


The pyrrole ring has found great use in the design and development of pharmaceuticals. Lipitor is the leading cholesterol-lowering drug and other bioactive pyrroles shown below include the opioid antagonist norbinaltorphimine, a broad-spectrum insecticide, a sodium-independent dopamine receptor antagonist, a DNA cross-linking agent, and an antipsychotic agent.

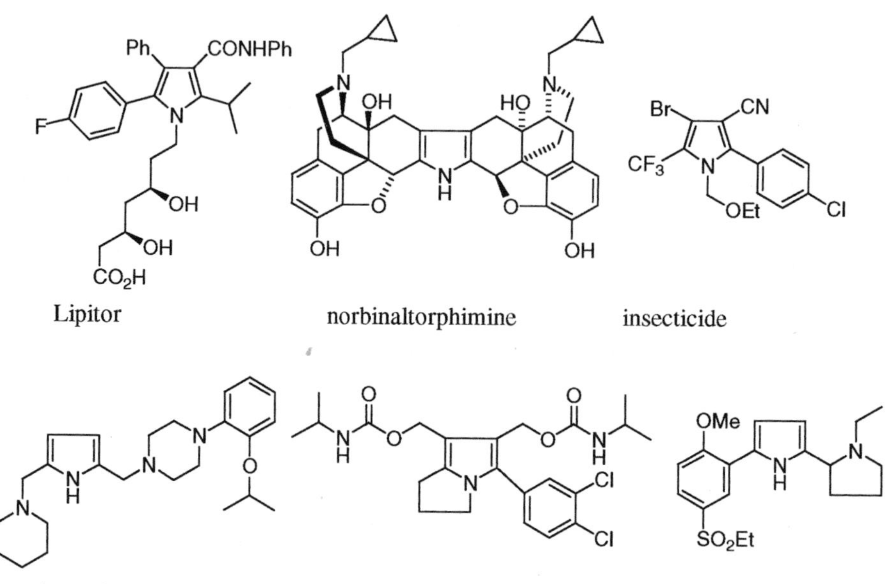

## 2.1 Synthesis of pyrrolyl halides

Although pyrrolyl halides are well-known compounds, their instability to acid, alkali, and heat precludes their commercial availability. Since pyrrole is a very reactive, π-excessive heterocycle, it undergoes halogenation extremely readily [6, 7]. For example, the labile 2-bromopyrrole, which decomposes above room temperature, is a well-known compound, as are *N*-alkyl-2-halopyrroles, readily prepared by direct halogenation, usually with NBS for the synthesis of bromopyrroles [8, 9]. The 2-halopyrrole is usually the kinetic product but the 3-halopyrrole is often the thermodynamic product, and this property of halopyrroles can be exploited in synthesis. For example, *N*-benzylpyrrole (**1**) can be dibrominated to give **2** as the kinetic product, which rearranges to **3** upon treatment with acid [10, 11]. Other *N*-alkyl-2,5-dibromopyrroles are available in this fashion.

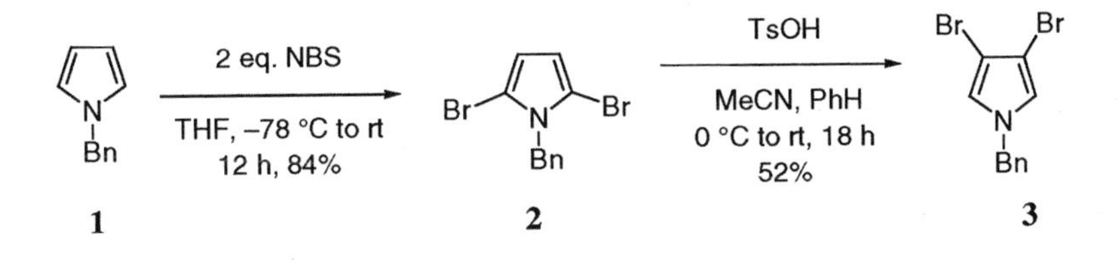


Bromination of **1** with $Br_2$ ($CCl_4$, 0°C) affords 1-benzyl-3-bromopyrrole directly in 55–66% yield [12]. Since these bromopyrroles are still very labile, other *N*-protecting groups, which are electron-withdrawing, have been employed to provide access to more robust halogenated pyrroles. The bromination of *N*-BOC-pyrrole, which is readily available [13], with 1,3-dibromo-5,5-dimethylhydantoin (NBH) or NBS, can be adjusted to afford either the 2-bromo derivative [14] or the 2,5-dibromo derivative in good yields [14, 15]. Not surprisingly, a bulky *N*-protecting group directs halogenation to the C-3 position of pyrrole. Thus, the bromination of *N*-tritylpyrrole (**4**) affords either mono- (**5**) or 3,4-dibromination products depending on conditions [16]. The trityl protecting group is readily removed under Birch conditions.

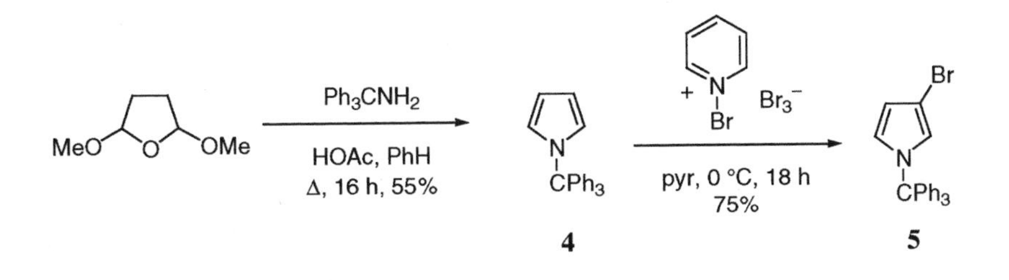


A more versatile *N*-protection strategy is seen in Muchowski's *N*-TIPS-pyrrole (**6**) [17–19]. Whereas bromination of *N*-trimethylsilylpyrrole gives almost exclusively C-2 attack, the NBS bromination of **6** affords the C-3 product **7** in a 96:4 ratio (C-3/C-2) in 93% yield. Deprotection is readily achieved with fluoride to give the unstable **8**. The C-3, C-4 dibromo derivative can be prepared by tribromination of **6** and then selective C-2 bromine-lithium exchange. Similar chemistry affords 3-iodo-*N*-TIPS pyrrole.

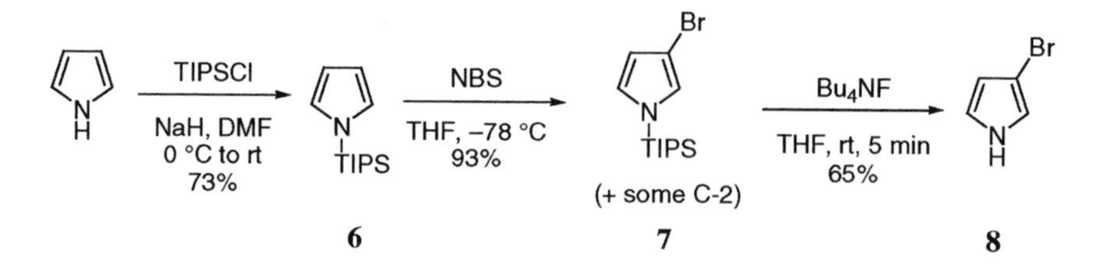


The power of Muchowski's method is seen by the fact that these bromopyrroles can be subjected to bromine-lithium exchange to afford the versatile 3-lithio species that can be quenched with a variety of electrophiles in good to excellent yields [18–21]. This is illustrated by a synthesis of verrucarin E (**11**) [19].

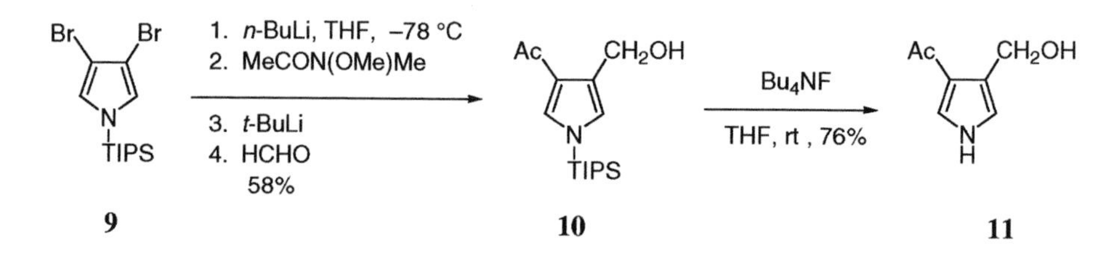


If the pyrrole is substituted with an electron-withdrawing group, then more vigorous halogenation conditions are required, but the products are usually more stable than simple halogenated pyrroles. For example, the bromination of pyrrole-2-carboxylic acid (**12**) yields the 4,5-dibromo isomer (**13**) in excellent yield [22]. Similarly, the bromination of 4-chloropyrrole-2-carboxylic acid furnishes 5-bromo-4-chloropyrrole-2-carboxylic acid in 90% yield [23].

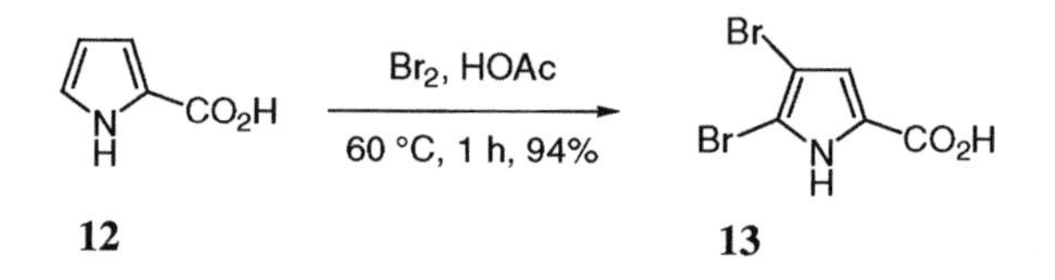


Various iodinated pyrroles have been prepared by direct iodination [19, 24] or *via* thallation [25]. For example, 3-iodo-*N*-TIPS-pyrrole is prepared in 61% yield from **6** [19], and 3,4-diiodo-2-formyl-1-methylpyrrole is available in 54% yield *via* a bis-thallation reaction [25].

Although *N*-protected 2-lithiopyrroles are readily generated and many types are known [6, 26], these intermediates have not generally been employed to synthesize halogenated pyrroles. One exception is the synthesis of the two natural seabird hexahalogenated bipyrroles **15** and **17**, which were prepared as shown from bipyrrole **14** [27]. The presumed inductive effect of the nitrogen atoms is responsible for the regiochemistry of the halogen-metal exchange reaction in going from **15** to **17**.

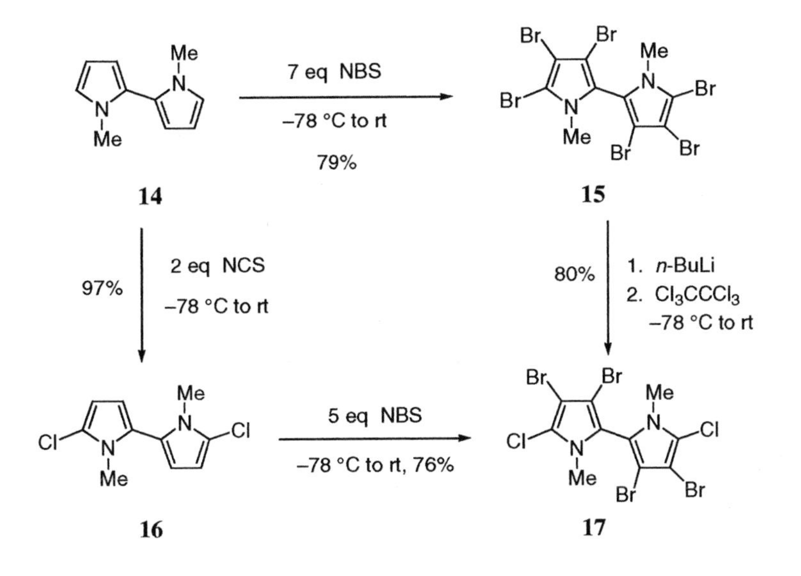


Other halopyrroles and related triflates will be cited as appropriate in the subsequent sections.

## 2.2 Oxidative coupling/cyclization

In most cases, the oxidative addition process consumes stoichiometric amount of $Pd(OAc)_2$. One of the earliest examples of the use of palladium in pyrrole chemistry was the $Pd(OAc)_2$ induced oxidative coupling of *N*-methylpyrrole with styrene to afford a mixture of olefins **18** and **19** in low yield based on palladium acetate [28].

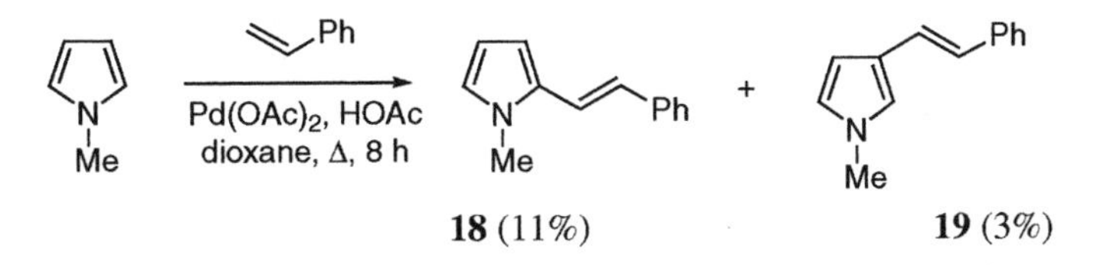


A similar reaction of pyrroles **20** with acrylates provides the C-2 substituted α-alkenyl derivatives **21** in 24–91% yield [29]. The 2,6-dichlorobenzoyl protecting group is noteworthy as it prevents cyclization of the phenyl group onto the pyrrole ring (*vide infra*).

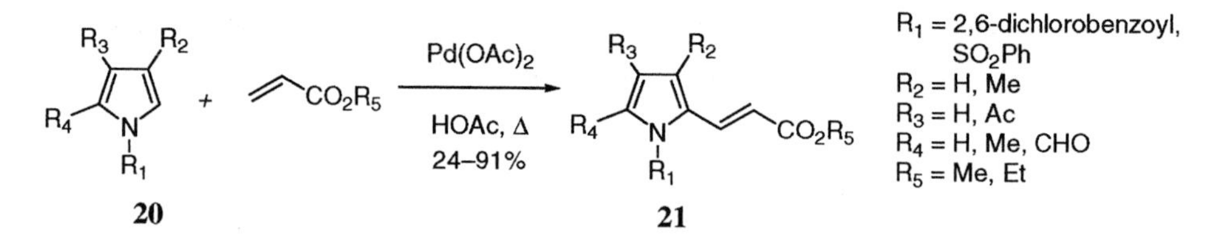


Itahara has also found that the phenylation of *N*-aroylpyrroles can be achieved using $Pd(OAc)_2$ [30, 31]. Although *N*-benzoylpyrrole (**22**) yields a mixture of diphenylpyrrole **23**, cyclized pyrrole **24**, and bipyrrolyl **25** as shown, 1-(2,6-dichlorobenzoyl)pyrrole **26** gives the diphenylated pyrrole **27** in excellent yield. The *N*-aroyl groups are readily cleaved with aqueous alkali, and the arylation reaction also proceeds with *p*-xylene and *p*-dichlorobenzene. Unfortunately, *N*-methyl-, *N*-acetyl-, and *N*-(phenoxycarbonyl)pyrroles give complex mixtures of products.

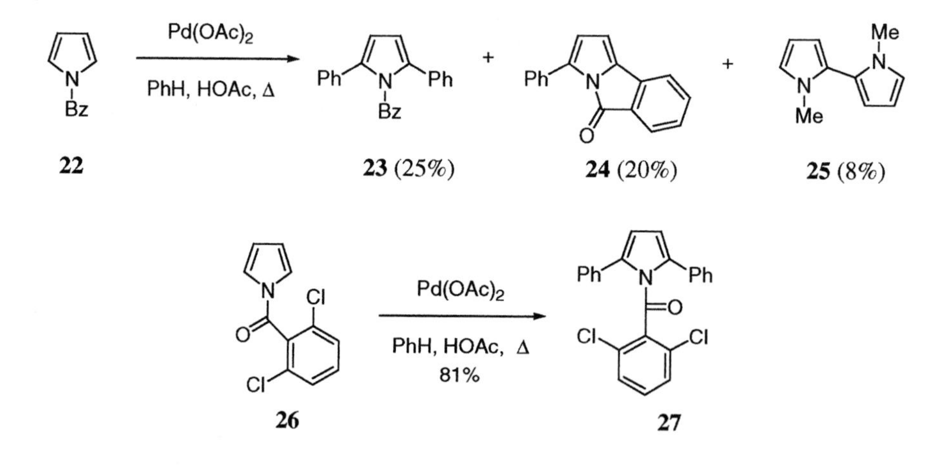


As expected, reaction of *N*-aroylpyrroles **28** in the absence of added arene affords the bipyrroles **29** or cyclized product **30** [32, 33]. Bipyrrole **31** was prepared *via* this oxidative coupling reaction [32].

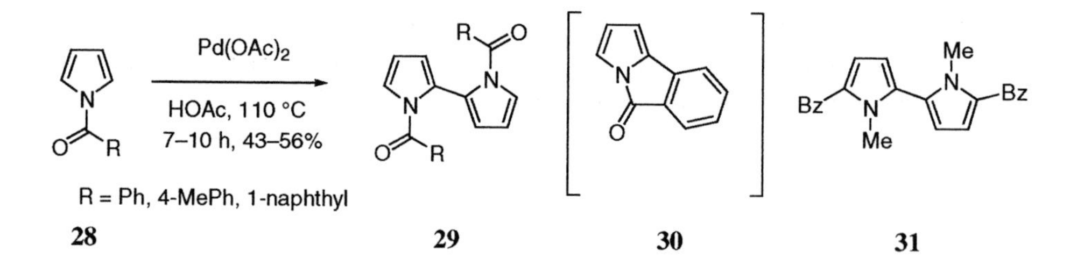

Itahara has extended these stoichiometric $Pd(OAc)_2$ induced reactions to the coupling of *N*-(phenylsulfonyl)pyrrole (**32**) and 1,4-naphthoquinone (**33**) to afford **34** [34].

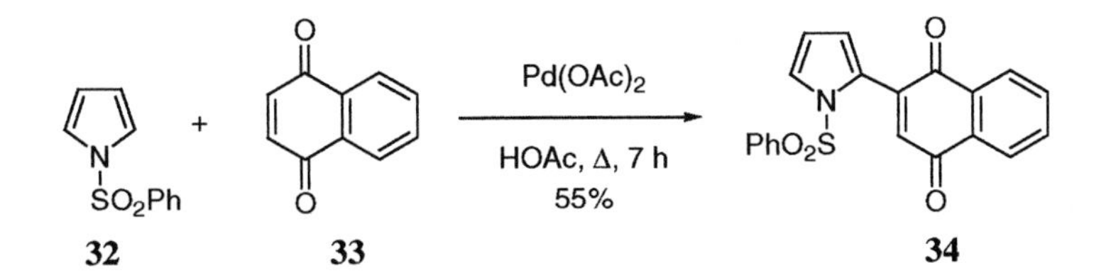


A key step in Boger's synthesis of prodigiosine and related compounds is the oxidative cyclization of dipyrrolyl ketones such as **35** to give **36** in excellent yield [35, 36]. Noteworthy is the use of polymer-supported $Pd(OAc)_2$ in this chemistry.

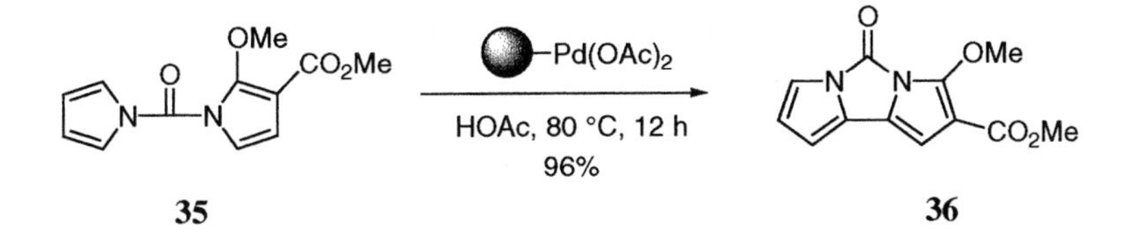


Pyrrole rings frequently serve as precursors to indole rings [37] and $PdCl_2$ induces the oxidative cyclization of pyrrole **37** to a mixture of **38** and **39** [38]. Since the oxidation of tetrahydroindoles to indoles, such as **38** to **39**, is usually straightforward, this transformation can be viewed as a novel and efficient indole ring synthesis.

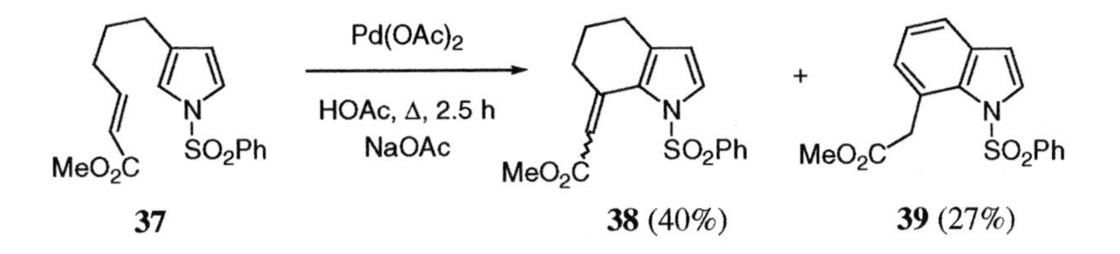


A new pyrrole ring synthesis developed by Arcadi involves the addition of ammonia or benzylamine to 4-pentynones, the latter of which are conveniently prepared *via* a palladium oxidative coupling sequence as shown below for the synthesis of **40** [39, 40].

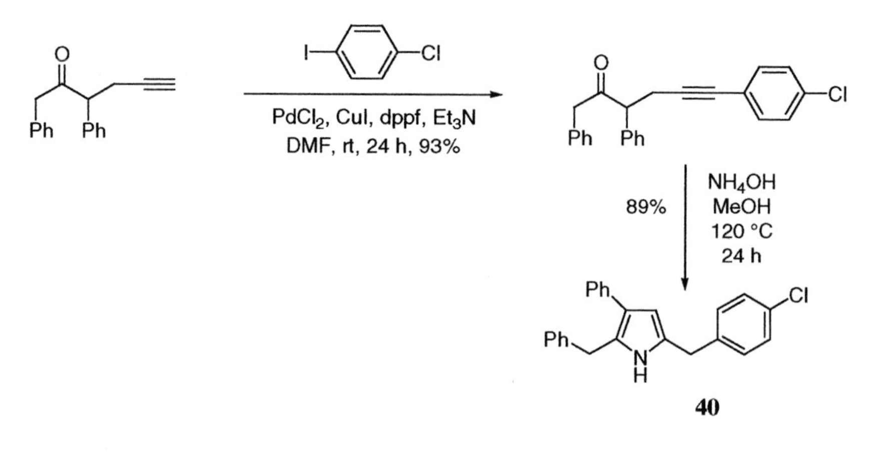


Ohta has developed a facile and efficient synthesis of pyrroles **42** that involves the Pd-catalyzed oxidative cyclization of hydroxy enamines such as **41** [41]. Fused pyrroles **43** and **44** were also synthesized in similar fashion.

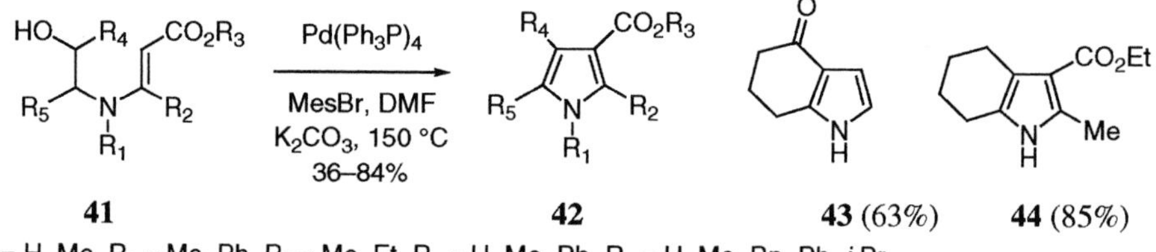


## 2.3 Coupling reactions with organometallic reagents

### 2.3.1 Kumada coupling

Pyrrole Grignard reagents can be generated either by bromine-magnesium exchange on bromopyrroles [42–44] or by transmetalation of lithiopyrroles with magnesium bromide [45–47]. The resulting pyrrole Grignard reagents undergo a variety of coupling reactions (the Kumada coupling [48, 49]) with aryl, alkyl, and heteroaryl halides as catalyzed by palladium [42, 43, 45–47]. Several examples are shown below leading to pyrroles **45–47** [43, 45, 46].

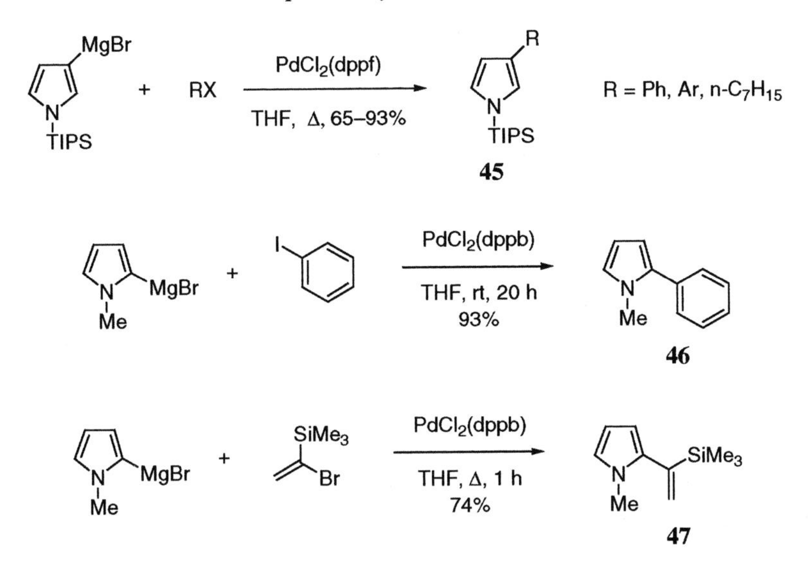


Noteworthy is the fact that one can utilize the appropriate bromopyrrole (e.g., **48**) in conjunction with the desired Grignard reagent in a one-step operation to afford the corresponding substituted pyrroles (e.g., **49**) [43]. The mixed pyrrole-pyridine heterocycle **50** was made in this fashion [45].

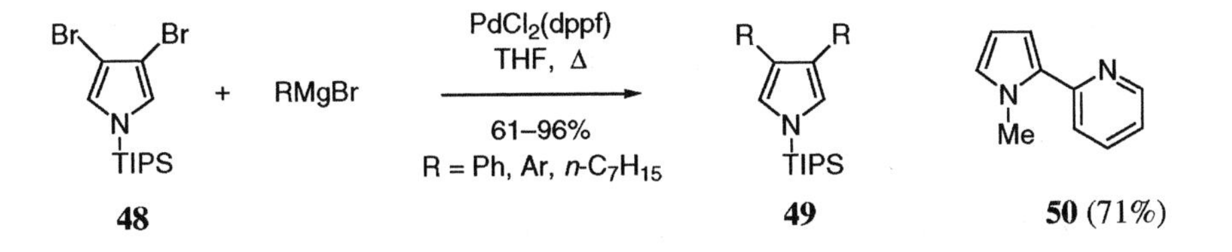


### 2.3.2 Negishi coupling

Pyrrolylzinc reagents are normally generated by transmetalation of lithiopyrroles with zinc chloride. As shown in the examples below, the resulting pyrrolylzinc species undergo the Negishi coupling with a range of aryl and acyl halides leading to pyrroles **46**, **51**, and **52** [23, 45, 50, 51]. Pyrrole **51** was synthesized in a search for novel near-infrared dyes [45].

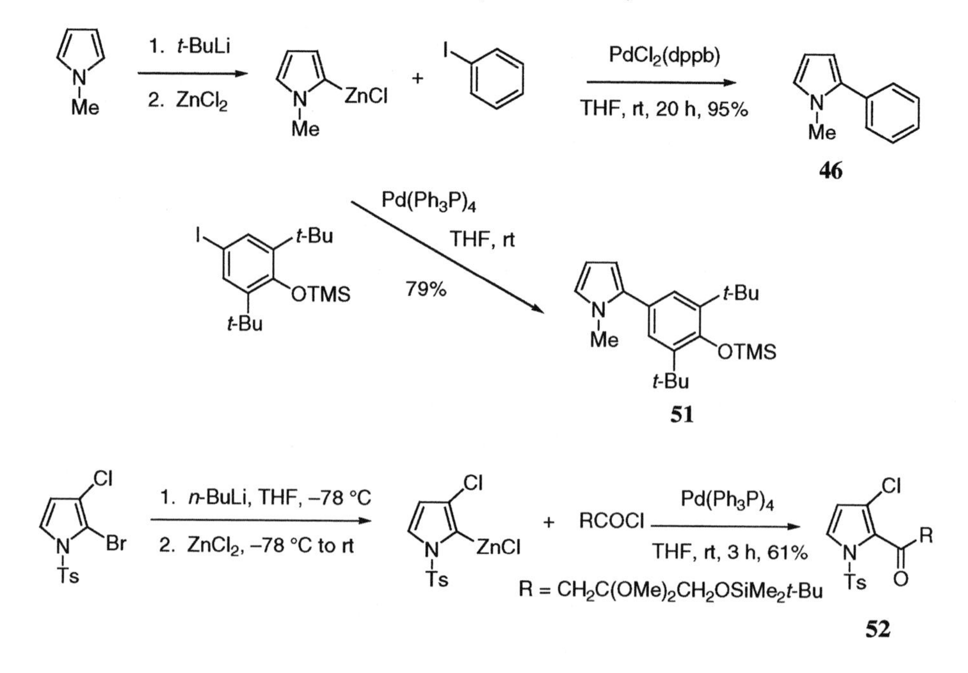


The *N*-pyrrolylzinc chloride **53** undergoes Pd-catalyzed coupling with perfluoroalkyl iodides to afford the 2-substituted pyrroles **54** in good yield [52]. Smaller amounts (15–20%) of 2-perfluoroalkanoyl pyrroles, which presumably arise by hydrolysis of the benzylic difluoromethylene group, are also formed. This reaction, which is performed in one pot, also affords 2-phenylpyrrole (75%) and 3-phenylpyrrole (5%) with iodobenzene. Some biphenyl (15%) is also formed.

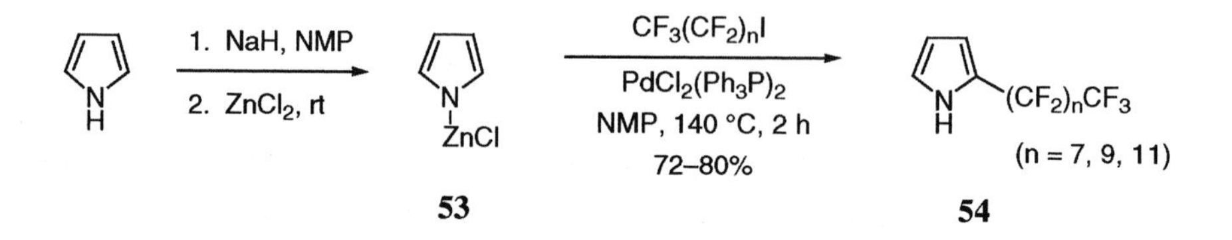


Dihydropyrrole triflate **55** is smoothly converted to the 2-phenyl derivative **56** under Negishi conditions [53]. This sequence represents a useful alternative to conventional pyrrole Negishi couplings.

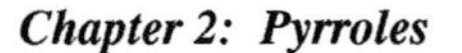

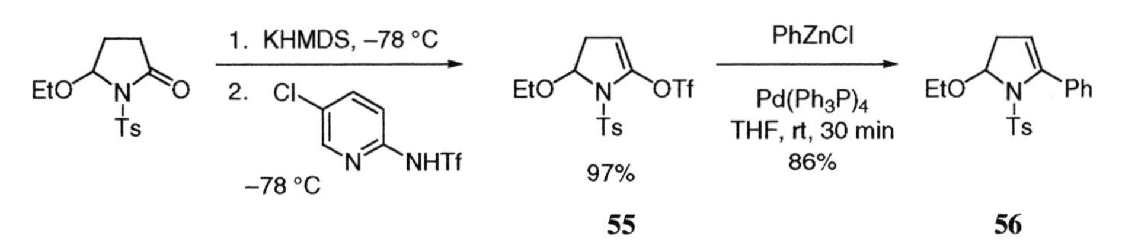


The Negishi coupling of halogenated porphyrins has also been described [54, 55].

### 2.3.3 Suzuki coupling

As we have seen in Chapter 1, the Suzuki coupling reaction is a powerful method for preparing biaryls and several applications in pyrrole chemistry have been described. Schlüter reported the first pyrrole boronic acid **57**, but, surprisingly, the related pyrrole borate **58** could not be hydrolyzed [15].

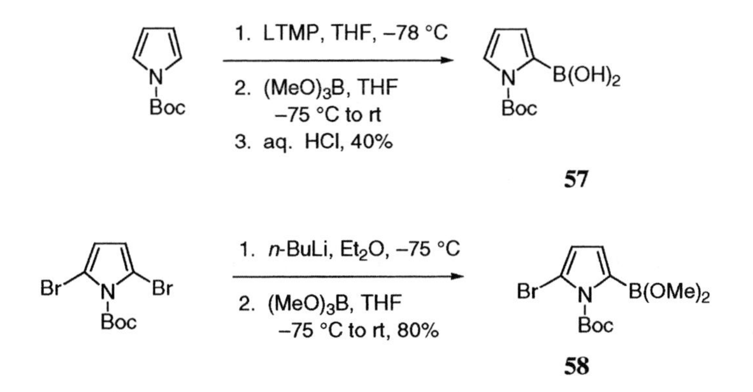


Gallagher has effected the Suzuki coupling reaction of **57** with a wide variety of aryl and heteroaryl halides to give pyrroles **59** as shown [56]. In all cases the major side product is the bipyrrole. Iodobenzene gives a higher yield than bromobenzene, but all of the other examples are aryl and heteroaryl bromides. This study included the synthesis of new, selective dopamine D3 receptor antagonist intermediates **60** and **61**, among others. The BOC group is readily cleaved in these compounds with methoxide in methanol at room temperature.

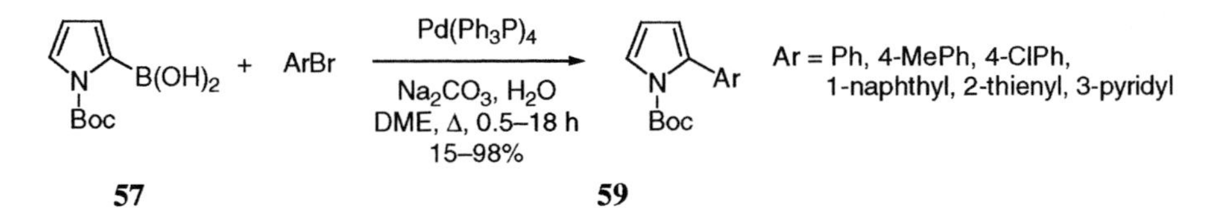

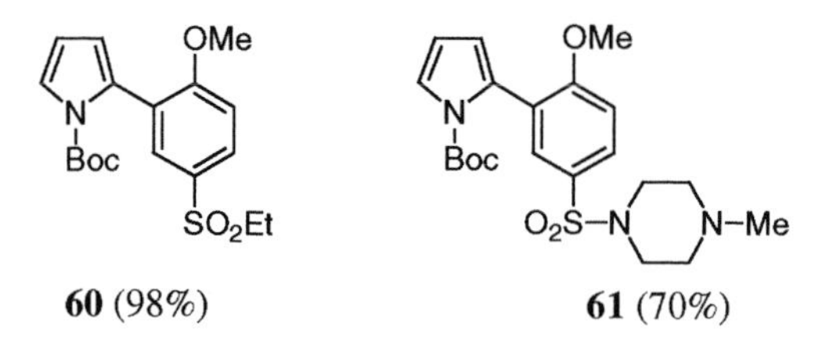


A synthesis of the immunosuppressive agent undecylprodigiosine involved a Suzuki coupling between **57** and triflate **62** to give tripyrrole **63** [57]. The corresponding Stille reaction was unsuccessful.

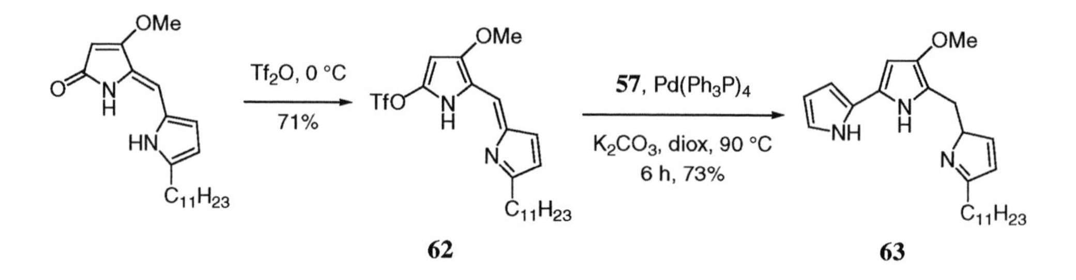


Ketcha has prepared 1-(phenylsulfonyl)pyrrole-2-boronic acid (**65**) in low yield and effected Suzuki coupling reactions to afford the corresponding 2-aryl derivatives **66** [58].

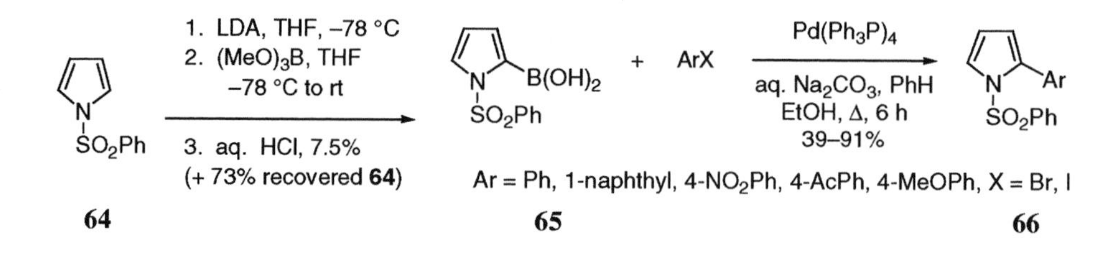


The pyrrole-3-boronic acid **67** has been prepared from the 3-iodopyrrole by Muchowski and subjected to a range of Suzuki couplings to afford 2-arylpyrroles **68** [59]. Subsequent fluoride deblocking to give **69** occurs in excellent yield.

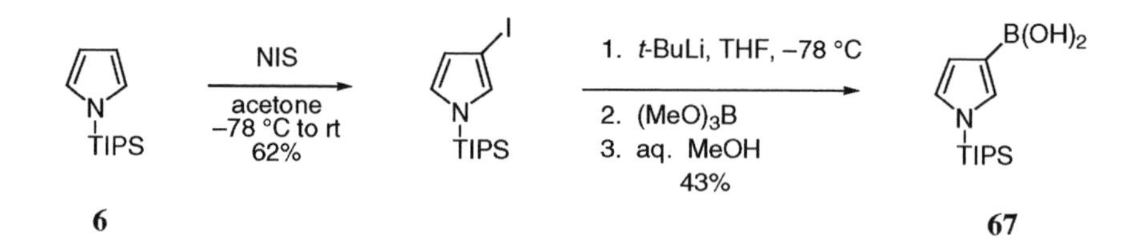

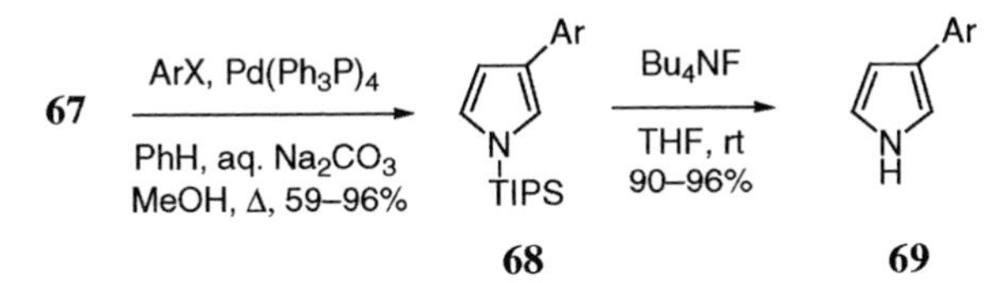


Ar = Ph, 4-PhMe, 4-PhOMe, 4-PhNHCOMe, 4-ClPh, 4-$NO_2$Ph, 4-$CO_2$MePh, 4-PhCHO, 4-PhCN, 2-PhMe, 3-PhOMe, 3-pyridyl; X = Br, I

The pyrrole component can also be employed as the aryl halide in Suzuki coupling with aryl boronic acids. Thus, Chang has effected several such reactions using phenylboronic acid and halopyrroles such as **70** and **71** [60].

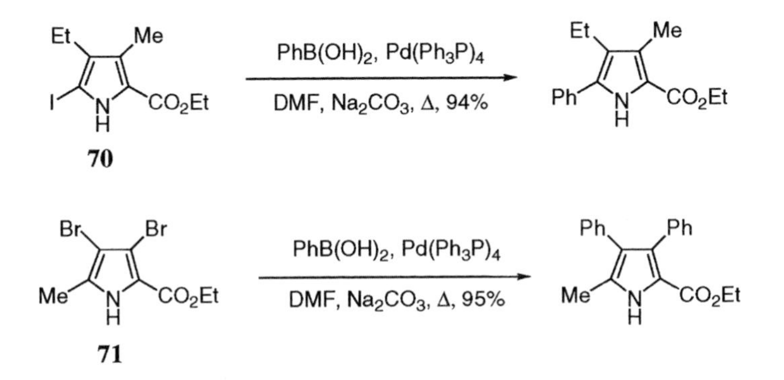


Using 2-bromopyrrole **72**, Burgess has synthesized 2-arylpyrroles **73** in excellent yield [61]. The resulting hydrolyzed pyrroles **74** were used to prepare 3,5-diaryl BODIPY® dyes.

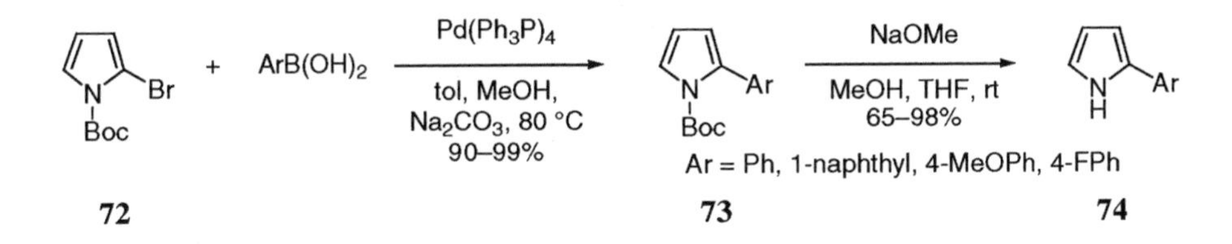


The Suzuki coupling has been utilized to craft β-octasubstituted tetramesitylporphyrins using various arylboronic acids [62], and Schlüter has adopted this reaction to prepare phenyl-pyrrole mixed polymers **75** [63]. The BOC group is easily removed by heating [64] and polymers with molecular weights of up to 23,000 were synthesized. These polymers are potentially interesting for their electrical and nonlinear optical properties [65].

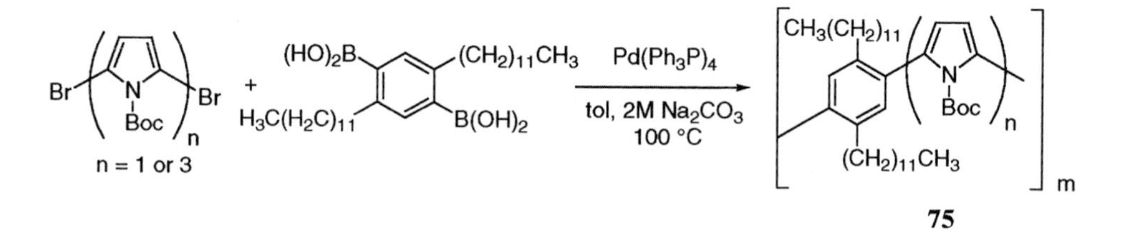


### 2.3.4 Stille coupling

Bailey described the first application of the Stille coupling to pyrroles, and one of the earliest examples of any such reaction involving heterocycles [66]. Lithiation of *N*-methylpyrrole and quenching with trimethylstannyl chloride gives 2-(trimethylstannyl)pyrrole (**76**), and palladium-catalyzed coupling with iodobenzene affords 1-methyl-2-phenylpyrrole (**46**) in good yield.

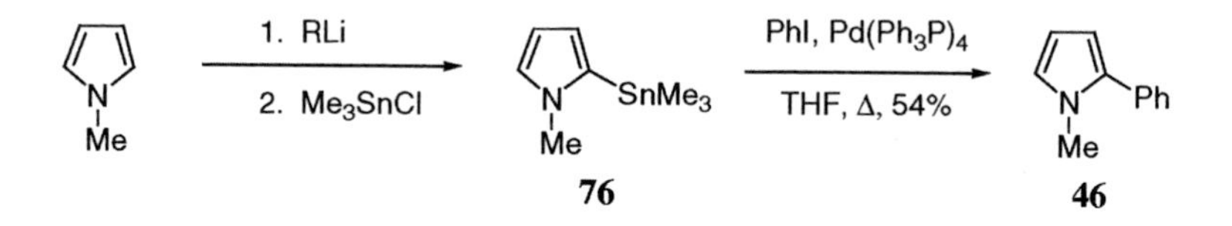


Mono- or bis-bromine-lithium exchange on dibromopyrrole **77** affords stannylpyrroles **78** or **79**, respectively, and *N*-BOC-2-trimethylstannylpyrrole is obtained in 75% yield from *N*-BOC pyrrole by lithiation with LTMP and quenching with $Me_3SnCl$ [15].

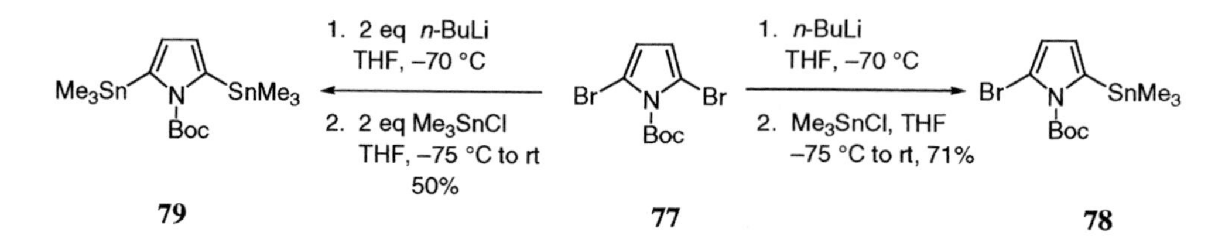


Direct lithiation of *N*-(dimethylamino)pyrrole (**80**) and subsequent quenching with trimethylstannyl chloride affords **81** in excellent yield [67]. The *N*-dimethylamino group can be removed with $Cr_2(OAc)_4{\bullet}2H_2O$.

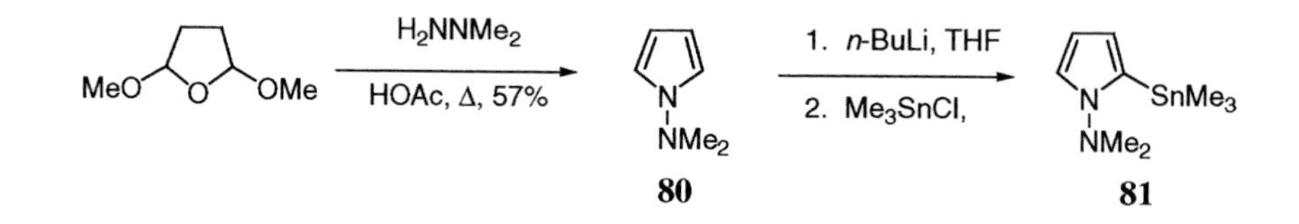

Although the stannylated pyrroles are normally obtained *via* lithiation, two other methods to prepare these Stille precursors have been devised. Caddick has found that the addition of tri-*n*-butylstannyl radical to pyrrole **82** affords stannylpyrrole **83** in good yield [68].

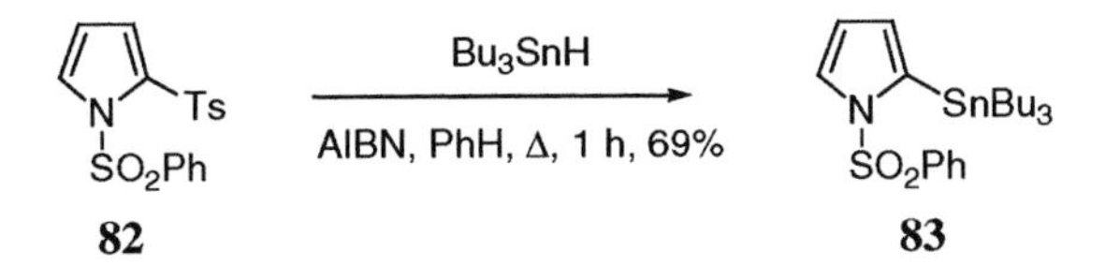


The second alternative synthesis of stannylpyrroles involves the versatile van Leusen pyrrole ring synthesis, illustrated below for the synthesis of **85** [69, 70]. Several examples were prepared in these studies, although at present the scope of this reaction is limited to pyrroles having an electron-withdrawing group at the 3-position. The structure of a BOC derivative of **85** ($R_1$ = Ph, $R_2$ = Bz) was established by X-ray crystallography [69].

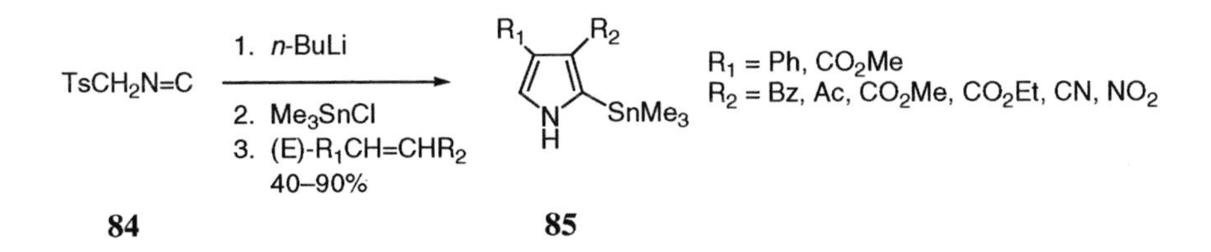


Stille couplings on *N*-protected derivatives of **85** proceed well, as shown below for the synthesis of pyrroles **86** [70].

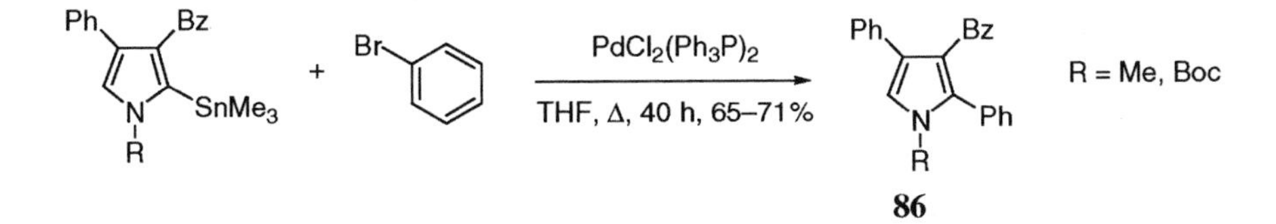


Dubac has employed a Stille coupling reaction to synthesize the pyrroles **88** and **89** from stannylpyrrole aldehyde **87** [71]. The latter tin compound was prepared as shown, and related stannylpyrroles were synthesized similarly [72] or using Muchowski's 6-dimethylamino-1-azafulvene dimer lithiation methodology [73].

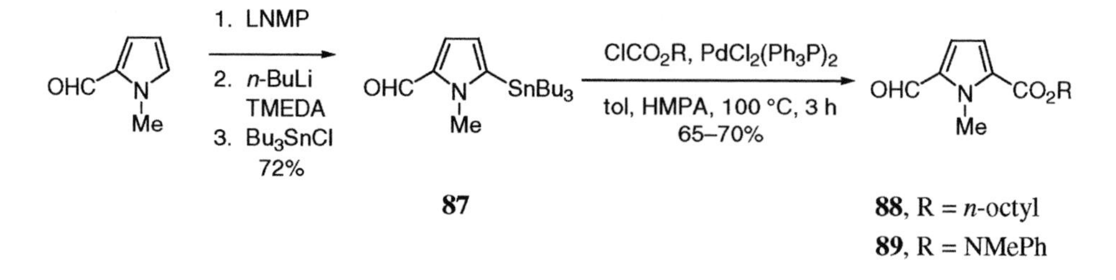


The related stannylpyrrole **90**, which was reported by Dubac [72], has been used to synthesize the sponge metabolite mycalazol 11 (**91**) and related compounds, which have activity against the P388 murine leukemia cell line [74].

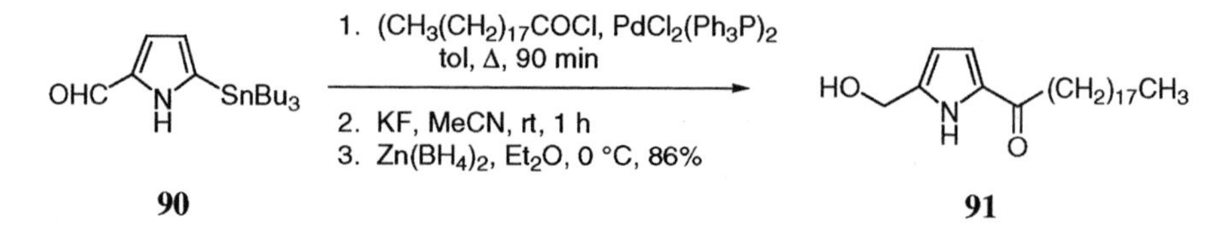


Stille couplings between stannylpyrrole **92** and siloles **93** and **94** have afforded the silole-pyrrole **95** [75]. These workers also prepared dimers and trimers of **95**.

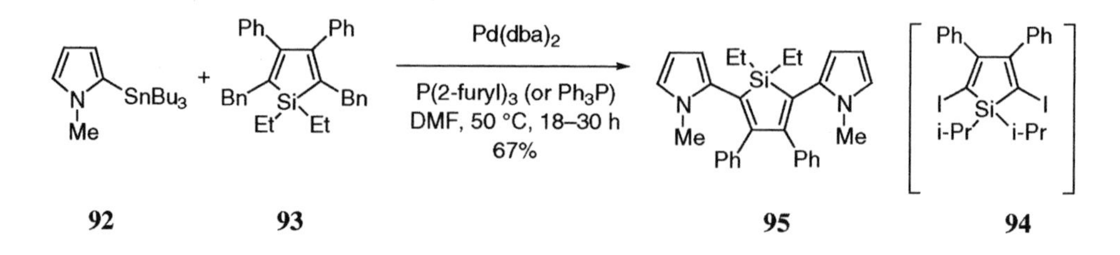


The Pd-catalyzed cross coupling reaction of 3-stannylated pyrroles is also known. Muchowski has thus prepared and utilized **96** to effect Stille couplings leading to **97** [59].

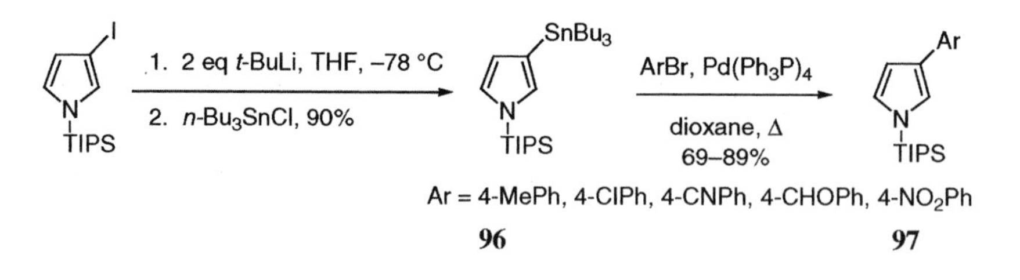

Pyrrole **98** has been employed in Stille couplings with dibromobenzoquinone **99** [76]. The product **100** can be subjected to a second Stille coupling to afford unsymmetrical diheteroarylquinones. Similar couplings between **98** and 2,3-dibromo-1,4-naphthoquinone were also described.

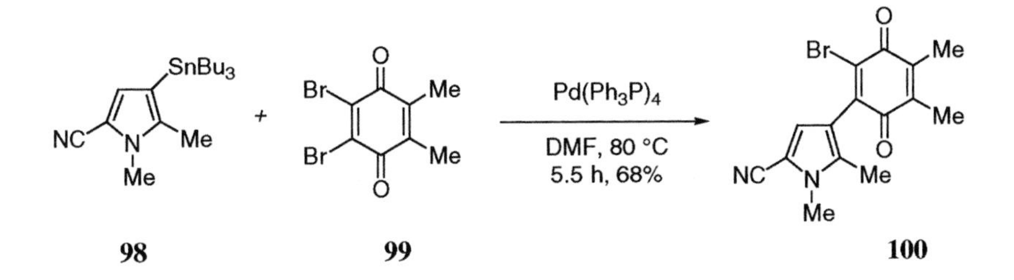


Halogenated pyrroles can serve as the aryl halide in Stille couplings with organotin reagents. Scott has used this idea to prepare a series of 3-vinylpyrroles, which are important building blocks for the synthesis of vinyl-porphyrins, bile pigments, and indoles [77]. Although 3-chloro- and 3-bromopyrroles fail completely or fared poorly in this chemistry, 3-iodopyrroles **101** work extremely well to yield 3-vinylpyrroles **102**.

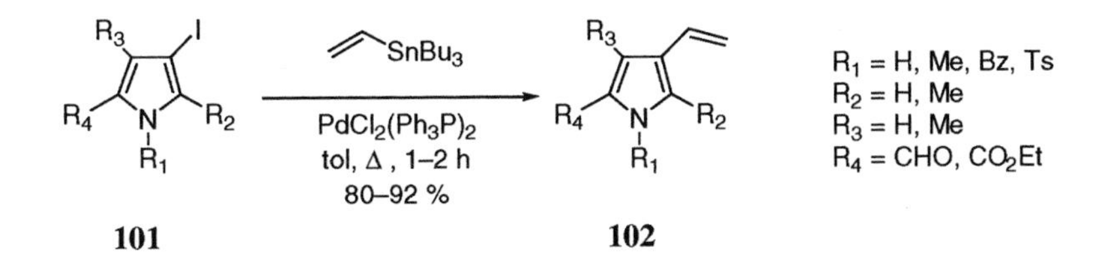


These workers have also used 3-iodopyrrole **103** to prepare the corresponding tin derivative **104** for Stille couplings to furnish 3-arylpyrroles **105** [78]. These pyrrole derivatives are important for the synthesis of β-aryl-substituted porphyrins for studies of heme catabolism. Also synthesized in this study were compounds **106** and **107** [78].

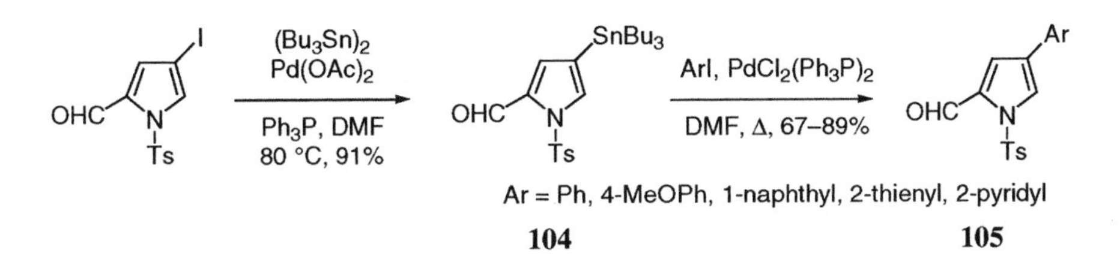

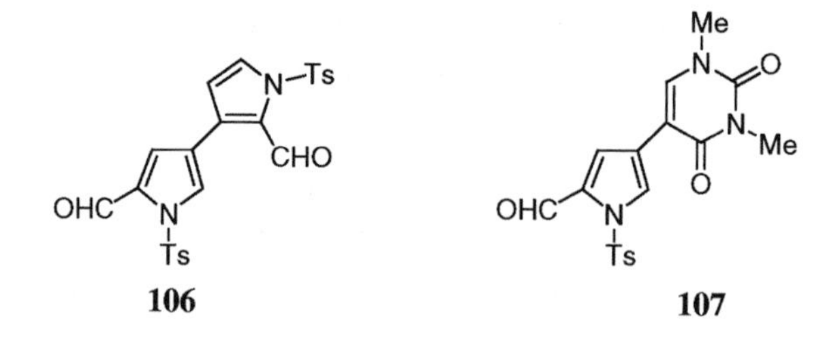


Tour has utilized dibromopyrrole **108** to prepare zwitterionic diiodopyrrole **110**, which in turn was employed in a synthesis of diphenyl derivative **111** and pyrrole polymers [10, 11].

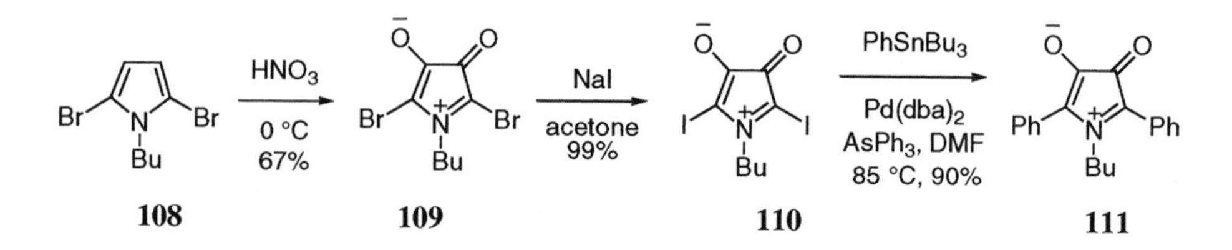


Stille has employed the Pd-catalyzed coupling that bears his name in syntheses of anthramycin and analogues [79, 80]. Thus, enol triflate **112** is smoothly coupled with acrylates to provide **113–115**.

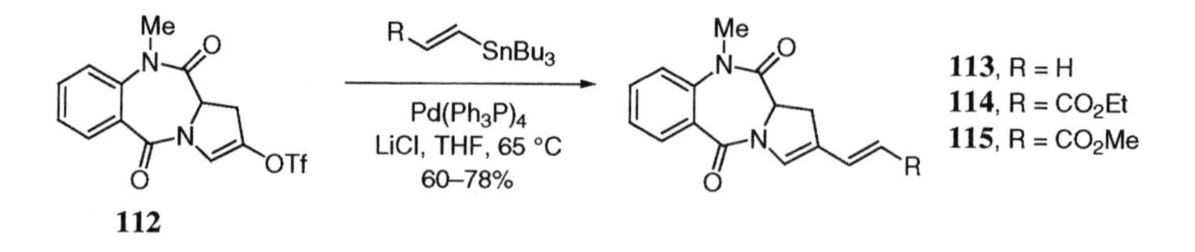


Dihydropyrrole triflate **55**, which was mentioned in Section 2.3.2, undergoes facile Stille couplings to yield, for example, 2-acylpyrrole derivative **116** after hydrolysis. Also prepared in these studies were the corresponding 2-vinyl (77%) and 2-phenyl (56%) derivatives [53, 81].

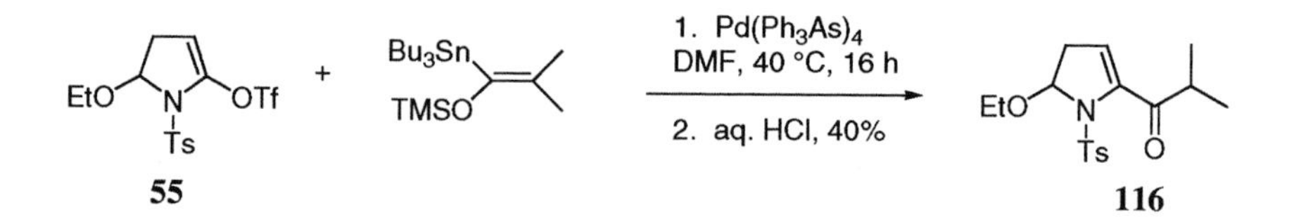


The Stille Pd-catalyzed cross coupling has been employed in the synthesis of modified porphyrins [54, 55, 82]. For example, the union of dihaloporphyrins with tri-*n*-butylvinylstannane affords protoporphyrin IX in excellent yield [82].

Pyridine **117** undergoes a Stille coupling with 1-methyl-2-trimethylstannylpyrrole to give **118** in 36% yield [83], and the same stannane has been joined with bromopyrimidines [84].

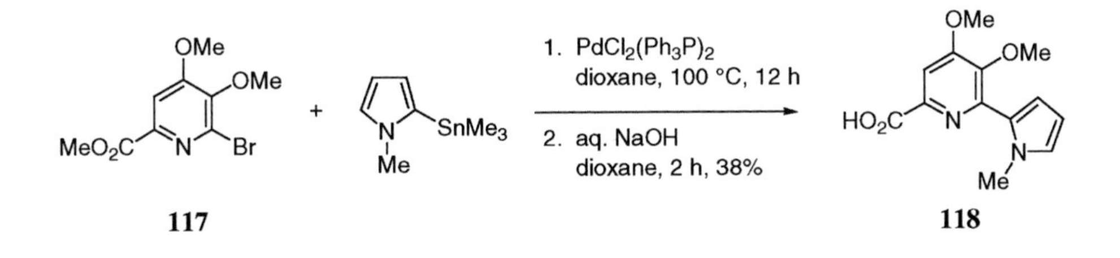


## 2.4 Sonogashira reaction

The Sonogashira reaction frequently serves as a platform for the construction of indoles, and we will explore this application in Chapter 3, but it also is a valuable method for the preparation of alkynyl pyrroles.

Muchowski has utilized *N*-TIPS-3-iodopyrrole to prepare a series of 3-alkynyl pyrroles **119** using standard Sonogashira conditions [59]. Pyrroles **120** were obtained after fluoride cleavage of the TIPS group. Similarly, 3,4-bis(alkynyl)pyrroles were prepared from *N*-TIPS-3,4-diiodopyrrole [59].

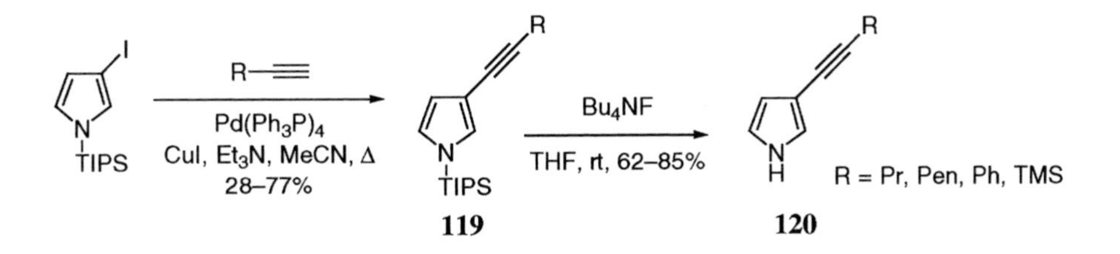


The coupling of trimethylsilylacetylene with 2,5-diiodo-1,3,4-trimethylpyrrole (**121**) affords the corresponding bis-acetylene after cleavage of the TMS groups [85].

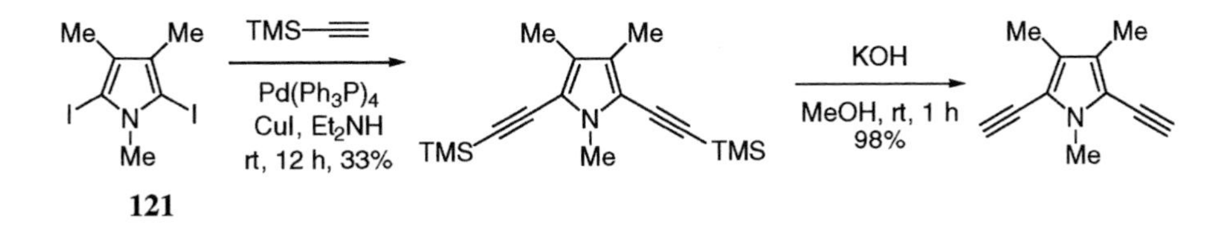


Tandem Sonogashira reactions starting with iodopyrrole **122** have furnished the dipyrrolylacetylene **124**, and dimerization of **123** gives **125** [86]. The Sonogashira coupling has also been reported for enol triflate **55** reacting with an *N*-protected propargyl amine [81].

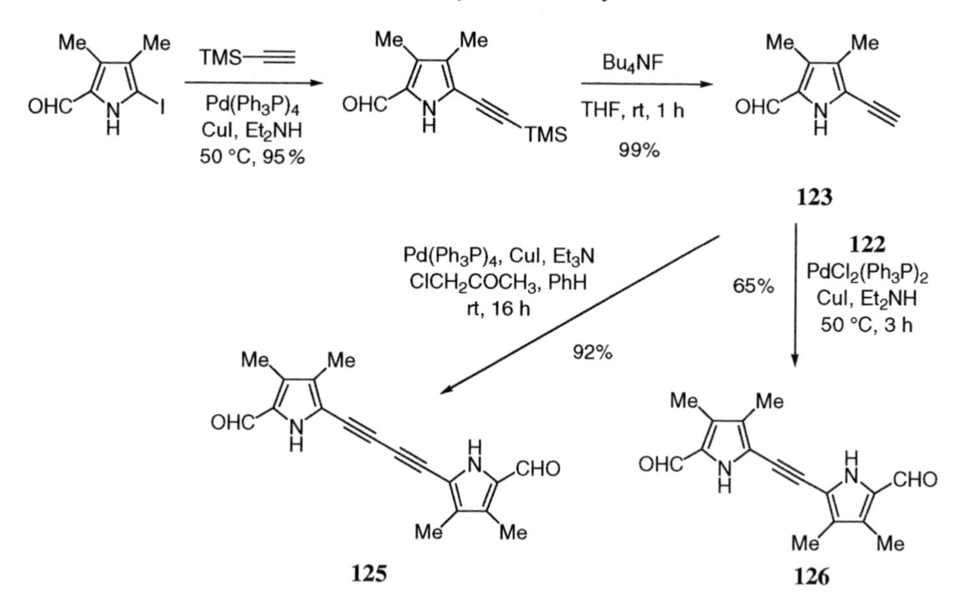


Dolphin has employed the Sonogashira coupling of terminal alkynes with zinc(II)-10-iodo-5,15-diphenylporphyrin (**126**) to prepare a series of alkynyl derivatives **127** [87], and similar syntheses of β-alkynyl porphyrins have been described [88].

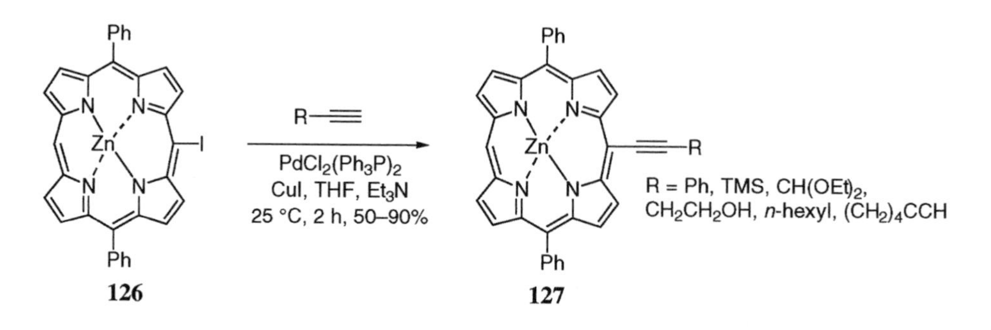


## 2.5 Heck and intramolecular Heck reactions

As will be seen in this chapter and in the rest of the book, the Heck reaction and its numerous variations represent a fantastically powerful set of tools available to the heterocyclic chemist. Although most Heck chemistry that involves pyrroles is intramolecular or entails synthesis of the pyrrole ring, a few intermolecular Heck reactions of pyrroles are known. Simple pyrroles (pyrrole, *N*-methylpyrrole, *N*-(phenylsulfonyl)pyrrole) react with 2-chloro-3,6-dialkylpyrazines under Heck conditions to give mixtures of C-2 and C-3 pyrrole-substituted pyrazines in low yields [89]. For example, chloropyrazine **128** reacts with *N*-(phenylsulfonyl)pyrrole to give a 5:2 mixture of coupled products (**129**, **130**). After alkali cleavage of the phenylsulfonyl group in **129** and **130**, the isomeric heterocycles could ultimately be separated.

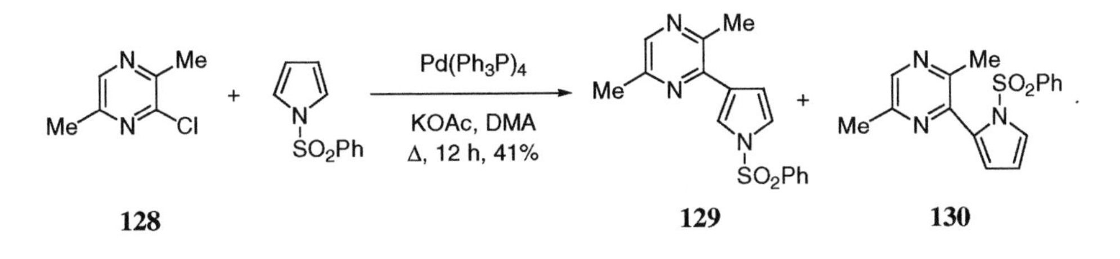


Pyrrole fails to undergo a Heck reaction with 1-bromoadamantane [90]. However, dihydropyrroles such as **131** and **132** undergo Heck reactions with ease, although the yields are variable [91–93]. Some examples are illustrated below. Since dihydropyrroles can be oxidized to pyrroles with a variety of reagents [94, 95], these Heck reactions of dihydropyrroles should constitute viable routes to pyrroles.

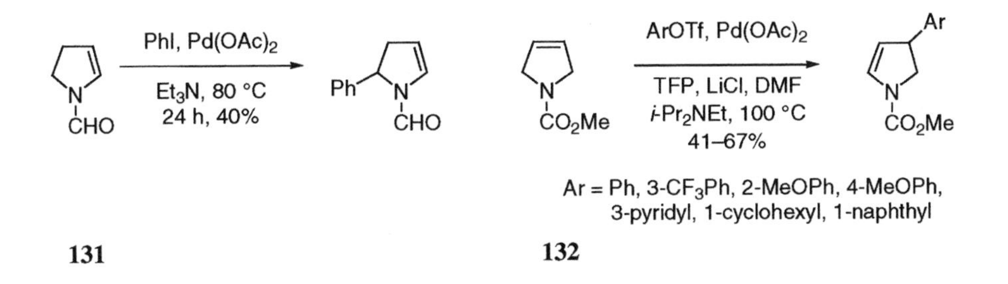


In these Heck reactions some degree of enantioselectivity (up to 83% *ee*) is achieved in the presence of (*R*)-BINAP, although the yields of Heck products are often very low in the highest degree of enantioselectivity (e.g., 19% isolated yield at 83% *ee*) [93]. An example of a tandem Heck reaction is shown below involving the arylation of dihydropyrrole **132** with 1-naphthyl triflate (**133**) [92]. Complete chirality transfer is observed for the arylation of **134** to **135**.

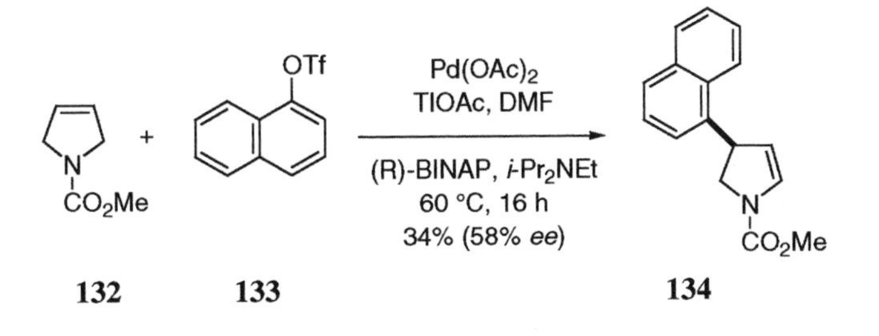

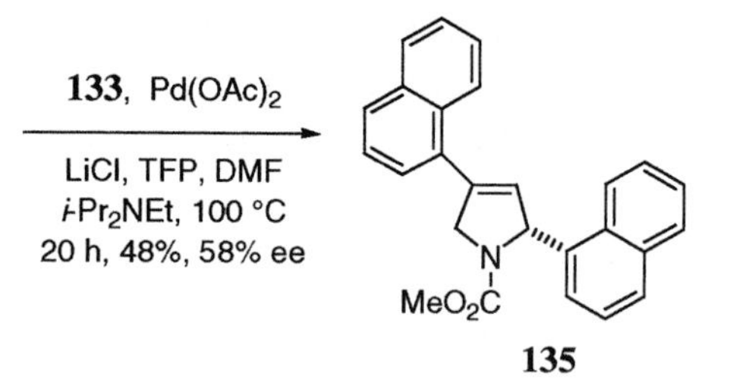


Stille employed a Heck reaction on triflate **136** to install the acrylamide side chain in his synthesis of anthramycin and analogues, as illustrated below [79, 80]. Ironically, a Stille reaction was less efficient in this transformation.

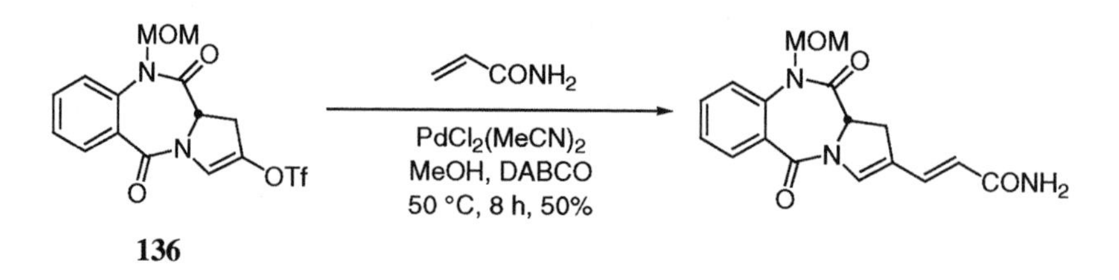


In a series of papers, rich in chemistry, Natsume has parlayed the intramolecular Heck reaction with **137** and **138** into an elegant construction of (+)-duocarmycin SA and related compounds [96–99].

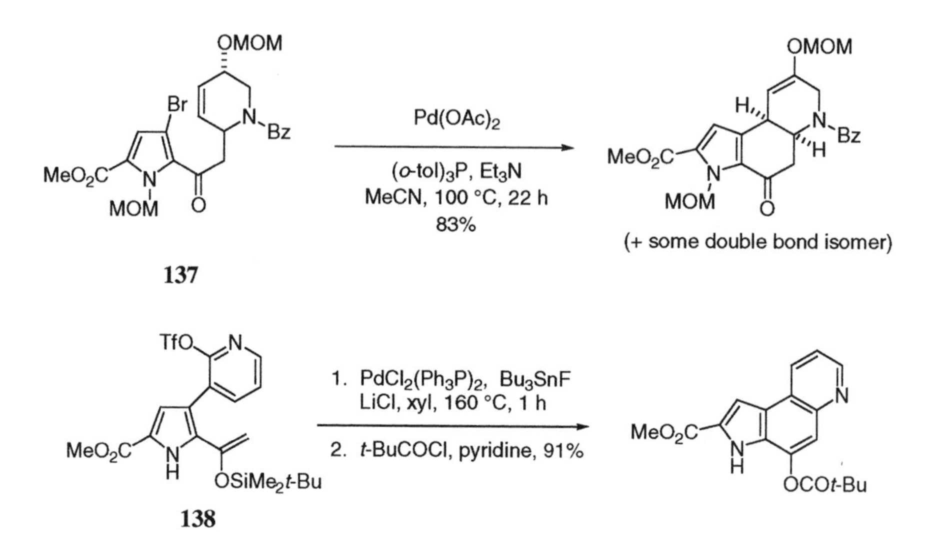


Grigg has utilized the Heck reaction in several ways, from the simple cyclization of *N*-(2-iodobenzoyl)pyrrole (**139**) to afford tricyclic lactam (**140**) [100] to the complex cascade transformation of pyrrole **141** with diphenylacetylene to tetracycle **144** [101]. The initial intermediate **142** undergoes a second Heck reaction with the pendant alkene to furnish **143**. A final insertion into the appropriate phenyl C-H bond completes the sequence.

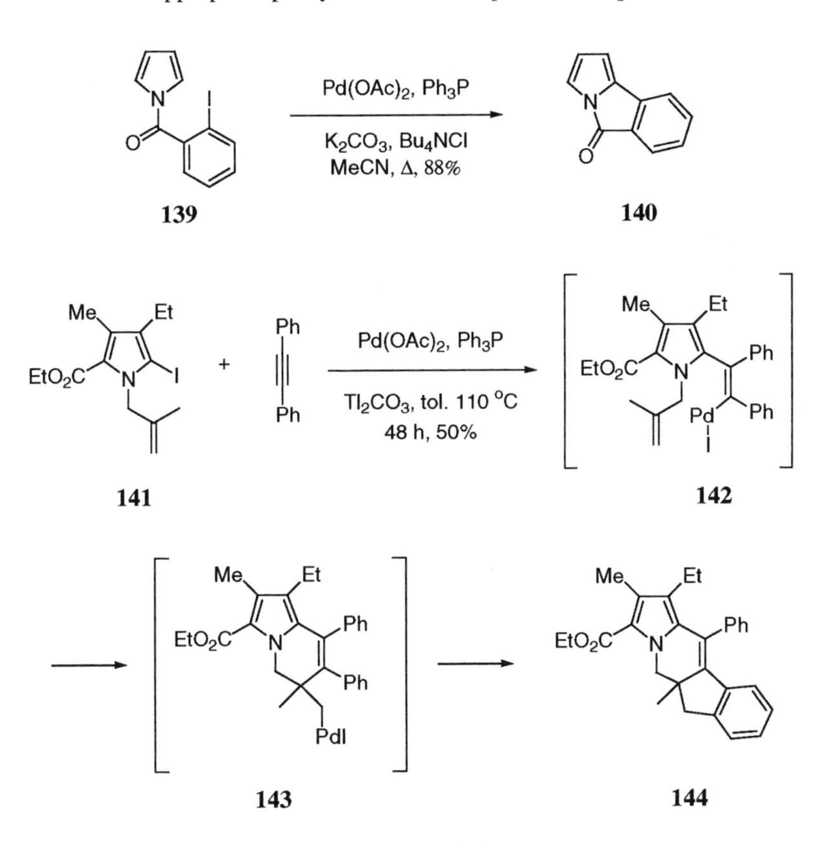


A related tandem Heck reaction is seen in the conversion of **145** to **146**, wherein the pyrrole ring is the site of termination [102].

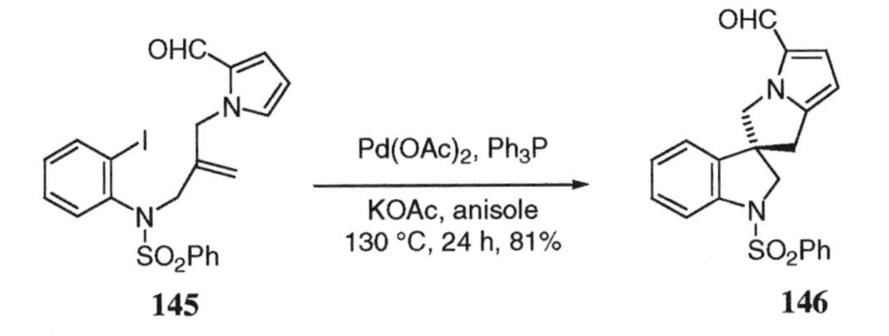


We will encounter more of these fantastic tandem cascade Pd-induced cyclizations in Chapter 3.

Several pyrrole-ring syntheses have been developed that utilize an intramolecular Heck reaction, but, since these transformations involve C-N bond formation, they are covered in Section 2.7. Grigg has employed the intramolecular Heck reaction to craft a series of spiro-pyrrolidines (**147**) [103], and Gronowitz has prepared several thienopyrroles, such as **148**, *via* a Heck strategy [104]. The *N*-BOC group in **148** is readily removed on mild heating after adsorption on silica gel (92% yield).

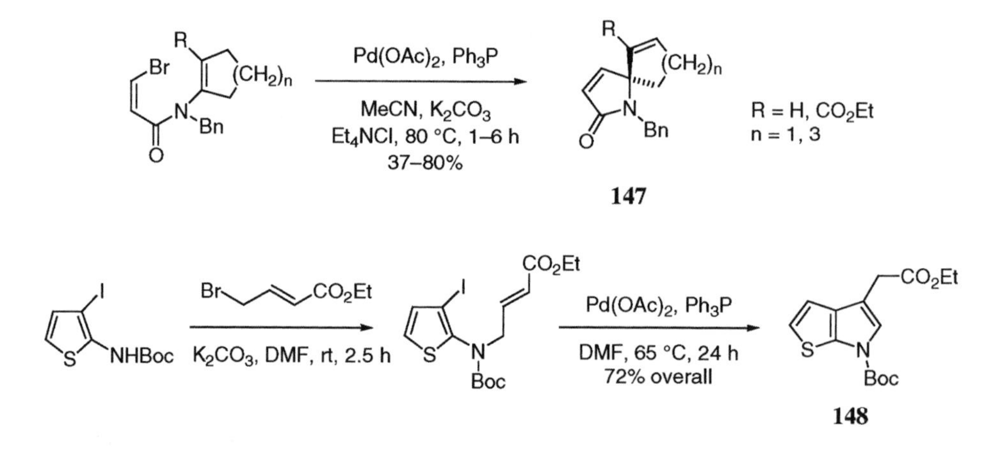


Since Heck reactions on metalated substrates are known (e.g., with organomercurials [105]), applications of these transformations to pyrrole chemistry have been reported. For example, mercuration of pyrrole **149** followed by exposure to methyl acrylate under Heck reaction conditions leads to **150** [106]. This Heck variation has been extended by Smith to mercurated porphyrins [107]. Pyrrolylacrylates like **150** have also been made using conventional Heck reactions on 3-iodopyrroles [108, 109].

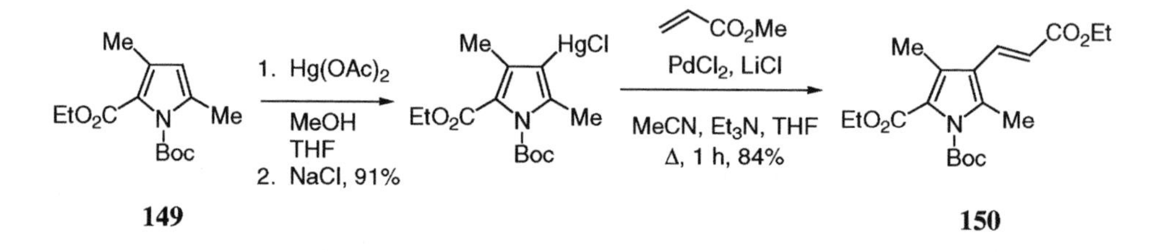


Pyrrole **151** can be thallated [110] and subjected to Heck (and Sonogashira) conditions to afford the anticipated products **152** [111].

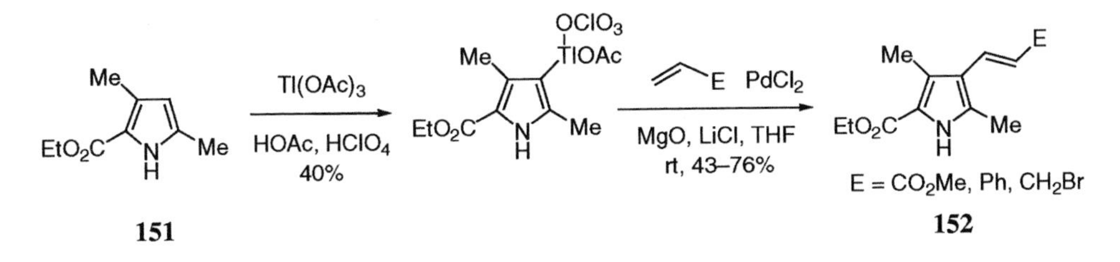


A Heck coupling of 2-lithio-1-methylpyrrole and bromohydrin **153** affords **154** [112].

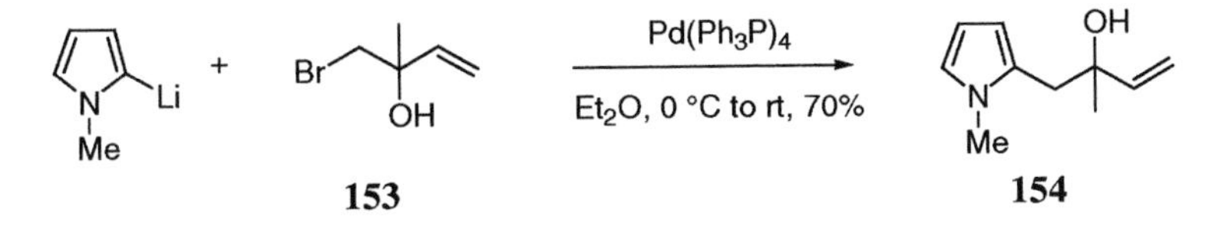


## 2.6 Carbonylation

As will be seen throughout the book, carbon monoxide has become an important player in the extension to carbonylation reactions. Since most of these involve CO at atmospheric pressure, this variation is quite accessible to the synthetic chemist.

Edstrom has utilized the carbonylation variation to engineer new routes to 3-substituted 4-hydroxyindoles, indolequinones, and mitosene analogs [113, 114]. For example, triflate **155** is converted to methyl ester **156** in high yield [114]. Subsequent oxidation affords indole **157**.

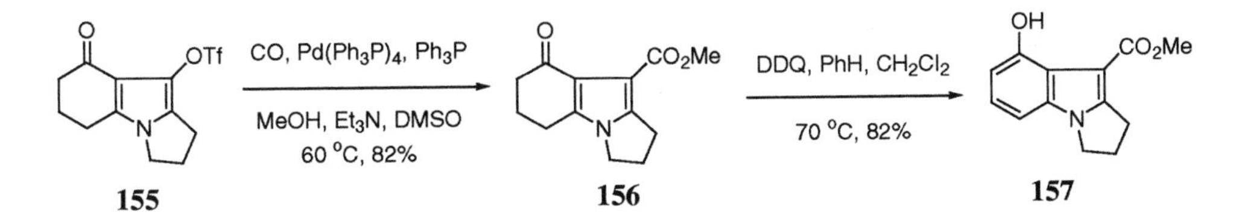


Dihydropyrrole triflate **55**, which we have encountered earlier in this chapter, undergoes a Pd-catalyzed carbonylation reaction to give ester **158** [53, 81]. A similar carbonylation sequence in the presence of the hydride donor *n*-$Bu_3SnH$ gives the corresponding aldehyde in 56% yield [81].

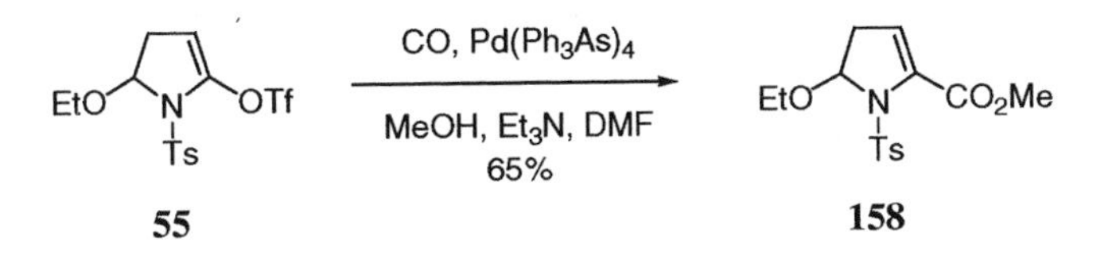

Under mercuration conditions, pyrrole itself reacts with a mixture of $Hg(OAc)_2$, $PdCl_2$, LiBr, CO, EtOH, and $Cu(OAc)_2$ to give 2-(ethoxycarbonyl)pyrrole, but in only 4% yield [115]. In contrast, using the thallation-palladium modification of the Heck reaction, Monti and Sleiter have prepared pyrrole ester **159** in high yield [111].

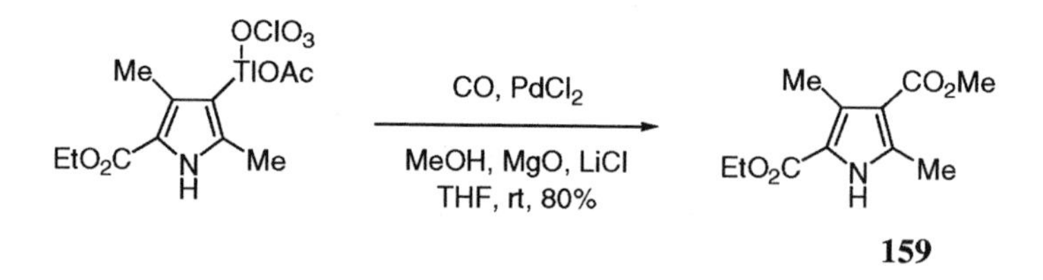


Cyclocarbonylation of pyrrole **160** leads in modest yield to 4-acetoxyindole **161** [116].

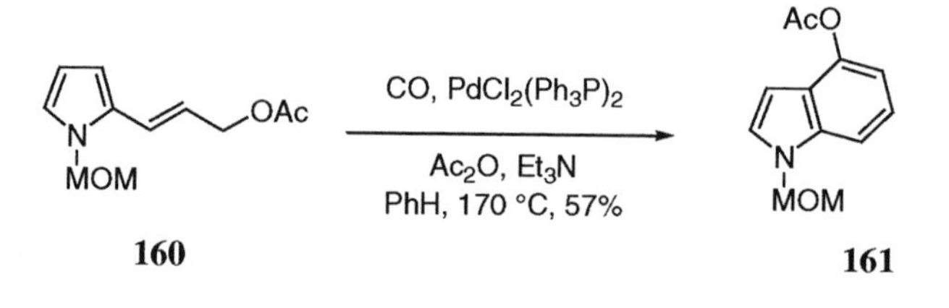


An interesting Pd-induced diyne cyclization-carbonylation sequence has been discovered by Chiusoli [117]. Thus, diyne **162** is transformed as shown into a mixture of (*E*)- and (*Z*)-bis(alkylidene)pyrrolidines **163** and **164**.

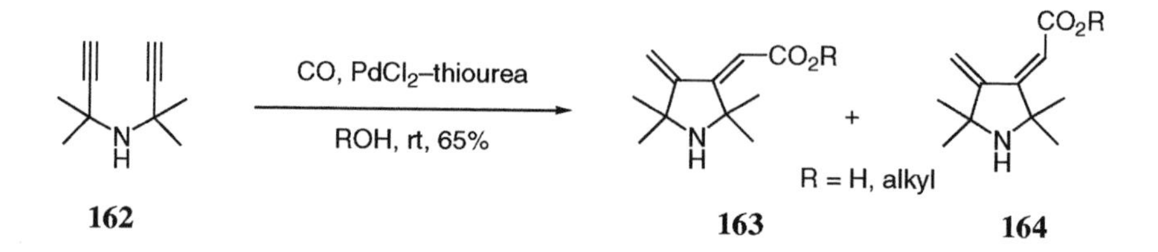


Ligand-directed palladation of 2-(dimethylaminomethyl)-1-(phenylsulfonyl)pyrrole (**165**) leads to the isolable palladium complex **166**. Exposure of **166** to CO yields pyrrole ester **167** in good yield [118].

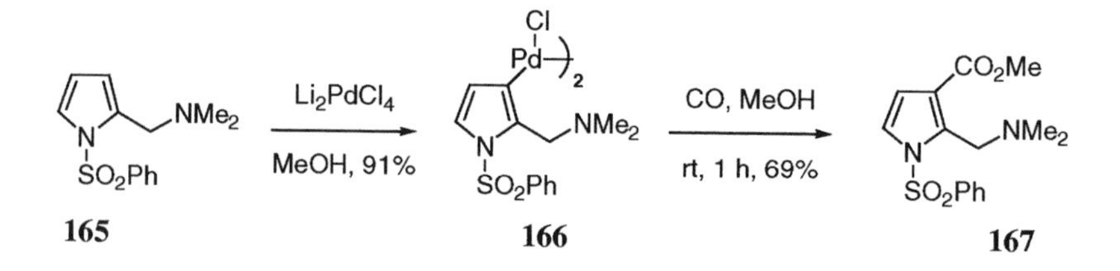

## 2.7 C—N bond formation reactions

Palladium-catalyzed reactions of alkenes containing nitrogen nucleophiles have proven to be a powerful methodology for C—N bond formation leading to pyrroles.

An early example of this strategy is the palladium black catalyzed conversion of (*Z*)-2-buten-1,4-diol with primary amines (cyclohexyl amine, 2-aminoethanol, *n*-hexyl amine, aniline) at 120 °C to give *N*-substituted pyrroles in 46–93% yield [119]. Trost extended this amination to the synthesis of a series of *N*-benzyl amines **169** from the readily available α-acetoxy-α-vinylketones **168** [120]. This methodology allowed for the facile preparation of pyrrolo-fused steroids.

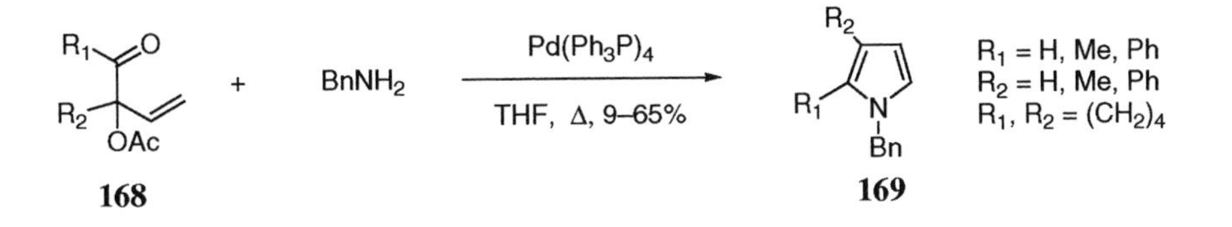


Fürstner has employed the Trost pyrrole synthesis in the first total synthesis of roseophilin, wherein this *N*-benzylpyrrole-ring forming step occurred in 70% yield [23]. Bäckvall has found that primary amines react with dienes under the guidance of Pd(II) to form pyrroles **170** in variable yields [121]. The intermediate π-allyl-palladium complexes are quite stable.

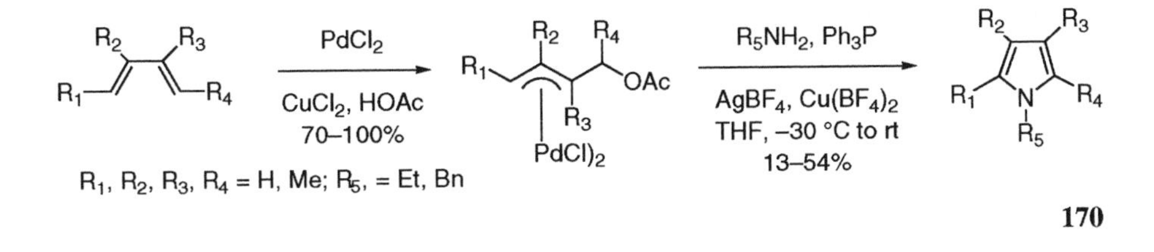


Most of the Pd-induced pyrrole ring forming reactions involve precursors having the requisite nitrogen nucleophile and alkene in the same molecule. Thus, as part of a general study of the amination of allylic systems, Genet and Bäckvall have observed that amine **171** cyclizes to pyrrole **172** [122]. Similar conditions and appropriate substrate design leads to bicyclic systems such as **173** and **174** where an amide nitrogen is a nucleophile [123].

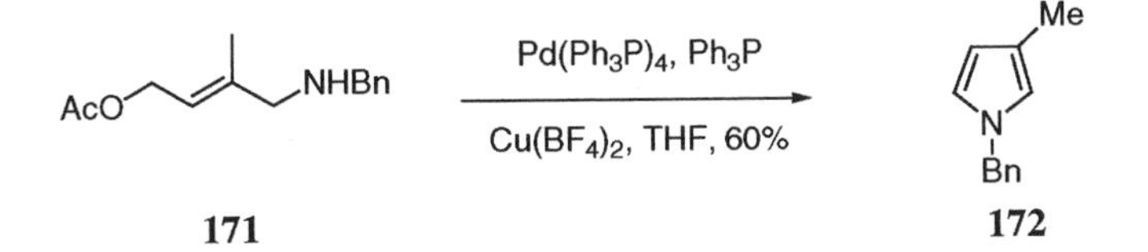

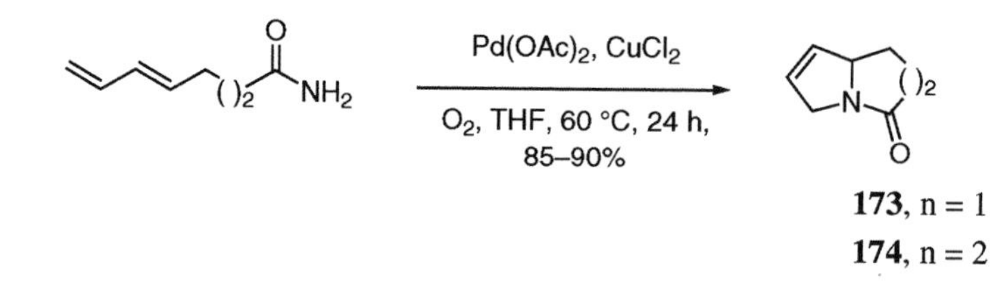


Likewise, the nitrogen in sulfonamides, such as **175** and **176**, serves very well in pyrrole-forming reactions, as evidenced by the two examples illustrated below [124, 125]. In the second reaction the hydroxyl group is essential for success, as is chloride (i.e., $Pd(OAc)_2$ does not work) [125]. It would seem that additional work is necessary to tame this latter reaction.

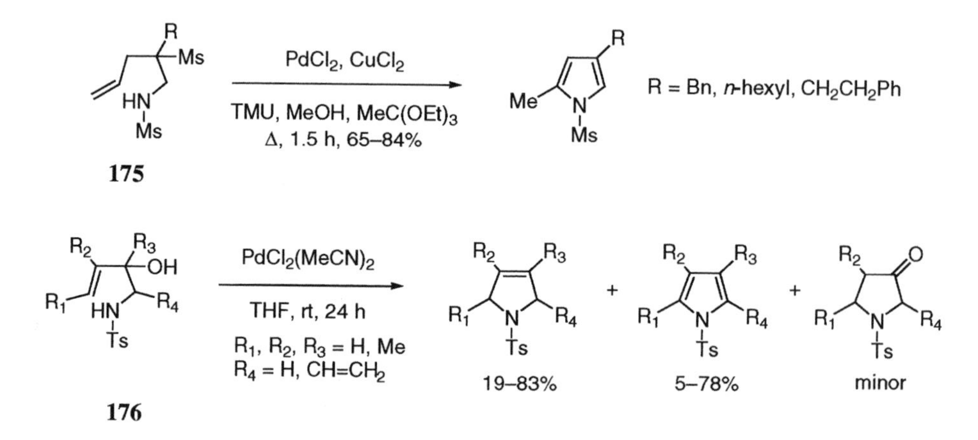


Cyclization of amino alcohol **177** leads to efficient chirality transfer into product dihydropyrrole **178** [126].

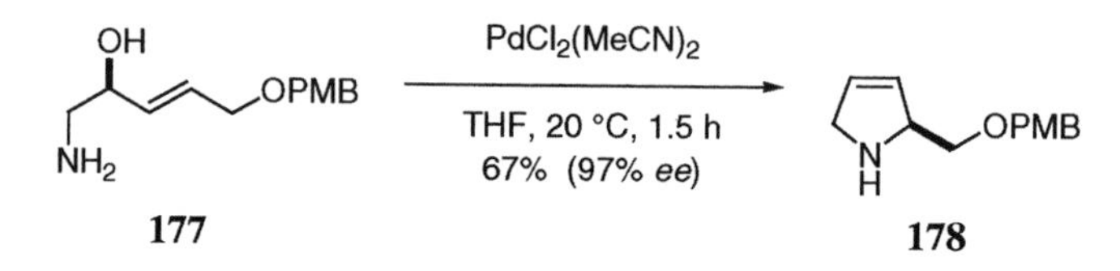


Alkynes can also serve as the recipient of cycloamination protocols. For example, 2,4-disubstituted pyrroles **180** are formed in high yields from Pd-catalyzed cyclization of aminoalkynes **179** [127]. Less effective is $Pd(Ph_3P)_4$ but $Pd(OAc)_2$ works as well as $PdCl_2$.

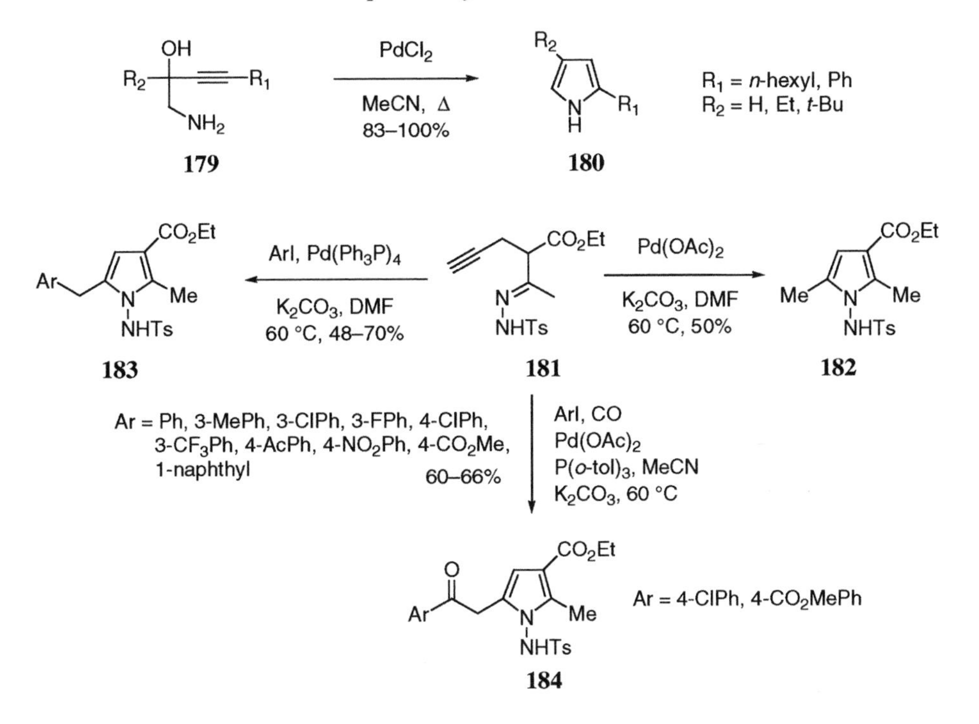


Arcadi has described a series of pyrrole ring-forming reactions leading to *N*-aminopyrrole derivatives **182–184** from common precursor **181**, depending on conditions [128]. Since *N*-aminopyrroles are widely used as pharmaceutical precursors, this chemistry could find wide appeal for the synthesis of medicinal agents.

The oxime nitrogen can also participate in cycloamination reactions to give pyrroles. Thus, treatment of oxime esters such as **185** with $Pd(Ph_3P)_4$ readily affords **186** [129]. The pentafluorophenyl group is necessary for good results; otherwise a Beckmann rearrangement can unfavorably enter the picture. The oxime stereochemistry makes no difference on the outcome of the reaction. In addition to **186**, pyrroles **187** and **188** were also prepared in this study (among others) [129].

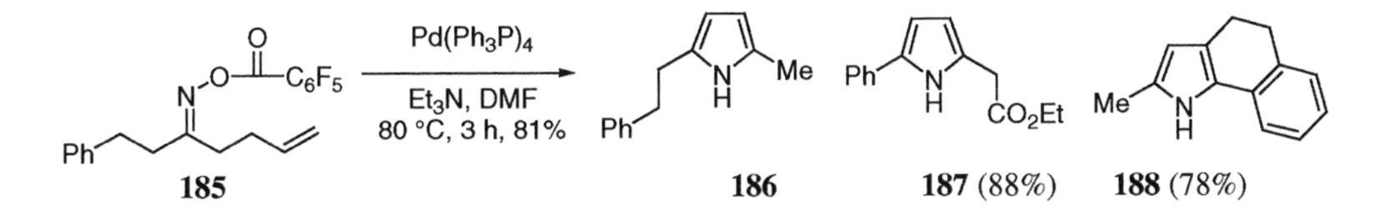


Larock has described the reaction of diphenylacetylene with iodosulfonamides **189** to give alkylidene dihydropyrroles **190** [130]. This ring-forming reaction is similar to the large number of related indole syntheses we will see in the next chapter. For example, **191** affords **192** under these conditions.

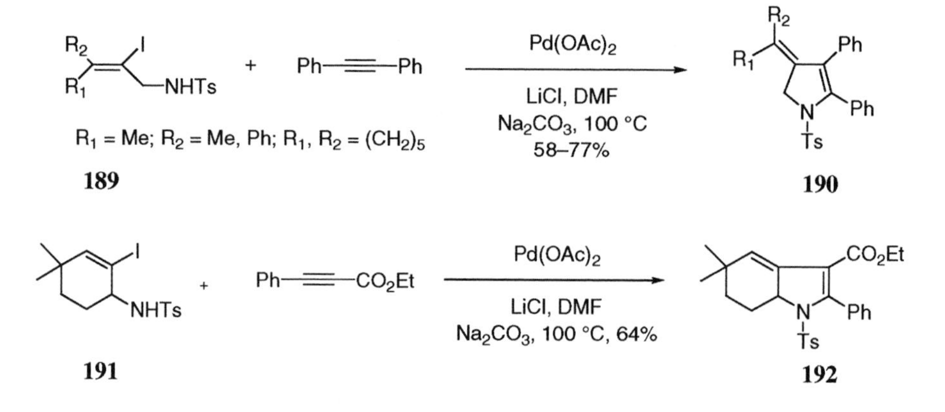


Using a similar Larock strategy, Gronowitz has synthesized a series of thienopyrroles, such as **193**, and other heterocyclic-fused pyrroles [131, 132]. Related reactions leading to indoles will be presented in the next chapter.

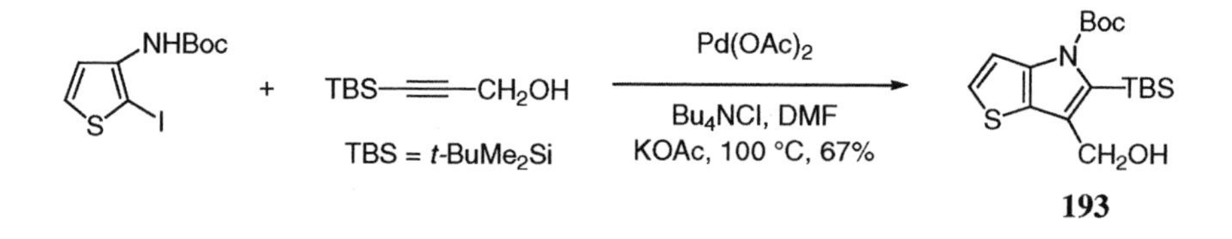


Two groups simultaneously found that trimethylsilyl cyanide reacts with disubstituted acetylenes in the presence of a Pd catalyst to form silylated 2-amino-5-cyanopyrroles **194** or **195** [133, 134]. These reactions are run neat and a variety of Pd species are successful in this transformation [133]. In the case of unsymmetrical diaryl acetylenes, the reaction is not regioselective [134].

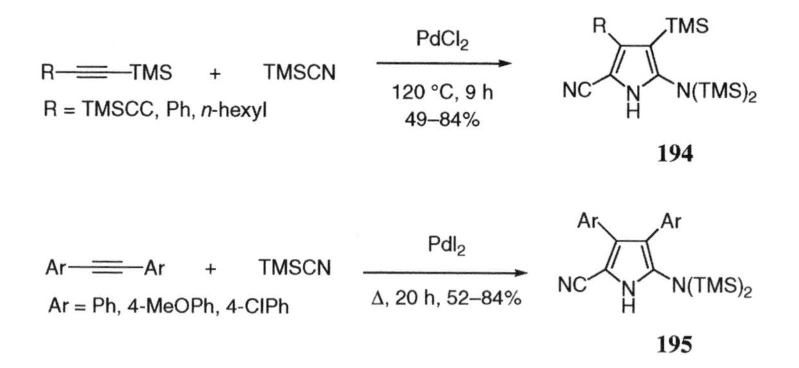

The powerful Buchwald-Hartwig aryl amination methodology [135–141] has been applied by Hartwig to the synthesis of *N*-arylpyrroles (**196**) [142, 143].

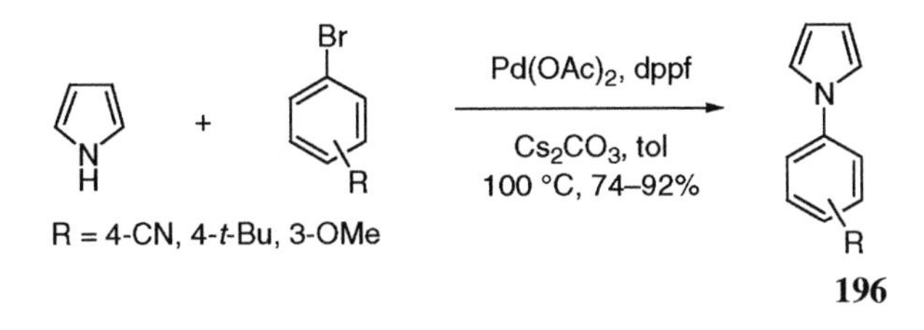


## 2.8 Miscellaneous

Several reactions and syntheses of pyrroles, which involve Pd catalysis, do not fall into the previous categories in this chapter, and therefore are presented here.

With regard to the Pd-catalyzed arylation of dihydropyrroles presented in Section 2.5, it is noteworthy that the readily available 2,5-dihydropyrroles **197** can be smoothly isomerized to 2,3-dihydropyrroles **198** under the influence of Pd [144].

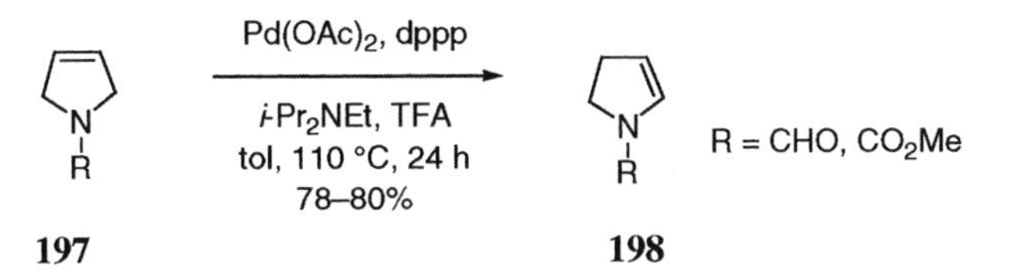


Iodopyrroles **199** can be conveniently deiodinated with formate as the hydride donor in the presence of Pd(0) [145]. This transformation is particularly important in the synthesis of dipyrromethanes for porphyrins and for linear pyrroles. Interestingly, no reaction occurs in refluxing THF.

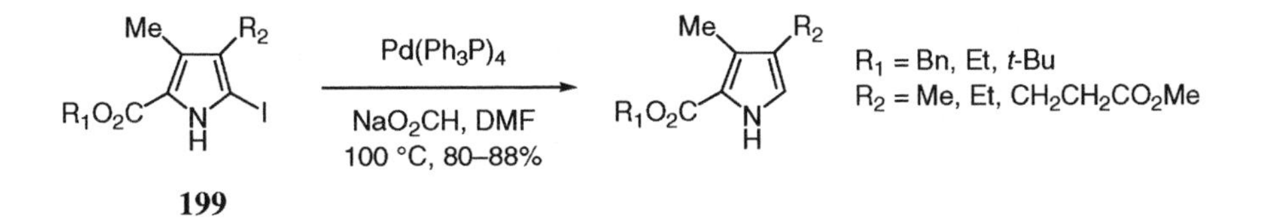


It is frequently necessary in synthesis to operate on the non-heterocycle portion of a molecule, and palladium technology often succeeds admirably in this regard. For example, *N*-(4-iodophenyl)pyrrole (**200**) is cyanated to **201** in excellent yield [146].

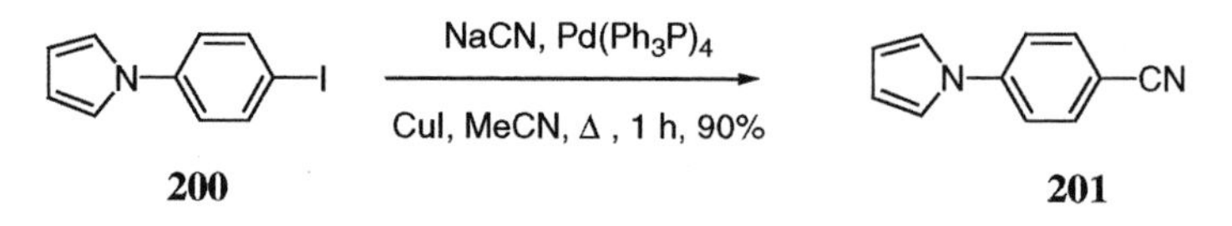


An efficient and novel route to 3,3'-bipyrroles, such as **203**, involves the oxidative-cyclization of cyclodiyne **202** [147].

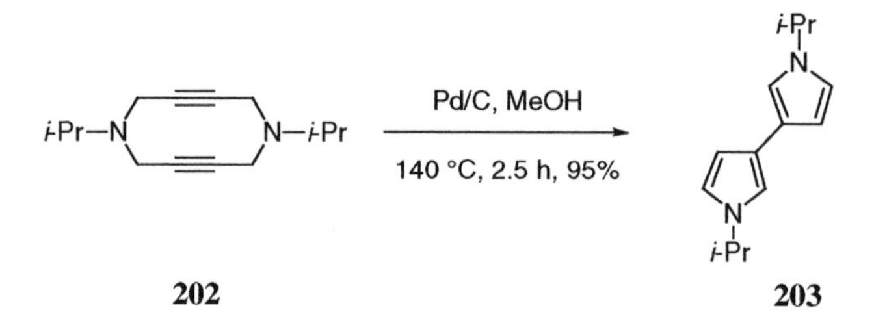


The use of Pd/C to effect decarbonylation reactions, which has long been known, is equally successful with pyrrole aldehydes, e.g., **204** to **205** [108].

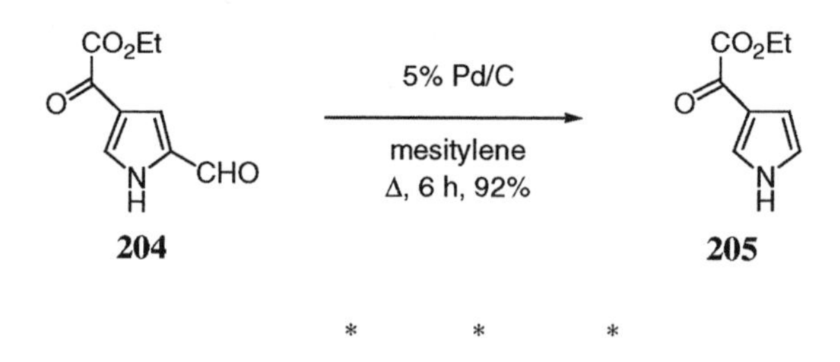


* * *

In conclusion, the already rich chemistry of pyrroles is greatly expanded by the palladium reactions presented in this chapter. The abundance of both 2- and 3-pyrrolyl halides and triflates has led to many examples of high-yielding Negishi, Suzuki, Stille, Sonogashira, and Heck reactions. Noteworthy are the excellent approaches to alkynyl pyrroles and porphyrins using Sonogashira, Stille, and Suzuki reactions.

## 2.9 References

1. Gribble, G. W. "*Pyrroles and their Benzo Derivatives*" In *Comprehensive Heterocyclic Chemistry* — Second Edition, Vol. 2; Katritzky, A. R.; Rees, C. W.; Scriven, E. F. V.; Pergamon: New York; Chapter 2.04, **1996**.
2. Gribble, G. W. *Prog. Chem. Org. Nat. Prod.* **1996**, *68*, 1–498.

3. (a) Gribble, G. W. *Chem. Soc. Rev.* **1999**, *28*, 335–46. (b) Gribble, G. W. *Environ. Sci. Pollut. Res.* **2000**, *7*, 37–49.
4. Tittlemier, S. A.; Simon, M.; Jarman, W. M.; Elliott, J. E.; Norstrom, R. J. *Environ. Sci. Technol.* **1999**, *33*, 26–33.
5. For a leading reference, see Martina, S.; Enkelmann, V.; Wegner, G.; Schlüter, A.-D. *Synth. Metals* **1992**, *51*, 299–305.
6. For a review of pyrrole chemistry, see Jones, R. A.; Bean, G. P. "The Chemistry of Pyrroles," Academic Press: New York, **1977**.
7. Anderson, H. J.; Loader, C. E. *Synthesis* **1985**, 353–64.
8. Cordell, G. A. *J. Org. Chem.* **1975**, *40*, 3161–9.
9. Gilow, H. M.; Burton, D. E. *J. Org. Chem.* **1981**, *46*, 2221–5.
10. Brockmann, T. W.; Tour, J. M. *J. Am. Chem. Soc.* **1995**, *117*, 4437–47.
11. Choi, D.-S.; Huang, S.; Huang, M.; Barnard, T. S.; Adams, R. D.; Seminario, J. M.; Tour, J. M. *J. Org. Chem.* **1998**, *63*, 2646–55.
12. Anderson, H. J.; Griffins, S. J. *Can. J. Chem.* **1967**, *45*, 2227–34.
13. Grehn, L.; Ragnarsson, U. *Angew. Chem., Int. Ed. Eng.* **1984**, *23*, 296–7.
14. Chen, W.; Cava, M. P. *Tetrahedron Lett.* **1987**, *28*, 6025–6.
15. Martina, S.; Enkelmann, V.; Wegner, G.; Schlüter, A.-D. *Synthesis* **1991**, 613–5.
16. Chadwick, D. J.; Hodgson, S. T. *J. Chem. Soc., Perkin Trans. 1* **1983**, 93–102.
17. Muchowski, J. M.; Solas, D. R. *Tetrahedron Lett.* **1983**, *24*, 3455–6.
18. Muchowski, J. M.; Naef, R. *Helv. Chim. Acta* **1984**, *67*, 1168–72.
19. Bray, B. L.; Mathies, P. H.; Naef, R.; Solas, D. R.; Tidwell, T. T.; Artis, D. R.; Muchowski, J. M. *J. Org. Chem.* **1990**, *55*, 6317–28.
20. Kozikowski, A. P.; Cheng, X.-M. *J. Org. Chem.* **1984**, *49*, 3239–40.
21. Stefan, K.-P.; Schuhmann, W.; Parlar, H.; Korte, F. *Chem. Ber.* **1989**, *122*, 169–74.
22. Ponasik, J. A.; Conova, S.; Kinghorn, D.; Kinney, W. A.; Rittschof, D.; Ganem, B. *Tetrahedron* **1998**, *54*, 6977–86.
23. Fürstner, A.; Weintritt, H. *J. Am. Chem. Soc.* **1998**, *120*, 2817–25.
24. Farnier, M.; Fournari, P. *Bull. Soc. Chim. Fr.* **1973**, 351–9.
25. Hollins, R. A.; Colnago, L. A.; Salim, V. M.; Seidl, M. C. *J. Heterocycl. Chem.* **1979**, *16*, 993–6.
26. For an excellent summary, see Gharpure, M.; Stoller, A.; Bellamy, F.; Firnau, G.; Snieckus, V. *Synthesis* **1991**, 1079–82.
27. Gribble, G. W.; Blank, D. H.; Jasinski, J. P. *Chem. Commun.* **1999**, 2195–6.
28. Asano, R.; Moritani, I.; Fujiwara, Y.; Teranishi, S. *Bull. Chem. Soc. Jpn.* **1973**, *46*, 663–4.
29. Itahara, T.; Kawasaki, K.; Ouseto, F. *Bull. Chem. Soc. Jpn.* **1984**, *57*, 3488–93.
30. Itahara, T. *J. Chem. Soc., Chem. Commun.* **1981**, 254–5.
31. Itahara, T. *J. Org. Chem.* **1985**, *50*, 5272–5.
32. Itahara, T. *J. Chem. Soc., Chem. Commun.* **1980**, 49–50.

33. Itahara, T. *Heterocycles* **1986**, *24*, 2557–62.
34. Itahara, T. *J. Org. Chem.* **1985**, *50*, 5546–50.
35. Boger, D. L.; Patel, M. *Tetrahedron Lett.* **1987**, *28*, 2499–502.
36. Boger, D. L.; Patel, M. *J. Org. Chem.* **1988**, *53*, 1405–15.
37. For reviews of indole ring syntheses, see (a) Gribble, G. W. *Contemp. Org. Synth.* **1994**, *1*, 145–72. (b) Gribble, G. W. *Perkin Trans. 1* **2000**, 1045–75.
38. Yokoyama, Y.; Suzuki, H.; Matsumoto, S.; Sunaga, Y.; Tani, M.; Murakami, Y. *Chem. Pharm. Bull.* **1991**, *39*, 2830–6.
39. Arcadi, A.; Rossi, E. *Synlett* **1997**, 667–8.
40. Arcadi, A.; Rossi, E. *Tetrahedron* **1998**, *54*, 15253–72
41. Aoyagi, Y.; Mizusaki, T.; Ohta, A. *Tetrahedron Lett.* **1996**, *37*, 9203–6.
42. Bumagin, N. A.; Sokolova, A. F.; Beletskaya, I. P.; Wolz, G. *Russ. J. Org. Chem.* **1993**, *29*, 136–7.
43. Bumagin, N. A.; Nikitina, A. F.; Beletskaya, I. P. *Russ. J. Org. Chem.* **1994**, *30*, 1619–29.
44. Abarbri, M.; Dehmel, F.; Knochel, P. *Tetrahedron Lett.* **1999**, *40*, 7449–53.
45. Minato, A.; Tamao, K.; Hayashi, T.; Suzuki, K.; Kumada, M. *Tetrahedron Lett.* **1981**, *22*, 5319–22.
46. Minato, A.; Suzuki, K.; Tamao, K.; Kumada, M. *Tetrahedron Lett.* **1984**, *25*, 83–6.
47. Minato, A.; Suzuki, K.; Tamao, K.; Kumada, M. *J. Chem. Soc., Chem. Commun.* **1984**, 511–3.
48. Hayashi, T.; Konishi, M.; Kumada, M. *Tetrahedron Lett.* **1979**, 1871–4.
49. Widdowson, D. A.; Zhang, Y.-Z. *Tetrahedron* **1986**, *42*, 2111–6.
50. Sakamoto, T.; Kondo, Y.; Takazawa, N.; Yamanaka, H. *Heterocycles* **1993**, *36*, 941–2.
51. Takahashi, K.; Gunji, A. *Heterocycles* **1996**, *43*, 941–4.
52. Filippini, L.; Gusmeroli, M.; Riva, R. *Tetrahedron Lett.* **1992**, *33*, 1755–8.
53. Luker, T.; Hiemstra, H.; Speckamp, W. N. *Tetrahedron Lett.* **1996**, *37*, 8257–60.
54. DiMagno, S. G.; Lin, V. S.-Y.; Therien, M. J. *J. Org. Chem.* **1993**, *58*, 5983–93.
55. DiMagno, S. G.; Lin, V. S.-Y.; Therien, M. J. *J. Am. Chem. Soc.* **1993**, *115*, 2513–5.
56. Johnson, C. N.; Stemp, G.; Anand, N.; Stephen, S. C.; Gallagher, T. *Synlett* **1998**, 1025–7.
57. D'Alessio, R.; Rossi, A. *Synlett* **1996**, 513–4.
58. Grieb, J. G.; Ketcha, D. M. *Synth. Commun.* **1995**, *25*, 2145–53.
59. Alvarez, A.; Guzmán, A.; Ruiz, A.; Velarde, E.; Muchowski, J. M. *J. Org. Chem.* **1992**, *57*, 1653–6.
60. Chang, C.K.; Bag, N. *J. Org. Chem.* **1995**, *60*, 7030–2.
61. Thoresen, L. H.; Kim, H.; Welch, M. B.; Burghart, A.; Burgess, K. *Synlett* **1998**, 1276–8.
62. Zhou, X.; Zhou, Z.; Mak, T. C. W.; Chan, K. S. *J. Chem. Soc., Perkin Trans. 1* **1994**, 2519–20.

63. Martina, S.; Schlüter, A.-D. *Macromolecules* **1992**, *25*, 3607–8.
64. Rawal, V. H.; Cava, M. P. *Tetrahedron Lett.* **1985**, *26*, 6141–4.
65. Brédas, J. L.; Chance, R. R., Eds. Conjugated Polymeric Materials: Opportunities in Electronics, Optoelectronics, and Molecular Electronics; Kluwer Academic Publishers: Dordrecht, The Netherlands, **1990**.
66. Bailey, T. R. *Tetrahedron Lett.* **1986**, *27*, 4407–10.
67. Martinez, G. R.; Grieco, P. A.; Srinivasan, C. V. *J. Org. Chem.* **1981**, *46*, 3760–1.
68. Caddick, S.; Joshi, S. *Synlett* **1992**, 805–6.
69. Meetsma, A.; Dijkstra, H. P.; Ten Have, R.; van Leusen, A. M. *Acta Cryst.* **1996**, *C52*, 2747–50.
70. Dijkstra, H. P.; Ten Have, R.; van Leusen, A. M. *J. Org. Chem.* **1998**, *63*, 5332–8.
71. Jousseaume, B.; Kwon, H.; Verlhac, J.-B.; Denat, F.; Dubac, J. *Synlett* **1993**, 117–8.
72. Denat, F.; Gaspard-Iloughmane, H.; Dubac, J. *J. Organomet. Chem.* **1992**, *423*, 173–82.
73. Muchowski, J. M.; Hess, P. *Tetrahedron Lett.* **1988**, *29*, 777–80.
74. Nabbs, B. K.; Abell, A. D. *Bioorg. Med. Chem. Lett.* **1999**, *9*, 505–8.
75. Tamao, K.; Ohno, S.; Yamaguchi, S. *Chem. Commun.* **1996**, 1873–4.
76. Yoshida, S.; Kubo, H.; Saika, T.; Katsumura, S. *Chem. Lett.* **1996**, 139–40.
77. Wang, J.; Scott, A. I. *Tetrahedron Lett.* **1995**, *36*, 7043–6.
78. Wang, J.; Scott, A. I. *Tetrahedron Lett.* **1996**, *37*, 3247–50.
79. Peña, M. R.; Stille, J. K. *Tetrahedron Lett.* **1987**, *28*, 6573–6.
80. Peña, M. R.; Stille, J. K. *J. Am. Chem. Soc.* **1989**, *111*, 5417–24.
81. Bernabé, P.; Rutjes, F. P. J. T.; Hiemstra, H.; Speckamp, W. H. *Tetrahedron Lett.* **1996**, *37*, 3561–4.
82. Minnetian, O. M.; Morris, I. K.; Snow, K. M.; Smith, K. M. *J. Org. Chem.* **1989**, *54*, 5567–74.
83. Massa, M. A.; Patt, W. C.; Ahn, K.; Sisneros, A. M.; Herman, S. B.; Doherty, A. *Bioorg. Med. Chem. Lett.* **1998**, *8*, 2117–22.
84. Peters, D.; Hörnfeldt, A.-B.; Gronowitz, S. *J. Heterocycl. Chem.* **1990**, *27*, 2165–73.
85. Ortaggi, G.; Scarsella, M.; Scialis, R.; Sleiter, G. *Gazz. Chim. Ital.* **1988**, *118*, 743–4.
86. (a) Cho, D. H.; Lee, J. H.; Kim, B. H. *J. Org. Chem.* **1999**, *64*, 8048–50. (b) Mártire, D. O.; Jux, N.; Aramendía, P. F.; Negri, R. M.; Lex, J.; Braslavsky, S. E.; Schaffner, K.; Vogel, E. *J. Am. Chem. Soc.* **1992**, *114*, 9969–78.
87. Boyle, R. W.; Johnson, C. K.; Dolphin, D. *J. Chem. Soc., Chem. Commun.* **1995**, 527–8.
88. Ali, H.; van Lier, J. E. *Tetrahedron* **1994**, *50*, 11933–44.
89. Aoyagi, Y.; Inoue, A.; Koizumi, I.; Hashimoto, R.; Tokunaga, K.; Gohma, K.; Komatsu, J.; Sekine, K.; Miyafuji, A.; Kunoh, J.; Honma, R.; Akita, Y.; Ohta, A. *Heterocycles* **1992**, *33*, 257–72.
90. Bräse, S.; Waegell, B.; de Meijere, A. *Synthesis* **1998**, 148–52.
91. Nilsson, K.; Hallberg, A. *J. Org. Chem.* **1990**, *55*, 2464–70.
92. Sonesson, C.; Larhed, M.; Nyqvist, C.; Hallberg, A. *J. Org. Chem.* **1996**, *61*, 4756–63.

93. Ozawa, F.; Hayashi, T. *J. Organometal. Chem.* **1992**, *428*, 267–77.
94. Shim, Y. K.; Youn, J. I.; Chun, J. S.; Park, T. H.; Kim, M. H.; Kim, W. J. *Synthesis* **1990**, 753–4.
95. Yagi, T.; Aoyama, T.; Shioiri, T. *Synlett* **1997**, 1063–4.
96. Muratake, H.; Abe, I.; Natsume, M. *Tetrahedron Lett.* **1994** *35*, 2573–6.
97. Muratake, H.; Abe, I.; Natsume, M. *Chem. Pharm. Bull.* **1996**, *44*, 67–79.
98. Muratake, H.; Tonegawa, M.; Natsume, M. *Chem. Pharm. Bull.* **1996**, *44*, 1631–3.
99. Muratake, H.; Tonegawa, M.; Natsume, M. *Chem. Pharm. Bull.* **1998**, *46*, 400–12.
100. Grigg, R.; Sridharan, V.; Stevenson, P.; Sukirthalingam, S.; Worakun, T. *Tetrahedron* **1990**, *46*, 4003–18.
101. Grigg, R.; Loganathan, V.; Sridharan, V. *Tetrahedron Lett.* **1996**, *37*, 3399–402.
102. Grigg, R.; Fretwell, P.; Meerholtz, C.; Sridharan, V. *Tetrahedron* **1994**, *50*, 359–70.
103. Grigg, R.; Sridharan, V.; Stevenson, P.; Sukirthalingam, S. *Tetrahedron* **1989**, *45*, 3557–68.
104. Wensbo, D.; Annby, U.; Gronowitz, S. *Tetrahedron* **1995**, *51*, 10323–42.
105. Heck, R. F. *J. Am. Chem. Soc.* **1971**, *93*, 6896–901.
106. Ganske, J. A.; Pandey, R. K.; Postich, M. J.; Snow, K. M.; Smith, K. M. *J. Org. Chem.* **1989**, *54*, 4801–7.
107. Morris, I. K.; Snow, K. M.; Smith, N. W.; Smith, K. M. *J. Org. Chem.* **1990**, *55*, 1231–6.
108. Demopoulos, B. J.; Anderson, H. J.; Loader, C. E.; Faber, K. *Can. J. Chem.* **1983**, *61*, 2415–22.
109. Faber, K.; Anderson, H. J.; Loader, C. E.; Daley, A. S. *Can. J. Chem.* **1984**, *62*, 1046–50.
110. Monti, D.; Sleiter, G. *Gazz. Chim. Ital.* **1990**, *120*, 587–90.
111. Monti, D.; Sleiter, G. *Gazz. Chim. Ital.* **1994**, *124*, 133–6.
112. Araki, S.; Ohmura, M.; Butsugan, Y. *Bull. Chem. Soc. Jpn.* **1986**, *59*, 2019–20.
113. Edstrom, E. D. *Synlett* **1995**, 49–50.
114. Edstrom, E. D.; Yu, T.; Jones, Z. *Tetrahedron Lett.* **1995**, *36*, 7035–8.
115. Jaouhari, R. Dixneuf, P. H.; Lécolier, S. *Tetrahedron Lett.* **1986**, *27*, 6315–8.
116. Iwasaki, M.; Kobayashi, Y.; Li, J.-P.; Matsuzaka, H.; Ishii, Y.; Hidai, M. *J. Org. Chem.* **1991**, *56*, 1922–7.
117. Chiusoli, G. P.; Costa, M.; Masarati, E.; Salerno, G. *J. Organomet. Chem.* **1983**, *255*, C35–8.
118. Cartoon, M. E. K.; Cheeseman, G. W. H. *J. Organomet. Chem.* **1982**, *234*, 123–36.
119. Murahashi, S.; Shimamura, T.; Moritani, I. *J. Chem. Soc., Chem. Commun.* **1974**, 931–2.
120. Trost, B. M.; Kernan, E. *J. Org. Chem.* **1980**, *45*, 2741–6.
121. Bäckvall, J.-E.; Nyström, J.-E. *J. Chem. Soc., Chem. Commun.* **1981**, 59–60.
122. Genet, J. P.; Balabane, M.; Bäckvall, J. E.; Nyström, J. E. *Tetrahedron Lett.* **1983**, *24*, 2745–8.

123. Anderson, P. G.; Bäckvall, J.-E. *J. Am. Chem. Soc.* **1992**, *114*, 8696–8.
124. Igarashi, S.; Haruta, Y.; Ozawa, M.; Nishide, Y.; Kinoshita, H.; Inomata, K. *Chem. Lett.* **1989**, 737–40.
125. Kimura, M.; Harayama, H.; Tanaka, S.; Tamaru, Y. *J. Chem. Soc., Chem. Commun.* **1994**, 2531–3.
126. Saito, S.; Hara, T.; Takahashi, N.; Hirai, M.; Moriwake, T. *Synlett* **1992**, 237–8.
127. Utimoto, K.; Miwa, H.; Nozaki, H. *Tetrahedron Lett.* **1981**, *22*, 4277–8.
128. Arcadi, A.; Anacardio, R.; D'Anniballe, G.; Gentile, M. *Synlett* **1997**, 1315–7.
129. Tsutsui, H.; Narasaka, K. *Chem. Lett.* **1999**, 45–6.
130. Larock, R. C.; Doty, M. J.; Han, X. *Tetrahedron Lett.* **1998**, *39*, 5143–6.
131. Wensbo, D.; Eriksson, A.; Jeschke, T.; Annby, U.; Gronowitz, S.; Cohen, L. A. *Tetrahedron Lett.* **1993**, *34*, 2823–6.
132. Wensbo, D.; Gronowitz, S. *Tetrahedron* **1996**, *52*, 14975–88.
133. Kusumoto, T.; Hujama, T.; Ogata, K. *Tetrahedron Lett.* **1986**, *27*, 4197–200.
134. Chatani, N.; Hanafusa, T. *Tetrahedron Lett.* **1986**, *27*, 4201–4.
135. Hartwig, J. F. *Synlett* **1997**, 329–40.
136. (a) Marcoux, J.-F.; Wagaw, S.; Buchwald, S. L. *J. Org. Chem.* **1997**, *62*, 1568–9. (b) Sadighi, J. P.; Harris, M. C.; Buchwald, S. L. *Tetrahedron Lett.* **1998**, *39*, 5327–30.
137. Hartwig, J. F. *Angew. Chem. Int. Ed.* **1998**, *37*, 2046–67.
138. (a) Wolfe, J. P.; Wagaw, S.; Marcoux, J.-F.; Buchwald, S. L. *Acc. Chem. Res.* **1998**, *31*, 805–18. (b) Yang, B. H.; Buchwald, S. L. *J. Organometal. Chem.* **1999**, *576*, 125–46.
139. Hartwig, J. F. *Angew. Chem. Int. Ed.* **1998**, *37*, 2090–2.
140. Belfield, A. J.; Brown, G. R.; Foubister, A. J. *Tetrahedron* **1999**, *55*, 11399–428.
141. Huang, J.; Grasa, G.; Nolan, S. P. *Org. Lett.* **1999**, *1*, 1307–9.
142. Mann, G.; Hartwig, J. F.; Driver, M. S.; Fernandez-Rivas, C. *J. Am. Chem. Soc.* **1998**, *120*, 827–8.
143. Hartwig, J. F.; Kawatsura, M.; Hauck, S. I.; Shaughnessy, K. H.; Alcazar-Roman, L. M. *J. Org. Chem.* **1999**, *64*, 5575–80.
144. Sonesson, C.; Hallberg, A. *Tetrahedron Lett.* **1995**, *36*, 4505–6.
145. Leung, S. H.; Edington, D. G.; Griffith, T. E.; James, J. J. *Tetrahedron Lett.* **1999**, *40*, 7189–91.
146. Anderson, B. A.; Bell, E. C.; Ginah, F. O.; Harn, N. K.; Pagh, L. M.; Wepsiec, J. P. *J. Org. Chem.* **1998**, *63*, 8224–8.
147. Gleiter, R.; Ritter, J. *Tetrahedron* **1996**, *52*, 10383–8.

# CHAPTER 3

## Indoles

Indole is perhaps the single most visible heterocycle in all of chemistry. It is embodied in a myriad of natural products, pharmaceutical agents, and a growing list of polymers. In addition to the hundreds of well-known indole plant alkaloids (e.g., yohimbine, reserpine, strychnine, ellipticine, lysergic acid, physostigmine), the indole ring is present in an array of other organisms. The indigo analog Tyrian Purple is the ancient Egyptian dye produced by Mediterranean molluscs, and a bromine-containing triindole is found in an *Orina* sponge. Sciodole is a *Tricholoma* fungal product and the plant growth hormone 4-chloroindole-3-acetic acid is utilized by peas, beans, lentils and other plants. Another chlorine-containing fungal metabolite is pennigritrem, produced by *Penicillium nigricans*. The funnel-web spider employs the toxic argiotoxin 659 in chemical defense.

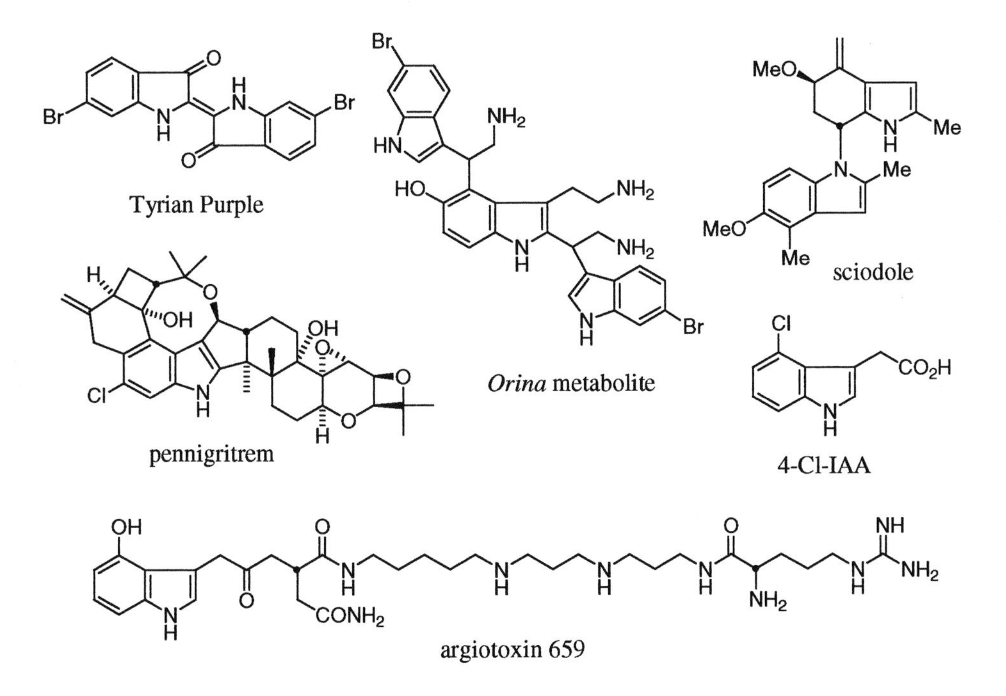


The central importance of the indole derivatives serotonin and the amino acid tryptophan in living organisms has inspired chemists to design and synthesize thousands of indole-containing pharmaceuticals. Some of the most recent indole-containing drugs are shown here. Three new antimigraine drugs — competitors of the highly effective Sumatriptan — are Naramig, Zomig, and Maxalt. The antiemetics Nasea and Anzemet are potent and highly selective 5-$HT_3$ receptor antagonists for the treatment of chemotherapy-induced nausea and vomiting. Rescriptor is a new HIV-1 reverse transcriptase inhibitor, and Serdolect is a neuroleptic for acute and chronic schizophrenia. Lescol is an HMG-CoA reductase inhibitor, and Somatuline (not shown) is an indole cyclic peptide for the treatment of acromegaly by acting as a growth hormone inhibitor. Accolate is a new antiasthma drug, and the simple oxindole ReQuip is effective in the treatment of early-stage Parkinson's disease.

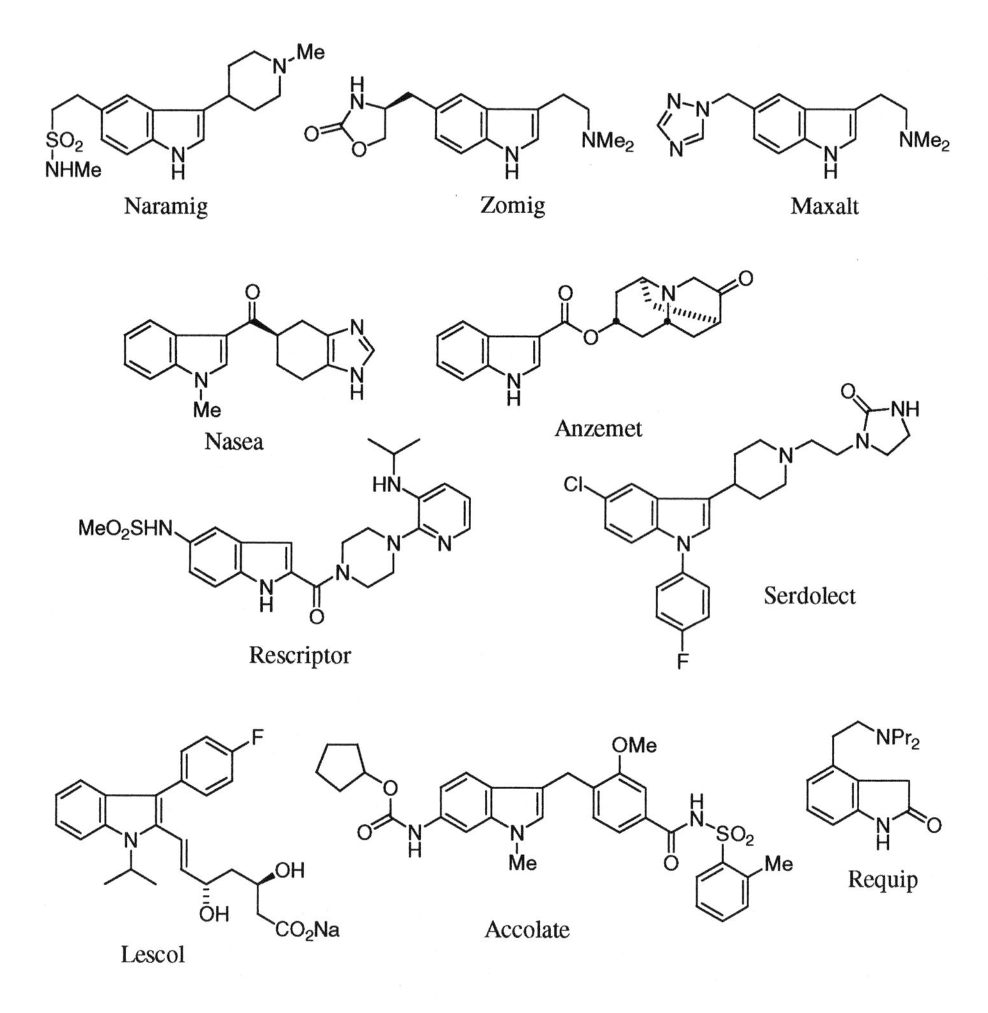

Although much less so than pyrrole polymers, indole polymers are beginning to be synthesized and studied as new materials. Electropolymerized films of indole-5-carboxylic acid are well-suited for the fabrication of micro pH sensors and they have been used to measure ascorbate and NADH levels. The three novel pyrroloindoles shown have been electrochemically polymerized, and the polymeric pyrrolocarbazole has similar physical properties to polyaniline.

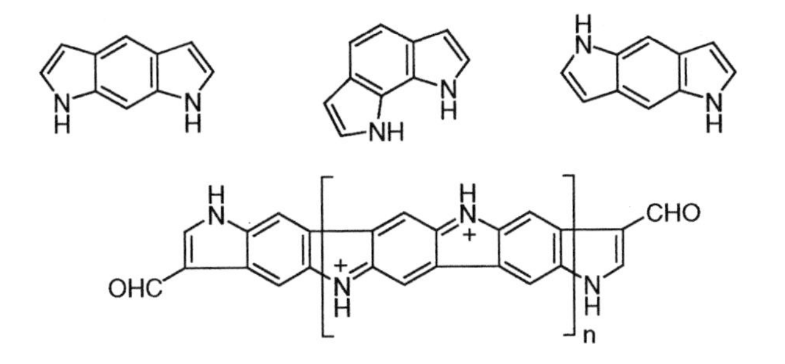


## 3.1 Synthesis of indolyl halides

Like pyrrole, indole is a very reactive π-excessive heterocycle and reacts with halogens and other electrophiles extremely rapidly [1]. Nature has exploited this property to produce more than 300 halogenated indoles, mainly bromoindoles in marine organisms. A few examples are illustrated below. Chapter 2 contains references to reviews of these natural organohalogens.

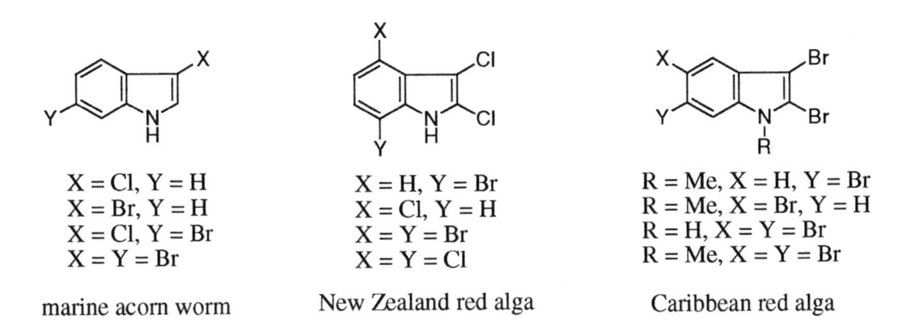


Early syntheses of haloindoles involved direct reactions of indoles with chlorine, bromine, or iodine. In some cases, this approach was reasonably successful, but the instability of the resulting 3-haloindoles made product isolation and further chemistry difficult. For example, although attempted preparations of 3-chloro-, 3-bromo-, and 3-iodoindole were described in the early 1900's [2], only recently have practical syntheses of these compounds and their *N*-protected derivatives become available. For example, 3-bromoindole (**2**) can be prepared in

64% yield by the treatment of indole (**1**) with pyridinium bromide perbromide. The product decomposes at 65 °C and should be stored at –20 °C [2]. Likewise, the unstable 3-iodoindole can be formed by treating **1** with $I_2$, KI, NaOH, MeOH [3]. The product was converted to indoxyl acetate (AgOAc, HOAc) in 28% overall yield.

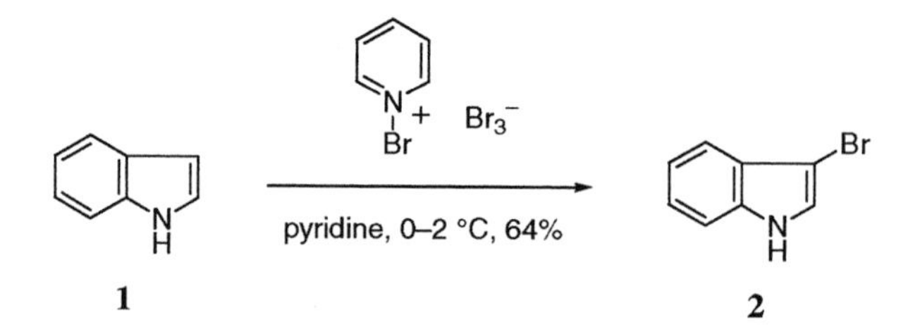


A vast improvement for the synthesis of both 3-bromo- and 3-iodoindole by using DMF as solvent was described by Bocchi and Palla, as summarized below [4]. This appears to be the method of choice for the preparation of simple 3-bromo- and 3-iodoindoles.

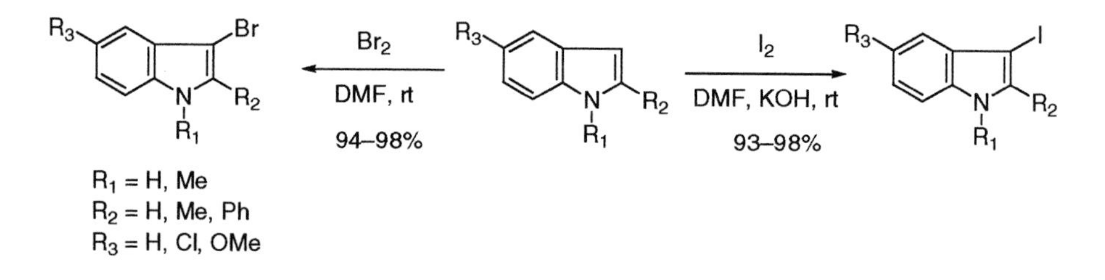


Erickson extended these reactions to useful preparations of both 3-chloroindole and several 2,3-dihaloindoles, many of which occur naturally [5]. When the C-3 position is already substituted, halogenation usually occurs at C-2. A summary of these halogenations is shown below. Erickson was able to improve Piers' synthesis of **2** to a yield of 82%. Interestingly, the action of sulfuryl chloride on 3-iodoindole gives the *ipso* product 3-chloroindole in 84% yield.

The notoriously unstable 2-chloroindole was first synthesized in pure form by Powers [6], and this procedure was later extended to the preparation of 2-bromoindole by Erickson [5]. The method involves reaction of oxindole with either $POCl_3$ or $POBr_3$ but yields are very low in both cases (26% for 2-chloroindole and 15% for 2-bromoindole). Powers also synthesized the more stable 1-benzyl-2-chloroindole from *N*-benzyloxyindole and $POCl_3$, but discontinued all work in this area after developing a severe skin rash from these haloindoles. This toxicity is consistent with the role of natural halogenated indoles in chemical defense by marine organisms.

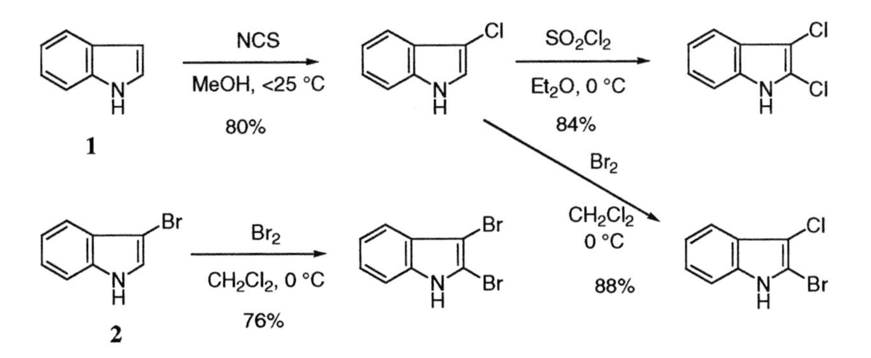


The most efficient route to *N*-unsubstituted 2-haloindoles is that developed by Bergman involving a Katritzky indole C-2 lithiation protocol [7] and quenching with selected sources of halogen as illustrated [8]. Bromine and iodine gave lower yields. Furthermore, subjecting 2-iodoindole to the Bocchi and Palla procedure with iodine affords 2,3-diiodoindole in 82% yield. Bergman observed that the order of stability of 2-haloindoles at –20 °C is 2-I >> 2-Cl > 2-Br. Nevertheless, simple *N*-unsubstituted 2- and 3-haloindoles would seem to be too labile to undergo consistently successful Pd-catalyzed chemistry.

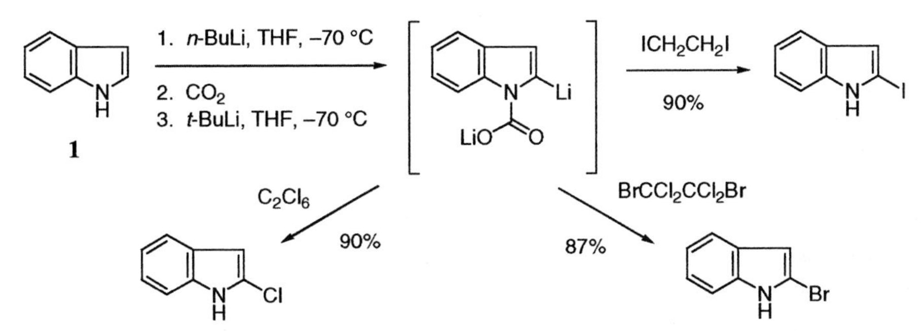


Several methods for synthesizing *N*-protected (usually with electron-withdrawing groups) 2- and 3-haloindoles have been developed and the resulting haloindoles are much less prone to decomposition than the unsubstituted compounds. Bromination of *N*-(phenylsulfonyl)indole (**3**), which is readily available *via* lithiation [9, 10] or phase-transfer chemistry [11, 12], affords 3-bromo-1-(phenylsulfonyl)indole (**4**) in nearly quantitative yield [12].

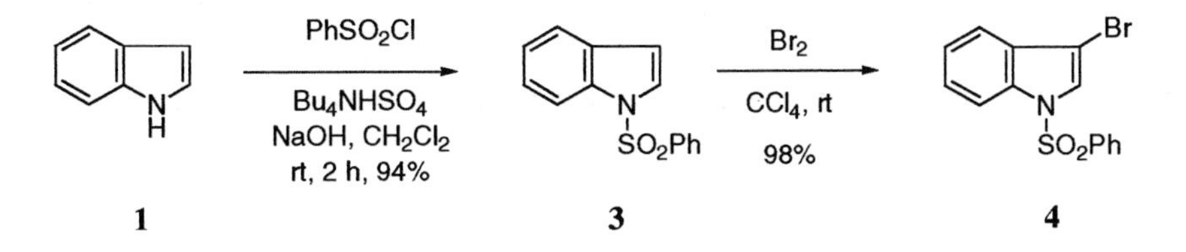

The synthesis of both 3-iodo-1-(phenylsulfonyl)indole (**5**) and 2,3-diiodo-1-(phenylsulfonyl)-indole (**6**) can be achieved in excellent overall yields as illustrated [10]. The preparation of **5** is done in one pot.

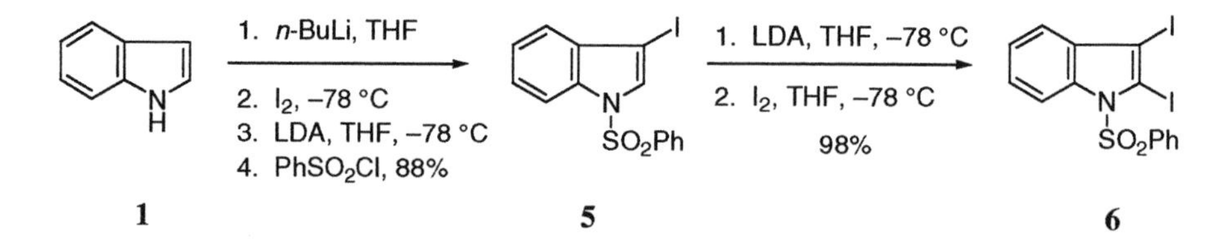


The lithiation of 3-haloindoles represents an excellent method for the preparation of other 2,3-dihaloindoles. For example, treatment of **4** with LDA followed by quenching with CNBr affords 2,3-dibromoindole **7** in good yield [13], and quenching with iodine furnishes 3-bromo-2-iodoindole **8** [14].

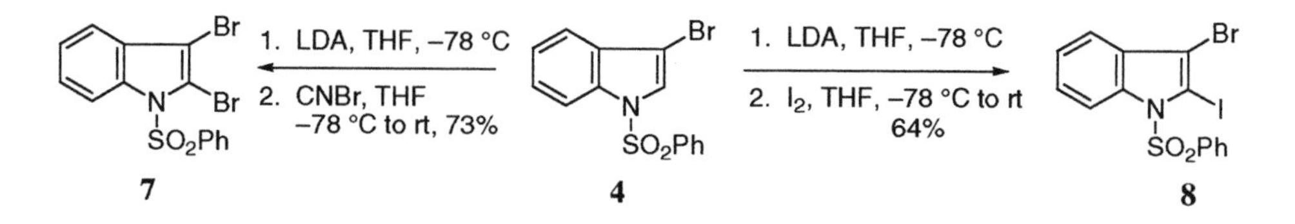


Likewise, quenching the 2-lithio species derived from **5** with CNBr gives 2-bromo-3-iodo-1-(phenylsulfonyl)indole in 80% yield. Lithiation of *N*-(phenylsulfonyl)indole (**3**) with LDA followed by quenching with CNBr or benzenesulfonyl chloride gives the corresponding 2-bromo and 2-chloro derivatives in 82% and 93% yields, respectively [14]. Similar lithiation methods have been used to prepare 2-iodo-1-methylindole [15] and 1-Boc-2-iodoindole, the latter of which can be converted to 2-iodotryptamine, and its 5-methoxy derivative can be similarly crafted [16]. Widdowson prepared 2- and 3-fluoro-1-(4-toluenesulfonyl)indoles from the corresponding trimethylstannylindoles by reaction with either cesium fluoroxysulfate or Selectfluor™ [17, 18]. Unfortunately, the C-F bond is too strong to undergo oxidative addition of Pd so these fluoroindoles will have to await the development of new technology to find applications in synthesis. Treatment of 2-(trimethylsilyl)azaindoles with ICl furnishes the corresponding 2-iodoazaindoles [19].

An entirely different approach to 3-haloindoles involves a mercuration/iodination sequence, which has been adopted by Hegedus to prepare 4-bromo-3-iodo-1-(4-toluenesulfonyl)indole for use in the synthesis of ergot alkaloids [20, 21]. We will discuss this chemistry later.

The synthesis of the benzene-ring substituted haloindoles entails more conventional aryl ring chemistry or *de novo* synthesis of the indole ring [1, 22–24]. As expected, these haloindoles are far more stable than the C-2 and C-3 haloindoles.

Hegedus also described an efficient synthesis of 4-bromoindole (**10**) and 4-bromo-1-(4-toluenesulfonyl)indole (**11**) starting from 2-amino-6-nitrotoluene (**9**) [20]. The synthesis is lengthy but yields are good, and the method involves the Hegedus indole ring synthesis discussed later in this chapter.

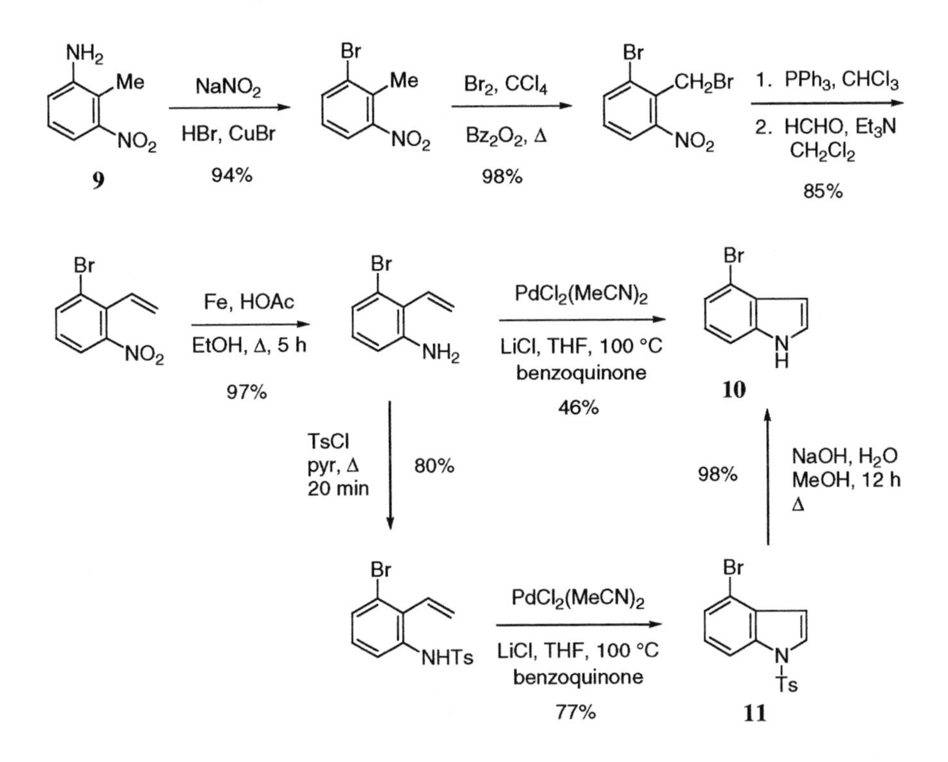


Conventional aryldiazonium salt chemistry on 4-aminoindole provides 4-iodo-1-(4-toluenesulfonyl)indole (**13**), 4-iodoindole (**14**), and 1-(*tert*-butyldimethylsilyl)-4-iodoindole (**15**) in excellent yields as shown [25, 26].

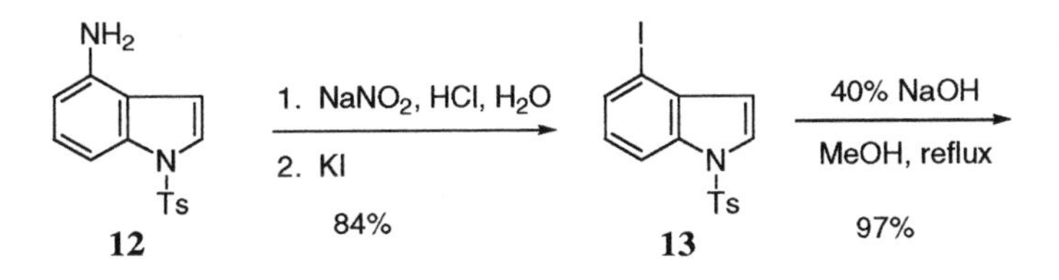

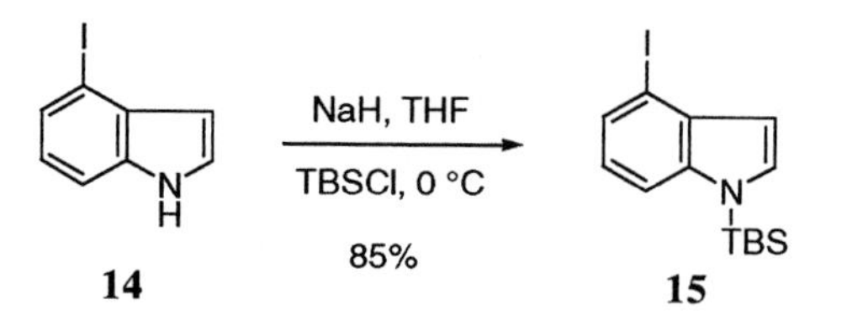


An excellent synthesis of 5-bromo- (**18**) and 5-iodoindole (**19**) involves protecting the indole double bond as sulfonate **16**, acetylation to **17**, and halogenation [27]. Indoline itself undergoes bromination at C-4 and C-7 [28].

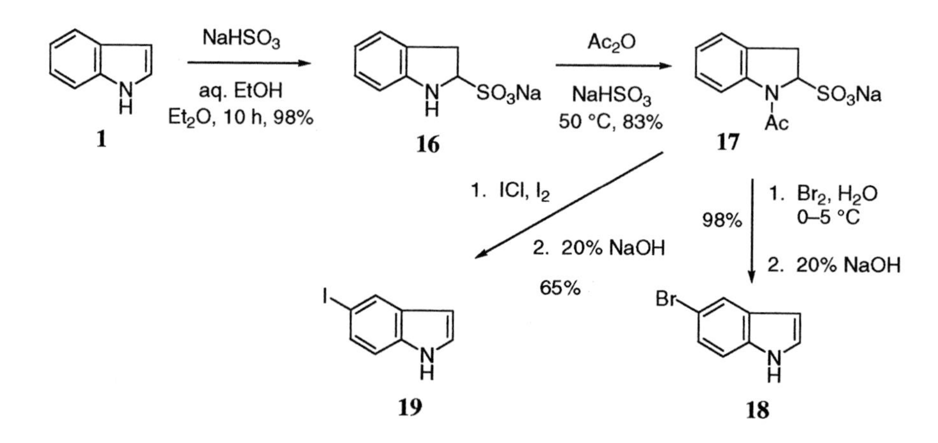


Carrera and Sheppard improved upon a Leimgruber-Batcho indole synthesis [24] to prepare 6-bromoindole (**20**) in excellent overall yield from 4-bromo-2-nitrotoluene [29a], and Rapoport utilized this method to synthesize 4-, 5-, 6-, and 7-bromoindole [29b].

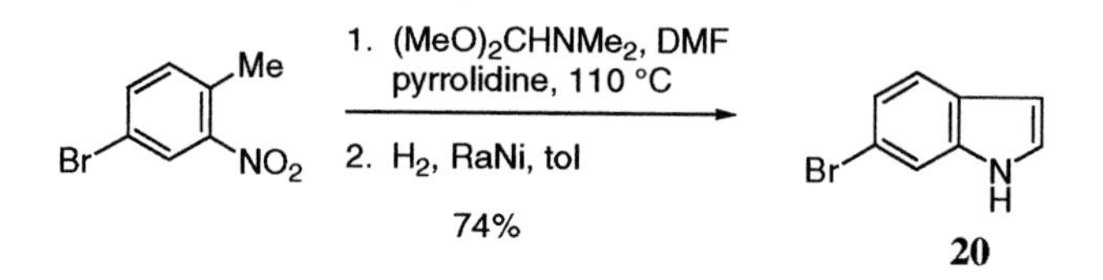


The Bartoli indole synthesis [24] is an excellent method for preparing 7-bromoindole (**21**) [29a, 30].

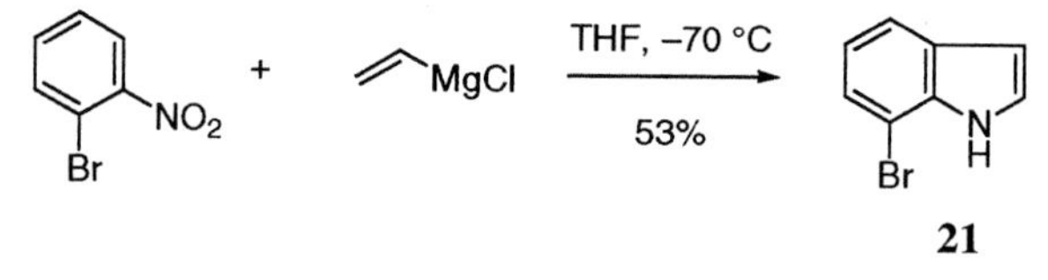

In addition, due to the proximity and inductive effect of the indoline nitrogen, a metalation strategy provides for a facile synthesis of 7-bromoindoline, as well as other 7-substituted indolines. Thus, Iwao and Kuraishi find that *N*-Boc-indoline (**22**) is smoothly lithiated at C-7 to give 7-lithio **23**. Quenching **23** with suitable electrophiles gives 7-haloindolines **24–26**, and quenching with tri-*n*-butylstannyl chloride affords the corresponding tin derivative in 65% yield [31].

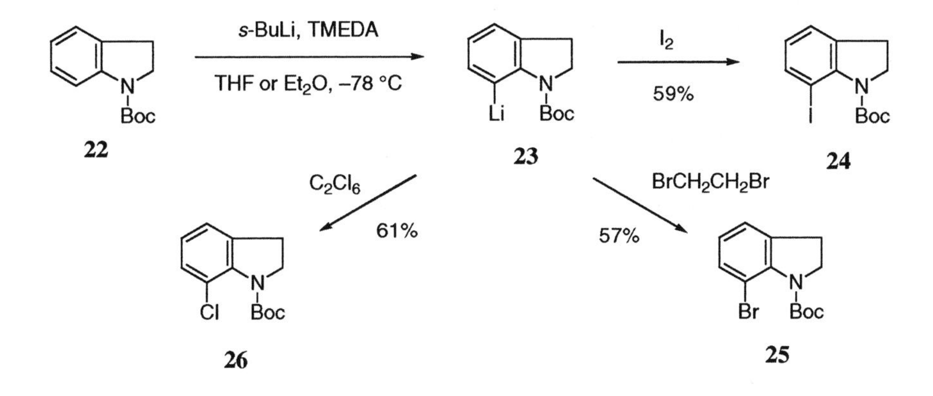


Meyers adopted and refined this method to prepare 1-benzyl-7-bromoindoline (60% overall yield from **22**) for use in Grignard coupling chemistry, and obtained a higher yield of 7-bromoindoline **25** by quenching with 1,2-dibromo-1,1,2,2-tetrafluoroethane (68%) [32, 33]. It is important to note that several methods are available for the oxidation of indolines to indoles [22–24], including one involving $PdCl_2$ as developed by Kuehne [34]. Interestingly, in the presence of $Ph_3P$, indoline reacts with $PdCl_2$ to give the stable complex **27**, confirmed by X-ray crystallography [35]. In any event, conditions can usually be found to oxidize halo-substituted and other indolines to the indoles (Pd/C, $MnO_2$, DDQ, NBS, *etc.*).

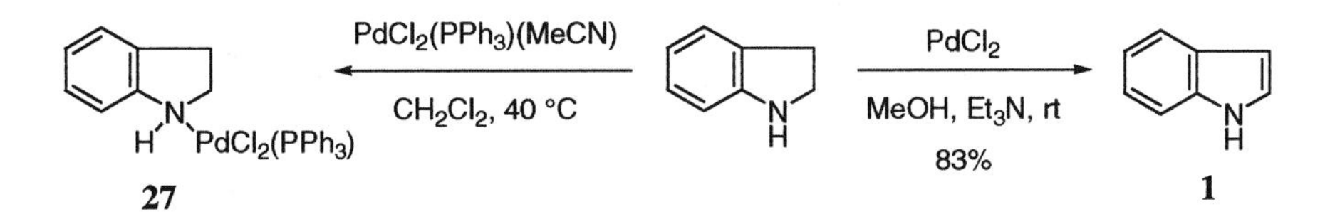


A third efficient synthesis of 7-bromoindole (**21**) involves the Stille stannylation of 2,6-dibromoaniline to give enol ether **28**, which, after hydrolysis and cyclization, affords **21** in 96% overall yield [36].

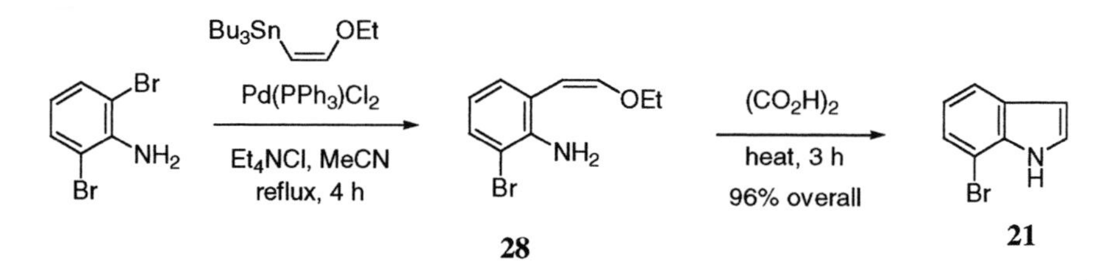


Widdowson expanded his hexacarbonylchromium chemistry to the synthesis and lithiation of $Cr(CO)_3$-*N*-TIPS indole (**29**), leading to 4-iodoindole **30** after oxidative decomplexation [37]. Stannylation at C-4 could also be achieved using this method (62% yield), and comparable chemistry with 3-methoxymethylindole leading to C-4 substitution was described.

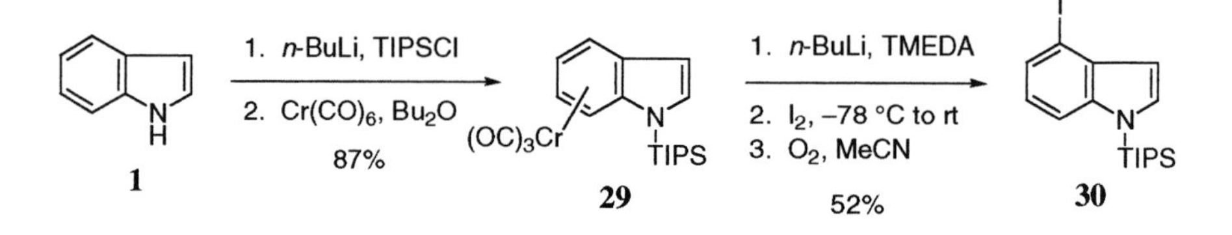


Another general approach to benzene-ring haloindoles involves thallation chemistry. Hollins and co-workers demonstrated that C-4 thallation occurs readily in a series of 3-acylindoles **31**, affording 4-iodoindoles **32** following treatment of the thallated intermediates with KI [38].

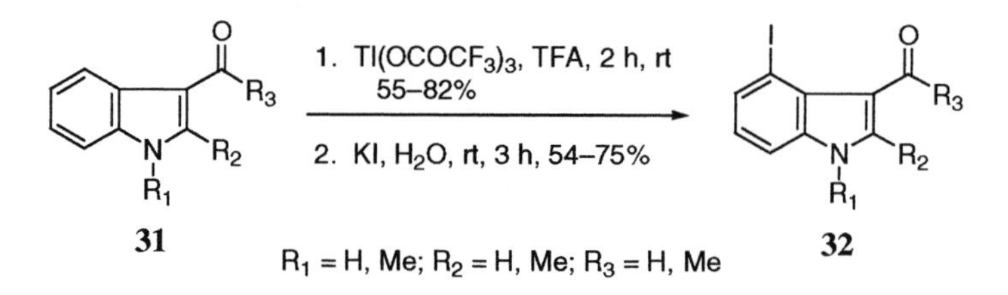


Somei improved this methodology by quenching the appropriate thallated intermediate with $I_2$/CuI/DMF to give 4-iodoindole-3-carboxaldehyde in 94% yield [39], and he extended this method to achieve efficient syntheses of C-7 haloindoles [40, 41]. For example, 7-iodoindole (**33**) was prepared in good overall yield from *N*-acetylindoline as illustrated. Thallation at C-5 is a minor (5%) pathway.

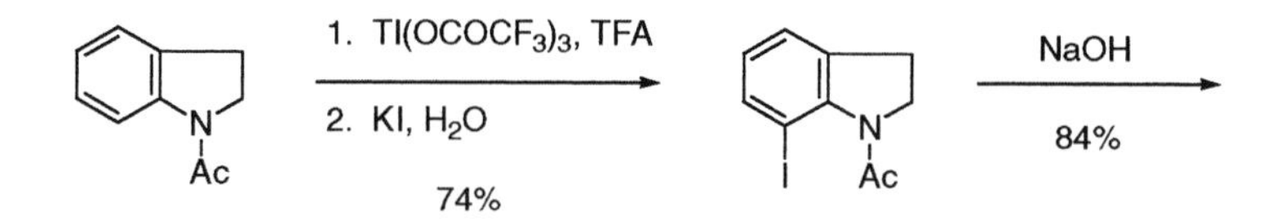

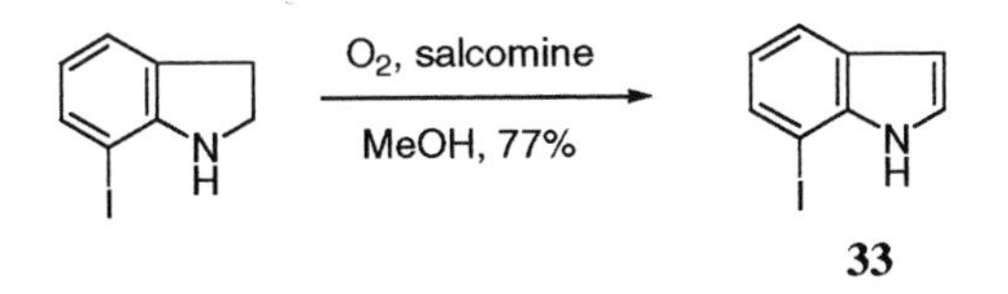


The invention of the triflate (trifluoromethylsulfonyl) group — one of the world's best leaving groups — has led to its use in palladium chemistry [42]. Conway and Gribble described the synthesis of 3-indolyl triflate **34** [12] and 2-indolyl triflate **35** from oxindole [43]. Mérour synthesized the *N*-phenylsulfonyl derivative **36** by employing a Baeyer-Villiger oxidation of the appropriate indolecarboxaldehyde [44].

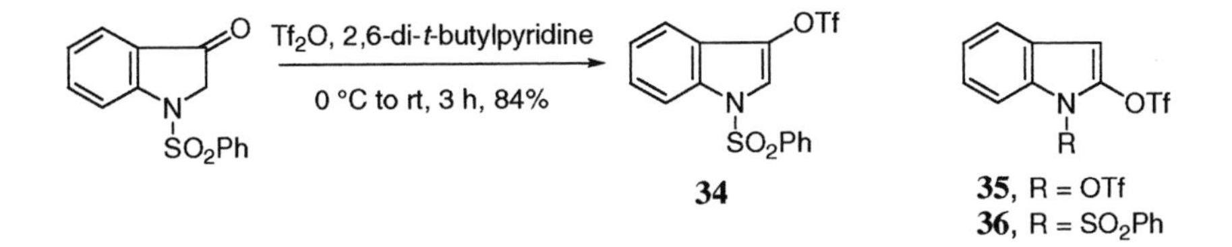


## 3.2 Oxidative coupling/cyclization

Most of the early applications of palladium to indole chemistry involved oxidative coupling or cyclization using stoichiometric Pd(II). Åkermark first reported the efficient oxidative coupling of diphenyl amines to carbazoles **37** with $Pd(OAc)_2$ in refluxing acetic acid [45]. The reaction is applicable to several ring-substituted carbazoles (Br, Cl, OMe, Me, $NO_2$), and 20 years later Åkermark and colleagues made this reaction catalytic in the conversion of arylaminoquinones **38** to carbazole-1,4-quinones **39** [46]. This oxidative cyclization is particularly useful for the synthesis of benzocarbazole-6,11-quinones (e.g., **40**).

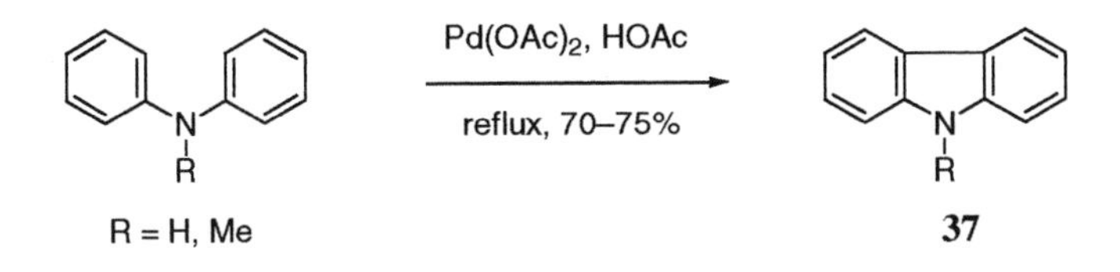

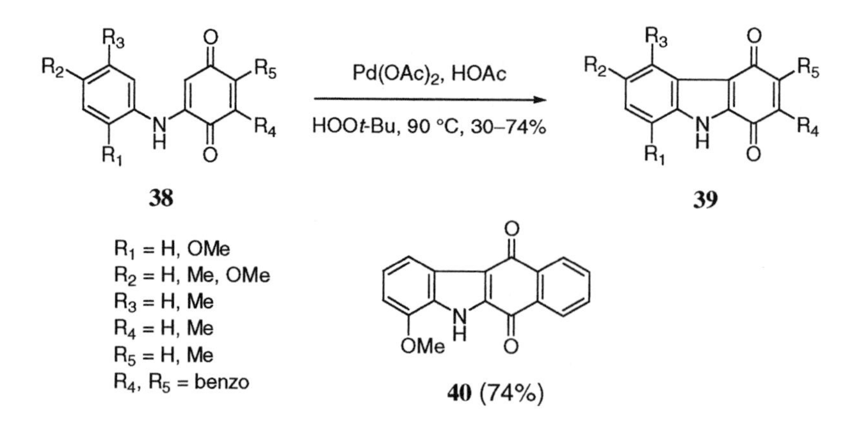


Such oxidative cyclizations with $Pd(OAc)_2$ have been used to synthesize carbazole and carbazolequinone alkaloids [47], ellipticines [48], 6*H*-pyrazolo[4,3-*c*]carbazole [49], and a series of 3-acetyl-1-methoxy-2-methylcarbazoles [50]. Knölker and co-workers exploited this reaction in the synthesis of numerous naturally occurring carbazolequinones including prekinamycin analogues (benzo[*b*]carbazolequinones) [51, 52], carbazomycins G and H [53], carbazoquinocin C [54, 55], and (±)-carquinostatin A [55]. Noteworthy is the use of $Cu(OAc)_2$ as a reoxidant to make these oxidative cyclizations catalytic [53, 54]. Illustrative is the synthesis of **41**, the acetate derivative of (±)-carquinostatin A [55]. In other cases, Knölker achieved yields of carbazolequinones higher than 90%.

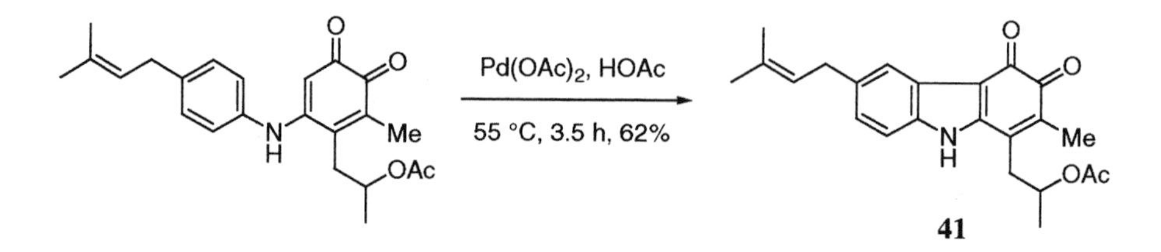


Trost was the first to apply a palladium-catalyzed cyclization reaction to the synthesis of indole alkaloids [56–59]. Thus, exposure of indole isoquinuclidine **42** to $PdCl_2(MeCN)_2$ gave ibogamine (**44**) (racemic and optically active) after reductive cleavage of the Pd-C bond in **43** [57]. Other iboga alkaloids were similarly crafted by Trost, and Williams adopted this strategy in a key step in his synthesis of (+)-paraherquamide B [60]. By reducing the palladium intermediate with $NaBD_4$, Trost deduced that this cyclization mechanism involves initial indole C-2 palladation following by ring closure to **43** [57].

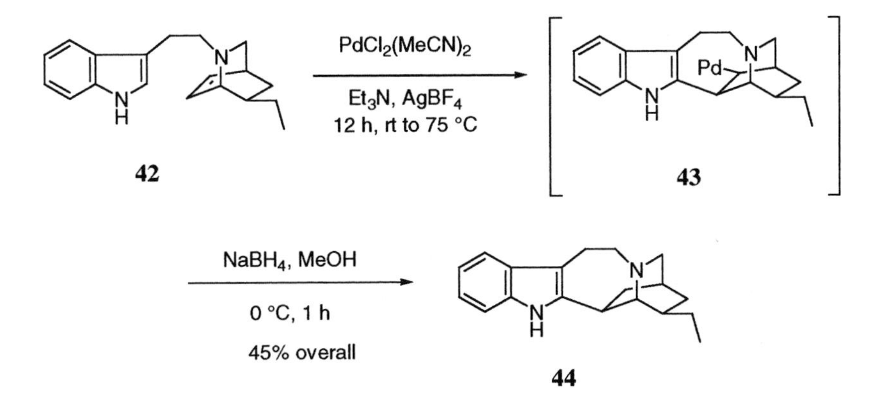


In a series of papers, Itahara established the utility of $Pd(OAc)_2$ in the oxidative cyclization of *C*- and *N*-benzoylindoles, and two examples are shown [61–63]. Itahara also found that the cyclization of 3-benzoyl-1,2-dimethylindole proceeds to the C-4 position (31% yield) [61]. Under similar conditions, both 1-acetylindole and 1-acetyl-3-methylindole are surprisingly intermolecularly arylated at the C-2 position by benzene and xylene (22–48% yield) [64, 65].

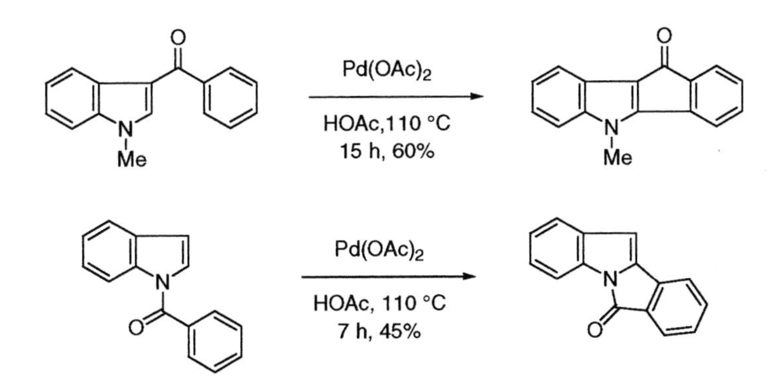


Itahara's work has paved the way for several $Pd(OAc)_2$-oxidative cyclizations leading to indole alkaloids. Thus, Black synthesized a series of pyrrolophenanthridone alkaloids (e.g., **45** to **46**) [66, 67] and, following the pioneering work of Bergman on the parent system [15], he effected the cyclization of diindole urea **47** to **48** [68]. The presence of a 3-methyl group in each indole ring blocks cyclization. *N*-Acylindolines corresponding to **45** also undergo this cyclization, and DDQ introduces the indole double bond quantitatively [66]. However, the cyclization of *N*-benzylindoles corresponding to **45** fails.

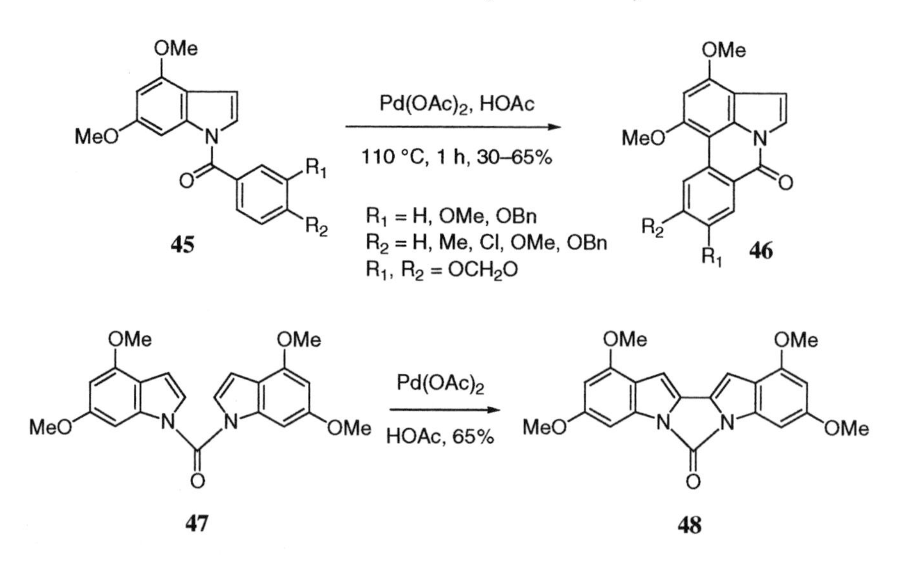


Hill described the $Pd(OAc)_2$-oxidative cyclization of bisindolylmaleimides (e.g., **49**) to indolo[2,3-*a*]pyrrolo[3,4-*c*]carbazoles (e.g., **50**) [69], which is the core ring system in numerous natural products, many of which have potent protein kinase activity [70]. Other workers employed this Pd-induced reaction to prepare additional examples of this ring system [71, 72]. Ohkubo found that $PdCl_2$/DMF was necessary to prevent acid-induced decomposition of benzene-ring substituted benzyloxy analogues of **49**, and the yields of cyclized products under these conditions are 85–100% [71].

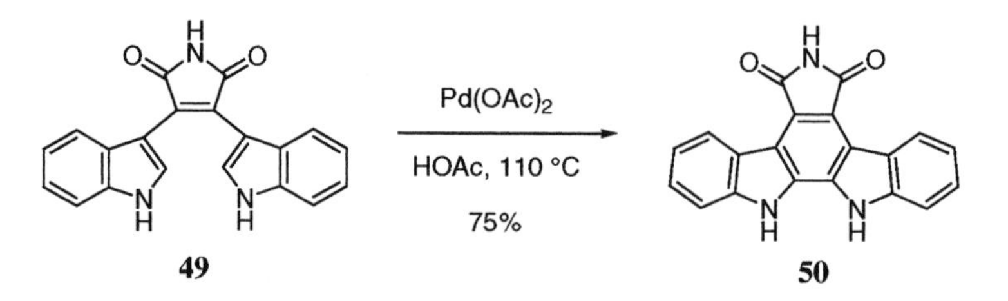


Srinivasan found that the typical stoichiometric $Pd(OAc)_2$ conditions effect cyclization of 2-(*N*-arylaminomethyl)indoles to aryl-fused β-carbolines in low yield [e.g., **51** to **52**] [73]. Similar to the chemistry observed with *N*-(phenylsulfonyl)pyrrole, 1,4-naphthoquinone also undergoes $Pd(OAc)_2$ oxidative coupling with *N*-(phenylsulfonyl)indole to give **53** in 68% yield [74].

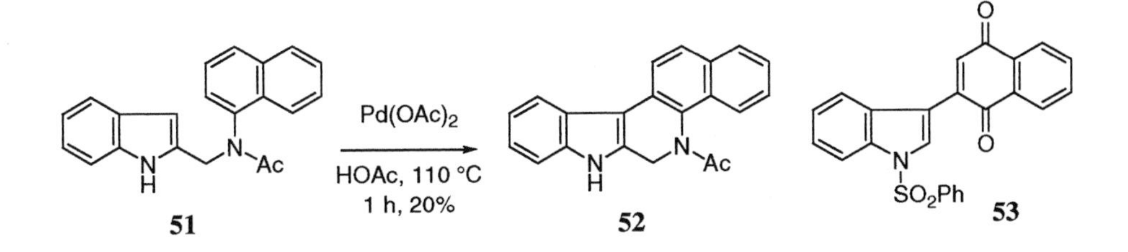


Intermolecular Pd(II) oxidative couplings with indoles are well established, although initial results were unpromising. For example, Billups found that indole reacts with allyl acetate (Pd(acac)/$Ph_3P$/HOAc) to give a mixture of 3-allyl- (54%), 1-allyl- (7%), and 1,3-diallylindole (11%) [75]. Allyl alcohol also is successful in this reaction but most other allylic alcohols fail. Likewise, methyl acrylate reacts with *N*-acetylindole ($Pd(OAc)_2$/HOAc) to give only a 20% yield of methyl (*E*)-3-(1-acetyl-3-indolyl)acrylate and a 9% yield of *N*-acetyl-2,3-bis(carbomethoxy)carbazole [76]. Itahara improved these oxidative couplings by employing both *N*-(2,6-dibenzoyl)indoles (e.g., **54**, **55**) and *N*-(phenylsulfonyl)indole as substrates [77, 78]. Reaction occurs at C-3 unless this position is blocked. The coupling can be made catalytic using AgOAc or other reoxidants [78]. Some examples are shown below and *E*-stereochemistry is the major or exclusive isomer. Acrylonitrile also reacts with **54** under these conditions (52%; *E*/*Z* = 3/1) [77], and methyl vinyl ketone, ethyl (*E*)-crotonate, and ethyl α-methylacrylate react with *N*-(phenylsulfonyl)indole under these oxidative conditions [78].

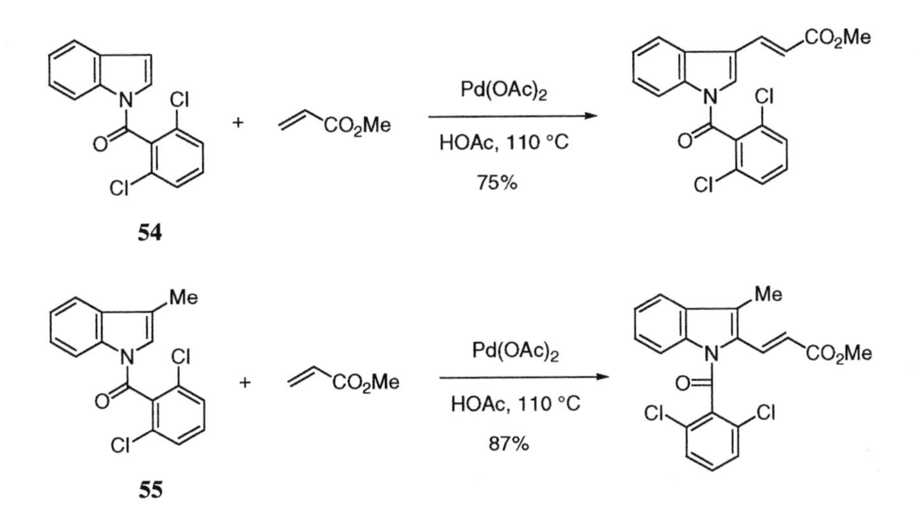

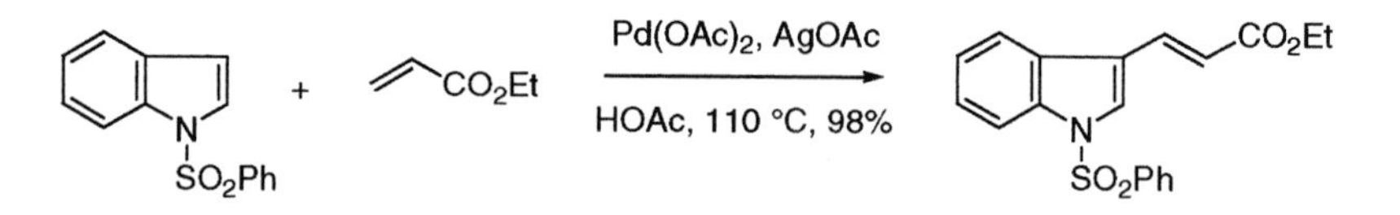


Hegedus found that 4-bromo-1-(4-toluenesulfonyl)indole (**11**) reacts with methyl acrylate to form the C-3 product in low yield under stoichiometric conditions [20]. Heck reactions of this substrate and related haloindoles are discussed in Section 3.5. Yokoyama, Murakami and co-workers also utilized **11** in total syntheses of clavicipitic acid and costaclavine, one key step of which is the oxidative coupling of **11** with **56** to give dehydrotryptophan derivative **57** [79]. The use of chloranil as a reoxidant to recycle Pd(0) to Pd(II) greatly improves the coupling over earlier conditions [80, 81]. For example, chloranil was more effective than DDQ, $MnO_2$, $Ag_2CO_3$, $Co(salen)_2/O_2$, and $Cu(OAc)_2$. In the absence of chloranil the yield of **57** is 31%.

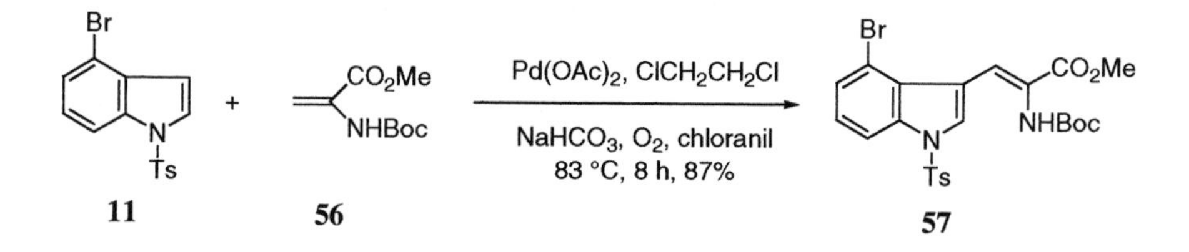


The reaction of *N*-protected dehydroalanine methyl esters (e.g. **56**, **59**) with other indoles **58** can also be effected to give the corresponding dehydrotryptophans **60**, invariably as the *Z*-isomers [81]. Murakami, Yokoyama and co-workers also studied oxidative couplings of acrylates, acrylonitrile, and enones with 2-carboethoxyindole, 1-benzylindole, and 1-benzyl-2-carboethoxyindole and $PdCl_2$ and $CuCl_2$ or $Cu(OAc)_2$ to give C-3 substitution in 50–84% yields [82, 83].

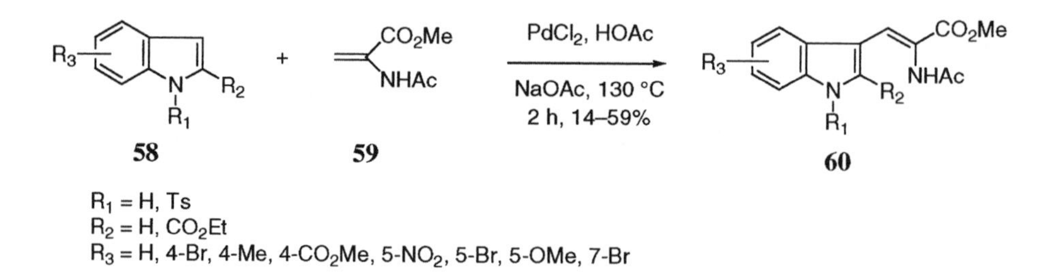


These workers have also found that 3-alkenylpyrroles, such as **61**, are cyclized to indoles [84].

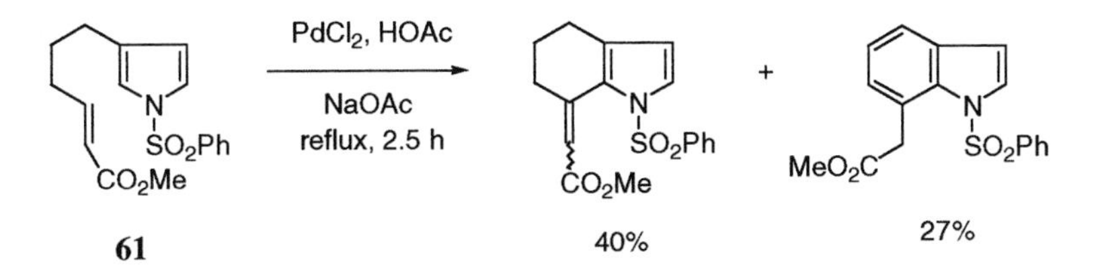


In addition to examining the vinylation of 1-methyl-2-indolecarboxaldehyde with methyl acrylate ($Pd(OAc)_2$/HOAc/AgOAc) to give methyl (*E*)-3-(2-formyl-1-methyl-3-indolyl)-acrylate in 60% yield, Pindur found that similar reactions of methyl 3-(1-methyl-2-indolyl)-acrylate afford bis(carbomethoxy)carbazoles albeit in low yield [85]. Fujiwara discovered that the combination of catalytic $Pd(OAc)_2$ with benzoquinone and *t*-butylhydroperoxide serves to couple indole with methyl acrylate to give methyl (*E*)-3-(3-indolyl)acrylate in 52% yield [86].
Abdrakhmanov and co-workers observed the cycloaddition of *N*-(1-methyl-2-butenyl)aniline or 2-(1-methyl-2-butenyl)aniline with $PdCl_2$/DMSO to give a 69% yield of a mixture of 2-ethyl-3-methylindole and 2,4-dimethylquinoline [87]. The authors propose that a Claisen rearrangement is initially involved. A similar oxidative cyclization of a 5-amino-indoleacrylate was the starting point for syntheses of CC-1065 and related compounds [88].

## 3.3 Coupling reactions with organometallic reagents

### 3.3.1 Kumada coupling

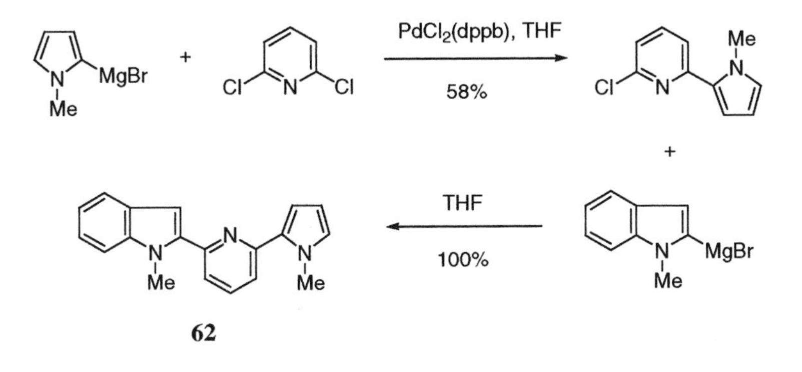


Of all the palladium-catalyzed coupling reactions, the Kumada coupling has been applied least often in indole chemistry. However, this Grignard-Pd cross-coupling methodology has been used to couple 1-methyl-2-indolylmagnesium bromide with iodobenzene and α-bromovinyltrimethylsilane to form 1-methyl-2-phenylindole and 1-methyl-2-(1-trimethylsilyl)vinylindole in 79% and 87% yields, respectively [89, 90]. Kumada constructed the triheterocycle **62** using a tandem version of his methodology [91].

Kondo employed the Kumada coupling using the Grignard reagents derived from 2- and 3-iodo-1-(phenylsulfonyl)indole to prepare the corresponding phenyl derivatives in 50% yield [92]. Widdowson expanded the scope of the Kumada coupling and provided some insight into the mechanism [93].

### 3.3.2 Negishi coupling

Although the Negishi coupling has been less frequently used in indole synthetic manipulations than either Suzuki or Stille couplings, we will see in this chapter that Negishi chemistry is often far superior to other Pd-catalyzed cross coupling reactions involving indoles. One of the first such examples is Pichart's coupling of 1-methyl-2-indolylzinc chloride (**63**) with iodopyrimidine **64** to give **65** [94].

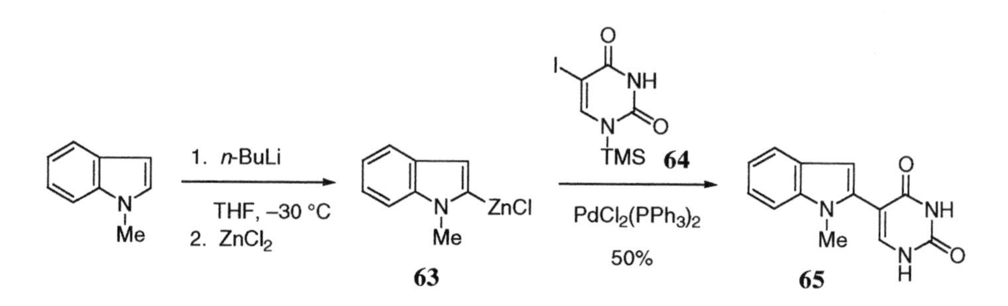


Sakamoto and co-workers studied extensively the generation and couplings of *N*-protected 2- and 3-indolylzinc halides [95–98]. The best *N*-protecting groups are Boc, $SO_2Ph$, and $CO_2^-$, and the zinc species can either be generated by transmetalation of a lithiated indole with $ZnCl_2$ or by oxidative addition of active Zn to iodoindoles. The latter technique is required for successful generation of 3-indolylzinc halides since 1-(phenylsulfonyl)-3-lithioindoles are known to rearrange to the more stable 2-lithioindoles at room temperature [10]. A selection of these reactions is shown that illustrates the range of compatible functional groups. These workers also generated 3-carbomethoxy-1-(phenylsulfonyl)-4-indolylzinc iodide and coupled it with iodobenzene under the same conditions to give the phenylated derivative in 43% yield.

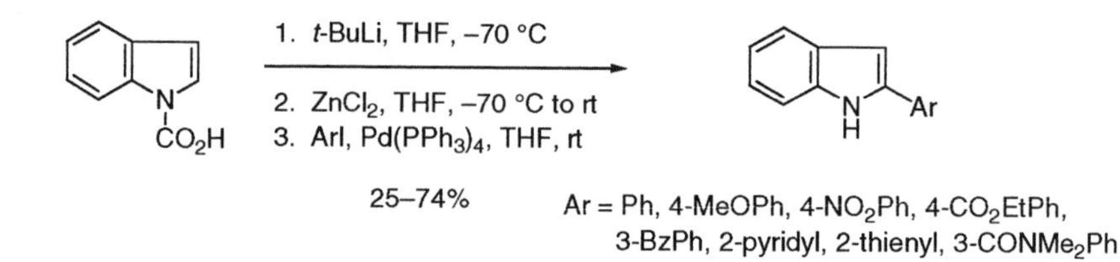

Independently and simultaneously, Bosch developed similar Negishi indole chemistry [99–101] to that described above by Sakamoto. An important discovery is that 1-(*tert*-butyldimethylsilyl)-3-lithioindole is stable at room temperature and does not rearrange to the 2-lithio isomer. Therefore, a halogen-lithium exchange sequence followed by transmetalation with $ZnCl_2$ conveniently generates the 3-indolylzinc species. This species and 1-(phenylsulfonyl)-2-indolylzinc chloride smoothly couple with halopyridines under Negishi conditions to give 3- and 2-(2-pyridyl)indoles, respectively, as illustrated below for the former (**66** to **67** to **68**) [100, 101]. The *tert*-butyldimethylsilyl group is readily removed with TsOH. Bosch and co-workers also coupled **67** with a range of other heteroaryl halides (2- and 3-bromothiophene, 3-bromofuran, 2-chloropyrazine, and 3-bromo-1-(phenylsulfonyl)indole) [100, 101].

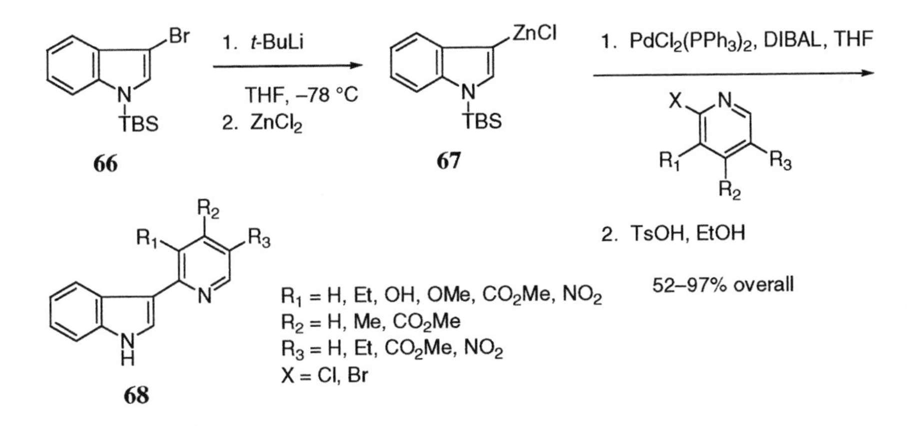


Danieli extended the Pd-catalyzed coupling of 2-indolylzinc chlorides to a series of halopyridin-2-ones and halopyran-2-ones [102]. This Negishi coupling is more efficient than a Suzuki approach but not as good as a Stille coupling. An example of the latter will be shown in Section 3.3.4. These workers also generated zinc reagents from 5-iodopyridin-2-one and 5-bromopyran-2-one but Negishi couplings were sluggish. Since direct alkylation of a 2-lithioindole failed, Fisher and co-workers utilized a Negishi protocol to synthesize 2-benzylindole **69** as well as the novel CNS agent **70** [103].

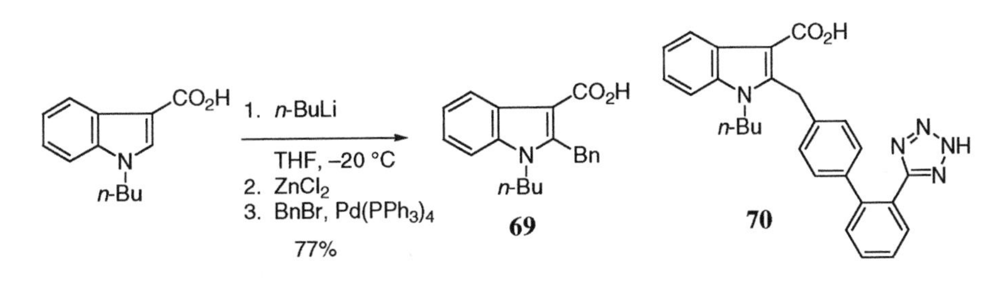

Cheng and Cheung also employed a 2-indolylzinc chloride **72** to couple with indole **71** in a synthesis of "inverto-yuehchukene" **73** [104]. Other Pd catalysts were no better in this low-yielding process.

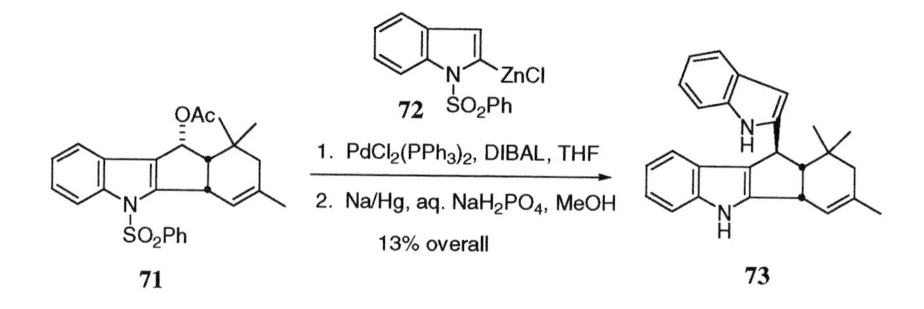


Maas and co-workers coupled 1-methyl-2-indolylzinc chloride (**63**) with several (*Z*)-bromostyrenes (e.g., **74**) to give the corresponding 2-(2-arylethenyl)-1-methylindoles (e.g., **75**) [105]. Although the (*Z*)-olefin is the kinetic product, this compound can be isomerized to the (*E*)-olefin. The β-bromostyrenes are conveniently synthesized by treating the requisite dibromostyrene with $Pd(PPh_3)_4/Bu_3SnH$ (61–79%). In addition to **74**, 1-[(*Z*)-2-bromo-ethenyl)]-4-nitrobenzene and 2-[(*Z*)-2-bromoethenyl)]thiophene were prepared using this method and successfully coupled with **63**.

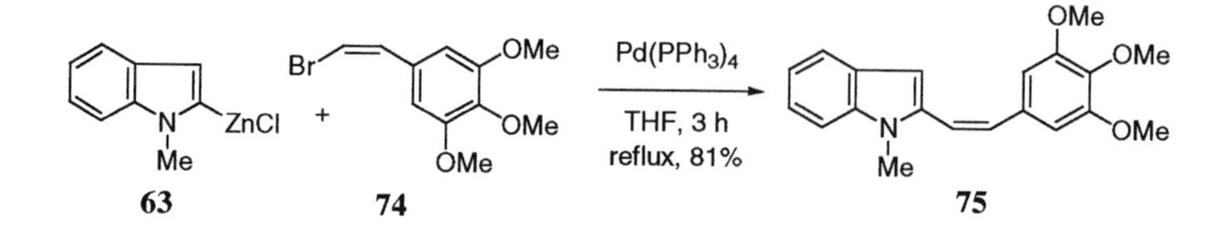


In a synthesis of polyketides, Kocienski crafted indole **78** from 2-iodo-1-methylindole and the appropriate organozinc reagent **77** derived from the corresponding stannane (**76**), which itself was reluctant to undergo a Stille coupling [106].

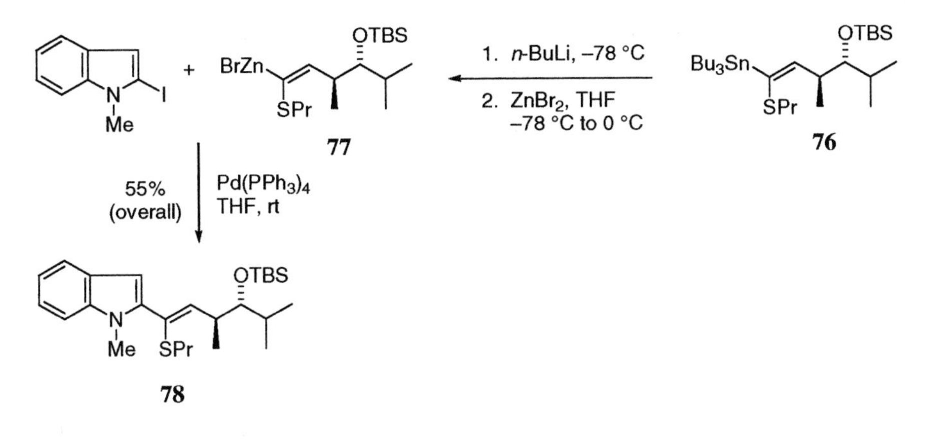


In his general studies aimed at the synthesis of ergot alkaloids, Hegedus effected the coupling of dihaloindole **79** with the zinc reagent **80**, generated *in situ* from 1-methoxy-1,2-propadiene, to afford ketene **81** [21].

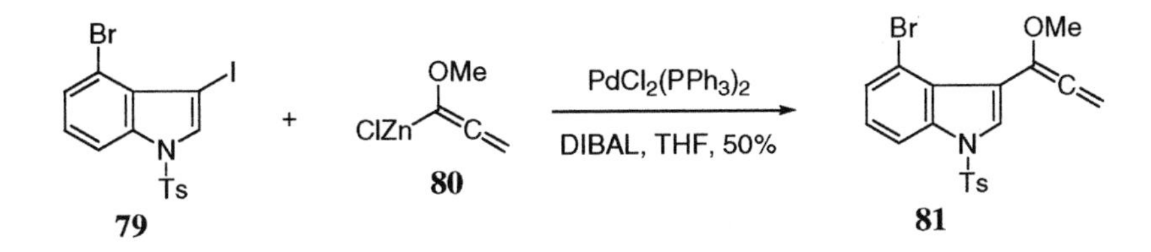


Negishi methodology can also be used to achieve the 3-acylation of indoles. Thus, Faul used this tactic to prepare a series of 3-acylindoles **83** from indole **82** [107]. Indole **82** could also be iodinated cleanly at C-3 with *N*-iodosuccinimide (78%).

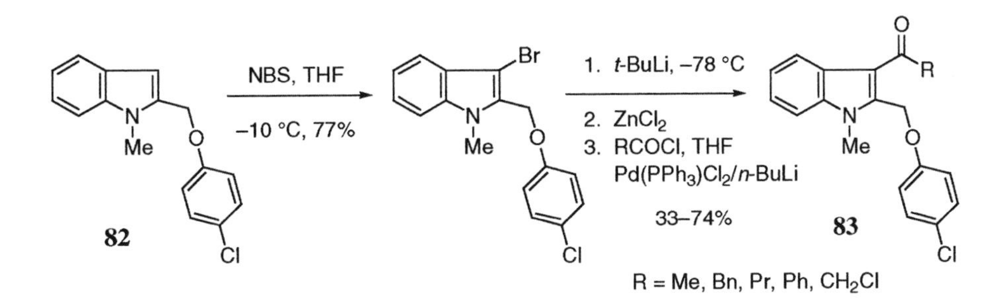


Bergman described indole C-3 acylation with acid chlorides *via* 1-indolylzinc chloride in the *absence* of palladium [108]. Davidsen and co-workers synthesized **86**, which is a potent antagonist of platelet activating factor-mediated effects, using this Bergman acylation sequence as shown [109]. As will be discussed in the next section, a Suzuki coupling was used to prepare **84**.

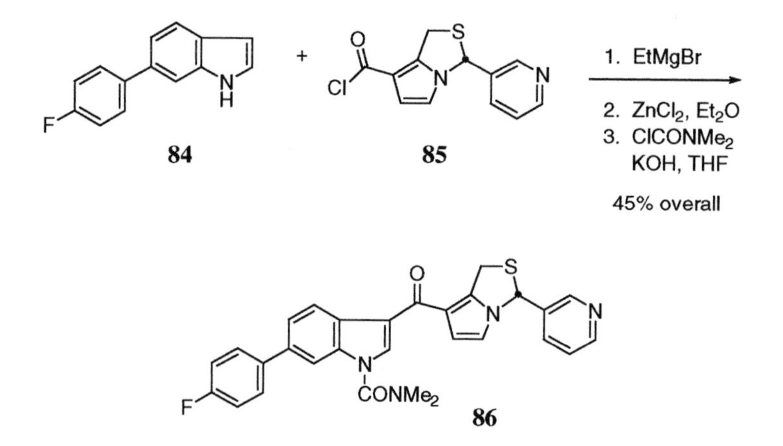


In continuation of his extraordinarily versatile and efficient directed-metalation technology, Snieckus employed indole **87** to selectively lithiate C-4 and to effect a Negishi coupling with 3-bromopyridine to give **88** in 90% yield [110]. In contrast, a Suzuki protocol gave **88** in only 19% yield (with loss of the TBS group).

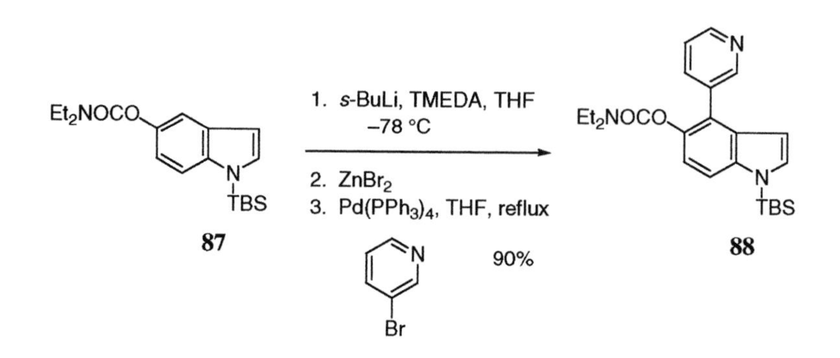


1,9-Dilithio-β-carboline (**89**), which was generated from 1-bromo-β-carboline, undergoes the Negishi coupling with 2-chloroquinoline to form the alkaloid nitramarine (**90**) [111].

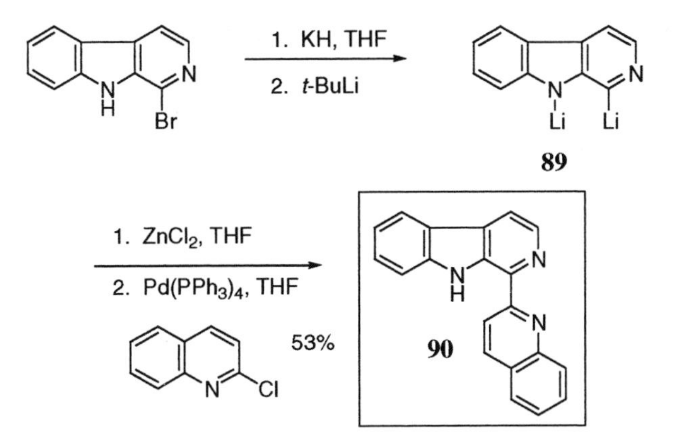


Grigg employed organozinc chemistry to construct 3-alkylidenedihydroindoles such as **91** *via* a tandem Pd-catalyzed cyclization-cross coupling sequence [112]. A similar route to such compounds was reported by Luo and Wang; e.g., **92** to **93** [113].

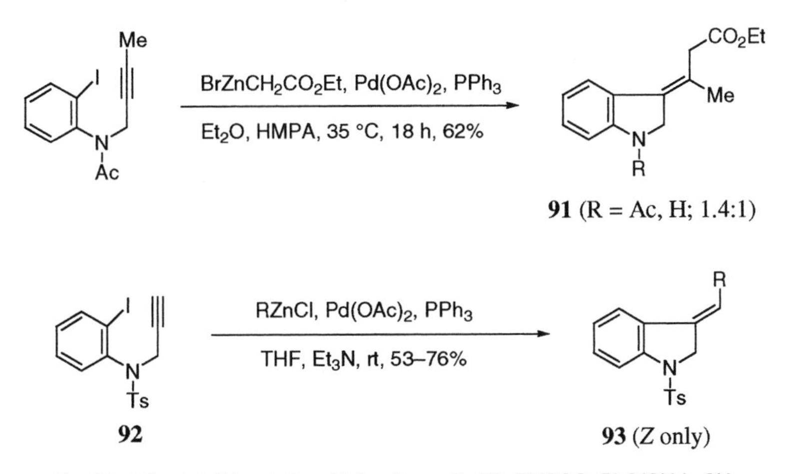


R = Ph, 2-furyl, 2-thienyl, 2-pyridyl, *n*-Bu, *n*-BuCC, TMSCC, $PhC(CH_2)=CH_2$

### 3.3.3 Suzuki coupling

Although the first report of an indoleboronic acid was by Conway and Gribble in 1990, this compound (**94**) was not employed in Suzuki coupling, but rather it was utilized *en route* to 3-indolyl triflate **34** as described in section 3.1 [12].

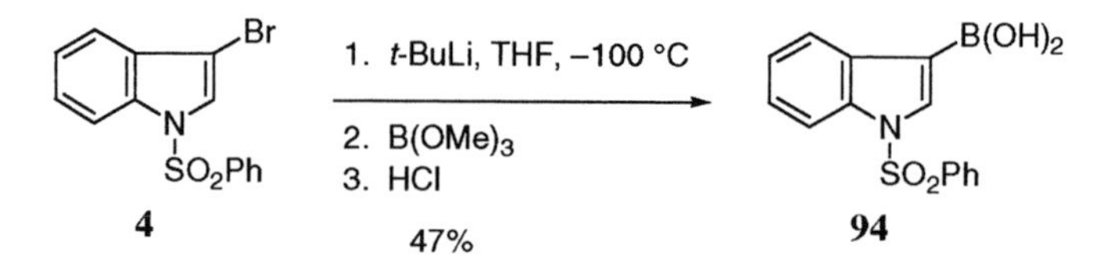


In the intervening years, indoleboronic acids substituted at all indole carbon positions have found use in synthesis. For example, Claridge and co-workers employed **94** in a synthesis of isoquinoline **95** under standard Suzuki conditions in high yield [114]. Compound **95** was subsequently converted to the new Pd-ligand 1-methyl-2-diphenylphosphino-3-(1'-isoquinolyl)indole.

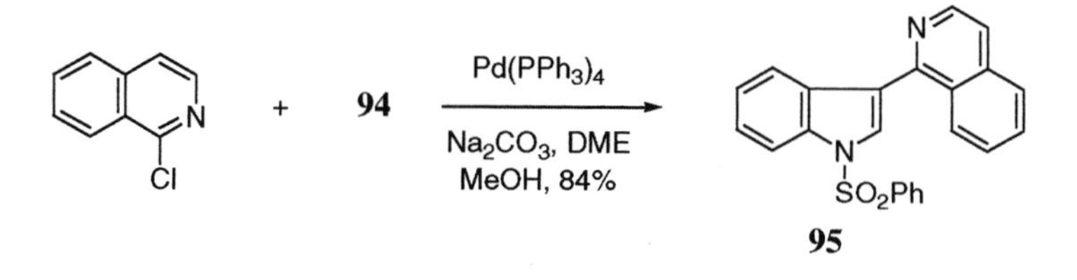


Martin utilized indoleboronic acids in Pd-catalyzed coupling to great effect, and has improved upon the halogen-metal exchange route to indole-3-boronic acids by adopting a mercuration-boronation protocol as illustrated below for the preparation of **96** and **97** [115, 116].

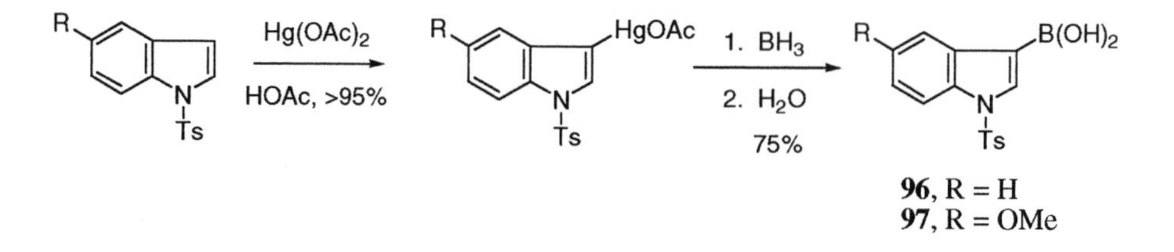


Boronic acids **96** and **97** couple very well with vinyl triflates **98** and **99** under typical Suzuki conditions ($Pd(PPh_3)_4/Na_2CO_3/LiCl/DME$) to give indoles **100** and **101**, respectively, in 76–92% yield [115, 116]. Enol triflates **98** and **99** were prepared in good yield (73–86%) from *N*-substituted 3-piperidones, wherein the direction of enolization (LDA/THF/–78 °C; $PhNTf_2$) is dictated by the *N*-substituent.

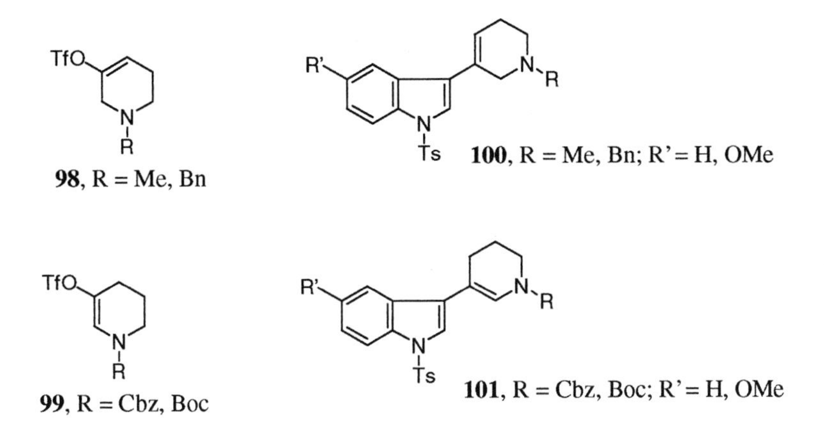


In the course of their successful syntheses of the marine alkaloids nortopsentins A–D, Kawasaki and co-workers were able to prepare selectively boronic acid **103** from 1-(*tert*-butyldimethylsilyl)-3,6-dibromoindole (**102**) and effect Suzuki couplings to give 3-arylindoles **104** in good yields [117]. Complementary to this chemistry is the direct Pd-catalyzed reaction of **102** with arylboronic acids to give 6-aryl-3-bromoindoles **105** [117].

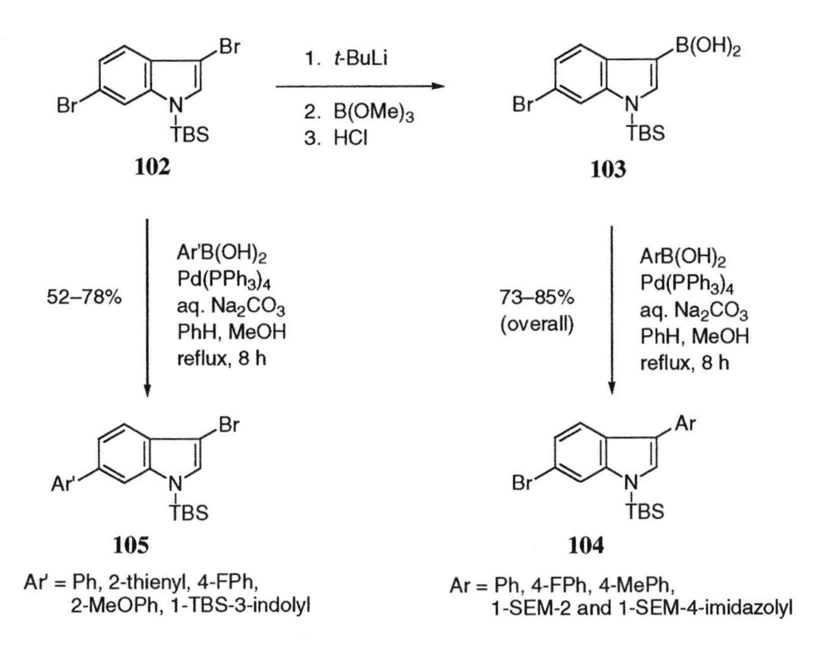


Hoerrner utilized *N*-TIPS boronic acid **106** to prepare dehydro-β-methyltryptophan **107**, *en route* to a synthesis of β-(2*R*, 3*S*)-methyltryptophan by asymmetric hydrogenation of **107** [118].

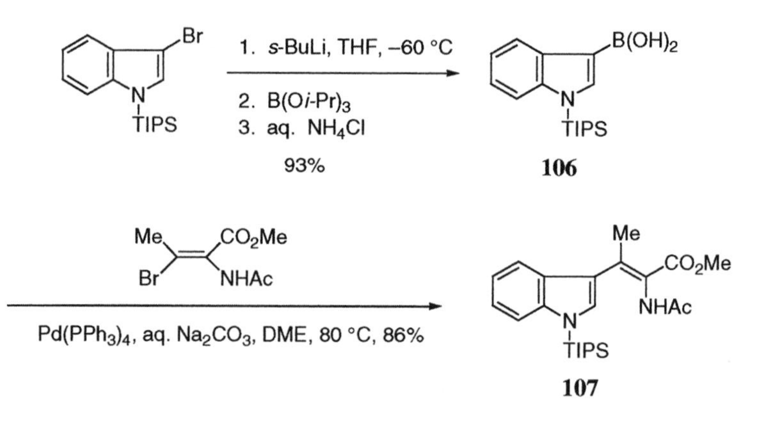


3-Indoleboronic acid **96** was employed by Neel to prepare bis(indolyl)maleimides such as **109** [119]. However, since the standard Suzuki conditions failed (triflate **108** apparently decomposing under the reaction conditions), the use of a phosphine-free Pd catalyst [120] and cesium fluoride [121] was necessary and gave **109** in an acceptable yield of 55%.

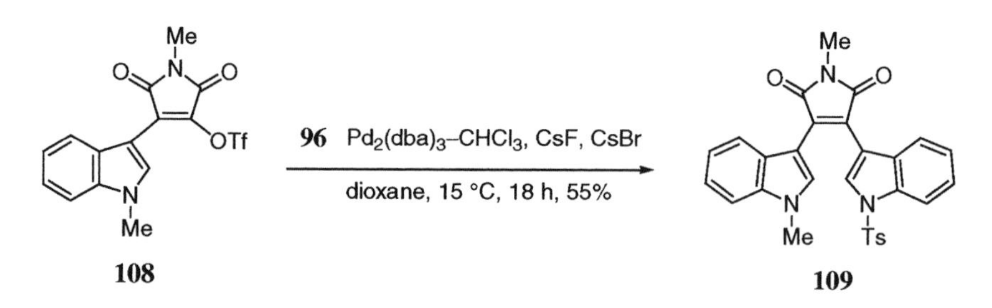


Merlic synthesized a series of *N*-substituted 2-iodoindoles (*N*-methyl, *N*-allyl, *N*-benzyl, *N*-phenylsulfonyl) and converted them to the corresponding boronic acids. Two of the latter (*N*-methyl and *N*-benzyl) undergo a Suzuki coupling with 2-iodoindole to give 2,2'-biindolyls in 51–67% yield *en route* to indolocarbazoles [122, 123]. By comparison, a Negishi protocol failed and a Stille coupling gave 2,2'-biindolyls in lower yields. Gallagher prepared *N*-Boc-2-indoleboronic acid **110** for use in Suzuki coupling with aryl bromides and iodobenzene in a search for new selective dopamine $D_3$ receptor antagonists [124].

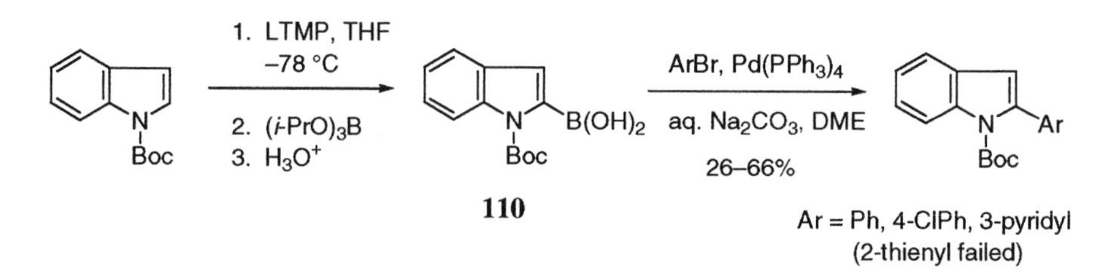

Martin prepared indole-5-boronic acid from 5-bromoindole by halogen-lithium exchange (44% yield) and performed Suzuki couplings with a wide range of aryl bromides (mainly) and heteroaryl bromides to give the expected compounds in 52–94% yield under the standard conditions [125, 126]. In addition to the typical substituted aryl iodides (fluoro, methoxy, nitro), the range of iodo partners included pyridines, pyrimidines, pyrazines, furans, thiophenes, thiazoles, and isoxazoles. Roussi and co-workers executed Suzuki couplings with a variety of 5-, 6-, and 7-indoleboronic acids and aryl bromides, and also with arylboronic acids and 5-, 6-, and 7-bromoindoles to give the biaryl products in yields up to 60% [127]. The first combination gave higher yields of coupling products. Unfortunately, these extensive model studies did not translate into a successful intramolecular version that the authors had devised as a synthesis of chloropeptin and kistamycin model compounds [128]. Thus, neither **111** nor **112** underwent the desired intramolecular Suzuki coupling (although a Ni(0)-mediated approach was successful).

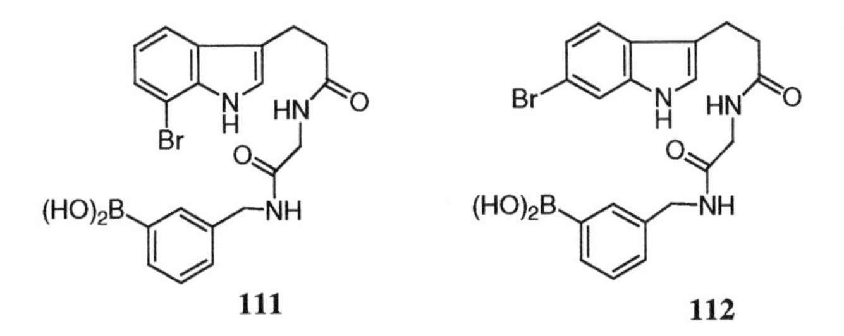


The "reverse" Suzuki coupling of indolyl halides (and triflates) with aryl- and heteroarylboronic acids has been pursued by a number of investigators. The choice of which coupling partner is the boronic acid and which is the halide (or triflate) is governed by relative substrate accessibility and availability. Zembower and Ames incorporated a Suzuki coupling into a general synthesis of 5-substituted tryptophans **115** *via* the iodinated cyclic tautomer **113** as illustrated [129]. The Suzuki product **114** (R = Ph) and **113** were unraveled by a sequence of $H_2SO_4$ (to regenerate the indole) (85–91%), TMSI carbamate cleavage (75–83%), and base hydrolysis of the *N*-acetyl and methyl ester (61–69%) to yield **115**. These Suzuki couplings were performed under conventional conditions with $RB(OH)_2$ or with the modified reagents derived from 9-BBN.

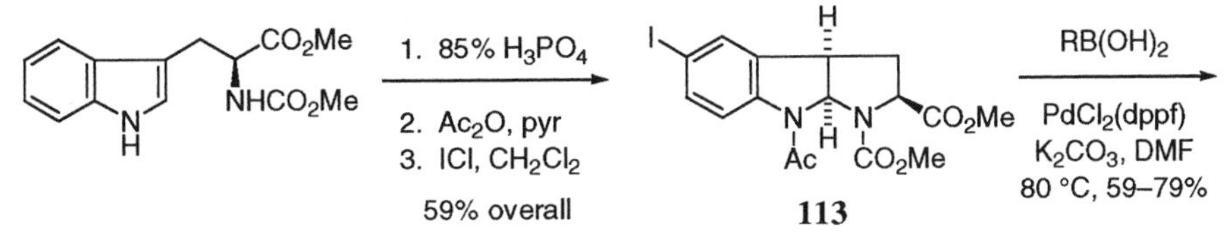

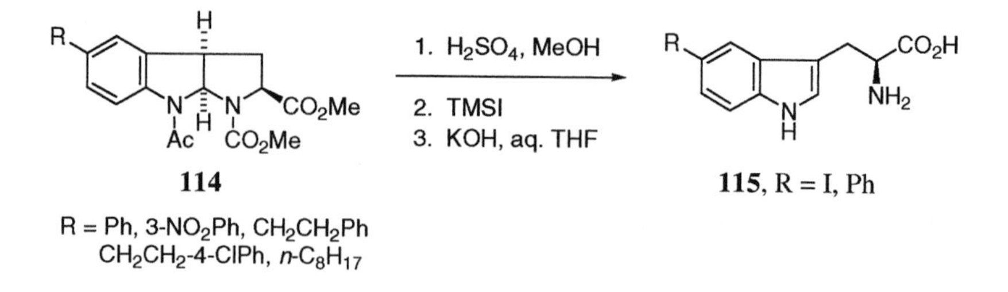


Snieckus described short syntheses of ungerimine (**121**) and hippadine by Suzuki couplings of boronic acid **118** with 7-bromo-5-(methylsulfonyloxy)indoline (**116**) and 7-iodoindoline (**117**), respectively [130]. Cyclization and aerial oxidation also occur. Treatment of **119** with Red-Al gave ungerimine (**121**) in 54% yield, and oxidation of **120** with DDQ afforded hippadine in 90% yield. Indoline **116** was readily synthesized from 5-hydroxyindole in 65% overall yield by mesylation, reduction of the indole double bond, and bromination. Indoline **117** was prepared in 67% yield from *N*-acetylindoline by thallation-iodination and basic hydrolysis.

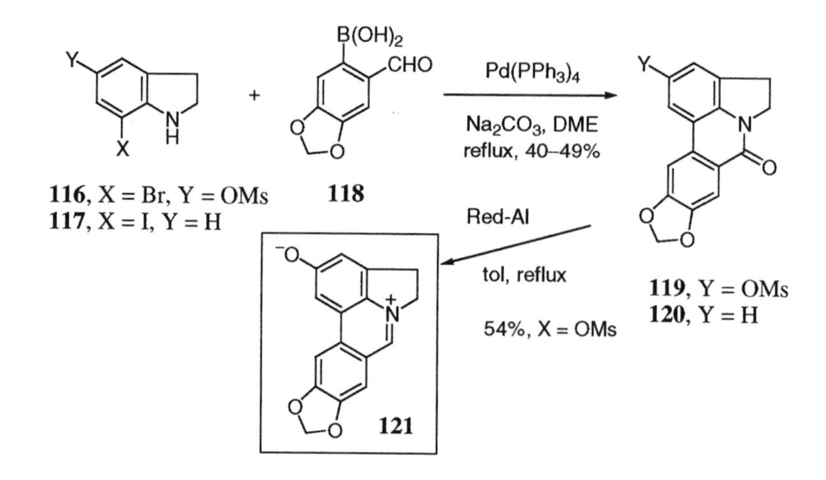


Martin effected the synthesis of several 3,5-diarylated indoles by a tandem Stille-Suzuki sequence [131]. The latter reaction involves exposure of 3-(3-pyridyl)-5-bromo-1-(4-toluenesulfonyl)indole with arylboronic acids (aryl = 3-thienyl, 2-furyl, phenyl) under typical conditions to give the expected products in 86–98% yield [131]. Carrera engaged 6- and 7-bromoindole in Pd-catalyzed couplings with 4-fluoro- and 4-methoxyphenylboronic acids to prepare 6- and 7-(4-fluorophenyl)indole (90% and 74% yield) and 6-(4-methoxyphenyl)indole (73% yield) [29]. Banwell and co-workers employed 7-bromoindole in a Suzuki coupling with 3,4-dioxygenated phenylboronic acids *en route* to the synthesis of Amaryllidaceae alkaloids [132]. Yields of 7-arylated indoles are 93–99%. Moody successfully coupled 4-bromoindole

**122** with boronic acid **123**, which was derived from the corresponding bromo compound, to give **124** in high yield [133].

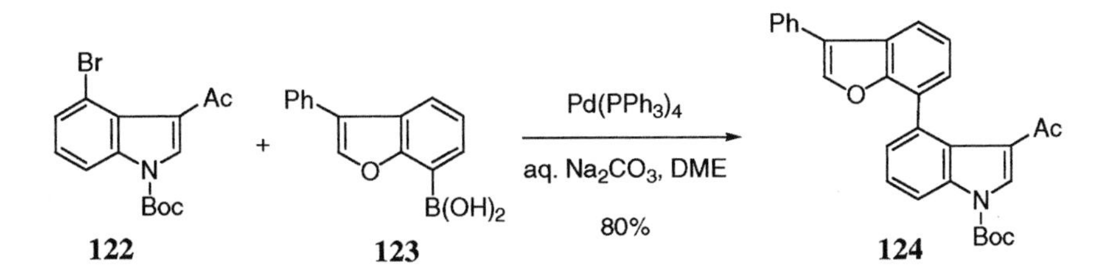


Terashima employed diethyl-(3-pyridyl)borane as the boron partner in a Suzuki coupling with both 5-bromo-1-(4-toluenesulfonyl)indole and 3-bromo-1-(mesitylenesulfonyl)indole to give the corresponding pyridylindoles in modest yields (47%, 39%) [134]. Somei effected the coupling of phenyl-, 2-furyl- and 1-hexenylboronic acids with 4-thallated indole-3-carboxaldehyde ($Pd(OAc)_2$/DMF) to give 4-substituted 3-formylindoles [135]. Regioselective thallation of indole-3-carboxaldehyde is achieved using thallium tris-trifluoroacetate in 77% yield. Indole **125**, which is available by the Buchwald zirconium indoline synthesis, was used by Buchwald to synthesize **127** *via* a Suzuki protocol [136]. Boronate ester **126** is prepared by the hydroboration of 3-methyl-1-butyne with catechol borane. Indole **127** had been used in earlier studies to synthesize the clavicipitic acids.

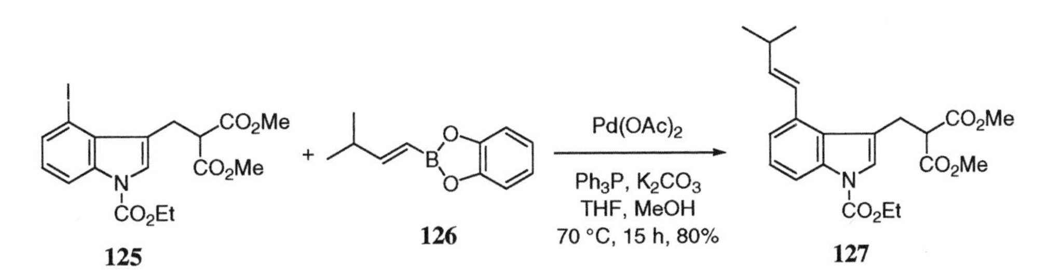


The medicinal importance of 2-aryltryptamines led Chu and co-workers to develop an efficient route to these compounds (**130**) *via* a Pd-catalyzed cross-coupling of protected 2-bromotryptamines **128** with arylboronic acids **129** [137]. Several Suzuki conditions were explored and only a partial listing of the arylboronic acids is shown here. In addition, boronic acids derived from naphthalene, isoquinoline, and indole were successfully coupled with **128**. The C-2 bromination of the protected tryptamines was conveniently performed using pyridinium hydrobromide perbromide (70–100%). 2-Phenyl-5-(and 7-)azaindoles have been prepared *via* a Suzuki coupling of the corresponding 2-iodoazaindoles [19].

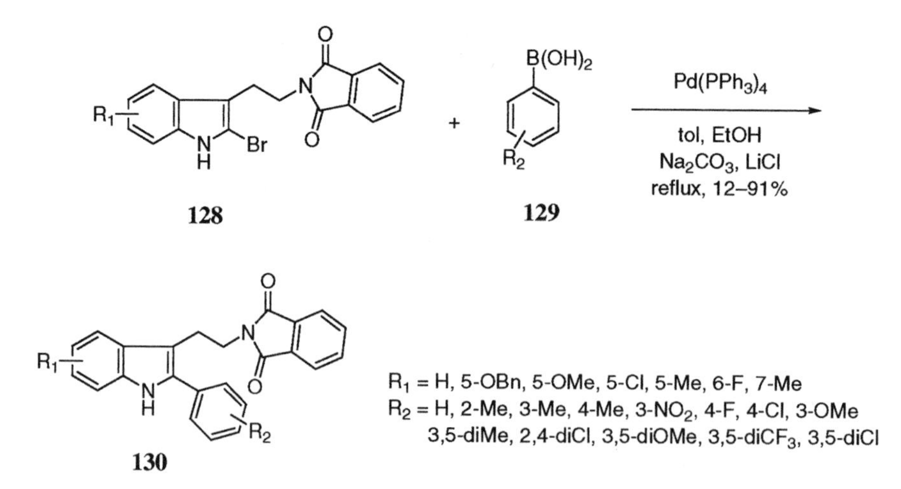


Indolyltriflates have been used in Suzuki couplings by Mérour [138, 139]. Thus, the readily available 1-(phenylsulfonyl)indol-2-yl triflate (**131**) smoothly couples with arylboronic acids in 65–91% yield. Similarly, Pd-catalyzed cross-coupling of phenylboronic acid with 1-benzyl-2-carbomethoxyindol-3-yl triflate affords the 3-phenyl derivative (62% yield) [139].

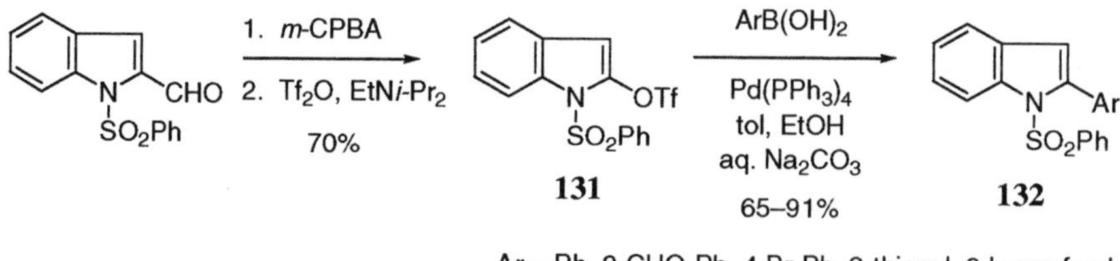


Doi and Mori prepared the enol triflate of 1-(4-toluenesulfonyl)-4-oxo-4,5,6,7-tetrahydroindole which couples with 3,4-methylenedioxyphenylboronic acid ($PdCl_2(PPh_3)_2$) to give the 4-substituted derivative in 52% yield. Dehydrogenation furnishes the indole in 59% yield [140]. Likewise, carbazole triflates have served as Suzuki partners with arylboronic acids and 9-alkyl-9-BBN derivatives in the synthesis of carazostatin, hyellazole, and carbazoquinocins B–F [141, 142a], and in the construction of several 4-aryl and 4-heteroaryl pyrrolo[3,4-*c*]carbazoles as models for future combinatorial studies of new protein kinase C inhibitors [143]. An aryl triflate was employed in the early stages of a Suzuki coupling route to novel naltrindoles [144]. Bracher and Hildebrand utilized 1-chloro-β-carboline in Suzuki methodology with the appropriate arylboronic acids to prepare 1-phenyl-β-carboline and the alkaloids komaroine (**133**) and perlolyrine (**134**) [145], and Mérour engineered the synthesis of benzo[5,6]cyclohepta[*b*]indole derivatives, such as **135**, *via* a Suzuki coupling of 5-methyl-11-bromo-5,6-dihydrobenzo[5,6]cyclohepta[*b*]indol-6-one with 2-thienylboronic acid [146].

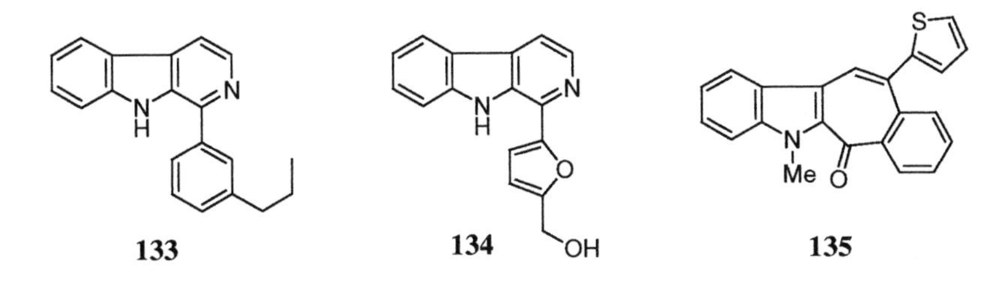


Ishikura and co-workers have done extensive work on the utility of indolylborates such as lithium triethyl(1-methylindol-2-yl)borate (**136**), prepared as shown from 1-methylindole, in Suzuki-like Pd-catalyzed reactions [147–157]. For example, **136** couples smoothly with aryl halides to afford 2-arylindoles **137** [147]. The amount of 2-ethyl-1-methylindole by-product, formed by ethyl group migration, can be minimized by refluxing the mixture. At room temperature 2-ethyl-1-methylindole is the major product. More recent work by Ishikura extended these couplings to the (removable) *N*-Boc analog of **136** with comparable yields to those obtained with **136** [157].

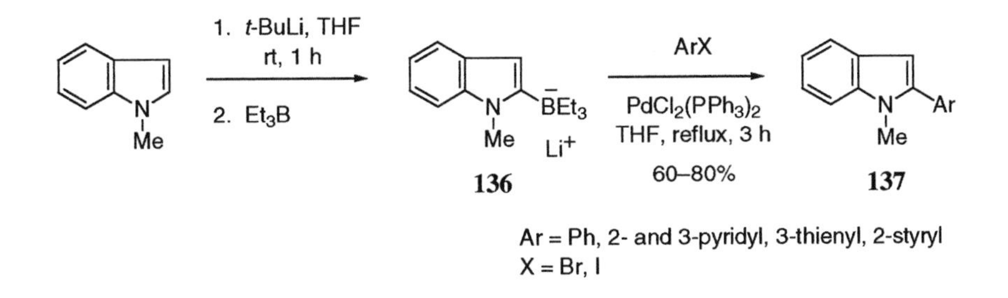


The extraordinary power of the Ishikura palladium-catalyzed couplings of indolylborates is revealed by the several examples shown below [148, 152, 154, 155]. The carbonylation version is discussed in Section 3.6. The formation of allenylindoles **139** *vis-á-vis* alkynylindoles **141** apparently depends on the equilibrium between an allenylpalladium complex and a propargylpalladium complex, and $S_N2$-like attack on the latter by **136** to give **141** is favored by the $Ph_3P$-ligated Pd complex.

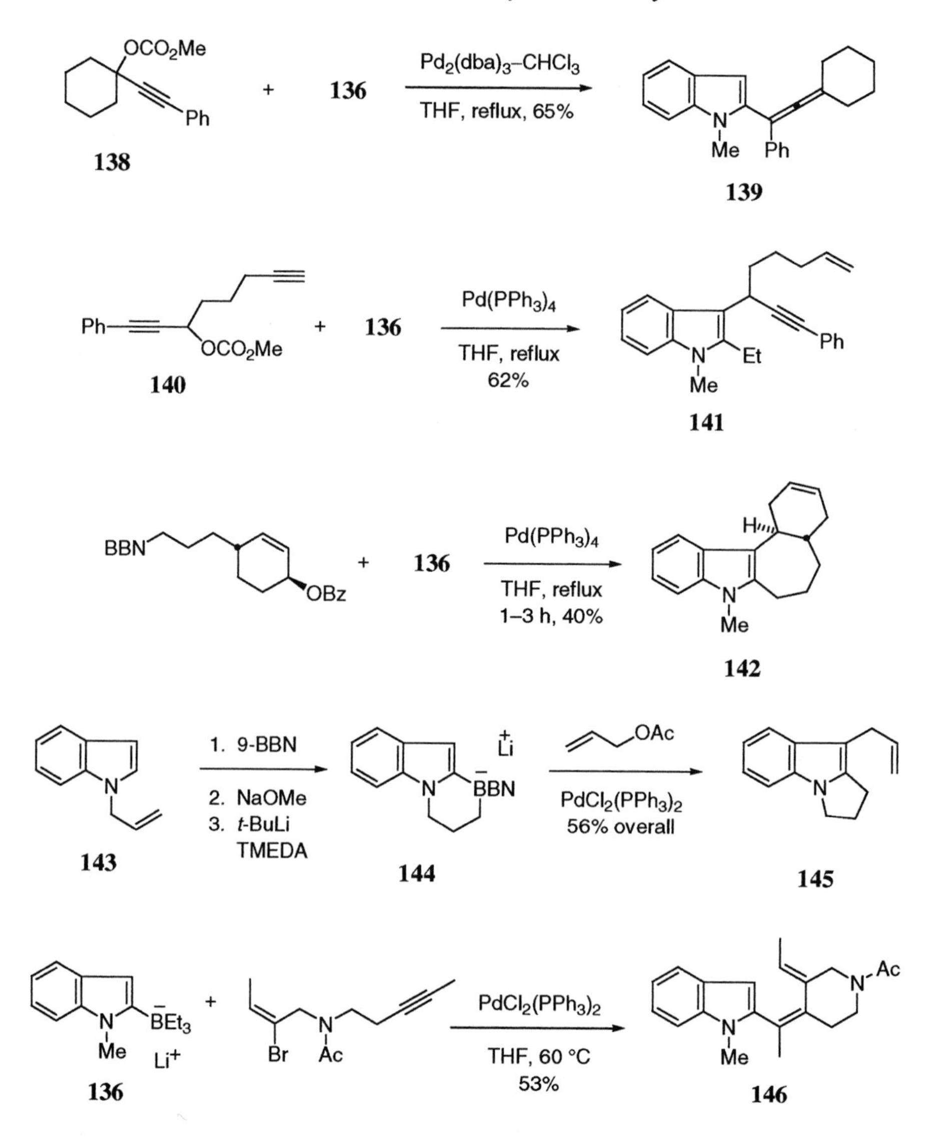


Murase and co-workers generated the *N*-methoxyindolylborate **147** and effected coupling with several indoles to give **148**, and, by reductive cleavage of the *N*-methoxyl group, arcyriacyanin A [25].

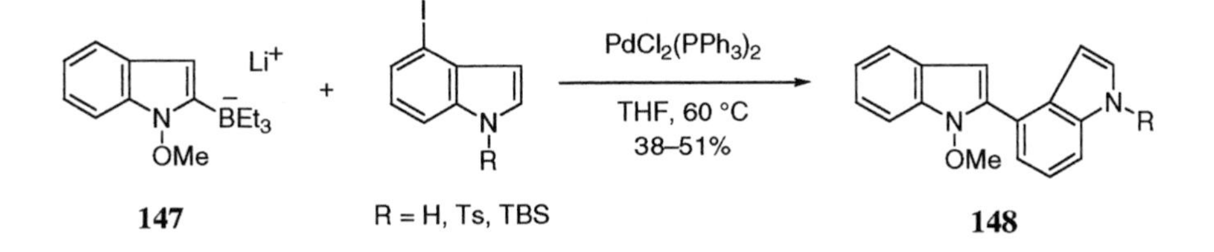


Palladium-catalyzed reactions of arylboronic acids have been utilized to craft precursors for constructing indole rings. Suzuki found that tris(2-ethoxyethenyl)borane (**149**) and catechol-derived boranes **150** readily couple with *o*-iodoanilines to yield **151**, which easily cyclize to indoles **152** with acid [158]. Kumar and co-workers used this method to prepare 5-(4-pyridinyl)-7-azaindoles from 6-amino-5-iodo-2-methyl-3,4'-bipyridyl [159]. A similar scheme with catechol-vinyl sulfide boranes also leads to indoles [160]. A Suzuki protocol has been employed by Sun and co-workers to synthesize a series of 6-aryloxindoles [161].

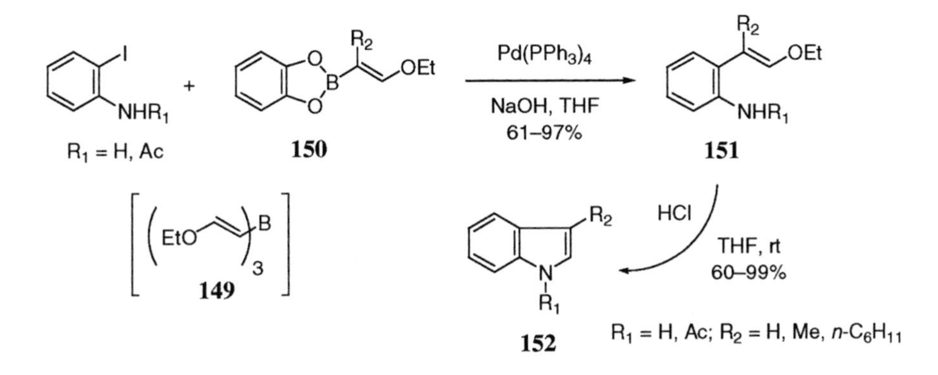


Abell utilized a Suzuki cross-coupling reaction on resin **153**. Subsequent acid treatment effected cyclization to indole **154**, which was readily cleaved with amines and alcohols to form potential libraries of amides and esters, respectively [162].

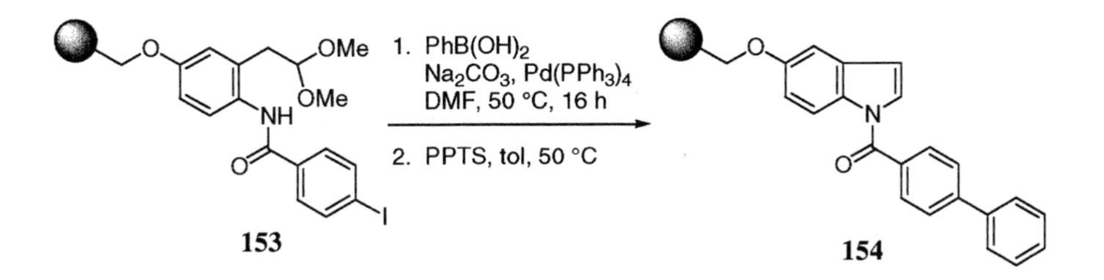


Quéguiner made great use of Suzuki methodology to prepare heterobiaryls for use in carboline synthesis [163]. This chemistry is discussed in Chapter 4. As explored in more detail in Section

3.5, Grigg developed several palladium-catalyzed tandem cyclization-anion capture processes, and these include organoboron anion transfer agents [112, 164, 206]. Two examples of this methodology are shown.

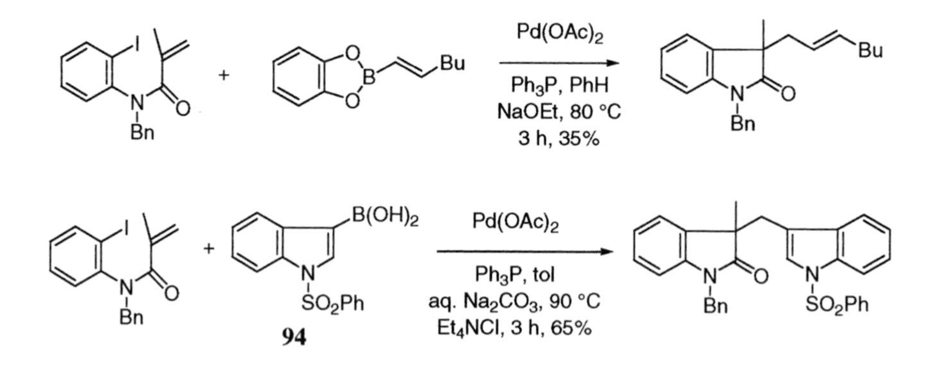


### 3.3.4 Stille coupling

Despite the well-documented toxicity of organotin compounds, the use of these reagents in Pd-catalyzed cross coupling reactions continues unabated, following the pioneering work of Stille. Indolylstannanes are usually prepared either by treating the appropriate lithioindole with a trialkyltin halide or by halogen-tin exchange with, for example, hexamethylditin. Typical procedures for the generation of (1-(4-toluenesulfonyl)indol-2-yl)trimethylstannane (**155**) and (1-(4-toluenesulfonyl)indol-3-yl)trimethylstannane (**156**) are illustrated [17, 18]. Bosch described an excellent route to the *N*-TBS-3-trimethylstannylindole [165].

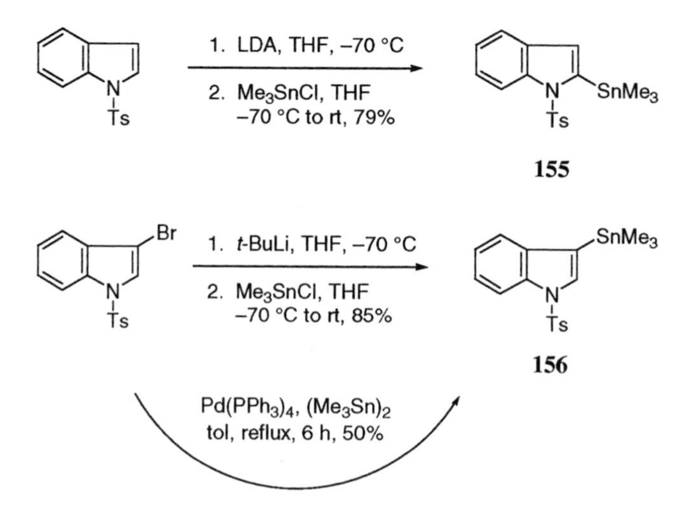

The indolyltributylstannanes, which are more robust than their trimethylstannyl counterparts, are prepared similarly [166, 167]. Labadie and Teng synthesized the *N*-Me, *N*-Boc, and *N*-SEM (indol-2-yl)tributylstannanes [167], and Beak prepared the *N*-Boc trimethyl- and tributyltin derivatives in high yield [166]. Caddick and Joshi found that tributylstannyl radical reacts with 2-tosylindoles to give the corresponding indole tin compounds as illustrated [168].

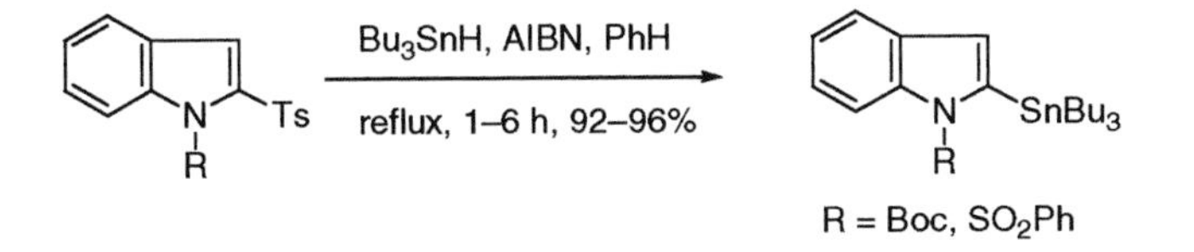


Stannylindoles that are substituted in the benzene ring can either be prepared by halogen-metal exchange or, in the case of the C-4 and C-7 positions, by directed lithiation. For example, *N*-TIPS-indole chromium complex **29** can be treated with trimethylstannyl chloride to give the C-4 substituted stannane (69% yield) [37]. Oxidative removal of the $Cr(CO)_3$ yields the *N*-TIPS indolylstannane (90% yield). Although indolylstannanes are often prone to premature destannylation, in some cases they can be manipulated prior to Pd-catalyzed cross-coupling reactions. For example, reaction of (1-methylindol-2-yl)trimethylstannane with tosyl isocyanate gives rise to **157** in high yield and a similar reaction with ethoxycarbonyl isothiocyanate gives **158**. In contrast, phenyl isocyanate gives *ipso* substitution [169]. Other syntheses of indolyltrialkylstannanes, including the elegant methodology of Fukuyama, will be presented later in this section.

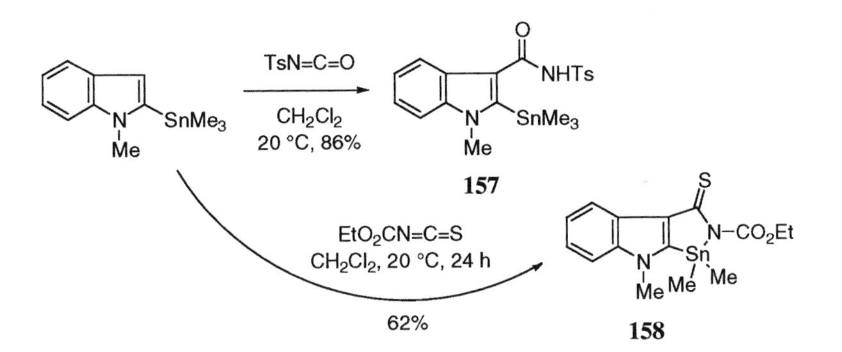


Palmisano and Santagostino first reported Stille reactions of indole-ring stannylindoles with their detailed studies of *N*-SEM stannane **159** [170]. Thus **159**, which is readily prepared by C-2 lithiation of *N*-SEM indole and quenching with $Bu_3SnCl$ (88%), couples under optimized Pd(0)-catalyzed conditions to give an array of cross-coupled products **160**. Some other examples and yields are shown. Stannane **159** could also be coupled with a (chloromethyl)cephem derivative (95%) and with a lysergic acid derivative (94%).

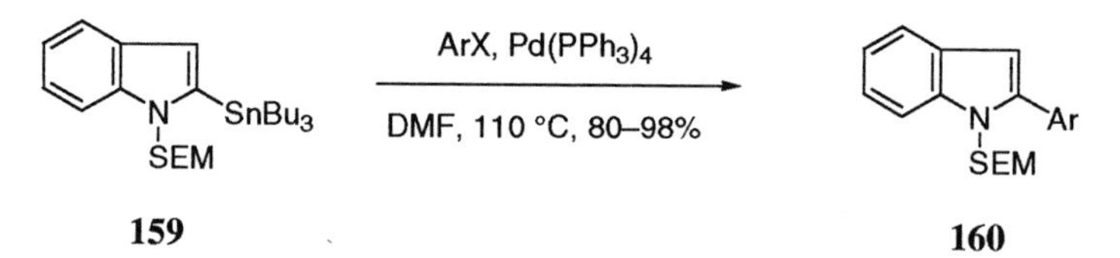


Ar = Ph, 2-$NO_2$-Ph, 2-Me-Ph, 4-OMe-Ph, 4-Ac-Ph, 2-pyridyl, 2-thienyl, Bn, allyl
X = Br, I

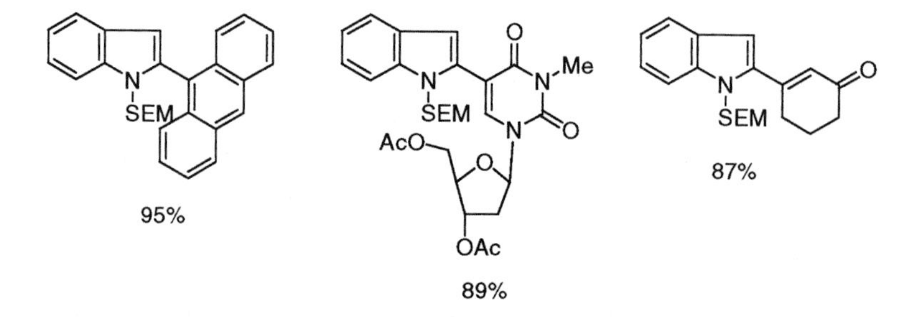


These workers also synthesized tryptamine stannane **161** and effected Stille couplings with this compound, including the intramolecular reaction **162** to **163** [171]. Eight- and 9-membered rings could also be fashioned in this manner. Other Pd catalysts were much less successful. The *N*-tosyl derivative of **162** was similarly prepared and used in Stille chemistry.

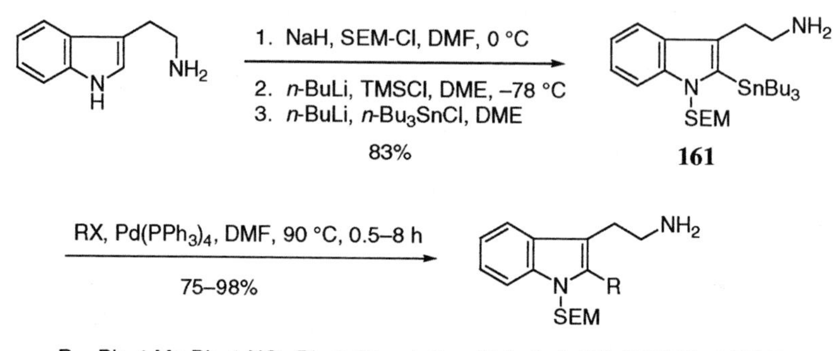


R = Ph, 4-Me-Ph, 4-$NO_2$-Ph, 2-thienyl, 2-pyridyl, vinyl, CH=CHTMS, CCTMS
X = Br, I

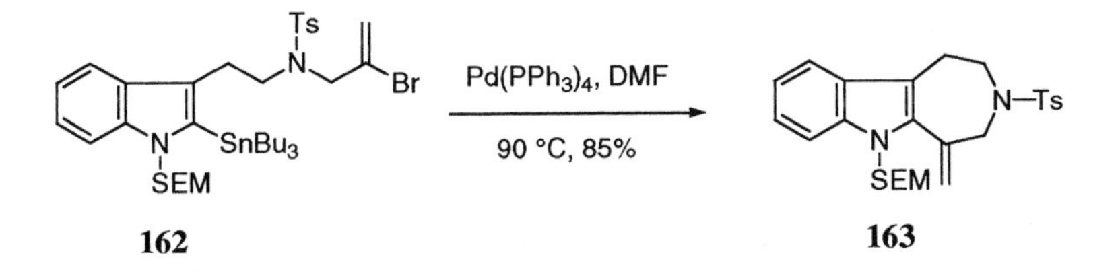


In a study complementary to that of Palmisano and Santagostino, Labadie and Teng published a thorough and careful exploration of the Pd-catalyzed cross-coupling ability of *N*-protected 2-tributylstannylindoles (*N* = Me, Boc, SEM) [167]. Although the *N*-Me and *N*-SEM stannanes are unstable on silica gel, these can be purified by distillation. These two independent studies clearly establish the power of Stille Pd-catalyzed cross coupling chemistry for synthesizing 2-substituted indoles. Using the Katritzky method of C-2 lithiation of indole-1-carboxylic acid, Hudkins synthesized 1-carboxy-2-tributylstannylindole and achieved cross coupling reactions with an array of aromatic and heterocyclic halides using $PdCl_2(Ph_3P)_2$ [172]. This stannane is stable for one month at –20°C. A synthesis of arcyriacyanin A features the coupling of *N*-Boc-2-trimethylstannylindole with 4-bromo-1-(4-toluenesulfonyl)indole (**11**) [173]. Joule employed stannane **159** in a Pd-catalyzed coupling with 2-iodonitrobenzene to set the stage for a synthesis of isocryptolepine [174]. Although 1-tosyl-2-tributylstannylindole did couple with 2-bromonitrobenzene, subsequent chemistry was less successful than that involving **159**. Danieli made use of **159** to effect couplings with 5-bromo-2-pyridones and 5-bromo-2-pyranone to give the C-2 coupled products in 44–71% yields [102]. Use of 1-(4-methoxyphenylsulfonyl)-2-tributylstannylindole led only to indole dimer. Fukuyama devised a novel tin-mediated indole ring synthesis leading directly to 2-stannylindoles that can capture aryl and alkyl halides in a Pd-catalyzed cross-coupling termination reaction [175–176]. The presumed pathway is illustrated and involves initial tributylstannyl radical addition to the isonitrile **164**, cyclization, and final formation of stannylindole **165**.

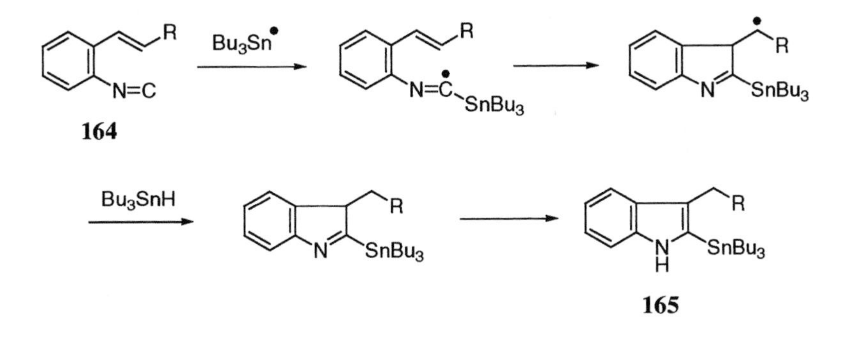

Moreover, the *in situ* reaction of **165** under Stille conditions affords a variety of coupled products **166**.

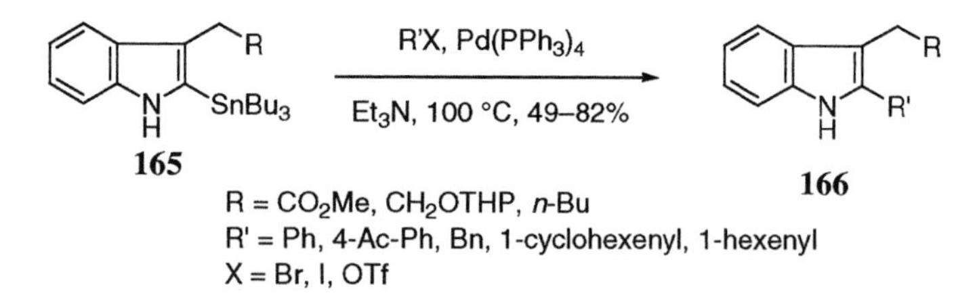


The potential power of Fukuyama's method is illustrated by the synthesis of biindolyl **168** which was used in a synthesis of indolocarbazoles [176]. The isonitriles (e.g., **167**) are generally prepared by dehydration of the corresponding formamides with $POCl_3$.

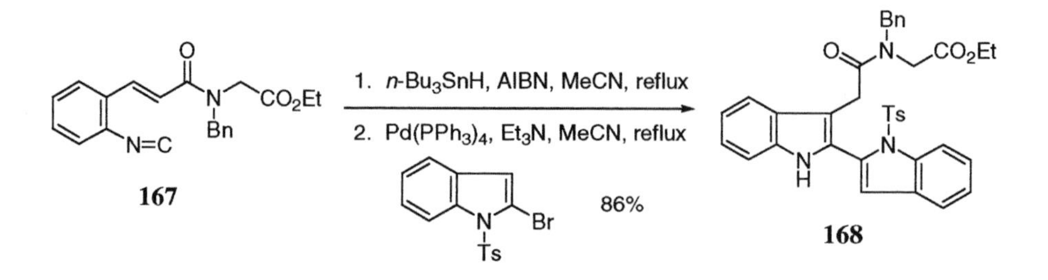


An application of Stille couplings to the solid phase using a traceless *N*-glycerol linker with 2-stannylindoles has been developed [177]. Only a few examples of the use of 3-stannylindoles in Stille reactions have been described. Ortar and co-workers prepared **169** and **170** and effected Pd-catalyzed cross coupling reactions with several aryl, heteroaryl, and vinyl substrates (bromides, iodides, triflates) to give the expected products **171** in high yields [178]. Enol triflates behave exceptionally well under the Ortar conditions, e.g., **172** to **173**.

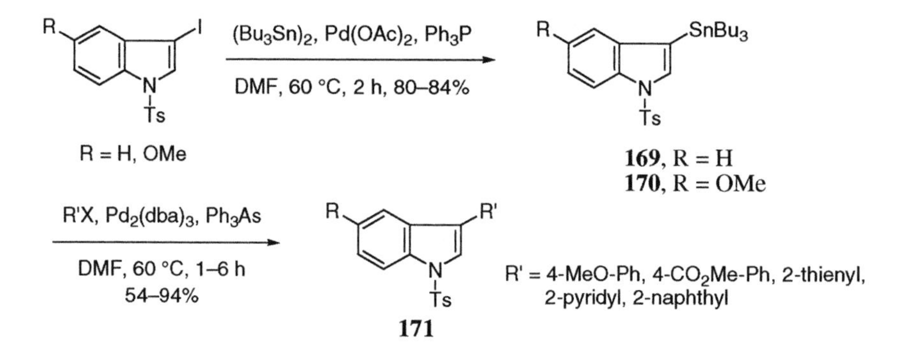

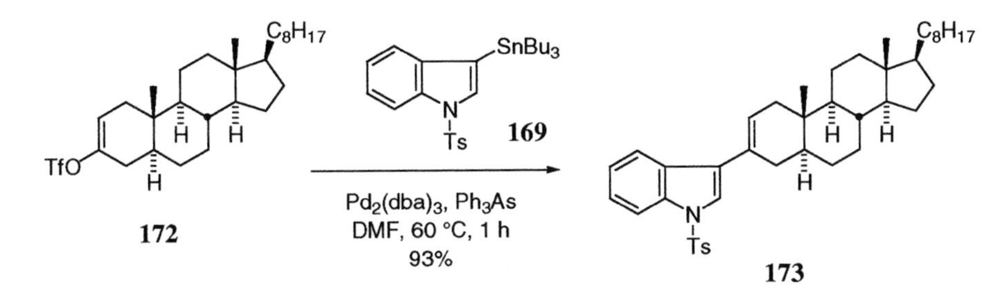


Murakami generated 3-tributylstannylindoles *in situ* (but also isolable) using 3-bromoindole **174**, allylic acetates and carbonates, and hexamethylditin [179]. A typical procedure is illustrated for the synthesis of **175**. The corresponding 5-bromo analog is allylated to the extent of 59%. 3-Stannylindoles couple smoothly in tandem fashion with 2,3-dibromo-5,6-dimethylbenzoquinone under Stille conditions [180].

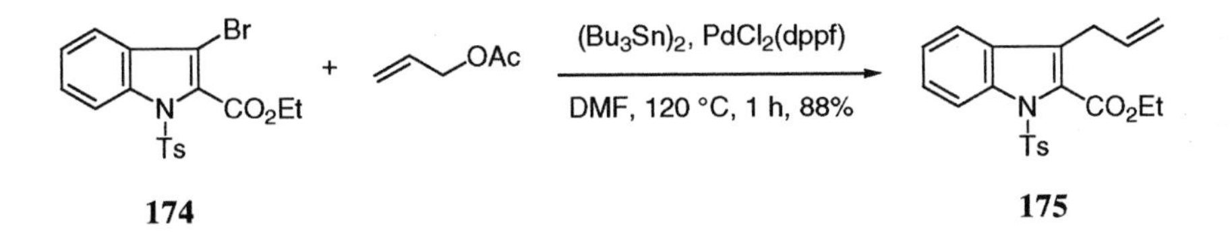


A number of investigators employed 2- or 3-haloindoles (or triflates) in combination with organotin compounds to effect Pd-catalyzed cross coupling. Somei used a Stille coupling of 2-bromo-3-indolecarboxaldehyde, which can be prepared from oxindole (77% yield), with several organotin reagents to synthesize the corresponding 2-substituted indoles [181]. One such reagent, (3-hydroxy-3-methyl-1-buten-1-yl)tributylstannane, was used in a synthesis of borrerine. Other substituents are methyl, phenyl, 2- and 3-pyridyl, and methyl acryloyl. Palmisano effected a similar coupling of the *N*-SEM derivative of 2-iodoindole with a stannylated pyrimidine (uridine) nucleoside [182]. Fukuyama successfully coupled methyl 2-iodoindol-3-ylacetate with methyl 2-tributylstannylacrylate in total syntheses of (±)-vincadifformine and (–)-tabersonine [176]. 2-Vinylazaindoles have been prepared in this fashion [19].

Indole triflates have proven to be very compatible with Pd-catalyzed methodologies, and Mérour applied triflate **36** in a Stille-type synthesis of 2-vinylindole **176** [138]. Gribble and Conway observed that the corresponding indole-3-triflate **34** gives the 3-vinylindole in 62% yield under similar conditions using $Pd(Ph_3P)_2Cl_2$ [183]. Mérour also studied similar Stille couplings of 2-carboethoxy-1-methyl-3-indolyltriflate with vinylstannane and ethoxyvinyl-stannane [139].

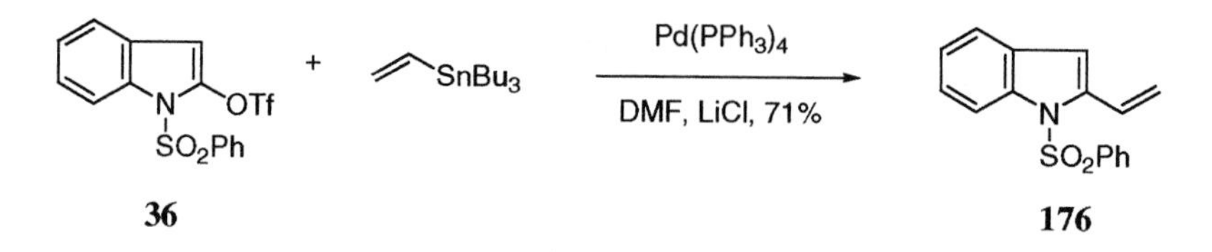


Sakamoto and co-workers employed a Stille coupling to prepare several 3-(2-ethoxyvinyl)indoles from (*Z*)-1-ethoxy-2-tributylstannylethene and 3-bromoindoles [184, 185]. For example, the reaction of 3-bromo-1-(methylsulfonyl)indole with this tin reagent affords the (*Z*)-indole in 83% yield. Although *Z*-isomer is the kinetic product, the *E*-isomer is often obtained after longer reaction periods. These workers also coupled 2-ethoxy-trialkylstannylacetylenes (alkyl = Me, Bu) with 3-iodoindole **5** to yield 3-ethoxyethynyl-1-(phenylsulfonyl)indole, which upon acid hydrolysis provides ethyl 1-(phenylsulfonyl)-3-indolylacetate (70%) [186]. Hibino and co-workers pursued similar chemistry with 2-formyl-3-iodo-1-tosylindole and vinylstannanes during their syntheses of several carbazole alkaloids (hyellazole, carazostatin, carbazoquinocins B-F) [141, 142]. The *N*-MOM indole protecting group was also employed in these studies, as was the unprotected indole. For example, 3-ethenyl-1-(methoxymethyl)indole-2-carboxaldehyde was prepared in 94% yield. In a tandem Stille-Suzuki operation, Martin was able to cross-couple 5-bromo-3-iodo-1-(4-toluene-sulfonyl)indole with 3-pyridyl, 5-pyrimidyl-, and 2-pyrazinyltrimethylstannanes to give the corresponding 5-bromo-3-heteroarylindoles in 55–62% yields [131]. In a beautifully engineered total synthesis of the marine alkaloid (+)-hapalindole Q, Albizati achieved the coupling of 3-bromoindole **177** with bromide **178** to give the key intermediate indole **179** [187]. No other Pd conditions were successful.

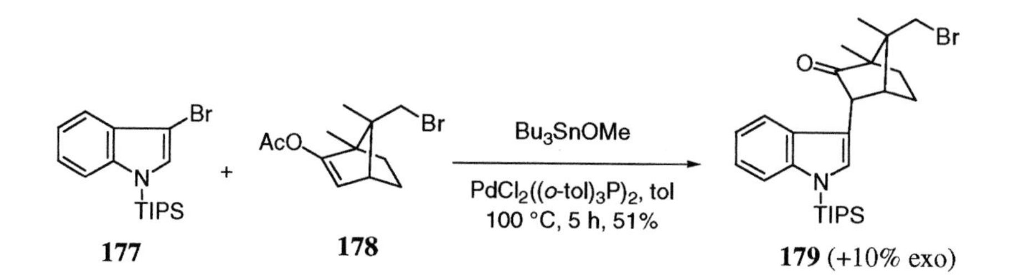


As a synthetic route to the grossularine natural products, Hibino and Potier independently studied the Stille coupling of ethyl 3-iodo-2-indolylcarboxylate, and *N*-protected analogs, with imidazolylstannanes [188–192]. An example is illustrated for **180** to **182** [191]. During the course of their studies in this area, Hibino and co-workers discovered an interesting case of *cine*-substitution, which seems to be the first such example in heterostannane reactions [190].

These workers also employed similar Stille reactions in syntheses of the antioxidant antiostatins and carbazoquinocins [192].

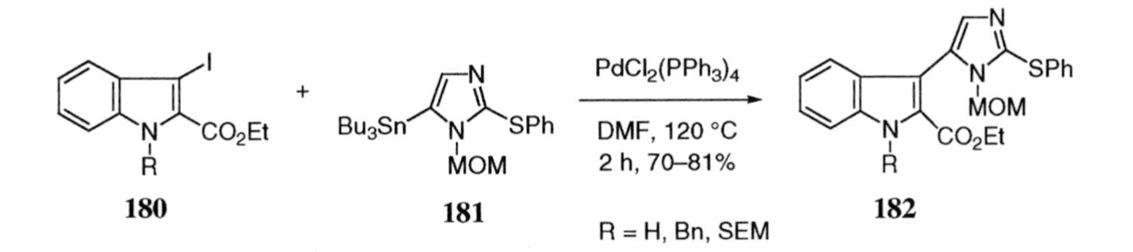


Palladium-catalyzed cross-coupling reactions involving the benzene ring positions in indoles have been the target of several investigations. Appropriately enough, Stille first studied the reactions of organotin compounds with C-4 substituted indoles. Thus, reaction of 1-tosyl-4-indolyltriflate (**183**) with stannanes **184** to give the coupled products **185** [193].

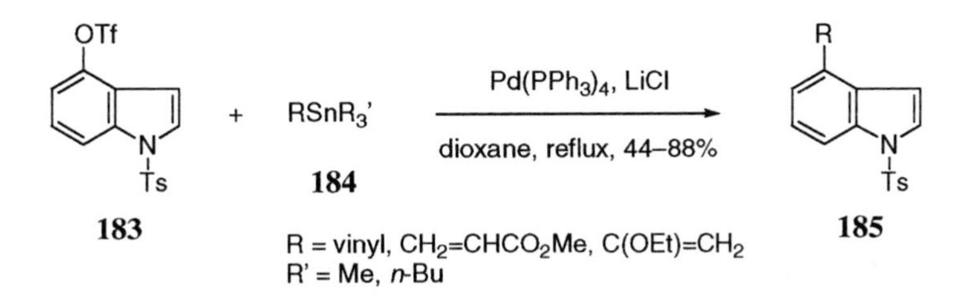


Buchwald effected a Stille reaction of 4-iodoindole **125** with vinyltributyltin to give the corresponding 4-vinylindole in 87% yield [136]. Widdowson and co-workers converted 4-iodo-1-triisopropylsilylindole (**30**) into the 4-methoxycarbonylmethylthio derivative in 98% yield by reaction with $Me_3SnSCH_2CO_2Me$ and $Pd(Ph_3P)_4$, at the beginning of a synthesis of (±)-chuangxinmycin methyl ester [194]. Somei employed thallated indole **186** in a cross-coupling reaction with pyridylstannane **187** to give **188** [195]. However, coupling of **186** with tetrabutylstannane was less successful (13% yield).

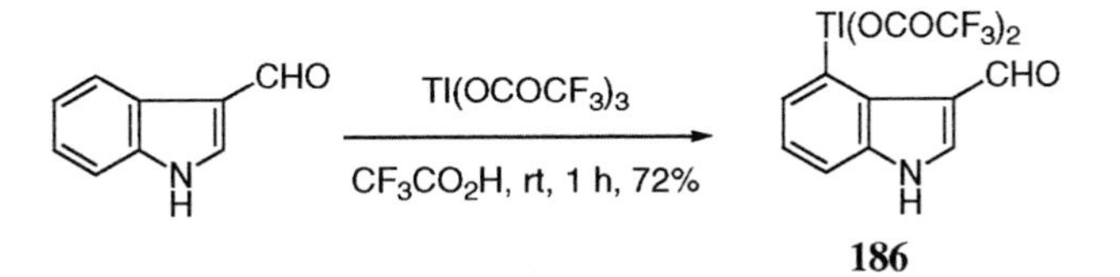

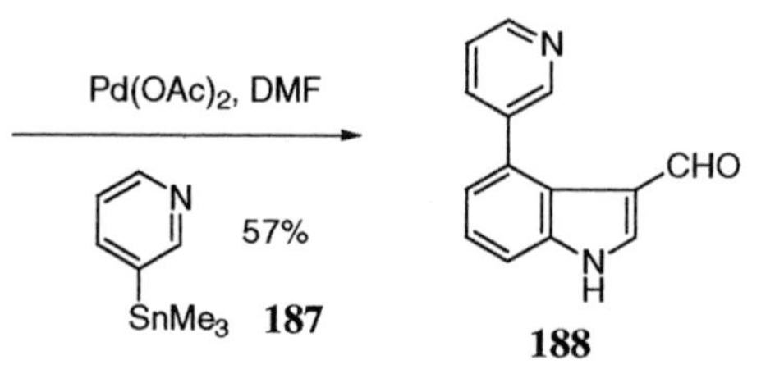


Doi and Mori made excellent use of dihydroindole triflate **189** in Pd-catalyzed cross-coupling reactions. This compound was discussed earlier in the Suzuki section, and it also undergoes Stille couplings as illustrated below [140]. A final dehydrogenation completes the sequence to indoles.

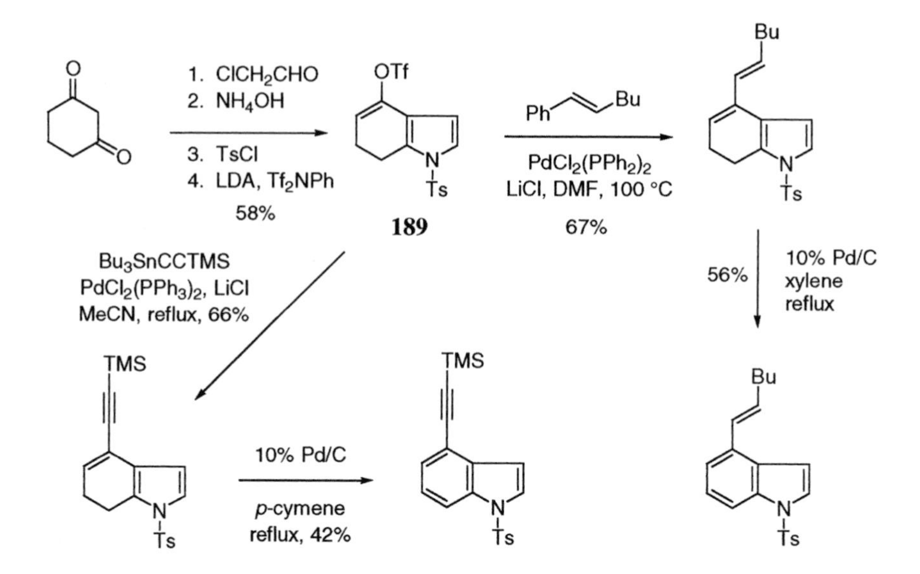


Martin prepared 5-trimethylstannylindole and effected coupling with bromobenzene to give 5-phenylindole [125]. In a search for new cAMP phosphodiesterase inhibitors, Pearce prepared the furylindole **190** from 5-bromoindole and 5-*tert*-butoxy-2-trimethylstannylfuran [196a]. Benhida and co-workers explored Stille couplings of 6-bromo- and 6-iodoindole, and methyl 6-iodoindol-2-ylacetate with a variety of heteroarylstannanes and vinylstannanes [196b].

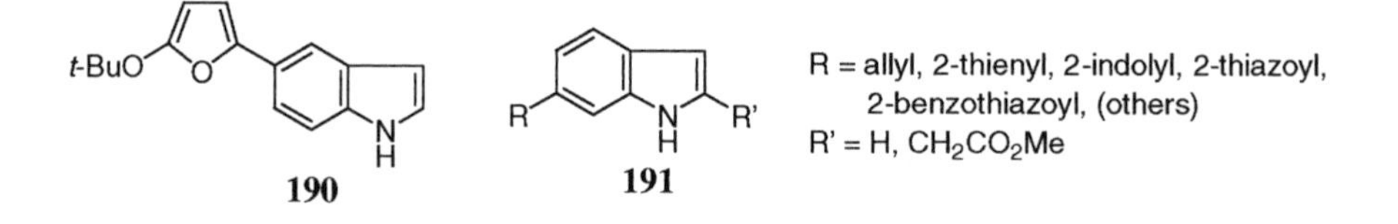

Miki effected Pd-catalyzed cross-coupling between dimethyl 7-bromoindole-2,3-dicarboxylate and both tributylvinyltin and tributyl-1-ethoxyvinyltin to yield the expected 7-vinylindoles [197]. Hydrolysis of the crude reaction product from using tributyl-1-ethoxyvinyltin gave the 7-acetylindole. Sakamoto used dibromide **192**, which was prepared by acylation of 7-bromoindole, in a very concise and efficient synthesis of hippadine [36]. The overall yield from commercial materials is 39%. Somewhat earlier, Grigg employed the same strategy to craft hippadine from the diiodoindoline version of **192** using similar cyclization reaction conditions ($(Me_3Sn)_2/Pd(OAc)_2$), followed by DDQ oxidation (90%) [198].

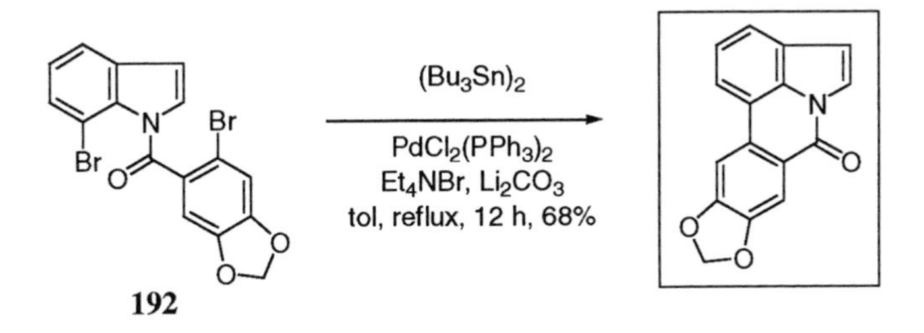


The propensity for *N*-protected indoles to undergo metalation at C-7 has inspired two research groups to pursue Stille couplings using this tactic. Somei and co-workers allowed tetramethyltin to react with (1-acetylindolin-7-yl)thallium bis(trifluoroacetate) to give a low yield of 1-acetyl-7-methylindoline (32%) [199]. Similarly, 1-acetyl-7-phenylindoline was obtained in 35% yield by reaction with tetraphenyltin. Iwao, Watanabe, and co-workers were able to lithiate and then stannylate the C-7 position of 1-*tert*-butoxycarbonylindoline. Coupling of this stannane with 6-bromopiperonal and 6-bromoveratraldehyde yielded the anticipated 7-arylindolines, which were converted into the pyrrolophenanthridone alkaloids hippadine, oxoassoanine, kalbretorine, pratosine, and anhydrolycorin-7-one [200]. Stille couplings have also been reported for carbazoles, carbolines, and other fused indoles. For example, Bracher and Hildebrand employed tributyl-1-ethoxyvinyltin in a reaction with 1-chloro-β-carboline ($PdCl_2(Ph_3P)_2$) to synthesize nitramarine and annomontine *via* the hydrolysis product 1-acetyl-β-carboline [201]. These workers also reported the reaction of tributylvinyltin with 1-chloro-β-carboline to give 1-vinyl-β-carboline in modest yield [145]. Bosch described tin couplings of a 1-bromonauclefine with tetraethyltin and tributylvinyltin to give the ethyl- and vinyl-substituted compounds in 87% and 95% yields, respectively [202]. In his researches towards the synthesis of grossularine analogs, Potier effected the Stille coupling of α-carboline triflate **193** with tributyl-4-methoxyphenyltin (**194**) to afford **195** [191]. Triflate **193** also couples readily with 1-(phenylsulfonyl)-3-tributylstannylindole under the same conditions (72% yield).

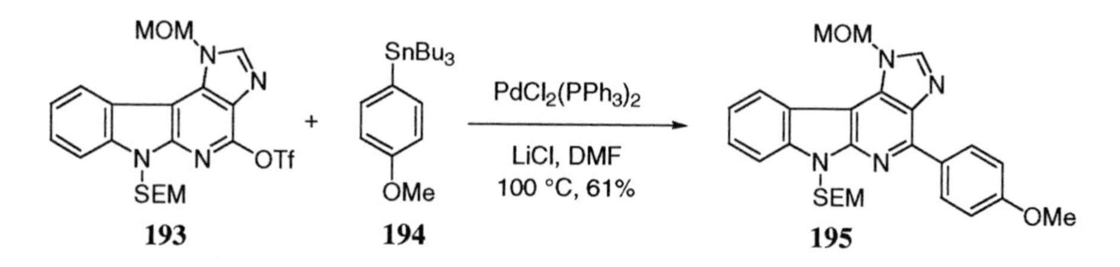


McCort employed Stille technology to join a SEM-protected 4-bromopyrrolo[3,4-*c*]carbazole to the 3-tributylstannyl derivatives of pyridine and quinoline [143]. Mérour effected coupling of tributylvinylstannane and tributylallylstannane with bromide **196** to afford the expected products **197** in excellent yield [146].

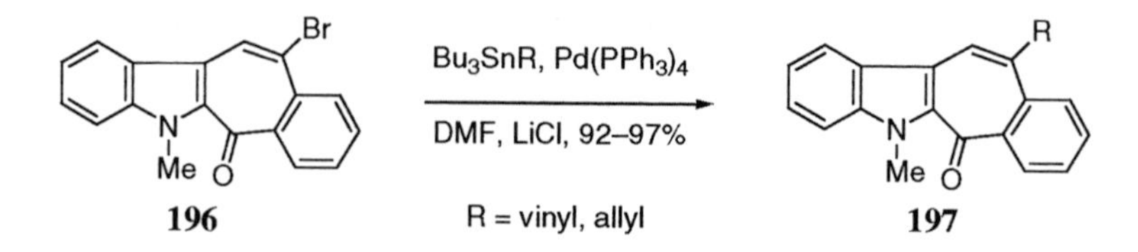


The Stille tin-coupling protocol was employed by Rice and co-workers to fashion novel naltrindole derivatives as potential new δ opioid receptor antagonists [144]. Thus, the triflate of naltrexone was allowed to react with tetramethyltin, tributylvinyltin, and tributyl-2-furyltin to give the requisite 3-desoxynaltrindole precursors. The reaction with trimethylphenyltin was less satisfactory. Several tin-mediated reactions leading directly to indole ring precursors have been described. Stille effected Pd-catalyzed couplings of *o*-bromoacetanilide (**198**) with tributylalkynyltins **199**. The resulting alkynylanilines **200** smoothly cyclize to indoles **201** under the influence of $PdCl_2(MeCN)_2$ [203]. We will encounter more of these Pd-catalyzed indole ring formations in Section 3.5. Likewise, many ring-substituted indoles (5-Me, 5-Cl, 5-OTf, 6-OMe, 6-$CO_2$Me) were prepared in this study. Much less successful were the *N*-Ts and *N*-$COCF_3$ analogs of **198**.

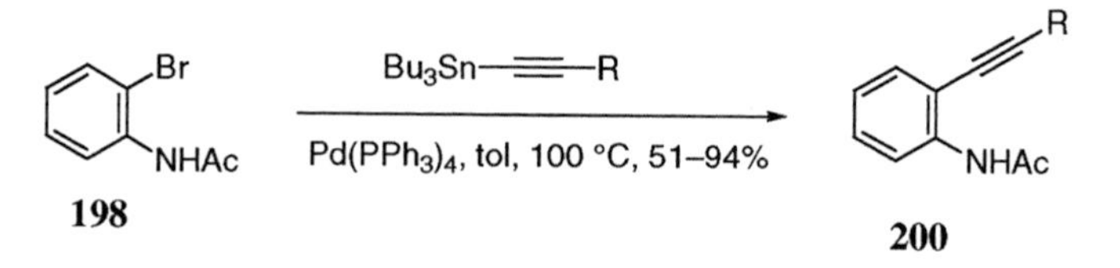

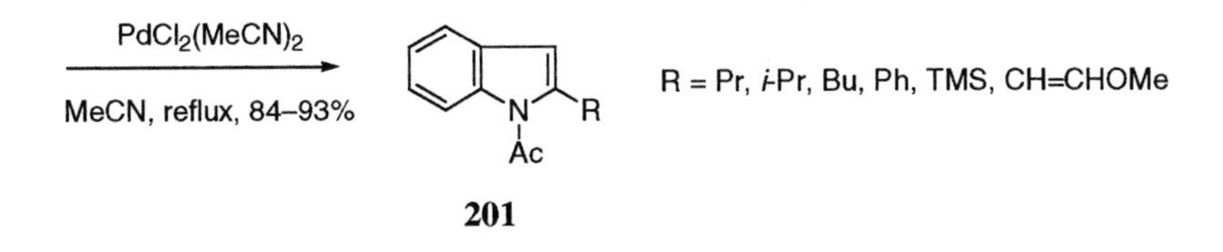


Sakamoto described similar reactions of *o*-bromoaniline derivatives with (*Z*)-tributyl-2-ethoxyvinyltin and subsequent cyclization of the coupled product with TsOH to yield, for example, *N*-acetylindole (29% yield overall) [185]. This research group also used this methodology to synthesize a series of azaindoles, an example of which is illustrated below [204]. Halonitropyridines were particularly attractive as coupling partners with tributyl-2-ethoxyvinyltin and precursors to azaindoles. Although the (*Z*)-isomer of **202** is obtained initially, it isomerizes to the (*E*)-isomer which is the thermodynamic product. This strategy represents a powerful method for the synthesis of all four azaindoles (1*H*-pyrrolopyridines). In fact, this method, starting with 2,6-dibromoaniline, is one of the best ways to synthesize 7-bromoindole (96% overall yield) [36].

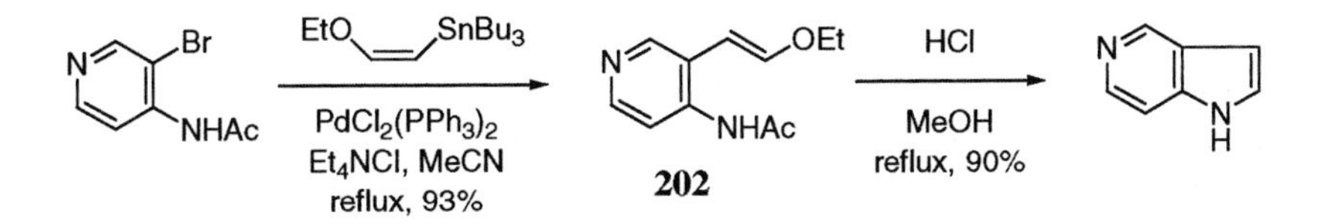


Iwao, Watanabe, and co-workers utilized tin-chemistry to construct a series of 2-amino-biphenyls for use in carbazole synthesis *via* arynic cyclization [205]. For example, treating *N*-(*t*-butoxycarbonyl)aniline with *t*-BuLi following by stannylation with $Bu_3SnCl$ gives the corresponding stannane. Coupling of this with 2-bromochlorobenzene yields *N*-(*t*-butoxycarbonyl)-2-(2'-chlorophenyl)aniline (74% yield). Subsequent treatment with excess $KNH_2$ affords carbazole in 99% yield. This protocol was successful in synthesizing glycozolinine and glycozolidine. Similarly, Quéguiner coupled heteroarylstannanes with phenylpyridines to give precursors for α-substituted δ-carbolines [163c]. A representative procedure is shown below.

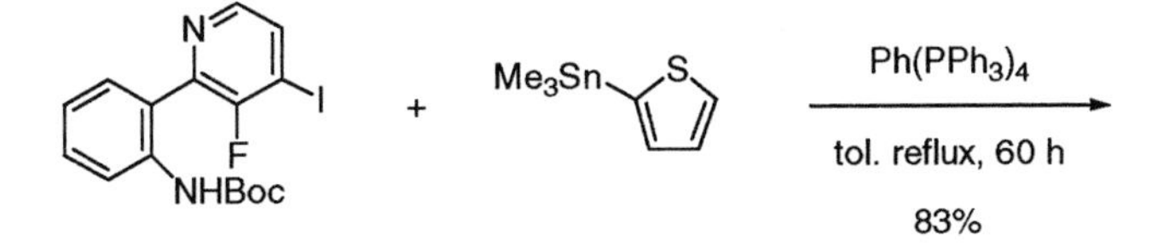

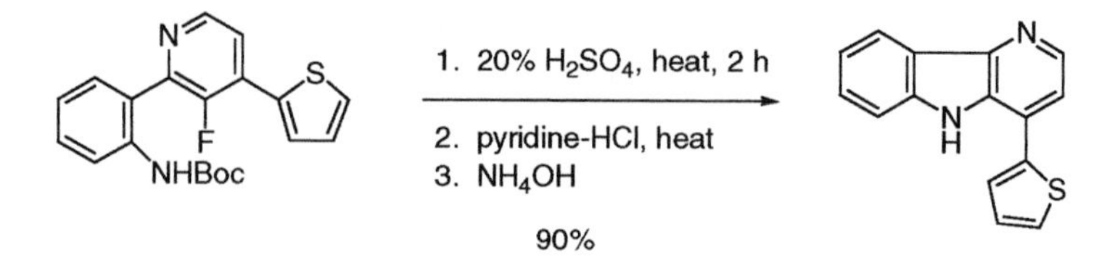


Grigg employed a tandem Pd-catalyzed cyclization to synthesize 3,3'-biindole **204** from symmetrical alkyne **203** [198]. The presumed intermediate exocyclic alkene is not detected.

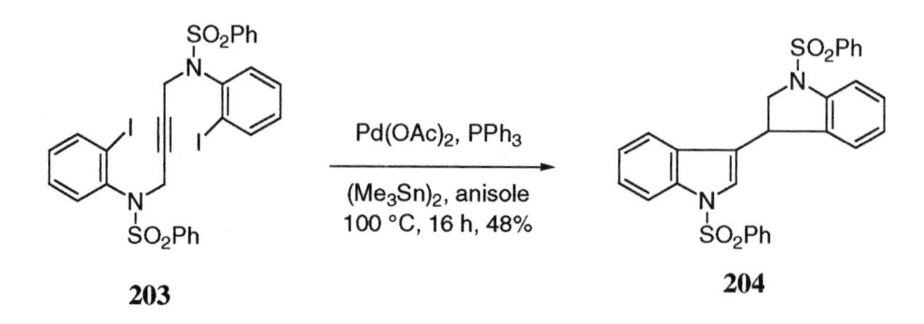


The combination of *o*-iodoaniline **205**, tributyl-2-thienylstannane (**206**), and bisalkyne **207** provides oxindole **208** in the presence of a Wilkinson's catalyst/Pd(0) system [206].

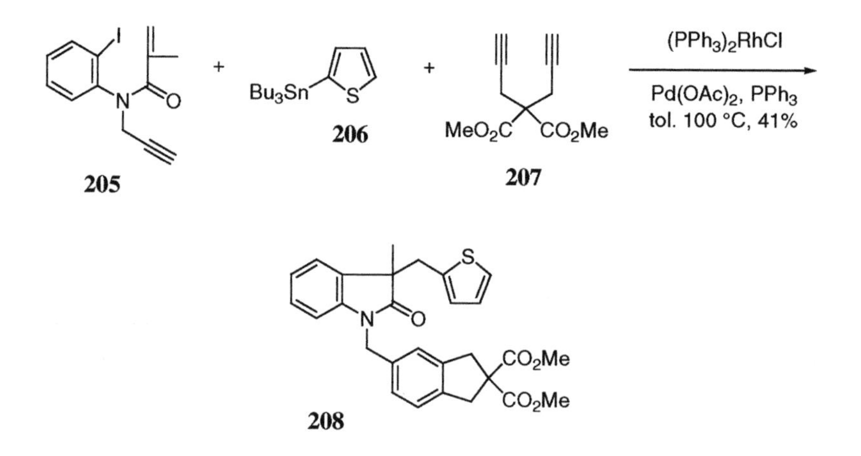


## 3.4 The Sonogashira coupling

The Sonogashira coupling is the Pd-catalyzed coupling of aryl halides and terminal alkynes [207], which, in the appropriate cases, can be followed by the spontaneous, or easily induced, cyclization to an indole ring. It is a sequel to the Castro acetylene coupling and subsequent cyclization to indoles in the presence of copper [208]. For example, Castro and co-workers found that copper acetylides react with *o*-iodoaniline to form 2-substituted indoles often in high yield. In the intervening years, the Pd-catalyzed cyclization of *o*-alkynylanilines to indoles has become a powerful indole ring construction. The related Larock indole ring synthesis is discussed in Section 3.5.

Yamanaka and co-workers were the first to apply the Sonogashira coupling reaction to an indole synthesis when they coupled trimethylsilylacetylene with *o*-bromonitrobenzene ($PdCl_2(Ph_3P)_2/Et_3N$). Treatment with NaOEt/EtOH gives *o*-(2,2-diethoxyethyl)nitrobenzene (39% overall), and hydrogenation and acid treatment affords indole (87%, two steps) [209–211]. The method is applicable to a variety of ring-substituted indoles and, particularly, to the synthesis of 4- and 6-azaindoles (pyrrolopyridines) from halonitropyridines. Taylor coupled thallated anilides **209** with copper(I) phenylacetylide to afford the corresponding *o*-alkynylanilides **210**. In the same pot catalytic $PdCl_2$ is then used to effect cyclization to *N*-acylindoles **211** [212]. Hydrolysis to the indoles **212** was achieved by base.

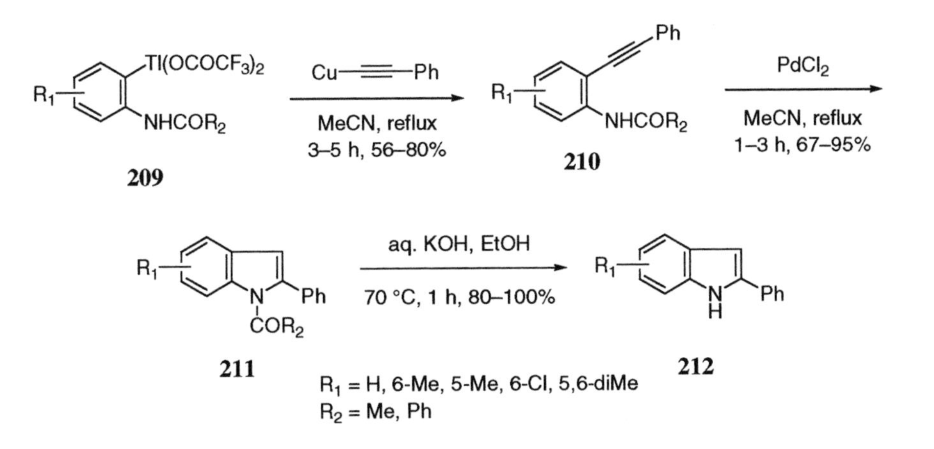


Tischler and Lanza effected coupling of several substituted *o*-chloro- and *o*-bromo-nitrobenzenes with trimethylsilylacetylene to give the *o*-alkynylnitrobenzenes **213** [213]. Further manipulation affords the corresponding indoles **214** in good to excellent yield.

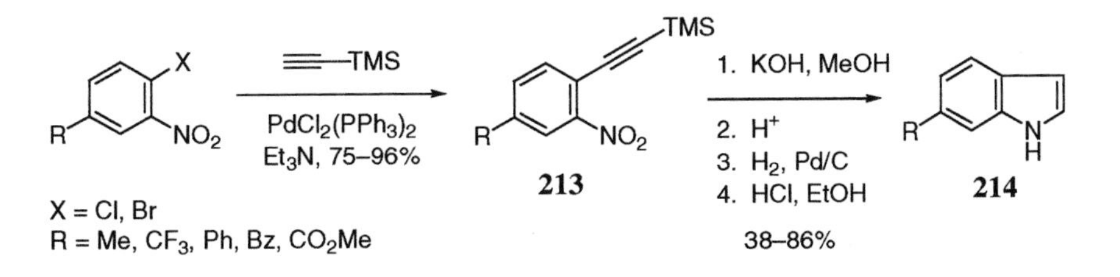

Yamanaka and co-workers studied Pd-catalyzed couplings of *o*-haloaniline derivatives with terminal alkynes, but cyclization to give indoles occurred upon treatment with mild base or CuI, often in one pot [211, 214, 215]. These workers also described the Pd-induced coupling of 1- and 2-substituted 2- and 3-iodoindoles with terminal alkynes to give the corresponding 2- and 3-alkynylindoles [217]. Similarly, Gribble and Conway coupled phenylacetylene with 3-indolyltriflate **34** to give the corresponding 3-(2-phenylethynyl)indole in 81% yield [183]. Monodendrons based on 9-phenylcarbazole were crafted by Zhu and Moore using the Sonogashira reaction. For example, the compound shown was prepared in high yield by these workers [217].

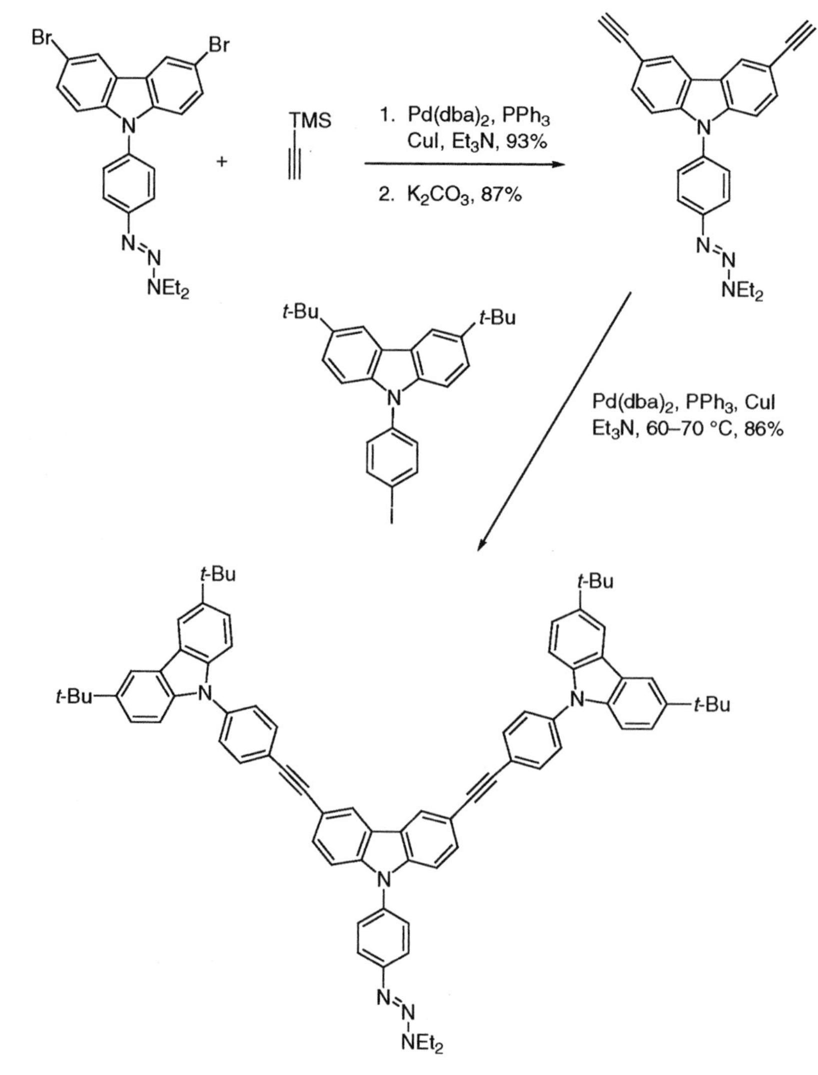

The combination of Pd-catalyzed coupling of terminal acetylenes with *o*-alkynylanilines or *o*-alkynylnitrobenzenes followed by base or CuI cyclization to an indole has been used in many situations with great success. Arcadi employed this methodology to prepare a series of 2-vinyl-, 2-aryl-, and 2-heteroarylindoles from 2-aminophenylacetylene and a subsequent elaboration of the acetylenic terminus. A final Pd-catalyzed cyclization completes the scheme [218].

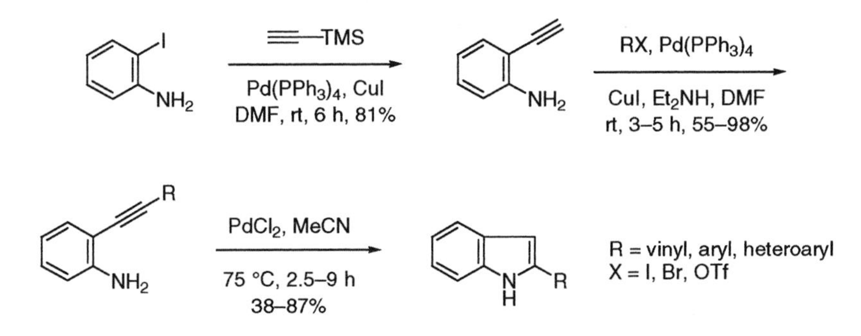


Other indoles that have been prepared using the Sonogashira coupling and cyclization sequence include 5,7-difluoroindole and 5,6,7-trifluoroindole [219], 4-, 5-, and 7-methoxyindoles and 5-, 6-, and 7-(triisopropylsilyl)oxyindoles [220], the 5,6-dichloroindole SB 242784, a compound in development for the treatment of osteoporosis [221], 5-azaindoles [222], 7-azaindoles [160], 2,2'-biindolyls [223, 176], 2-octylindole for use in a synthesis of carazostatin [224], chiral indole precursors for syntheses of carbazoquinocins A and D [225], a series of 5,7-disubstituted indoles [226], a pyrrolo[2,3-*e*]indole [226], an indolo[7,6-*g*]indole [227], pyrrolo[3,2,1-*ij*]quinolines from 4-arylamino-8-iodoquinolines [228], optically active indol-2-ylarylcarbinols [229], 2-alkynylindoles [176], 7-substituted indoles *via* the lithiation of the intermediate 2-alkynylaniline derivative [230], and a variety of 2,5,6-trisubstituted indoles [231]. This latter study employs tetrabutylammonium fluoride, instead of CuI or alkoxide, to effect the final cyclization of **215** to indoles **216** as summarized here.

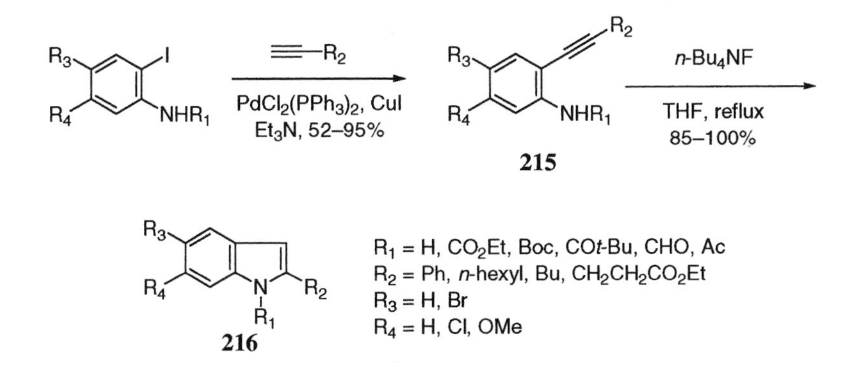

A new, water soluble palladium catalyst was used in the Sonogashira reaction ($Pd(OAc)_2$ triphenylphosphine-trisulfonate sodium salt) [232], and several groups adapted the Sonogashira coupling and subsequent cyclization to the solid-phase synthesis of indoles. Bedeschi and co-workers used this method to prepare a series of 2-substituted-5-indolecarboxylic acids [233]. Collini and Ellingboe extended the technique to 1,2,3-trisubstituted-6-indolecarboxylic acids [234]. Zhang and co-workers used the solid phase to prepare a series of 2-substituted-3-aminomethyl-5-indolecarboxamides, and, by manipulation of the resin-bound Mannich reaction intermediates, to synthesize 3-cyanomethyl-5-indole-carboxamide and other products of nucleophilic substitution [235]. This research team also employed a sulfonyl linker, as summarized below, to provide a series of substituted indoles [236]. The advantages of this particular approach are that the sulfonyl linker is "traceless," since it disappears from the final indole product, and the polystyrene sulfonyl chloride resin is commercially available.

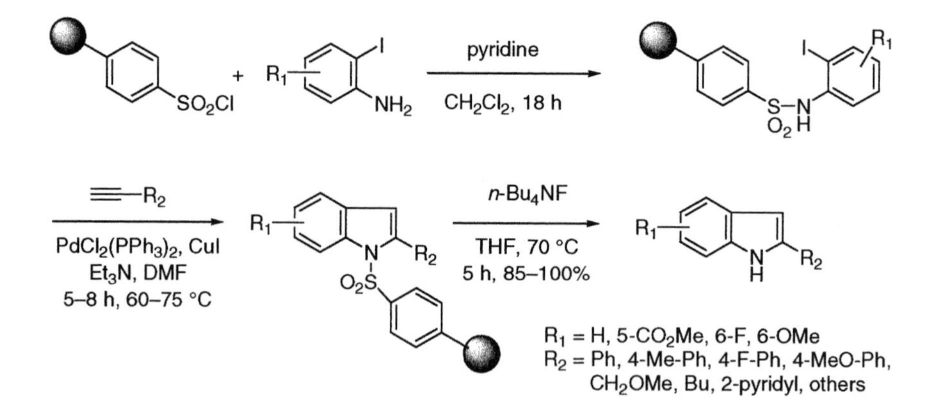


The Sonogashira coupling was used to prepare both *o*-alkynylformamides, for the synthesis of the corresponding isonitriles for the Fukuyama indole ring formation [177, 237], and several 4-alkynylindoles from the corresponding 4-iodoindoles [238]. Similarly, ethyl 3-triflyloxy-1-methylindole-2-carboxylate reacts with propargylic alcohol to form the corresponding indolyl-3-propynyl alcohol in 80% yield [139]. The Sonogashira coupling was also used to prepare arylalkyne substrates for subsequent cyclization under Heck-Hegedus-Mori or Larock conditions, and these are discussed in the appropriate sections later in this Chapter.

## 3.5 Heck couplings

The incredibly powerful and versatile Heck coupling reaction has found enormous utility in indole ring synthesis and in the elaboration of this important heterocycle. Due to the enormity of this topic, the section is divided into Heck reactions of indoles, the synthesis of the indole ring as developed by Hegedus, Mori-Ban, and Heck, and the Larock indole ring synthesis.

### 3.5.1 Heck coupling reaction

Both inter- and intramolecular Heck reactions of indoles have been pursued and these will be considered in turn. Appropriately, Heck and co-workers were the first to use Pd-catalyzed vinyl substitution reactions with haloindoles [239]. Thus, 1-acetyl-3-bromoindole (**217**) gave a 50% yield of 3-indolylacrylate **218**. A similar reaction with 5-bromoindole yielded (*E*)-methyl 3-(5-indolyl)acrylate (53% yield), but 3-bromoindole gave no identifiable product.

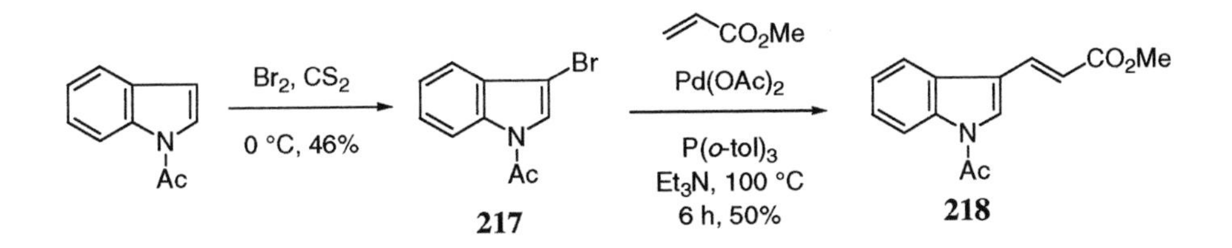


Somei and co-workers made extensive use of the Heck reaction with haloindoles in their synthetic approaches to ergot and other alkaloids [26, 40, 41, 240–249]. Thus, 4-bromo-1-carbomethoxyindole (69%) [26], 7-iodoindole (91%) (but not 7-iodoindoline or 1-acetyl-7-iodoindoline) [40, 41], and 1-acetyl-5-iodoindoline (96%) [41] underwent coupling with methyl acrylate under standard conditions ($Pd(OAc)_2/Ph_3P/Et_3N$/DMF/100 °C) to give the corresponding (*E*)-indolylacrylates in the yields indicated. The Heck coupling of methyl acrylate with thallated indoles and indolines is productive in some cases [41, 241, 246]. For example, reaction of (3-formylindol-4-yl)thallium bis-trifluoroacetate (**186**) affords acrylate **219** in excellent yield [241]. Similarly, this one-pot thallation-palladation operation from 3-formylindole and methyl vinyl ketone was used to synthesize 4-(3-formylindol-4-yl)-3-buten-2-one (86% yield).

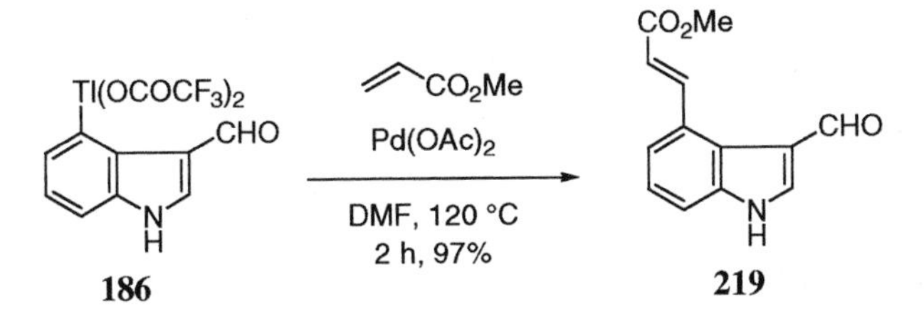


However, the most important of Somei's contributions in this area are Heck reactions of haloindoles with allylic alcohols [26, 240, 242–245, 247, 248]. For example, reaction of 4-

iodo-3-indolecarboxaldehyde with 2-methyl-3-buten-2-ol afforded alcohol **220** in high yield. This could be subsequently transformed to (±)-6,7-secoagroclavine (**221**) [242]. Interestingly, the one-pot thallation-palladation protocol failed in this case.

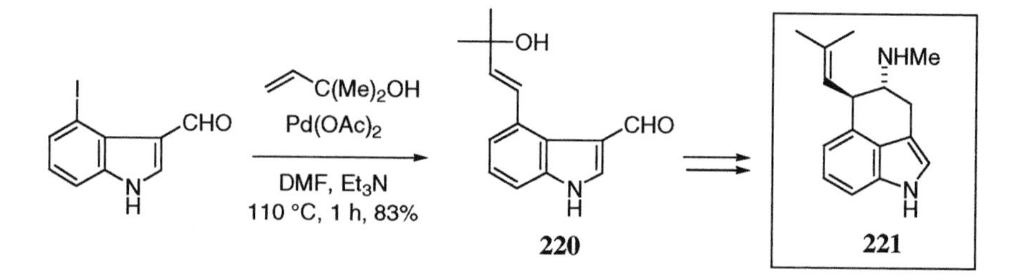


Somei adapted this chemistry to syntheses of (±)-norchanoclavine-I, (±)-chanoclavine-I, (±)-isochanoclavine-I, (±)-agroclavine, and related indoles [243–245, 248]. Extension of this Heck reaction to 7-iodoindoline and 2-methyl-3-buten-2-ol led to a synthesis of the alkaloid annonidine A [247]. In contrast to the uneventful Heck chemistry of allylic alcohols with 4-haloindoles, reaction of thallated indole **186** with 2-methyl-4-trimethylsilyl-3-butyn-2-ol affords an unusual 1-oxa-2-sila-3-cyclopentene indole product [249]. Hegedus was also an early pioneer in exploring Heck reactions of haloindoles [250–252]. Thus, reaction of 4-bromo-1-(4-toluenesulfonyl)indole (**11**) under Heck conditions affords 4-substituted indoles **222** [250]. Murakami described the same reaction with ethyl acrylate [83], and 2-iodo-5-(and 7-) azaindoles undergo a Heck reaction with methyl acrylate [19].

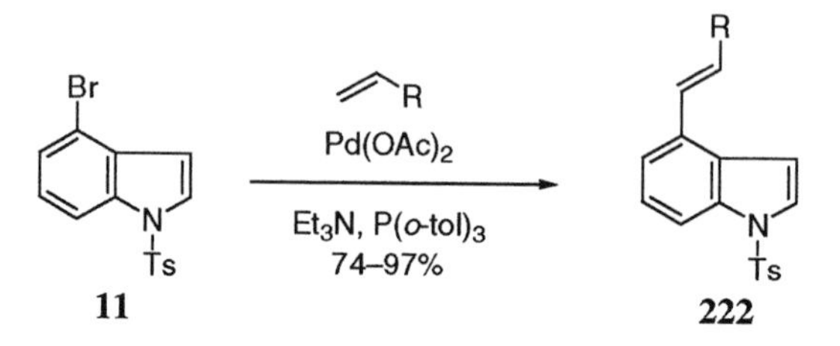


Hegedus also found that mercuration of **11** followed by a Heck reaction with methyl acrylate gives the corresponding 3-indolylacrylate in 70% yield [250]. More relevant to Hegedus's goal of ergot alkaloid synthesis was the observation that the appropriate dihaloindole reacted regioselectively at C-3 to give exclusively the (*Z*)-isomer **223** [250, 251].

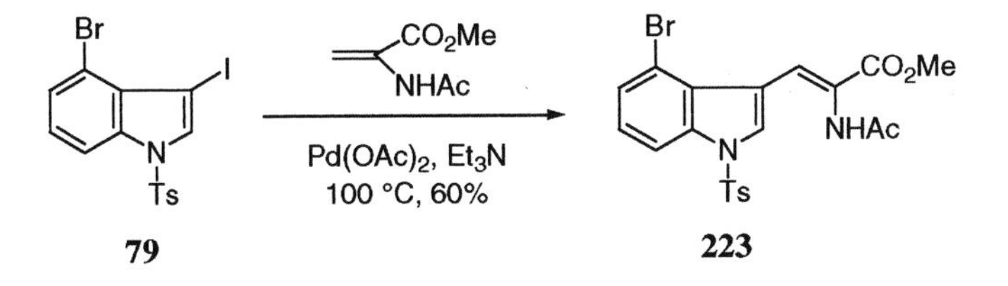


This reaction was used and extended by Murakami to the preparation of ring-substituted derivatives of **223** [80, 81], to a similar Heck reactions leading to the C-4 (*Z*)-products [80], and to syntheses of clavicipitic acid and costaclavine [79]. Semmelhack improved upon this Heck reaction with *N*-protected dehydroalanine esters by adding chloride to the reaction mixture in his synthesis of the 4-fluoro analog of **223** (77% *vs.* 40% yield) [253]. Using his methodology and a Suzuki coupling, Hegedus and co-workers synthesized the ergot alkaloid (±)-aurantioclavine [252], and the *N*-acetyl methyl ester of (±)-clavicipitic acid [251]. The key intramolecular Heck cyclization utilized in the latter synthesis is presented later. Also worthy of mention is the Heck reaction of 5-bromo-1-tosylindole with an *N*-vinyloxazolone leading to 5-(2-aminoethyl)indoline [254], an approach to lysergic acid involving vinylation of 4-bromo-1-tosyl-3-indolecarboxaldehyde [255], a synthesis of (–)-6-*n*-octylindolactam-V featuring a Heck reaction on the corresponding 6-bromoindolactam [256], and reaction of a 3-iodo-1-methylindole (but not the NH analog) with methyl acrylate as part of Sundberg's approach to iboga alkaloids [257]. The intramolecular version of this strategy is presented later. Blechert and co-workers employed a Heck reaction between 1-benzyl-7-bromo-4-ethylindole and methyl crotonate in the early stages of a synthesis of (±)-*cis*-trikentrin [265].

Triflates also undergo Heck reactions and Gribble and Conway reported several such couplings of 1-(phenylsulfonyl)indol-3-yl triflate (**34**) to afford 3-vinylindoles **224** [183]. Cyclohexene, allyl bromide, and methyl propiolate failed to react under these conditions, but triphenylphosphine afforded **225** in excellent yield (93%), and divinyl carbinol yielded the rearranged enal **226** (82% yield).

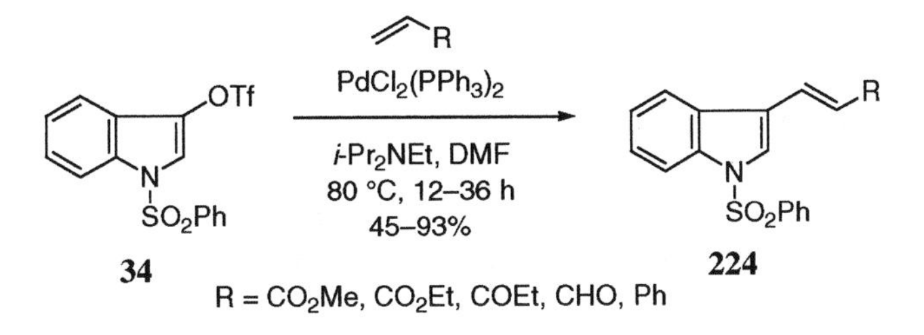

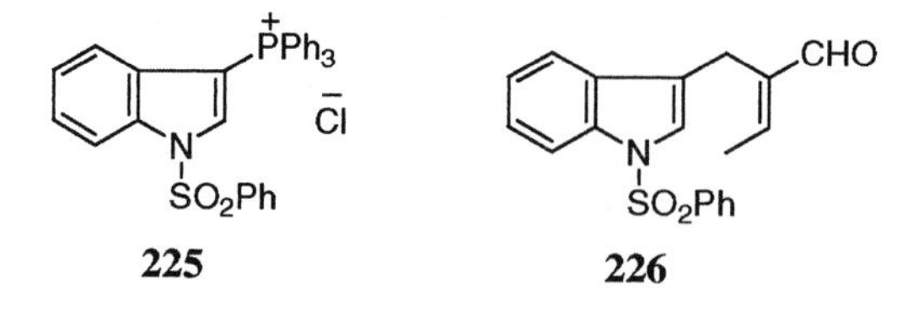


Mérour also explored the Heck reactions of indolyl triflates with allylic alcohols [139, 258]. For example, reaction of triflate **227** with allyl alcohol gives the rearranged allylic alcohol **228** [139].

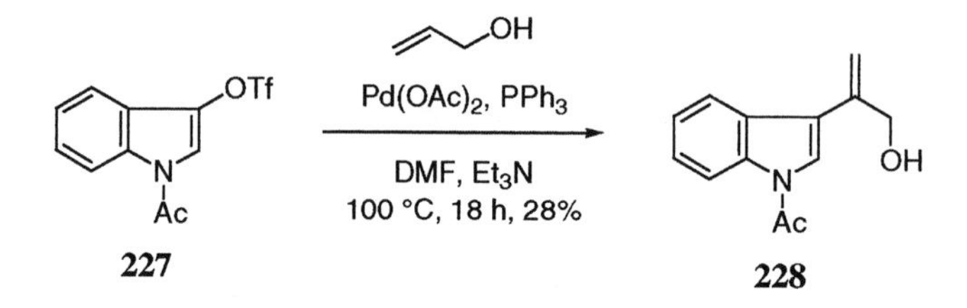


The vinyl triflate of Kornfeld's ketone has been subjected to Heck reactions with methyl acrylate, methyl methacrylate, and methyl 3-(*N*-*tert*-butoxycarbonyl-*N*-methyl)amino-2-methylenepropionate leading to a formal synthesis of lysergic acid [259]. A similar Heck reaction between 1-(phenylsulfonyl)indol-5-yl triflate and dehydroalanine methyl ester was described by this research group [260]. Chloropyrazines undergo Heck couplings with both indole and 1-tosylindole, and these reactions are discussed in the pyrazine Chapter [261]. Rajeswaran and Srinivasan described an interesting arylation of bromomethyl indole **229** with arenes [262]. Subsequent desulfurization and hydrolysis furnishes 2-arylmethylindoles **230**. Bis-indole **231** was also prepared in this study.

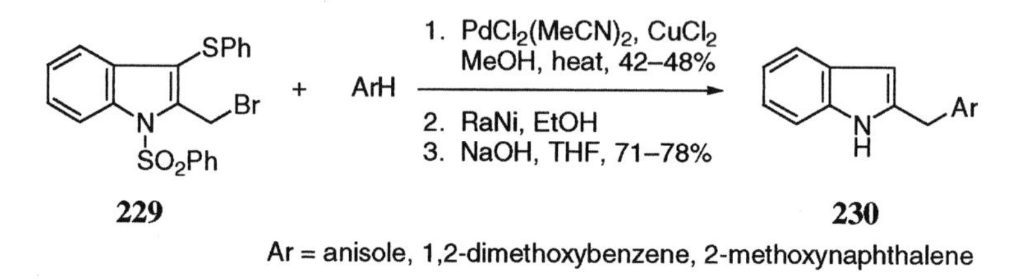

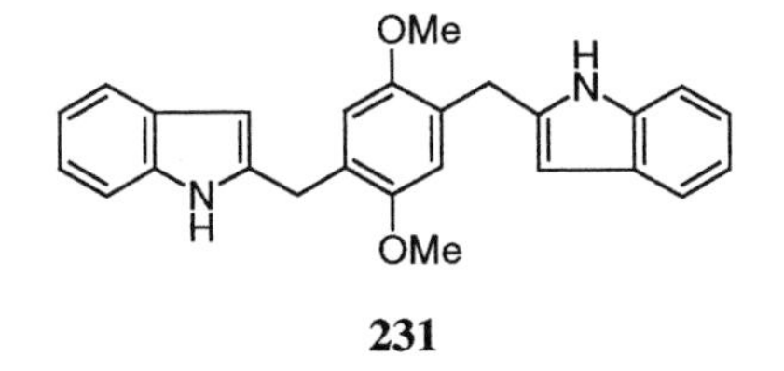


The Heck-mercuration modification, which was presented earlier, has been adapted to the solid phase by Zhang and co-workers, as illustrated below [236].

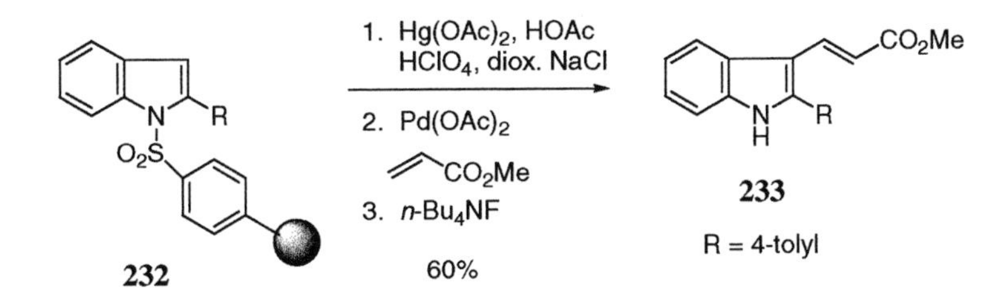


The intramolecular Heck reaction as applied to indoles has led to several spectacular synthetic achievements. As alluded to in the previous section, both Hegedus and Murakami exploited intramolecular Heck reactions to synthesize ergot alkaloids. In model studies, Hegedus noted that 3-allyl-4-bromo-1-tosylindole (**234**) cyclizes to **235** in good yield [250], and Murakami's group observed that, for example, **236** cyclizes to **237** [263]. Roberts effected similar cyclizations leading to 7- and 8-membered ring tryptophan surrogates [264].

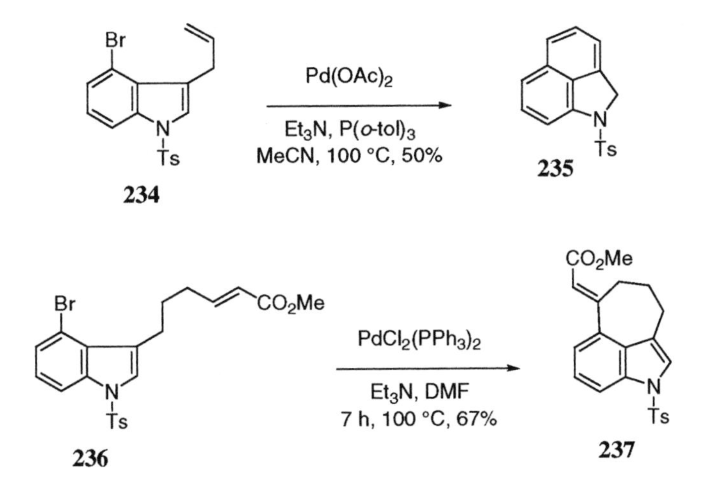

Hegedus' synthesis of (±)-clavicipitic acid *N*-acetyl methyl ester culminated in the Pd-induced cyclization of **238** to **239**, the latter of which was reduced to the target mixture [251]. Substrate **238** was prepared *via* a Heck reaction with the corresponding 4-bromo compound **223** and 2-methyl-3-buten-2-ol (83%). The cyclization also occurs with tosic acid (97%).

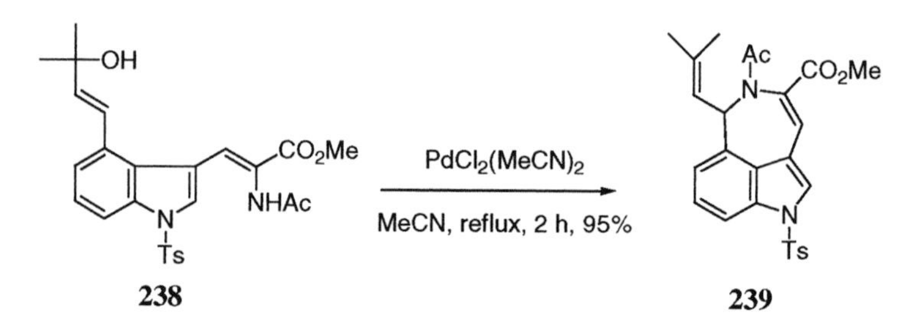


In a related approach, Murakami synthesized clavicipitic acid and costaclavine [79], and later extended this chemistry to a synthesis of chanoclavine-I featuring the intramolecular Heck vinylation **240** to **241** [266]. The corresponding enone failed to cyclize under these conditions. Noteworthy is that radical cyclizations, which often compete successfully with Heck reactions, were poor in this system.

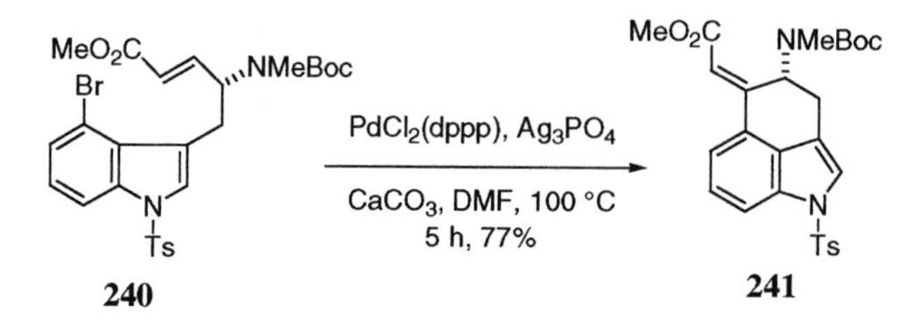


Several investigators have studied intramolecular Heck reactions on alkene-tethered haloindoles. Black prepared several 1-allyl-7-bromoindoles and found that they undergo cyclization in the presence of palladium as shown for **242** to **243** [267]. Although this new synthesis of pyrroloquinolines is reasonably general, some of the products are unstable. Substrate **242** was prepared by bromination of 4,6-dimethoxy-2,3-diphenylindole (92%) and *N*-alkylation.

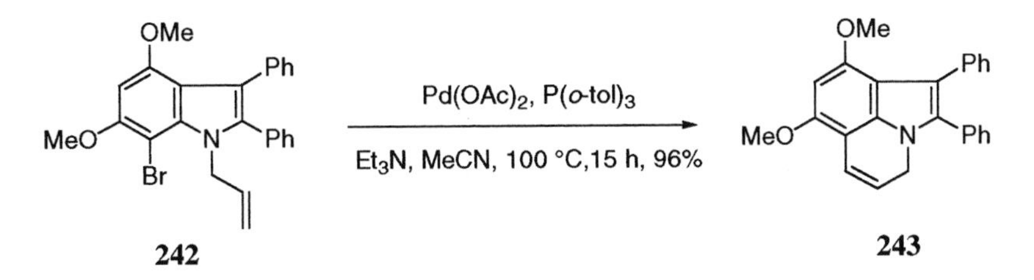


Gilchrist examined the cyclization of *N*-alkenyl-2-iodoindoles with palladium [268, 269]. For example, reaction of *N*-pentenylindole **244** under Heck conditions affords a mixture of **245** and **246** in very good yield. In the absence of TlOAc, **246** is the major product. Further exposure of **245** to $Pd(OAc)_2$ gives **246**. Reaction of 1-(4-butenyl)-2-iodoindole under similar conditions affords the pyrrolo[1,2-*a*]indole ring system in modest yield (35%).

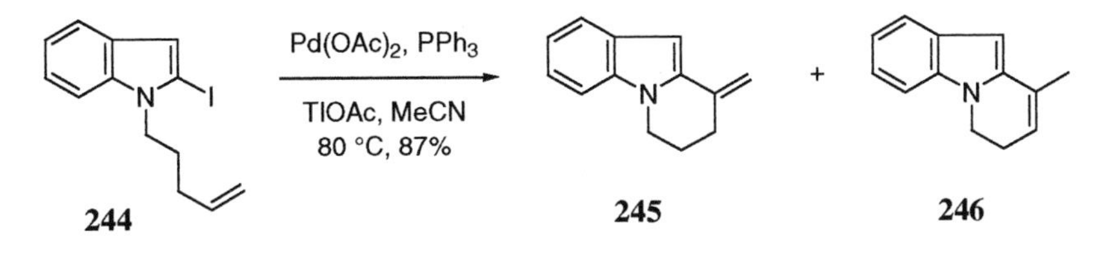


In his synthetic approaches to iboga alkaloids, Sundberg pursued several Heck cyclization strategies but found the best one to be **247** to **248** [257].

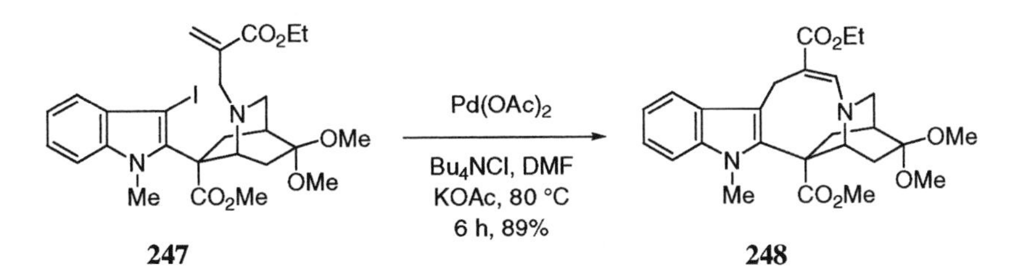


Merlic discovered the novel benzannulation of biindole **249** to **250** during studies to synthesize indolocarbazoles [123]. Several unsymmetrical biindoles were also prepared and their reactions with dimethyl acetylenedicarboxylate and related alkynes were studied. Yields of indolocarbazoles were 51–88% and some regioselectivity was observed in unsymmetrical cases (up to 80:20).

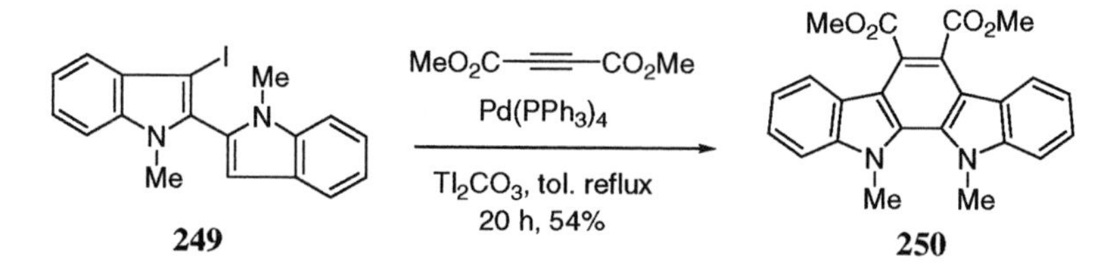


A particularly elegant domino Heck reaction involving 4-bromoindole and bromo(indolyl)maleimide **251** to give *N*-methylarcyriacyanin A (**252**) in one operation was reported by Steglich [173]. This alkaloid could also be prepared from triflate **253** in higher yield in a heteroaryl Heck reaction.

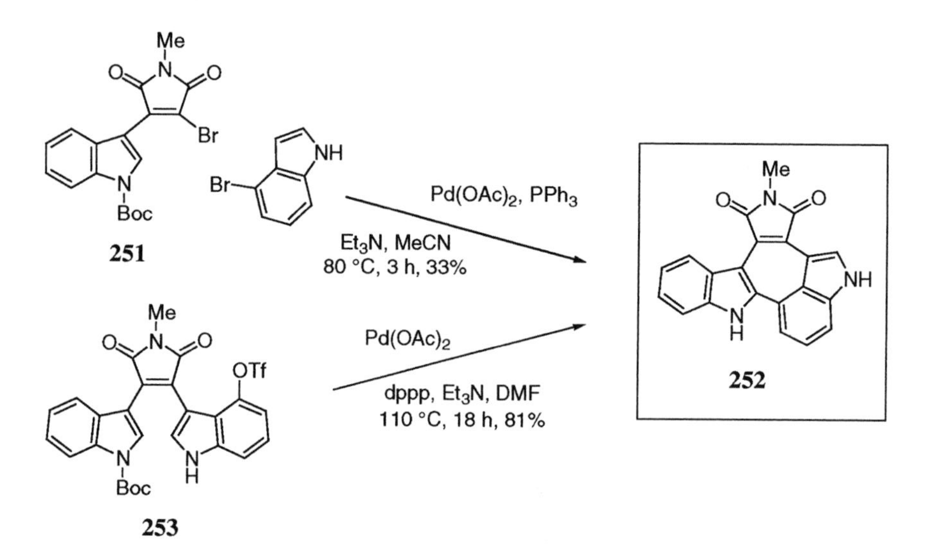


Several intramolecular Heck reactions involve aryl halides cyclizing onto indole rings. Grigg first described the simple Heck cyclizations of **254** and **255** [270], and this was followed by similar Heck reactions reported by Kozikowski and Ma on the bromide corresponding to **254** and the *N*-benzylindole **256** [271, 272]. These investigators also observed cyclization to the C-3 position in a Heck reaction of indole **257**, and they prepared a series of peripheral-type benzodiazepine receptors **258** using this chemistry. For example, **258** (n = 3, R = *n*-Pr) is obtained in 81% yield.

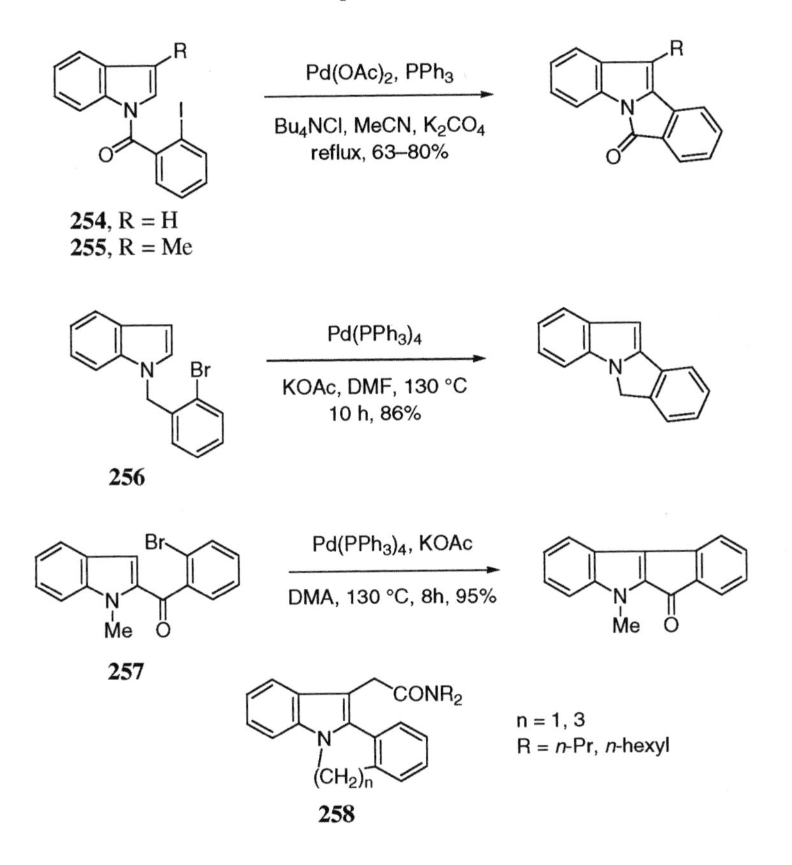


Kraus found that a Pd-catalyzed cyclization is superior to those involving tin-initiated radical cyclizations in the construction of pyrrolo[1,2-*a*]indoles such as **260** [273]. The bromide corresponding to **259** cyclizes in 48% yield, and *N*-(2-bromo-1-cyclohexenecarbonyl)indole-3-carboxaldehyde cyclizes in 60% yield. In contrast, the corresponding radical reactions afford these products in 35–53% yields. Substrate **261** failed to cyclize under these Heck conditions.

Mérour synthesized a series of indolo[2,1-*a*]isoquinolines and pyrrolophenanthridines *via* Heck cyclizations onto the C-2 and C-7 positions of indole, respectively [274]. Two examples are shown.

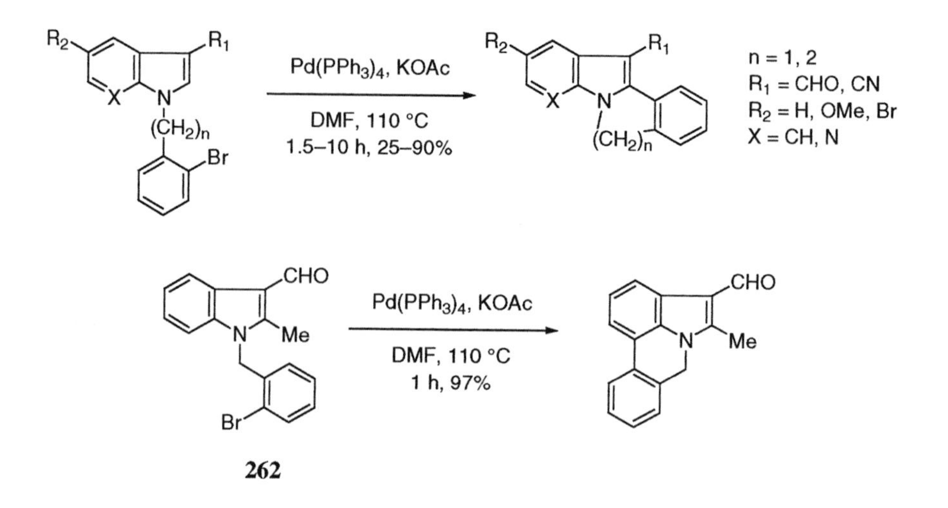


This indole C-7 Heck cyclization strategy was employed by Shao and Cai in a synthesis of anhydrolycorine-7-one from the requisite *N*-aroylindoline [275], by Miki in syntheses of pratosine and hippadine from substrates like **262** [276], and by Rigby to synthesize anhydrodehydrolycorine from an *N*-benzylhydroindolone [277, 278]. Thal and co-workers constructed examples of the new ring systems, pyrido[2',3'-*d'*]pyridazino[2,3-*a*]indole (**264**) and pyrido[2',3'-*d'*]diazepino[1,6,7-*h,i*]indole (**265**), by effecting Heck cyclizations on the appropriate 2-bromopyridine precursors (e.g., **263**) at C-2 or C-7, respectively [279, 280]. Compound **264** undergoes oxidative-addition with methyl acrylate at the C-3 position. This resulting product (not shown) can also be obtained from **263** in a tandem Heck sequence with methyl acrylate (62% yield).

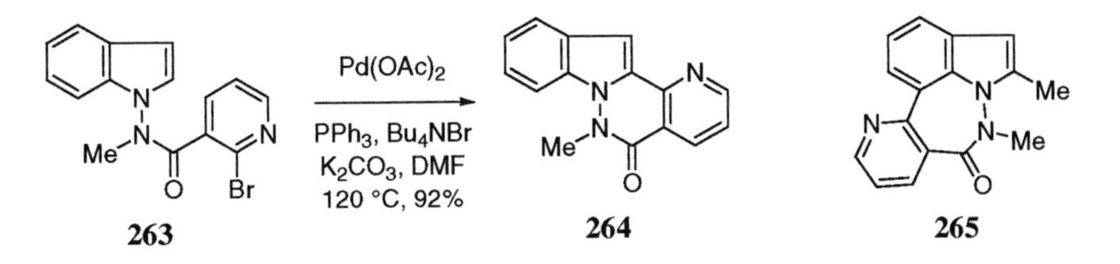


Several other Heck cyclizations that do not involve the indole ring directly have been developed. Kelly employed the Heck cyclization of **266** to synthesize (and revise the structure of) maxonine (**267**) [281]. The product resulting from attack at C-8 was also obtained.

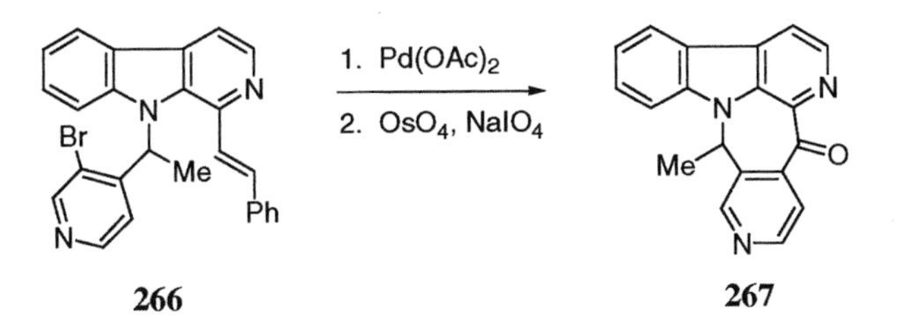


Gilchrist reported the conversion **268** to **269** [268, 269], and Grigg described syntheses of **270** and **271** using Heck conditions [270, 282].

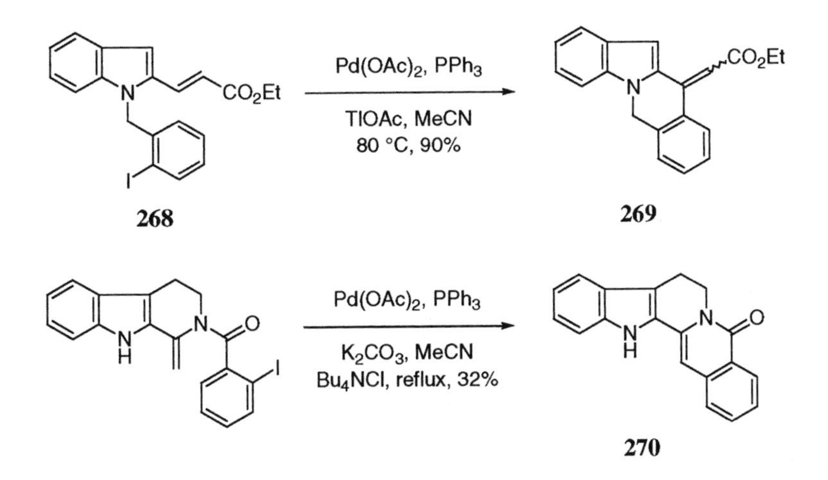

Mérour and co-workers achieved in excellent yield the cyclization of **272** to benzo[4,5]-cyclohepta[*b*]indole **273** [283, 284].

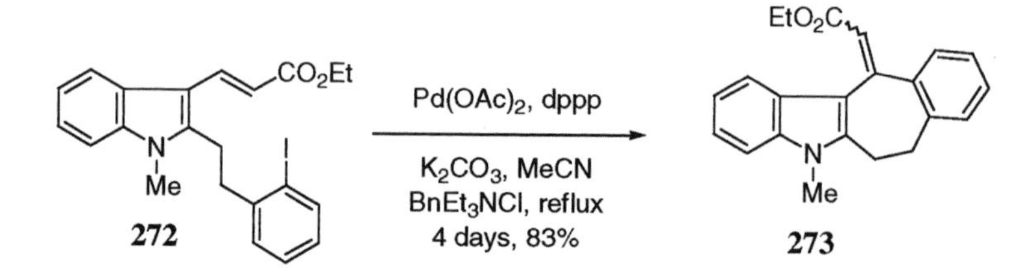


In a study of magallanesine analogs, Kurihara and co-workers effected the synthesis of **274** in low yield *via* Heck methodology [285].

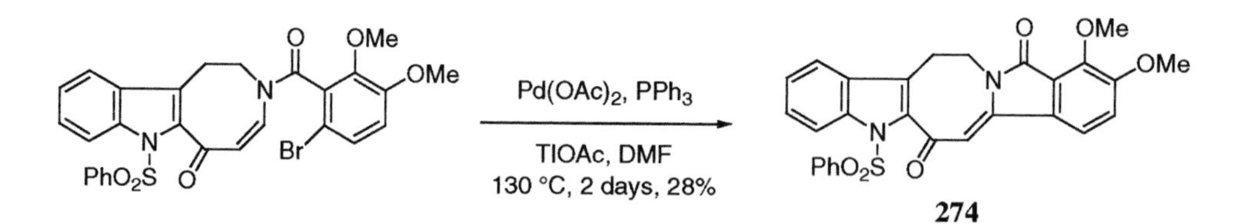


Rawal applied the Heck cyclization in elegant fashion to the construction of indole alkaloids. His route to geissoschizine alkaloids features a novel ring D formation, **275** to **276** [286]. Whereas classical Heck conditions favor the isogeissoschizal (**276b**) product, the "ligand-free" modification of Jeffery favors the geissoschizal (**276a**) stereochemistry.

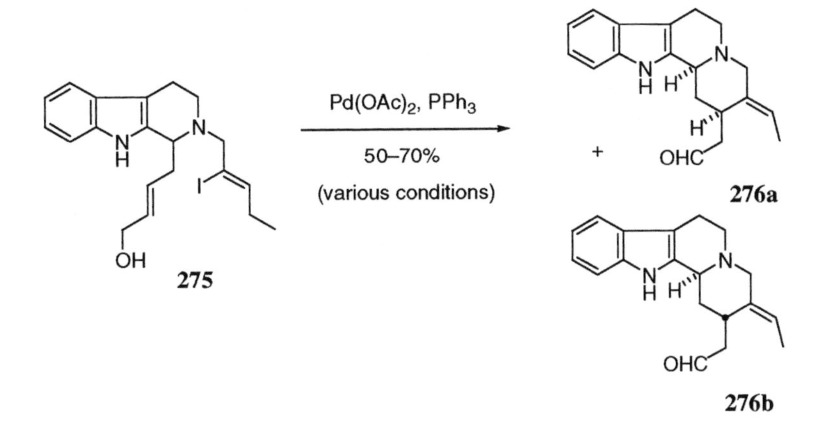


Following the application of a Heck cyclization to a concise synthesis of the *Strychnos* alkaloid dehydrotubifoline [287, 288], and earlier model studies [289], Rawal employed a similar strategy to achieve a remarkably efficient synthesis of strychnine [290]. Thus, pentacycle **277** is smoothly cyclized and deprotected to isostrychnine (**278**) in 71% overall yield — an appropriate finale to this section!

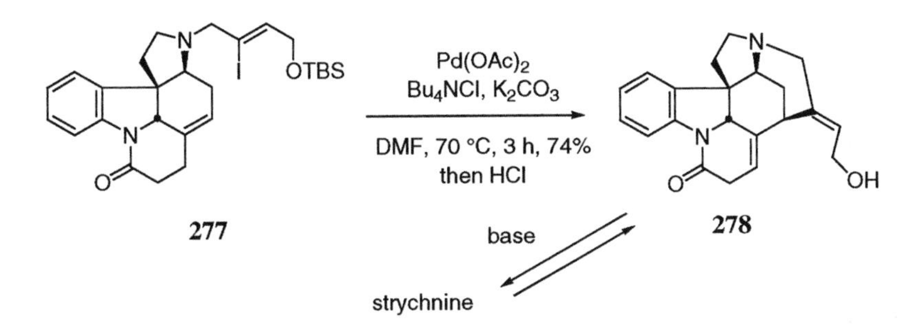


### 3.5.2 The Mori-Ban indole synthesis

The application of Heck cyclizations to the synthesis of indoles, indolines, and oxindoles was discovered independently by Mori-Ban's [296–298], and Heck's groups [299]. These investigators found that Pd can effect the cyclization of *o*-halo-*N*-allylanilines to indoles under Heck conditions [300]. The cyclization of *o*-halo-*N*-allylanilines to indoles is a general and efficient methodology, especially with the Larock improvements where he cyclized *o*-halo-*N*-allylanilines and *o*-halo-*N*-acryloylanilides into indoles and oxindoles [301]. For example, the conversion of **279** to **280** can be performed at lower temperature, shorter reaction time, and with less catalyst to give 3-methylindole (**280**) in 97% yield. Larock's improved conditions, which have been widely adopted, are catalytic (2%) $Pd(OAc)_2$, *n*-$Bu_4NCl$, DMF, base (usually

$Na_2CO_3$), 25 °C, 24 h. Larock extended his work in several ways [302–305], particularly with regard to Pd-catalyzed cross-coupling of *o*-allylic and *o*-vinylic anilides with vinyl halides and triflates to produce 2-vinylindoles [303–305], an example of which is shown [305]. The related "Larock indole synthesis" is discussed separately in the next section.

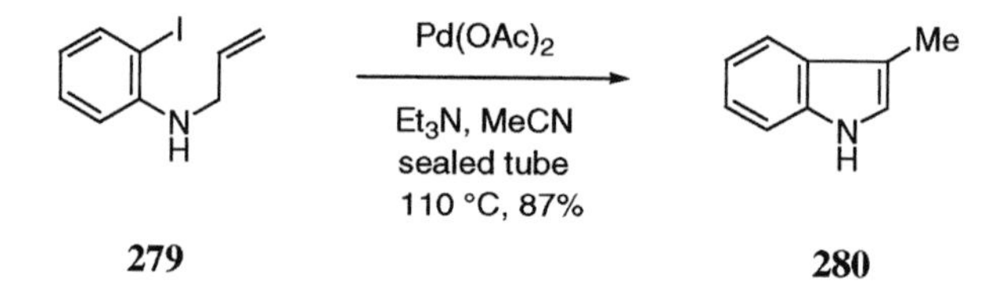


Genet showed that a water-soluble palladium catalyst system ($Pd(OAc)_2$-trisodium-phosphinetriyltribenzene-sulfonate) converts *o*-iodo-*N*-allylaniline to 3-methylindole in 97% yield [317]. The novel catalysts prepared from $Pd(OAc)_2$ and a fluorinated phosphine (e.g., $(C_6F_{13}CH_2CH_2)_2PPh$) in supercritical $CO_2$ also accomplish this cyclization to give 3-methylindole [318]. Hegedus extended his original work on the cyclization of *N*-allyl-*o*-bromoanilines to the synthesis of indoloquinones [319]. In a program to synthesize CC-1065 analogs, Sundberg prepared indole **281** in excellent yield [320]. Silver carbonate and sodium carbonate were less effective than triethylamine.

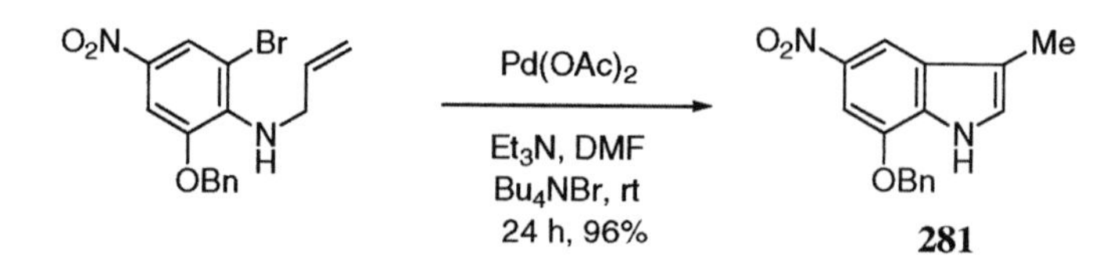


Likewise, Sakamoto has synthesized the CC-1065/duocarmycin pharmacophore **284** *via* the cyclization of **282** [321]. Silver carbonate prevented unwanted isomerization of the exocyclic double bond in **283**. Tietze and co-workers took full advantage of the power of these *N*-allyl-*o*-haloaniline Pd-catalyzed cyclizations in developing syntheses of the A-unit of CC-1065 and analogs [322–324].

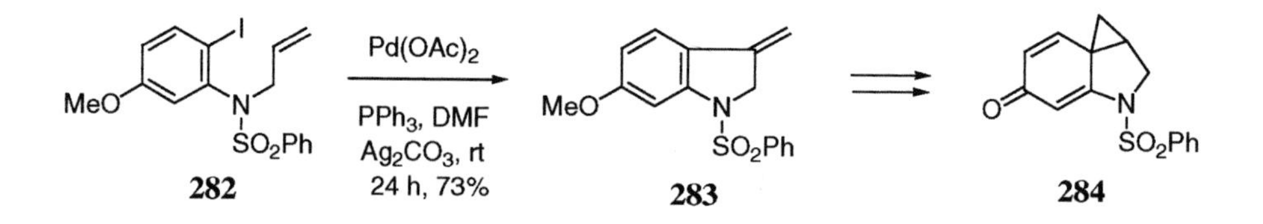


Hoffmann and co-workers crafted the desoxyeserolin precursor **286** from the *N*-pyrrolidone aniline derivative **285** [325].

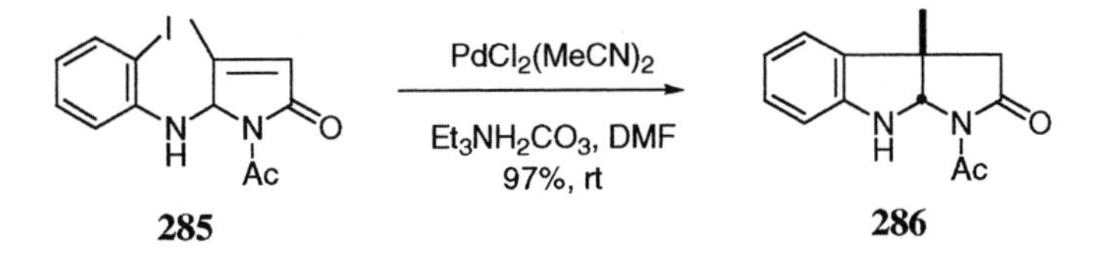


Macor exploited this methodology to synthesize several antimigraine analogs of Sumatriptan [326, 327]. An example is illustrated here.

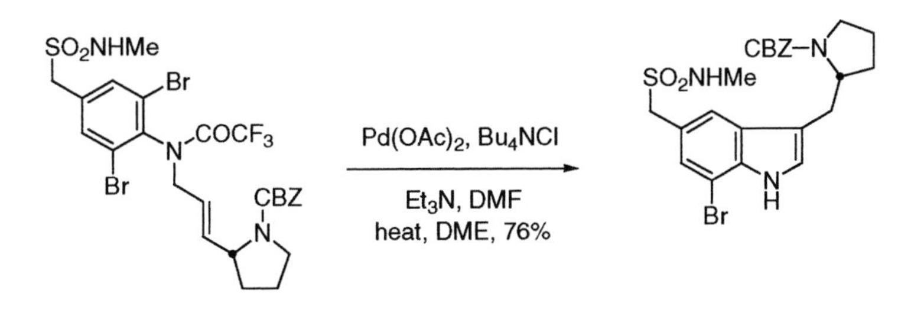


Gronowitz adapted this technology to one-pot syntheses of indole-3-acetic acids and indole-3-pyruvic acid oxime ethers from *N*-BOC protected *o*-iodoanilines [328, 329]. Rawal employed the Pd-catalyzed cyclization of *N*-(*o*-bromoallyl)anilines to afford 4- and 6-hydroxyindoles, and a 4,6-dihydroxyindole [330], and Yang and co-workers have used a similar cyclization to prepare δ-carbolines **287** and **288** as illustrated by the two examples shown [331]. The apparent extraneous methyl group in **288** is derived from triethylamine.

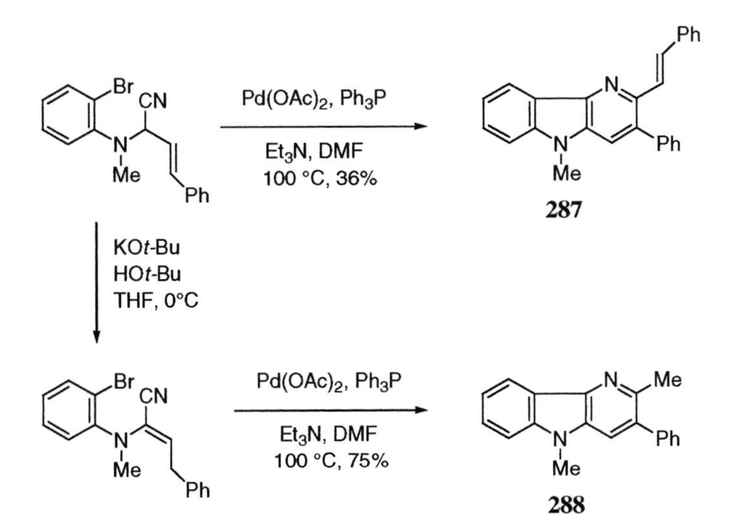

The cyclization of *N*-allyl-*o*-haloanilines was adapted to the solid phase for both indoles [332, 333] and oxindoles [334]. For example, as illustrated below, a library of 1-acyl-3-alkyl-6-hydroxyindoles is readily assembled from acid chlorides, allylic bromides, and 4-bromo-3-nitroanisole [332]. Zhang and Maryanoff used the Rink amide resin to prepare *N*-benzylindole-3-acetamides and related indoles *via* Heck cyclization [333], and Balasubramanian employed this technology to the synthesis of oxindoles *via* the palladium cyclization of *o*-iodo-*N*-acryloylanilines [334]. This latter cyclization route to oxindoles is presented later in this section.

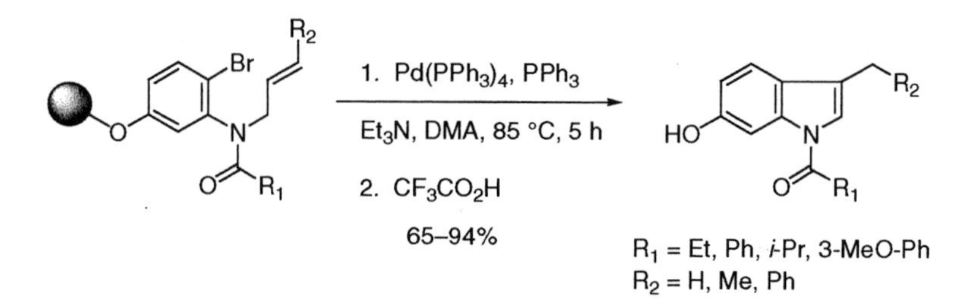


Whereas Hegedus [335] and Danishefsky [336] were the first to discover a tandem Heck reaction from *o*-allyl-*N*-acryloylanilines leading to tricyclic pyrrolo[1,2-*a*]indoles or pyridino[1,2-*a*]indoles [336], it has been the fantastic work of Grigg to unleash the enormous potential of this chemistry. Grigg and his co-workers parlayed their Pd-catalyzed tandem polycyclization-anion capture sequence into a treasure trove of syntheses starting with *N*-allyl-*o*-haloanilines [337–345]. Diels-Alder and olefin metathesis reactions can be interwoven into the sequence or can serve as the culmination step, as can a wide variety of nucleophiles. An example of the transformation of **289** to **290** is shown below in which indole is the terminating nucleophile [340].

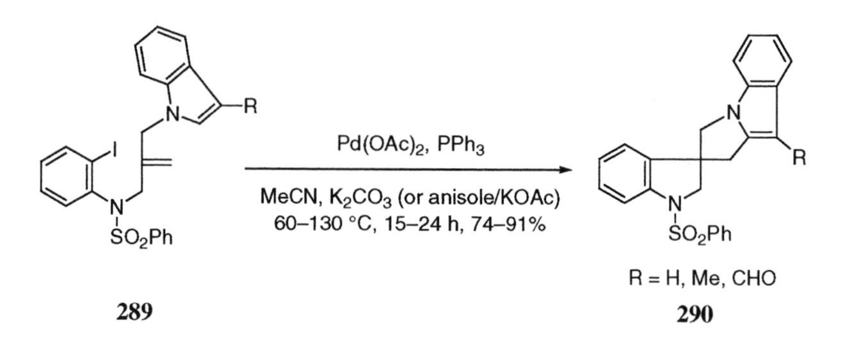


Grigg discovered that a 5-*exo-dig* Pd-catalyzed cyclization of *N*-acetylenic-*o*-haloanilines **291** to give 3-*exo*-alkylidene indolines **292** occurs in the presence of a hydride source, such as formic acid [346]. The reaction is stereoselective and regiospecific.

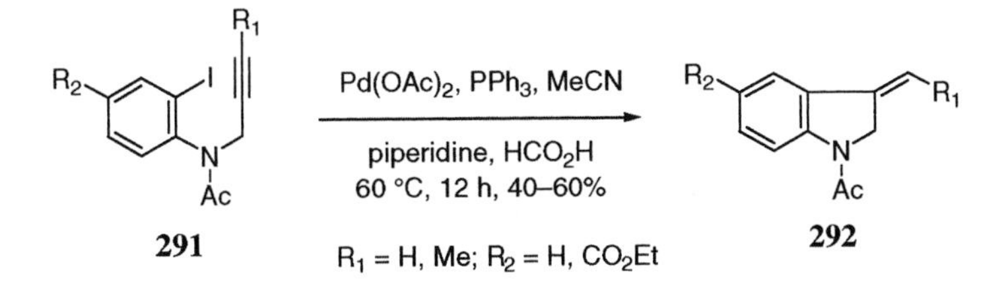


Grigg extended this alkyne cyclization to trapping with stannanes to give 3-*exo*-dienes [347], alkynes to afford tetracycles [338, 348], and alkenes leading to cyclopropanes [349], an example of which is illustrated. In his studies Grigg and co-workers have found that thallium and silver salts suppress direct capture of these palladium intermediates prior to capture [350].

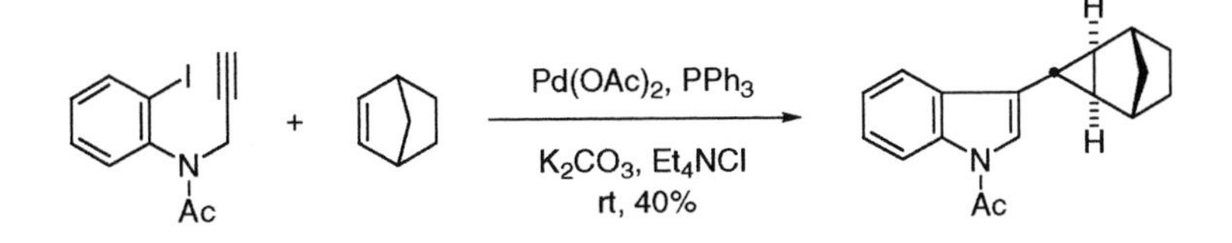


Overman's exhaustive study of the Pd-catalyzed cyclization of *o*-halo-*N*-acryloylanilines leading to spirooxindoles and related compounds has paid great dividends in advancing the art of organic synthesis [351–360]. Overman and his co-workers have developed this chemistry for the asymmetric synthesis of spirooxindoles leading to either enantiomer of physostigmine (**293**) [355, 359] and physovenine [359], for gelsemine studies [352, 353], and, *via* a spectacular bis-Pd-catalyzed cyclization, to total syntheses of chimonanthine and calycanthine [355], as summarized in the transformation of **294** to **295**. Hiemstra, Speckamp and co-workers have pursued similar studies of Pd-catalyzed cyclizations to spirooxindoles, culminating in total syntheses of (±)-gelsemine and (±)-21-oxogelsemine [361].

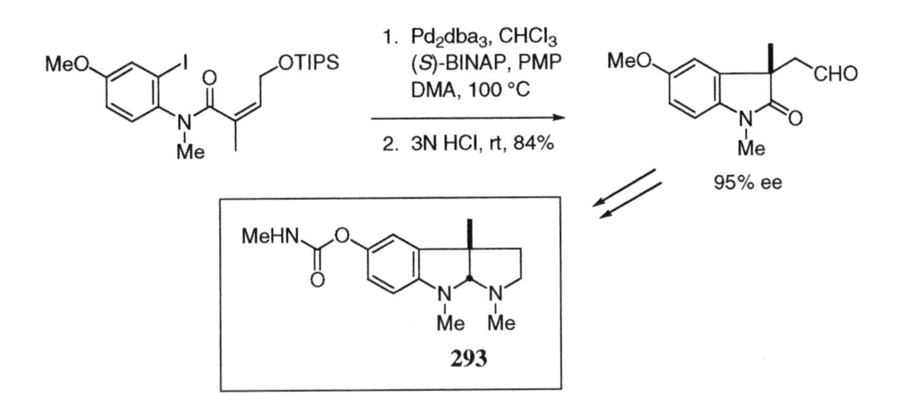

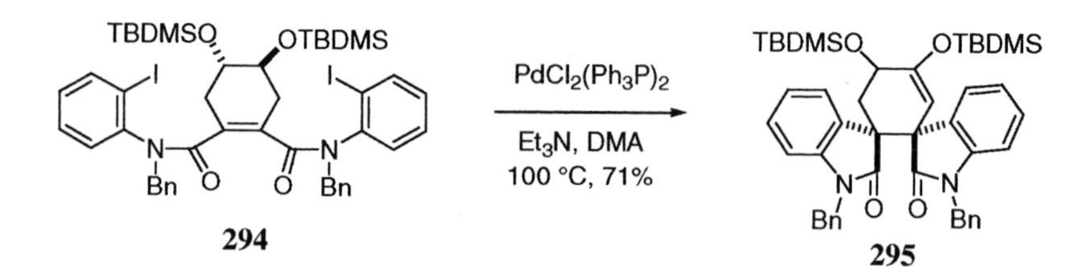


It has been known for sometime that 2-(2-bromoanilino)enones undergo Heck-type cyclizations to form indoles and carbazoles. Thus, Kibayashi reported the synthesis of 4-keto-1,2,3,4-tetrahydrocarbazoles in this manner [362], and Rapoport employed this reaction (**296** to **297**) to achieve an improved synthesis of 7-methoxymitosene [363]. A series of related mitosene analogs has been crafted using Pd chemistry by Michael and co-workers [364]. They found that P(*o*-tol)$_3$ was far superior to $PPh_3$ in conjunction with $Pd(OAc)_2$.

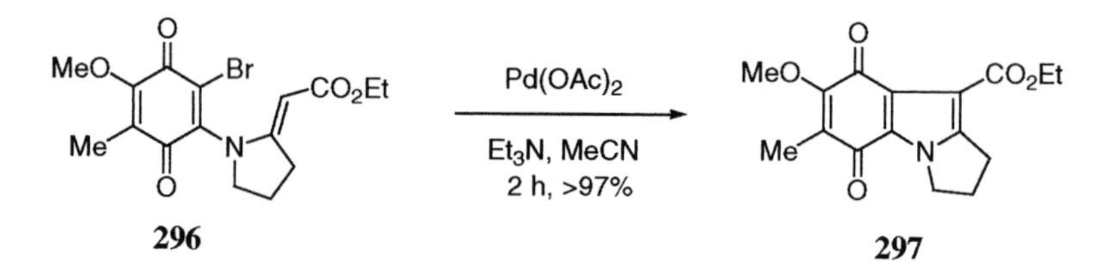


Kashara [365] and Sakamoto [366] have both shown that this Pd-cyclization is an excellent synthesis of a variety of 3-acylindoles, as shown for **298** to **299** [365]. Substrates **298** are also prepared using Pd. Sakamoto showed that 2-substituted 3-acylindoles are available in this manner [366].

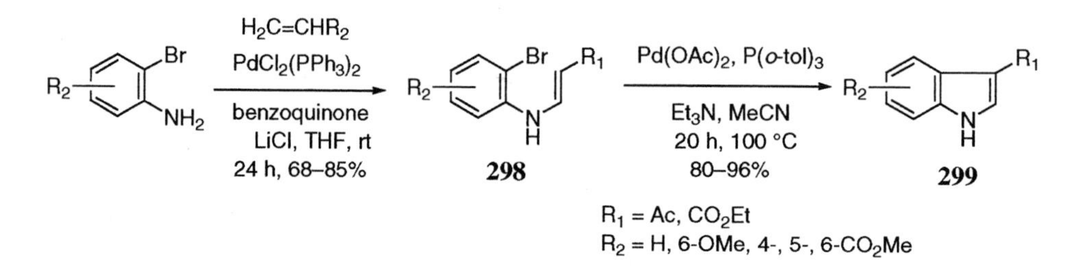


Related Pd-cyclizations have been applied to the synthesis of 3-carboethoxy-2-trifluoromethylindoles [367, 368] and 2-carbobenzyloxy-4-hydroxymethyl-3-methylindoles, a unit that is present in the antibiotic nosiheptide, from a 2-(2-iodoanilino) unsaturated ester [369]. A nice variation utilizes the *in situ* synthesis of 2-iodoanilino enamines and subsequent cyclization [370] as shown below.

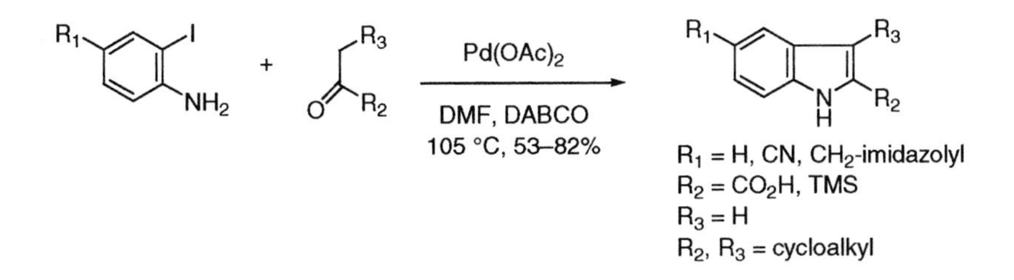


This indole synthesis has been extended to β-tetrahydrocarbolines (**300**) [371], azaketotetrahydrocarbazoles [372], carbolines, carbazoles, and pyrido[1,2-*a*]benzimidazoles [373]. Examples of the former two reaction types are illustrated. An early Heck cyclization of 2-carboxy-2'-iododiphenylamine to 1-carbazolecarboxylic acid (73% yield) [374] has been generally overlooked by subsequent investigators.

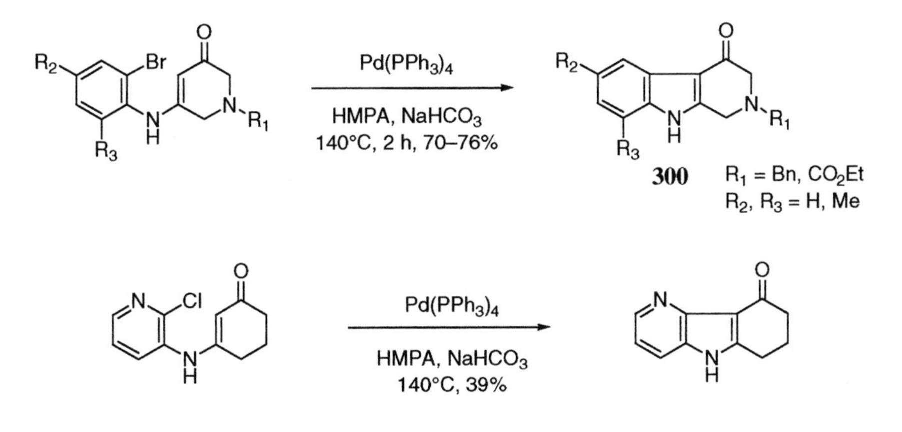


A few examples of Pd-catalyzed cyclization of *o*-allylanilines to indoles have been reported [375, 376], but only in the case of the cyclization of *N*-alkyl-*o*-siloxyallylanilines leading to 3-alkoxyindoles is the method useful [376]. Palladium has been often used to prepare *o*-vinylaniline derivatives for subsequent (non-Pd) cyclization. Although they do not fall in the Mori-Ban indole synthesis category for the indole ring-forming step, we have appended them at the end of this section because these reactions are still synthetically very useful. For example, the Pd-catalyzed reaction of *o*-iodoacetanilide with ethyl α-methoxyacrylate affords *o*-vinylacetanilide **301** [377]. Acid treatment gives 2-carboethoxyindole.

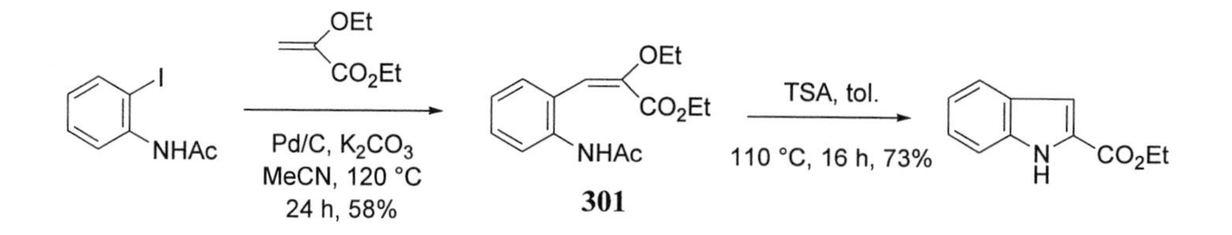


A related Heck reaction of substituted *o*-bromoacetanilides with styrenes followed by selenium-induced cyclization of the resulting *o*-styrylacetanilides gives 2-arylindoles [378]. Substituted *o*-bromonitrobenzenes react with ethyl vinyl ether under the influence of $Pd(OAc)_2$ to give the corresponding *o*-ethoxyethenylnitrobenzenes. Zinc reduction then yields indoles [379]. The one-step Pd-catalyzed conversion of *o*-bromoanilines to indoles **302** with enamines (or with *N*-vinyl-2-pyrrolidone) has been reported [380].

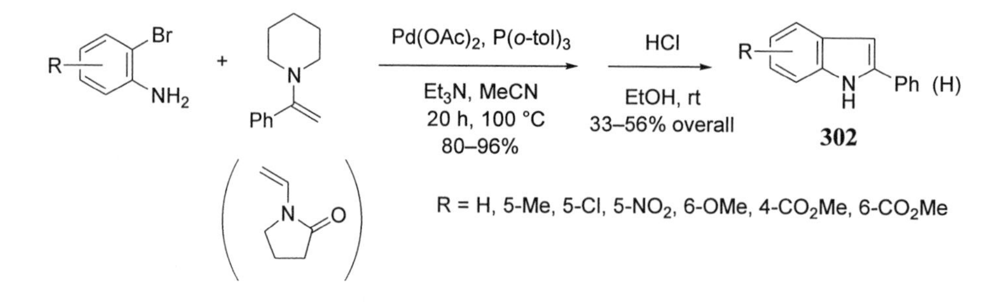


Ogasawara employed a Heck reaction of *o*-iodoaniline derivatives with dihydrodimethoxyfuran and vinylene carbonate to give intermediates that are readily cyclized to indoles with acid [381–383]. An example is shown below [381].

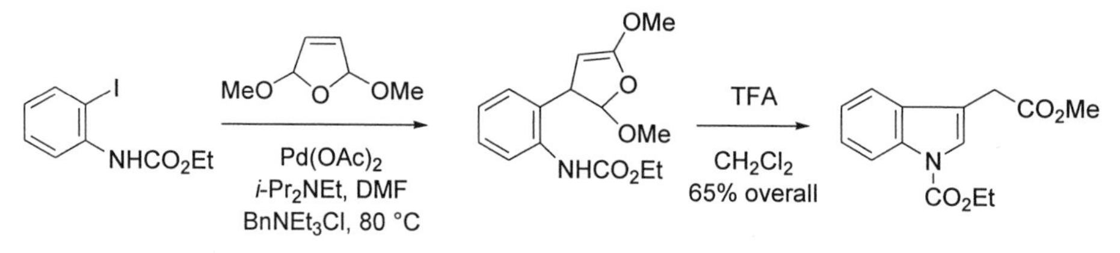


Carlström and Frejd described the one-step conversion of *o*-diiodobenzene (or iodobenzene) with *N*-Boc dehydroalanine methyl ester to give *N*-Boc-2-carbobenzyloxyindole in 30% yield [384].

### 3.5.3 The Larock indole synthesis

Larock and co-workers described the one-step Pd-catalyzed reaction of *o*-haloanilines with internal alkynes to give indoles [385, 386]. This excellent reaction, which is shown for the synthesis of indoles **303**, involves oxidative addition of the aryl halide (usually iodide) to Pd(0), *syn*-insertion of the alkyne into the ArPd bond, nitrogen displacement of the Pd in the resulting vinyl-Pd intermediate, and final reductive elimination of Pd(0).

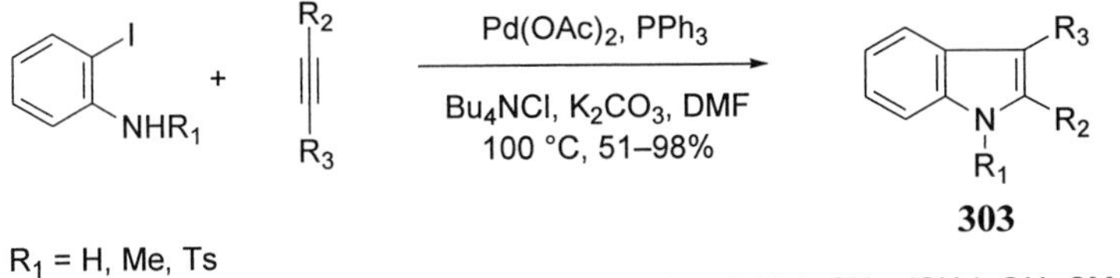


The reaction can be regioselective with unsymmetrical alkynes, and this is particularly true with silylated alkynes wherein the silyl group always resides at the C-2 indole position in the product. This is noteworthy because silyl-substituted indoles are valuable substrates for other chemistry (halogenation, Heck coupling). Gronowitz used the appropriate silylated alkynes with *o*-iodoanilines to fashion substituted tryptophans following desilylation with $AlCl_3$ [387]. Similarly, a series of 5-, 6-, and 7-azaindoles was prepared by Ujjainwalla and Warner from *o*-aminoiodopyridines and silylated (and other internal) alkynes using $PdCl_2$(dppf) [388]. Yum and co-workers also used a Larock indole synthesis to prepare 7-azaindoles **304** [389] and, from 4-amino-3-iodoquinolines, pyrrolo[3,2-*c*]quinolines **305**, which have a wide spectrum of biological activity [390].

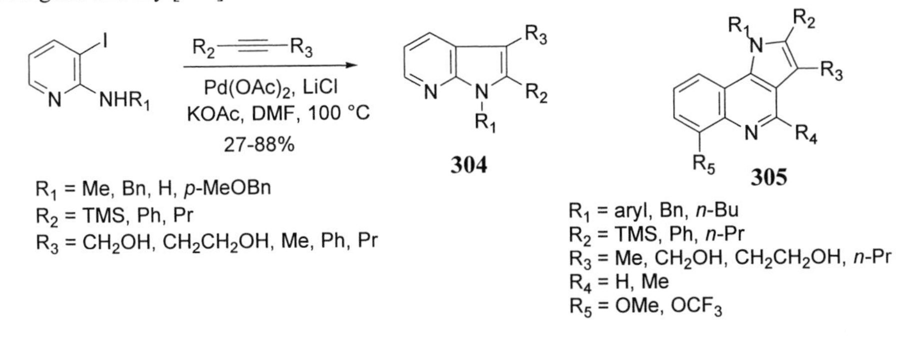


The Larock synthesis was used by Chen and co-workers to synthesize the 5-(triazolylmethyl)tryptamine MK-0462, a potent 5-$HT_{1D}$ receptor agonist, as well as a metabolite [391, 392]. Larock employed his methodology to prepare tetrahydroindoles [393], and Maassarani used this method for the synthesis of *N*-(2-pyridyl)indoles [394]. The latter study features the isolation of cyclopalladated *N*-phenyl-2-pyridylamines. Rosso and co-workers have employed this method for the industrial scale synthesis of an antimigraine drug candidate **306**. In this paper removal of spent palladium was best effected by trimercaptotriazine (**307**) although many techniques were explored [395].

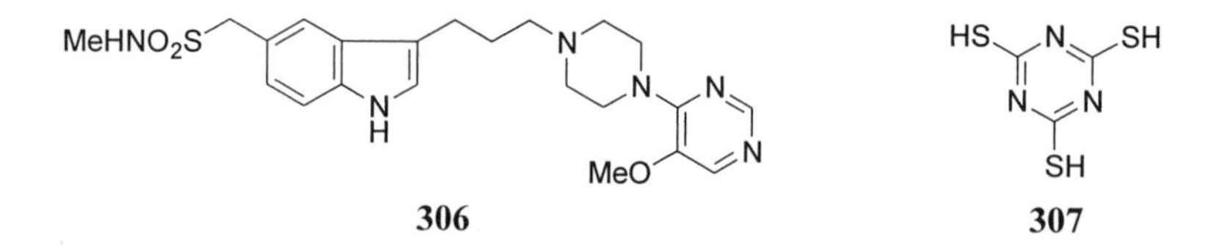


The Larock indole synthesis was adapted to the solid phase both for the synthesis of 1,2,3-trisubstituted indole-5-carboxamides [396] and, as illustrated, for the "traceless" synthesis of 2,3-disubstituted indoles **308** [397]. As seen earlier, the trimethylsilyl group is fastened to C-2 with complete regioselectivity. The TMS group is cleaved under the resin cleavage conditions. The original Larock conditions were not particularly successful.

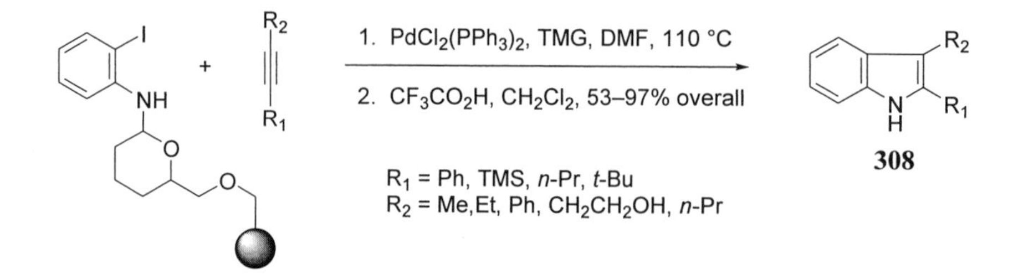


Similar Pd-catalyzed chemistry between *o*-iodoanilines and 1,3-dienes leading to 2-vinylindolines is also known, having been first described by Dieck and co-workers [398]. This reaction, which is shown for the synthesis of **309**, was discovered before Larock's work in this area. The same reaction with 1,3-cyclohexadiene gives the corresponding tetrahydrocarbazole in 70% yield.

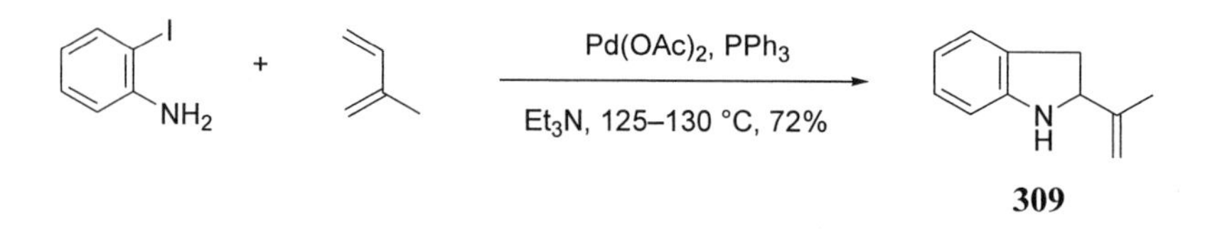

Larock extended this Pd-catalyzed diene heteroannulation to other dienes and anilines [399], including functionalized dienes leading to, for example, ketotetrahydrocarbazoles [400]. Back has employed 1-sulfonyl-1,3-dienes in this 2-vinylindoline synthesis [401], and the use of 1,3-dienes in constructing indolines has been adapted to the solid phase by Wang [402]. Interestingly, Larock has shown that the electronically-related vinylcyclopropanes undergo a similar cyclization with *o*-iodoanilines to form 2-vinylindolines, e.g., **310** [403, 404]. Vinylcyclobutane also reacts in a comparable manner.

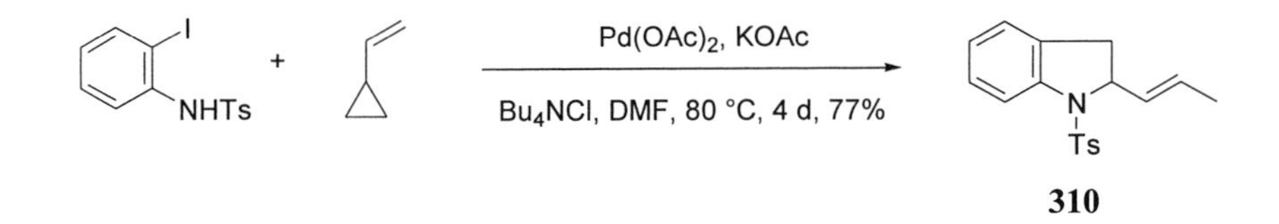


Larock found that allenes (1,2-dienes) undergo Pd-catalyzed reactions with *o*-iodoanilines to afford 3-alkylidene indolines, including examples using cyclic dienes, e.g., to give **311** [405], and ones leading to asymmetric induction, e.g., to give **312** [406, 407]. The highest enantioselectivities ever reported for any Pd-catalyzed intramolecular allylic substitution reactions were observed in this study. Mérour modified this reaction for the synthesis of 7-azaindolinones, following ozonolysis of the initially formed *exo*-methylene indoline [408].

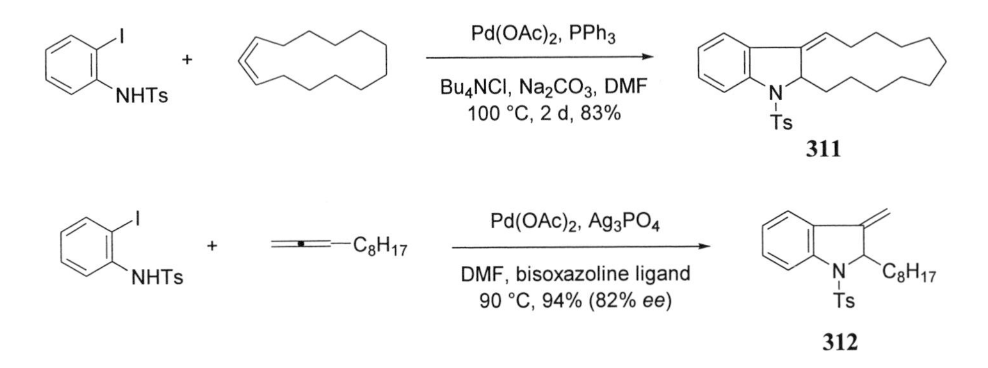


Prior to his work with internal alkynes, Larock found that *o*-thallated acetanilide undergoes Pd-catalyzed reactions with vinyl bromide and allyl chloride to give *N*-acetylindole and *N*-acetyl-2-methylindole each in 45% yield [409]. In an extension to reactions of internal alkynes with imines of *o*-iodoaniline, Larock reported a concise synthesis of isoindolo[2,1-*a*]indoles **313** and **314** [410]. The regioselectivity was excellent with unsymmetrical alkynes.

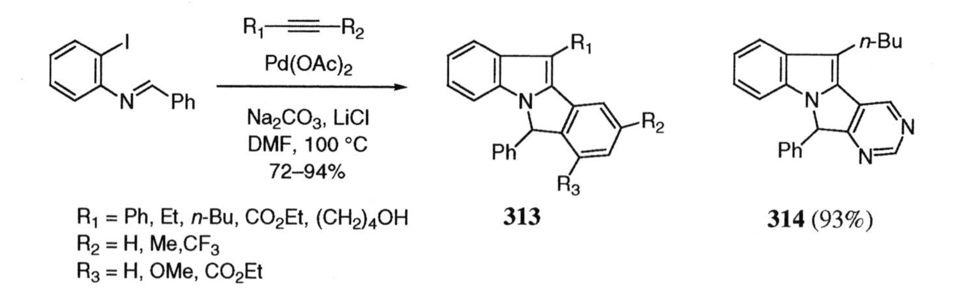


## 3.6 Carbonylation

The insertion of carbon monoxide into σ-alkylpalladium(II) complexes followed by attack by either alcohols or amines is a powerful acylation method. This carbonylation reaction has been applied in several different ways to the reactions and syntheses of indoles. Hegedus and co-workers converted *o*-allylanilines to indoline esters **315** in yields up to 75% [293]. In most of the examples in this section, CO at atmospheric pressure was employed.

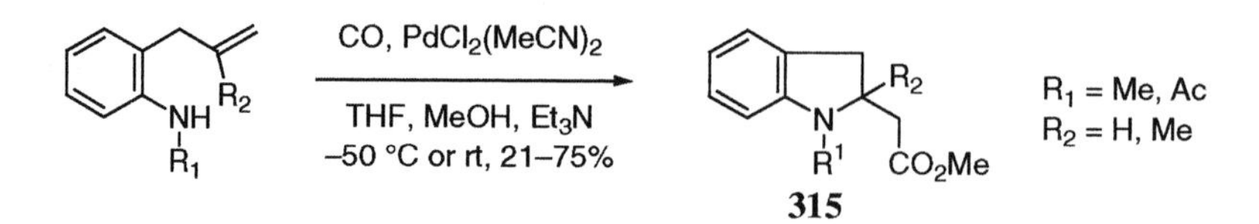


Hegedus applied this chemistry to the conversion of 1-lithio-3-methylindole to 1-(2-carbomethoxyethyl)-3-methylindole with CO and ethylene (53% yield) [411]. This interesting C—N bond formation reaction is revisited in Section 3.7. The amide precursor used by Overman in his gelsemine studies was prepared by a Pd-catalyzed carbonylation reaction between a vinyl triflate and *o*-bromoaniline [352]. A similar amination scheme was used to synthesize 21-oxoyohimbine from 1-(2-bromobenzyl)-1,2,3,4-tetrahydro-β-carboline [412]. Mori-Ban and co-workers employed a similar strategy to prepare the quinazoline alkaloid rutecarpine and related compounds [413]. Edstrom expanded his studies on the carbonylation of pyrroles (Chapter 2) to the methoxycarbonylation of 5-azaindolones leading to **316** [414, 415]. Doi and Mori performed a similar carbonylation on a dihydroindole 4-triflate leading, after dehydrogenation, to 4-carbomethoxy-1-tosylindole [140]. The direct carbonylation of *N*-acetyl- and *N*-benzoylindole leads to the corresponding C-3 carboxylic acids [416].

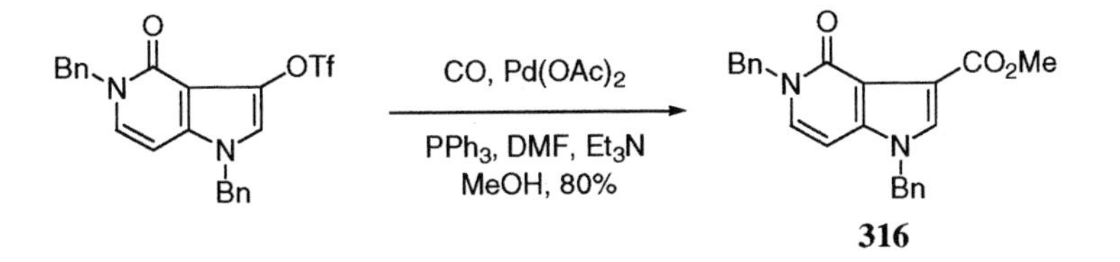


The tandem indolization-carbonylation of *o*-alkynylanilines has been explored by two groups. Kondo and co-workers have effected the synthesis of 3-carbomethoxyindoles **317** [417, 418], and Arcadi has modified this reaction to prepare 3-acylindoles **318** [419].

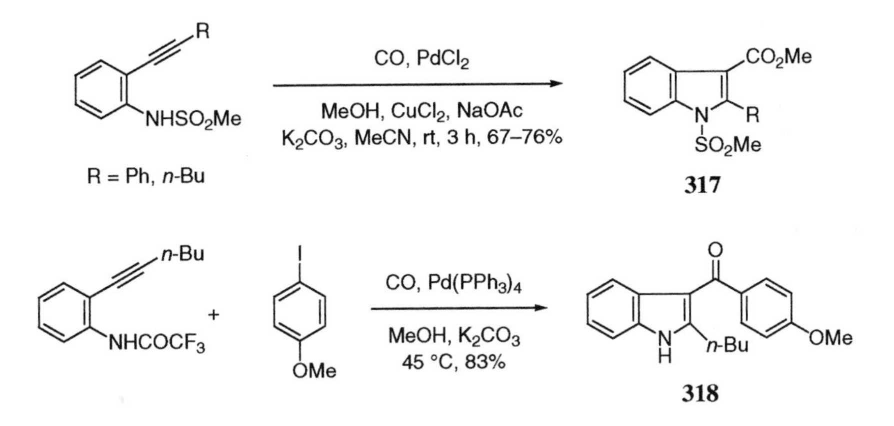


Herbert and McNeil have shown that the appropriate 2-iodoindole can be carbonylated in the presence of primary and secondary amines to afford the corresponding 2-indolecarboxamides in 33–97% yield. Further application of this protocol leads to amide **319**, which is a CCK-A antagonist (Lintitript) [420].

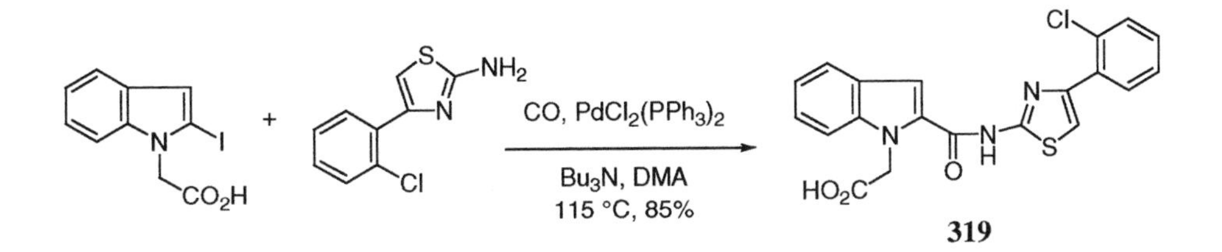


Fukuyama employed a vinyltin derivative in the carbonylation of 3-carbomethoxymethyl-2-iodoindole to afford **320** [176]. Buchwald effected the carbonylation of 4-iodoindole **321** to give lactam **322** [136], and a similar carbonylation reaction on 4-iodoindole malonate **125** gives ketone **323** in 68% yield.

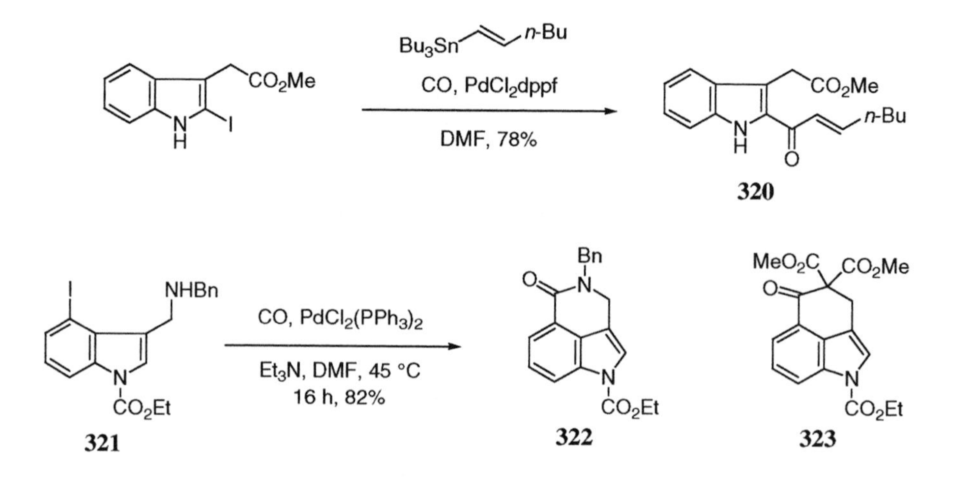


Somei carbonylated 7-thallated *N*-acetylindoline to give 7-carbomethoxy-*N*-acetylindoline as the major product ($CO/Pd(OAc)_2/MeOH/Cr(CO)_6$/52%) [199]. The carbonylation of C-2 in gramines and tryptamines has been achieved by Cenini and co-workers [421, 422]. Thus, treatment of gramine (3-dimethylaminomethylindole) with $Li_2PdCl_4$ ($PdCl_2$ + LiCl) followed by treatment of the palladated intermediate (isolable, 93% yield) with methanol and CO afforded 2-carbomethoxygramine in 92% yield. In their grossularine synthetic studies, Hibino and co-workers effected the carbonylation of an α-carboline triflate leading to methyl ester **324** [189]. The same triflate with an arylboronic acid and CO gave a low yield of an aryl ketone [188].

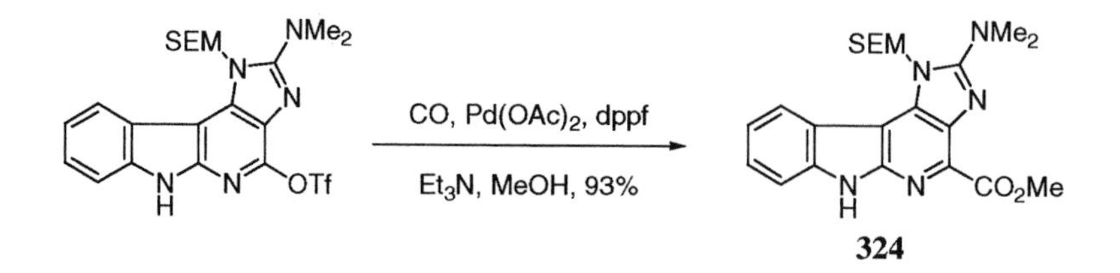


In the presence of tetramethyltin, 1-bromonauclefine reacts with CO in a Pd-catalyzed carbonylation to give the alkaloid naucletine [202]. Dong and Busacca effected a new synthesis of tryptamines and tryptophols *via* a Rh-catalyzed hydroformylation of functionalized anilines that are prepared by a standard Heck reaction, as shown for the preparation of tryptamine sulfonamide **325** [423]. This reaction is applicable to ring-substituted tryptamines (Cl, Br, F. OMe, $CF_3$). Likewise, the Rh-catalyzed carbonylation of *o*-alkynylanilines, which were prepared by a Pd-catalyzed Sonogashira coupling, leads to oxindoles (60–86% yields) [424].

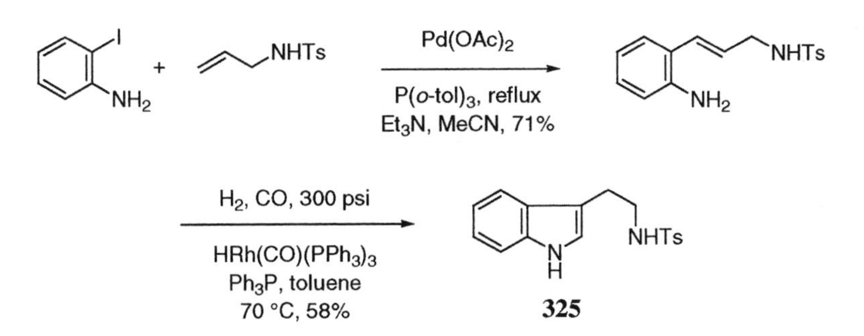


Hidai and co-workers found that 3-vinylindole **326** undergoes cyclocarbonylation to afford 1-acetoxycarbazole **327** [425]. The reaction of indole with allene and CO in the presence of catalytic Pd(0) leads to *N*-acylation (**328**) in good yield [426]. An analogous reaction with 5-hydroxyindole affords *N*- and *O*-acylation products (47% yield).

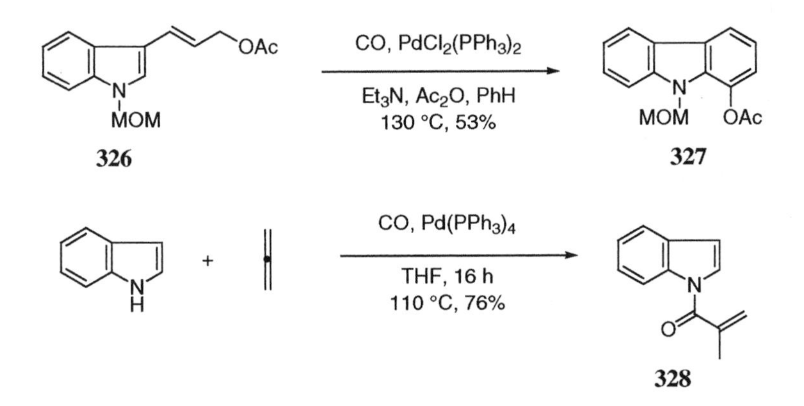


Grigg expanded his Pd-catalyzed cascade cyclization reactions to include carbonylation as the termination step [427]. Thus, indoline **329** is obtained in excellent yield and the spiroindoline **330** is secured as a single diastereomer. Thallium acetate results in significant improvement in these reactions by allowing for low-pressure carbonylation.

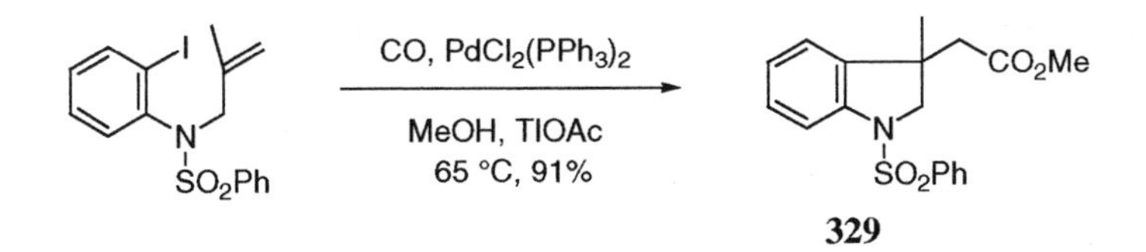

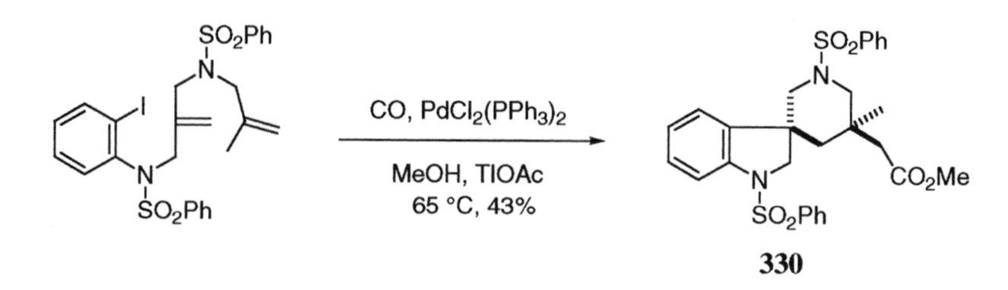


3-Spiro-2-oxindoles, such as **331**, are readily crafted from the Pd-catalyzed reactions of *o*-haloanilines with vinyl halides and triflates in the presence of CO [428]. The *o*-iodo enamide is presumed to form initially, followed by Heck cyclization.

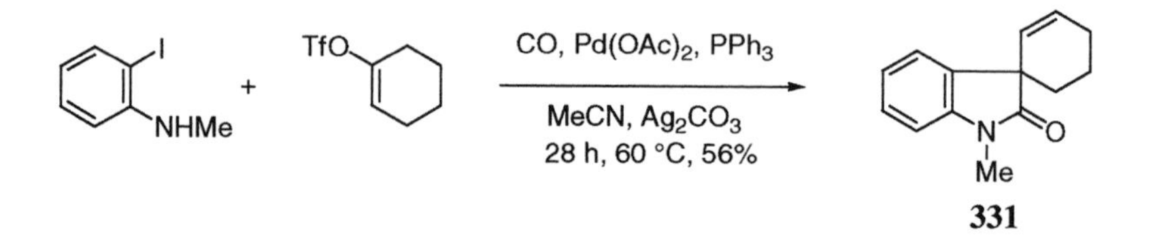


Grigg also extended these carbonylation reactions to a one- and two-pot protocols culminating in olefin metathesis [344]. For example, substrate **332** is converted to **334**, *via* **333**, under these conditions. *N*-Tosylindolines were constructed in like fashion.

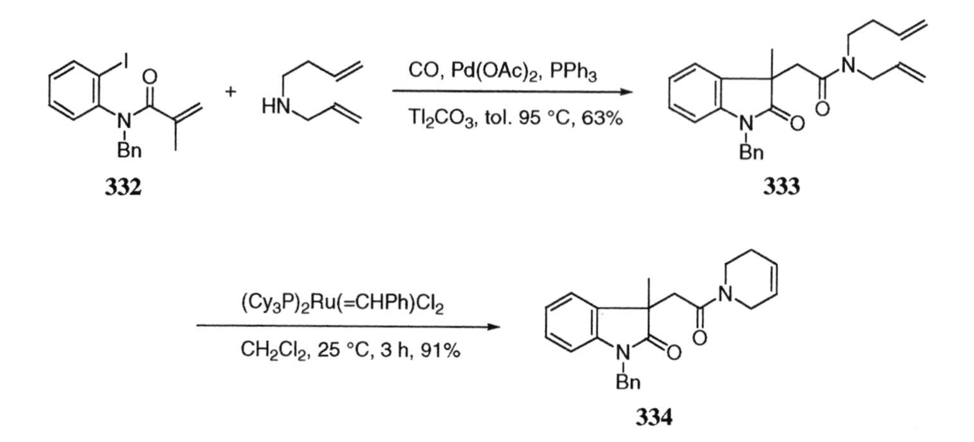


Ishikura has adapted his Pd-catalyzed cross-coupling methodology involving indolylborates to include carbonylation reactions [151, 153, 157, 429]. For example, **136** reacts with enol triflates in the presence of CO and Pd to give 2-acylindoles such as **335** [429]. This particular ketone was cyclized to the indole C-3 position and the resulting cyclopentanone was converted to yuehchukene analogs. Borate **136** also reacts with oxindole precursor **336** to give 2-

acylindole-oxindole **337** [151]. Noteworthy is that Ishikura also used the removable *N*-Boc indolylborates in this chemistry [157].

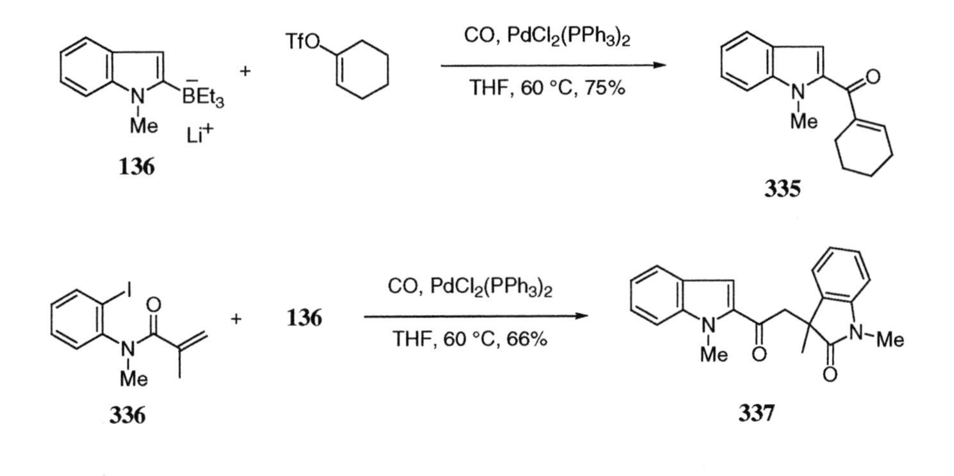


## 3.7 C—N bond formation reactions

While the Mori-Ban indole synthesis is catalyzed by a Pd(0) species, the Hegedus indole synthesis is catalyzed by a Pd(II) complex. In addition, the Mori-Ban indole synthesis is accomplished *via* a Pd-catalyzed vinylation (a Heck recation), whereas the Hegedus indole synthesis established the pyrrole ring *via* a Pd(II)-catalyzed amination (a Wacker-type process). Hegedus conducted the Pd-induced amination of alkenes [430] to an intramolecular version leading to indoles from *o*-allylanilines and *o*-vinylanilines [291–293, 295, 250, 251]. Three of the original examples from the work of Hegedus are shown below.

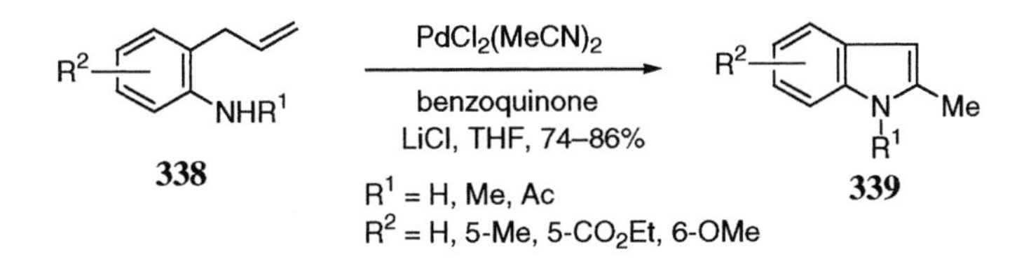

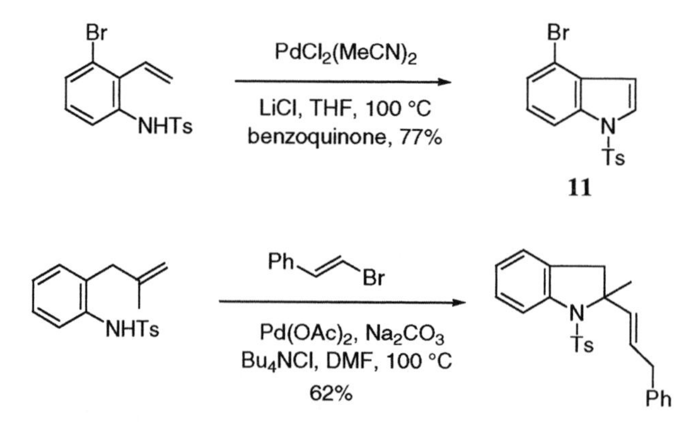


The Hegedus indole synthesis can be stoichiometric or catalytic and a range of indoles **340–342** was synthesized from the respective *o*-allylanilines in modest to very good yields (31–89%) [292].

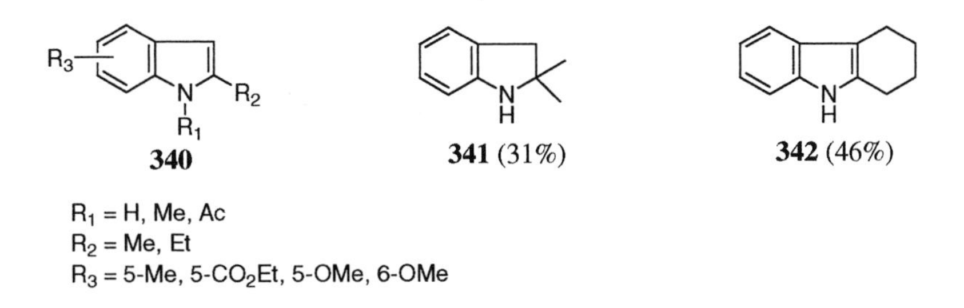


In addition, Hegedus and co-workers extended this chemistry to the *N*-alkylation of 1-lithioindole with alkenes ($PdCl_2(MeCN)_2$/THF/HMPA/$Et_3N$/then $H_2$) to afford *N*-alkylindoles in 28–68% yields [411]. Moreover, a similar reaction of *N*-allylindole **343** with nitriles leads either to **344a**, **344b**, or **344c** depending on how the intermediate Pd-alkyl (or acyl) complexes are treated [431]. The formation of these pyrazino[1,2-*a*]indoles is similar to a nitrile-Ritter reaction.

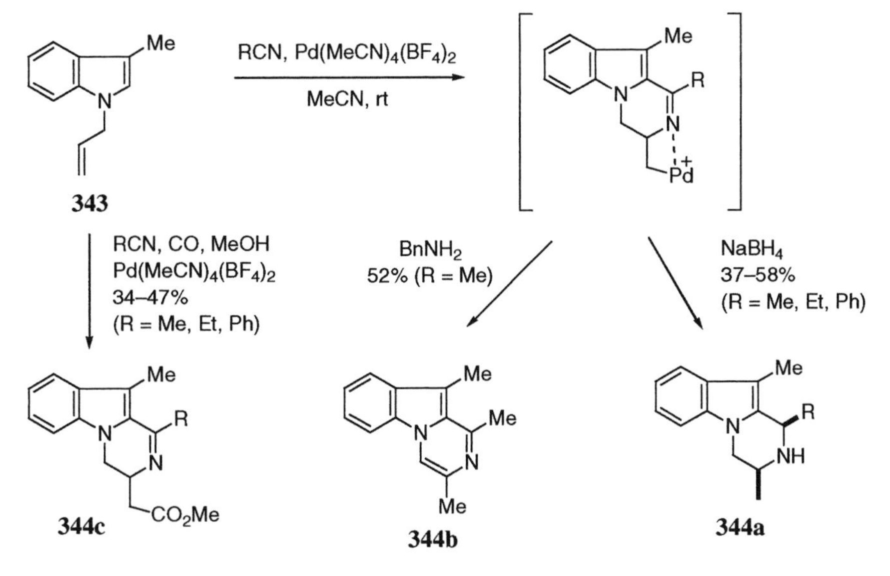


The Pd-catalyzed cyclization of *o*-vinylanilines to indoles has only been rarely utilized in synthesis, but Stille found this reaction to afford a variety of *N*-tosylindoles [193]. The requisite *o*-vinylaniline derivatives were conveniently prepared by a Stille coupling using tributylvinyltin and aryl bromides. A similar *ortho*-vinylation using $SnCl_4$–$Bu_3N$ and acetylene and subsequent Pd-catalyzed cyclization to give 5- and 7-methylindole, and 1,4- and 1,6-dimethylindole was described by Yamaguchi [306]. Kasahara reported the vinylation of *o*-bromoacetanilides **345** and the Pd-catalyzed cyclization of the resulting *o*-vinylacetanilides **346** to *N*-acetylindoles **347** [307].

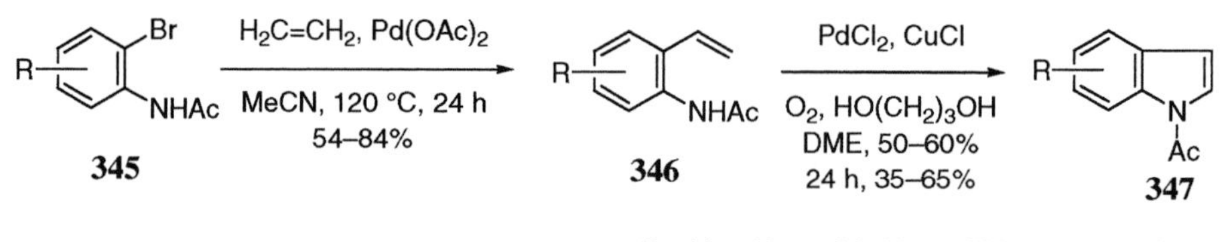


The cyclization of *o*-alkynylanilines to indoles, which usually does not require palladium, has been described in Section 3.4. In view of their extensive research with this transformation, this reaction is often referred to as the Sakamoto-Yamanaka indole synthesis [211, 214–216, 220, 230, 231]. Although the cyclization of *o*-alkynylanilines, which are often obtained by the Sonogashira coupling (Section 3.4), is usually accomplished with base, Kundu used $Pd(OAc)_2$ to effect the conversion of **348** to **349** [308].

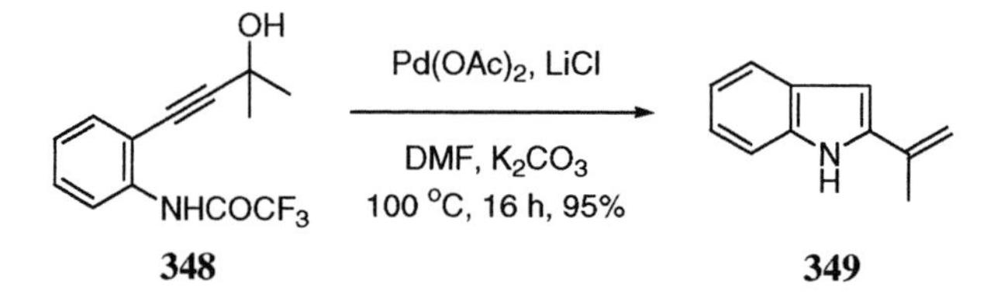


Trost used a Sonogashira coupling, followed by reduction of the nitro group, to prepare **350**, which in turn was converted to tricyclic lactam **351** by Pd-catalyzed cyclization (yield unreported) [309].

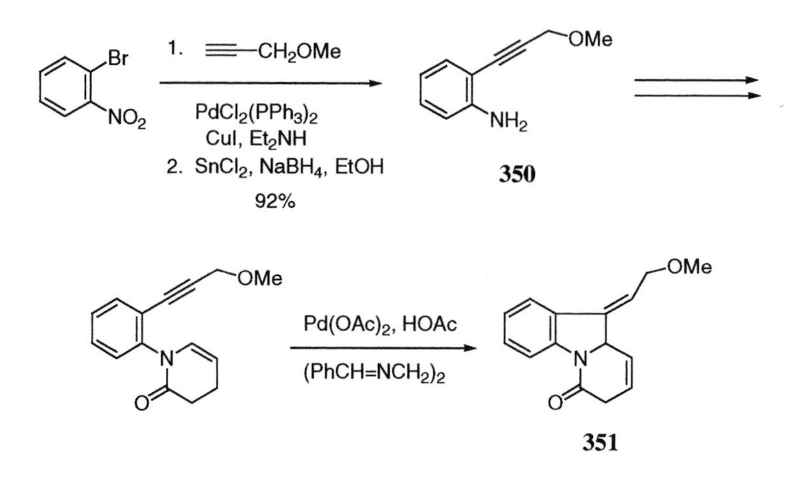

Illustrative of the mild conditions involved in these cyclizations, Fukuyama effected the transformation of **352** to biindole **353** [176a].

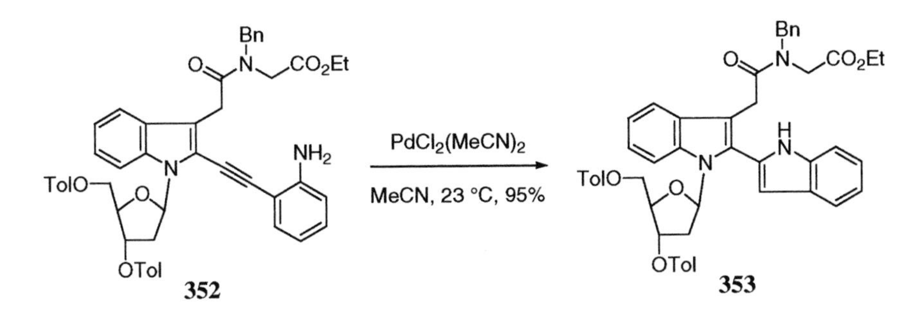


Several groups intercepted the indole-palladium complex that is initially obtained on cyclization by a subsequent Heck reaction. As will be seen, this can be a powerful elaboration of indoles. In the first example of this concept, Utimoto and co-workers ambushed intermediate **354** with a series of allylic chlorides to give **355**. Normal acid workup yields the corresponding C-3 unsubstituted indoles (52–83%) [310].

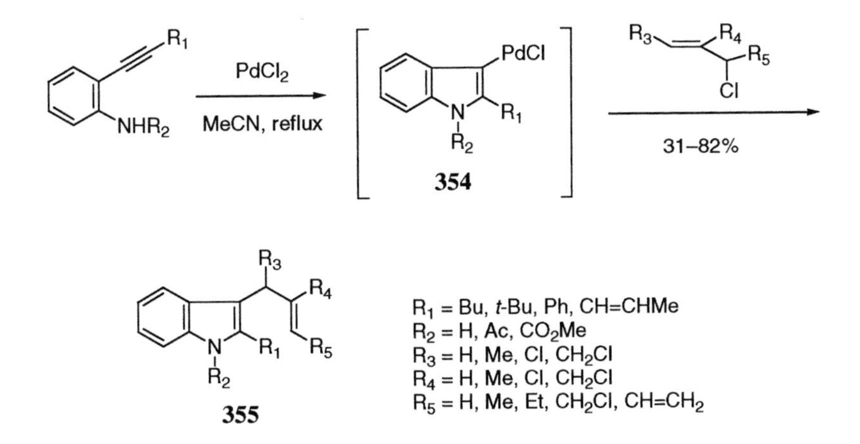


The research group of Cacchi made extensive use of these tandem cyclization-Heck reactions to prepare a wide variety of indoles [311–314]. For example, vinyl triflates react with *o*-aminophenylacetylene to afford an array of 2-substituted indoles in excellent yield, e.g., **356** to **357** [312], and a similar reaction of **358** with aryl iodides leads to an excellent synthesis of 3-arylindoles **359** [313].

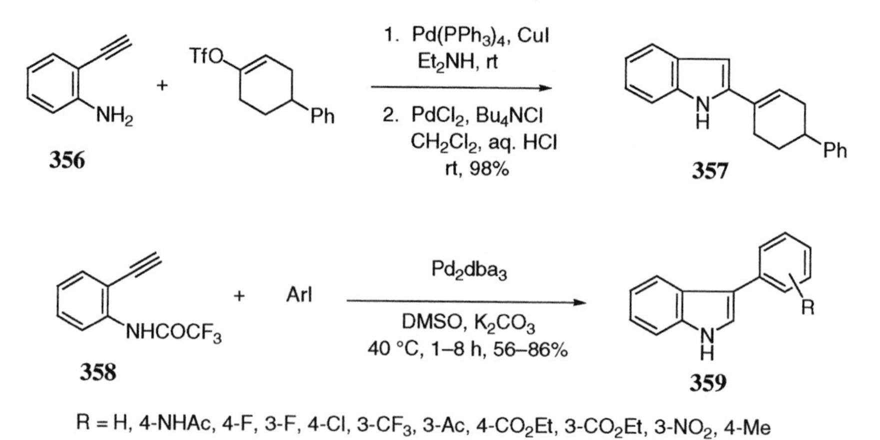


Saulnier expanded this reaction to an elegant synthesis of indolo[2,3-*a*]carbazole **355**, featuring a polyannulation from the diacetylene as shown [315].

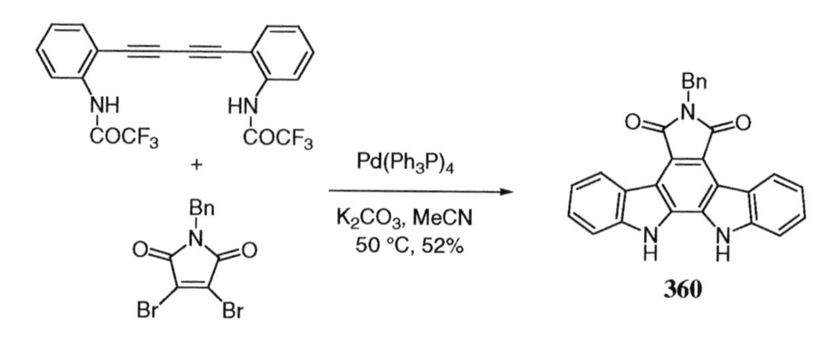


Yasuhara applied this methodology to the synthesis of 3-vinylindoles **361** [316].

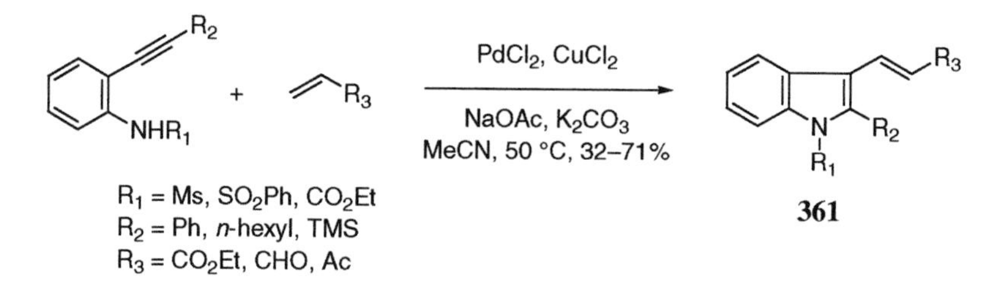


In a clever application of the hetero-Cope rearrangement, Martin used a Pd-catalyzed coupling of *N*-arylhydroxamates **362** with vinyl acetate to set up the [3,3] sigmatropic rearrangement **363** to **364** and final cyclization **364** to **365** [432–435]. Applications of this novel indole ring construction to the synthesis of the toxic fava bean metabolite 4-chloro-6-methoxyindole [434], marine acorn worm 4,6-dibromo- and 3,4,6-tribromoindoles [433, 434], CC-1065 subunits, and PDE-I and -II have been achieved. *N*-Arylhydroxamates **362** are readily prepared from the appropriate nitrobenzene by partial reduction and acylation.

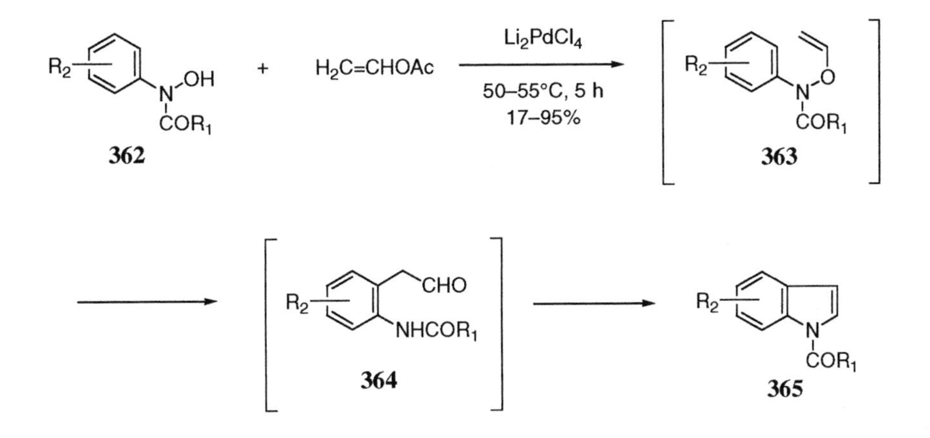


Boger and co-workers were the first to report the intramolecular amination of aryl halides in their synthesis of lavendamycin [436–438]. Thus, biaryl **366** is smoothly cyclized under the action of 1.5 equivalents of $Pd(Ph_3P)_4$ to β-carboline **367**, which comprises the CDE rings of lavendamycin.

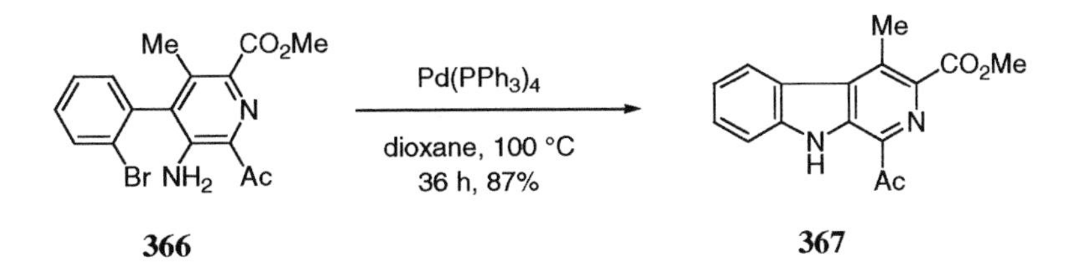


Buchwald parlayed the powerful Buchwald-Hartwig aryl amination technology [439–447] into a simple and versatile indoline synthesis [448–452]. For example, indole **368**, which has been employed in total syntheses of the marine alkaloids makaluvamine C and damirones A and B, was readily forged *via* the Pd-mediated cyclization shown below [448]. This intramolecular amination is applicable to the synthesis of *N*-substituted optically active indolines [450], and *o*-bromobenzylic bromides can be utilized in this methodology, as illustrated for the preparation of **369** [451]. Furthermore, this Pd-catalyzed amination reaction has been applied to the synthesis of arylhydrazones, which are substrates for the Fischer indole synthesis [453, 454].

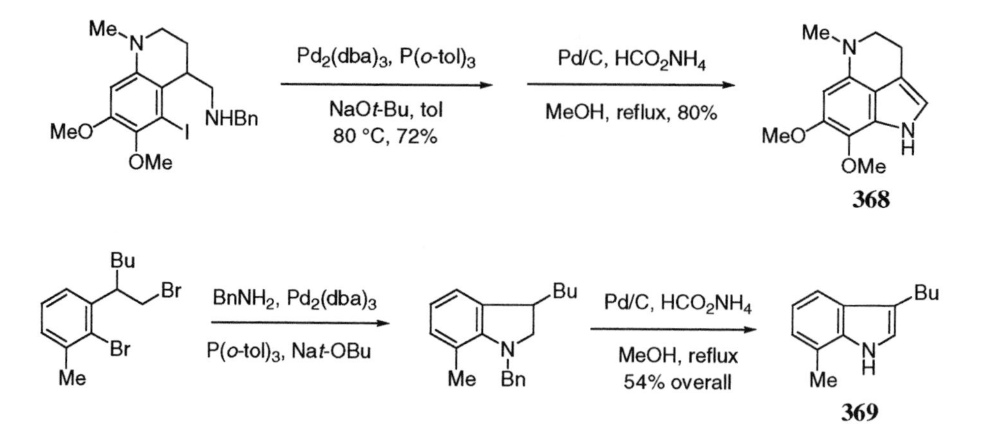


Snieckus and co-workers applied the Buchwald-Hartwig amination to the synthesis of *o*-carboxamido diarylamines, which can be elaborated to oxindoles [455]. Dobb synthesized α-carboline **370** *via* an intramolecular amination protocol [456]. These α-carbolines (pyrido[2,3-*b*]indoles) have been found to be modulators of the $GABA_A$ receptor, and this ring system is found in several natural products (grossularines, mescengricin). Snider achieved a similar cyclization of a 2-iodoindole leading to syntheses of (–)-asperlicin and (–)-asperlicin C as illustrated for the model reaction giving **371** [457]. The requisite 2-iodoindole was readily synthesized by a mercuration sequence ($Hg(OCOCF_3)_2$, $KI/I_2$/82%).

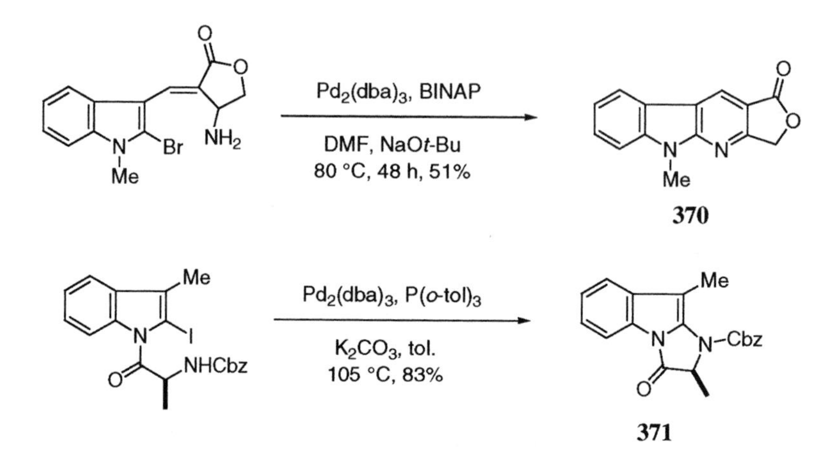


Several investigators have developed the reductive cyclization of *o*-nitrostyrenes into an efficient synthesis of indoles. Thus, research by the groups of Watanabe [458, 459], Söderberg [460, 461], and Cenini [462, 463] have established this reductive Pd-catalyzed *N*-heteroannulation reaction as a viable route to simple indoles and fused indoles (**372**) as shown below. Ohta described the related Pd-catalyzed cyclization of *o*-aminophenethyl alcohol to indole in 78% yield [464].

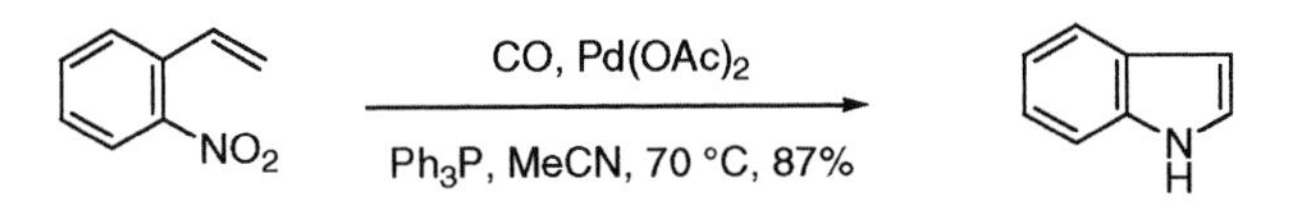


Also: 4-OH, 4-OMe, 5-OMe, 6-OMe, 4-Br, 4-$NO_2$, 4-$CO_2Me$, 5-$CO_2Me$, 6-$CO_2Me$, 7-$CO_2Me$, 2-Ph, 2-Me, and others

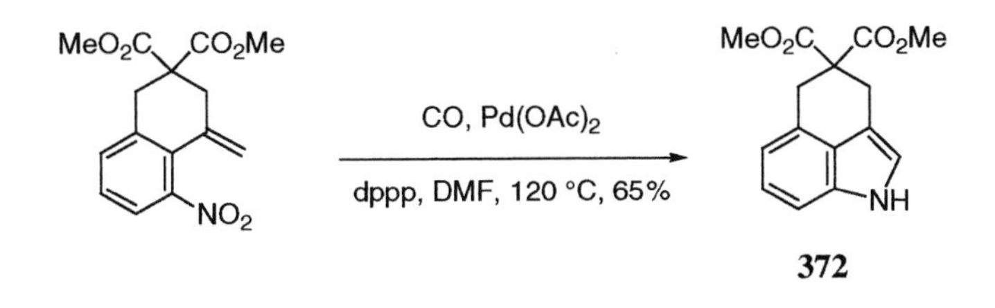


Izumi used $Pd(OAc)_2$ to effect the bis-alkoxylation of *o*-nitrostyrenes to form the corresponding *o*-nitrophenylacetaldehyde acetals, which, upon treatment with Fe/HOAc/HCl, give a variety of indoles (4-OMe, 5-OMe, 4-Me, 4-Cl, 4-$CO_2Me$, 6-Me) in 63–86% yields [465]. Mori and co-workers employed a titanium-isocyanate complex to construct tetrahydroindolones **373** from cyclohexane-1,3-diones [466].

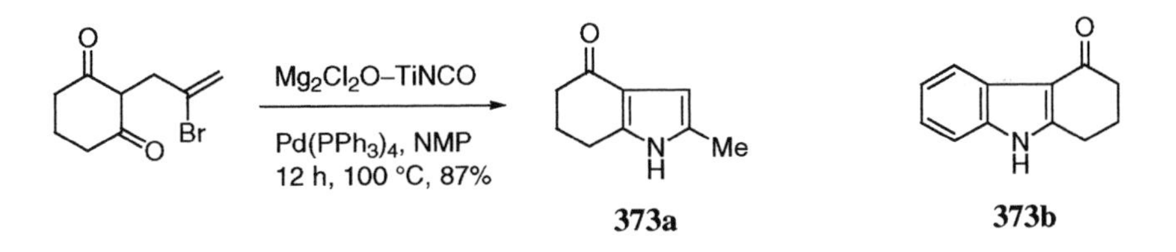


Mérour studied the reaction of indole triflates with diamines to afford pyrazino[2,3-*b*]indoles **374** and indolo[2,3-*b*]quinoxalines [467]. In the absence of palladium the yield of **374** is only 31% after 15 h. In some cases spiroindoxyls are formed.

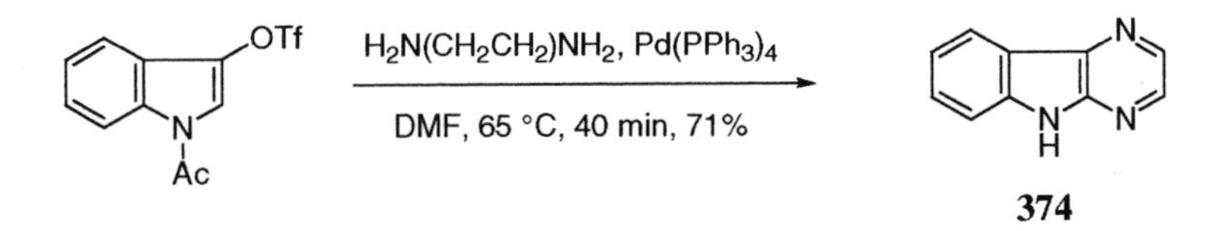

As we have seen earlier in this chapter, palladium is often employed to effect *N*-alkylation of indoles. Trost and Molander found that indole reacts with vinyl epoxide **375** to give indole **376** [468]. The utility of such *N*-alkylations remains to be established.

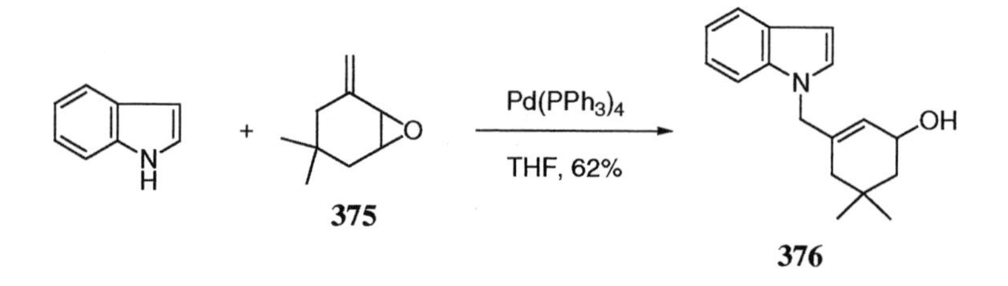


The Buchwald-Hartwig aryl amination methodology cited above in this section was engaged by Hartwig and others to synthesize *N*-arylindoles **377** [469]. Carbazole can be *N*-arylated under these same conditions with *p*-cyanobromobenzene (97% yield). Aryl chlorides also function in this reaction. The power of this amination method is seen by the facile synthesis of tris-carbazole **378** [469c].

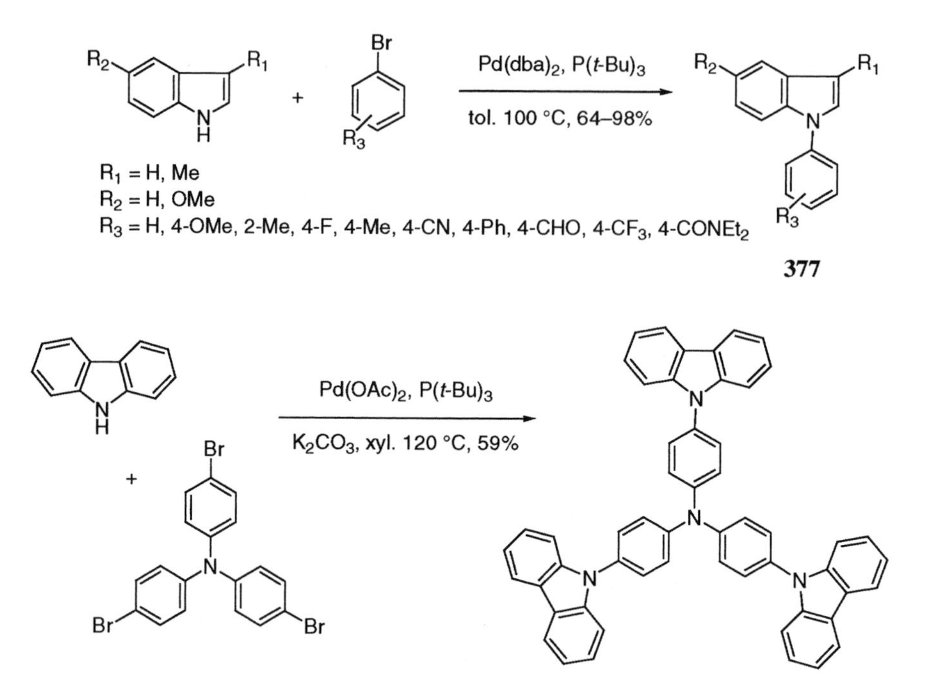

Although examples are sparse, a Pd-catalyzed carbon-sulfur bond formation leading to **379** was the penultimate reaction in a synthesis of (±)-chuangxinmycin [470, 194]. Earlier, Widdowson described the thiolation of 3-acetyl-4-iodoindole ($MeO_2CCH_2SsnMe_3$/$Pd(Ph_3P)_4$/83%) [471].

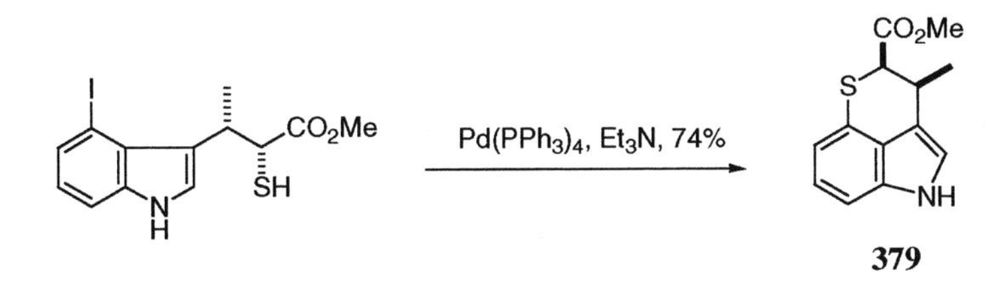


## 3.8 Miscellaneous

One of the more difficult transformations in organic synthesis is the deoxygenation of phenols to arenes. However, the use of palladium offers an attractive solution to this problem. The phenolic group in naltrindole can be deoxygenated *via* the derived triflate by treatment with formic acid and $PdCl_2(Ph_3P)_2$ good yield [144]. Triflate **380** is smoothly deoxygenated and further transformed to koumidine (**381**) under similar conditions by Sakai [472]. Noteworthy is the Pd-catalyzed isomerization of the ethylidene side chain. Sakai described a similar deoxygenation in the koumine series [473].

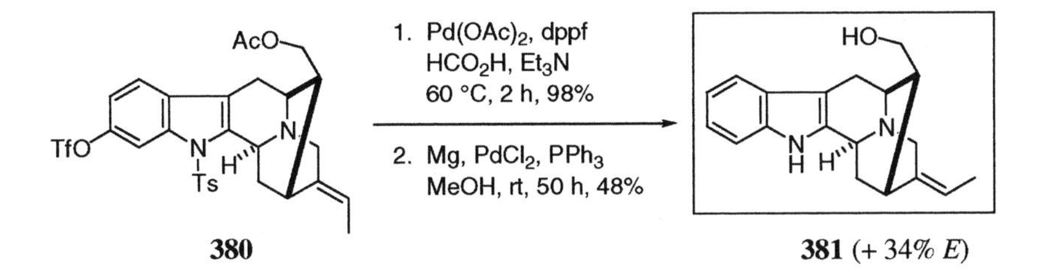


Halogen-containing indoles can be subjected to other nucleophilic substitution reactions under the guidance of palladium. Thus, 1-bromonauclefine is smoothly debrominated to nauclefine with $Pd(Ph_3P)_4$ (96%) [202], and 5-bromoindole is converted to 5-cyanoindole with NaCN and $Pd(Ph_3P)_4$ [474]. A similar reaction using potassium cyanide and the triflate of 1-hydroxy-β-carboline gives 1-cyano-β-carboline, which was transformed into the marine alkaloid eudistomin T upon reaction with benzyl Grignard reagent [475]. Palladium-catalysis has also been employed to deprotect *N*-allylindolines prepared by the Bailey-Liebeskind indoline synthesis [476–478] using the procedure of Genet [479]. Genet has also described the "ALLOC" deprotection of 4-formyl-*N*-(allyloxycarbonyl)indole using $Pd(OAc)_2$ and a water-soluble sulfonated $Ph_3P$ [480]. Takacs has employed indole as a trap for a Pd-mediated tetraene carbocyclization leading to **382** [481].

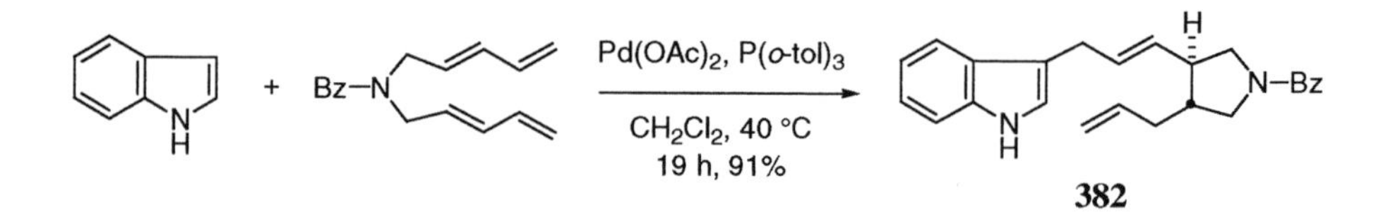


Fukuyama employed a $Pd(OAc)_2$ carbamate cleavage to trigger the final cyclization to (±)-catharanthine [482]. Godleski utilized palladium to close the D-ring in the key step of a synthesis of alloyohimbone [483]. In a search for novel rigid tryptamines, Vangveravong and Nichols used palladium acetate to catalyze the cyclopropanation of a 3-vinylindole chiral sultam leading to asymmetric syntheses of both enantiomers of *trans*-2-(indol-3-yl)cyclopropylamines and the corresponding cyclopropanecarboxylic acids [484]. Matsumoto and co-workers cyclized diazoindoles **383** to **384** with $Pd(OAc)_2$ in a system where $Rh_2(OAc)_4$ gives only benzene-ring cyclized products [485].

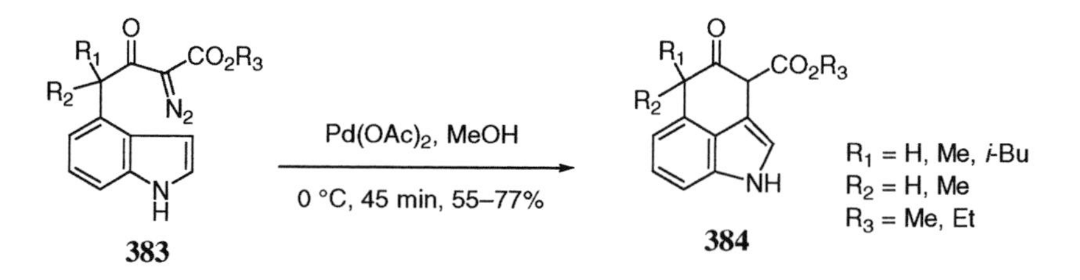


Although the yields are quite variable (6–100%), phenyl-1-azirines and allyl-1-azirines can be ring-enlarged to indoles under the action of Pd [486, 487]. In the example shown, **385** to **386**, the intermediate Pd-complex **387** was isolated [486]. The reaction of 2-phenylazirine with $Pd(Ph_3P)_4$ ($CO/BnNEt_3Cl/NaOH$) gives 2-styrylindole in 29% yield [487]. Interestingly, an atmosphere of $N_2$ is detrimental to the success of this reaction and CO is normally used.

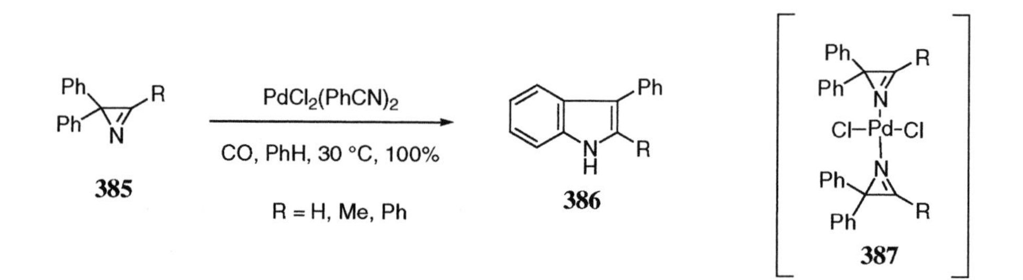

Yang described the Pd-induced cyclization of an aryl bromide onto a pendant cyano group leading to γ-carbolines and related compounds [488]. Genet studied the use of chiral palladium complexes in the construction of the C-ring of ergot alkaloids, a study that culminated in a synthesis of (–)-chanoclavine I [489–491]. For example, nitroindole **388** is cyclized to **389** in 57% yield and with enantioselectivities of up to 95% using $Pd(OAc)_2$ and (*S*)-(–)-BINAP.

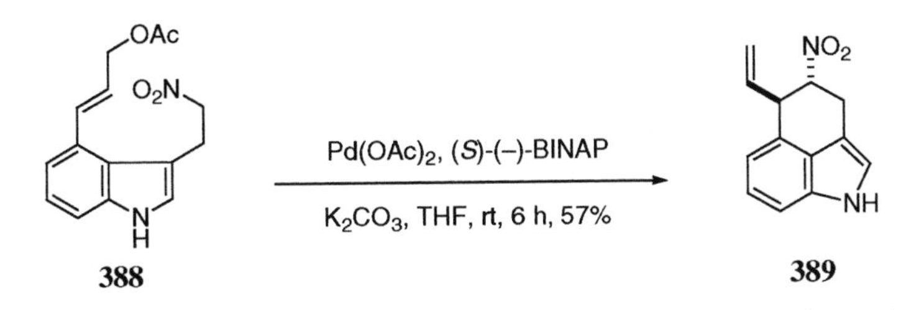


A fitting way to end this chapter is with Sakai's biomimetic syntheses of 11-methoxykoumine and koumine [492, 473]. Thus, the presumed biogenetic intermediate 18-hydroxy-taberpsychine (**390**), which was synthesized from 18-hydroxygardnutine, was acetylated and transformed into koumine (**391**) [473].

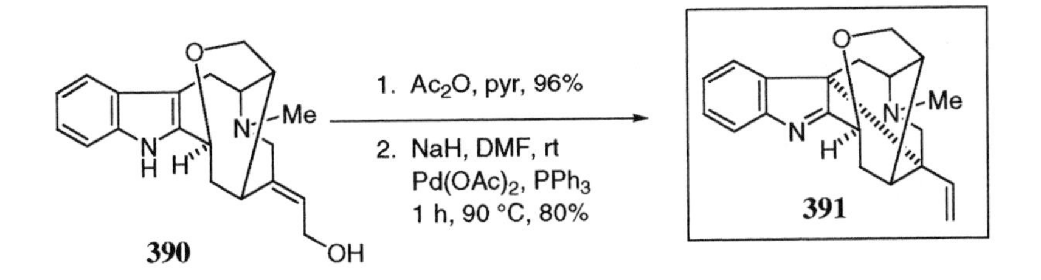


Sakai's elegant application of Pd-induced nucleophilic reactions of allylic acetates provides the first experimental support for the biogenesis of the koumine alkaloid skeleton and is an excellent concluding illustration of the power of palladium in indole chemistry.

* * *

In conclusion, the fantastically diverse chemistry of indole has been significantly enriched by palladium-catalyzed reactions. The accessibility of all of the possible halogenated indoles and several indolyl triflates has resulted in a wealth of synthetic applications as witnessed by the length of this chapter. In addition to the standard Pd-catalyzed reactions such as Negishi, Suzuki, Heck, Stille and Sonogashira, which have had great success in indole chemistry, oxidative coupling and cyclization are powerful routes to a variety of carbazoles, carbolines, indolocarbazoles, and other fused indoles.

## 3.9 References


1. For a review of haloindoles, see Powers, J. C. In "The Chemistry of Heterocyclic Compounds"; Houlihan, W. J., Ed.; J. Wiley and Sons: New York, 1972; Vol. 25, Part 2, p. 128.
2. For a leading reference to the early literature, see Piers, K.; Meimaroglou, C.; Jardine, R. V.; Brown, R. K. *Can. J. Chem.* **1963**, *41*, 2399–401.
3. Arnold, R. D.; Nutter, W. M.; Stepp, W. L. *J. Org. Chem.* **1959**, *24*, 117–8.
4. Bocchi, V.; Palla, G. *Synthesis* **1982**, 1096–7.
5. Brennan, M. R.; Erickson, K. L.; Szmalc, F. S.; Tansey, M. J.; Thornton, J. M. *Heterocycles* **1986**, *24*, 2879–85.
6. Powers, J. C. *J. Org. Chem.* **1966**, *31*, 2627–31.
7. Katritzky, A. R.; Akutagawa, K. *Tetrahedron Lett.* **1985**, *26*, 5935–8.
8. Bergman, J.; Venemalm, L. *J. Org. Chem.* **1992**, *57*, 2495–7.
9. Sundberg, R. J.; Russell, H. F. *J. Org. Chem.* **1973**, *38*, 3324–30.
10. Saulnier, M. G.; Gribble, G. W. *J. Org. Chem.* **1982**, *47*, 757–61.
11. Illi, V. O. *Synthesis* **1979**, 136.
12. Conway, S. C.; Gribble, G. W. *Heterocycles* **1990**, *30*, 627–33.
13. Conway, S. C.; Gribble, G. W. *Heterocycles* **1992**, *34*, 2095–108.
14. Gribble, G. W.; Allison, B. D.; Conway, S. C.; Saulnier, M. G. *Org. Prep. Proc. Int.* **1992**, *24*, 649–54.
15. Bergman, J.; Eklund, N. *Tetrahedron* **1980**, *36*, 1439–43.
16. Kline, T. *J. Heterocycl. Chem.* **1985**, *22*, 505–9.
17. Hodson, H. F.; Madge, D. J.; Widdowson, D. A. *Synlett* **1992**, 831–2.
18. Hodson, H. F.; Madge, D. J.; Slawin, A. N. Z.; Widdowson, D. A.; Williams, D. J. *Tetrahedron* **1994**, *50*, 1899–906.
19. Chi, S. M.; Choi, J.-K.; Yum, E. K.; Chi, D. Y. *Tetrahedron Lett.* **2000**, *41*, 919–22.
20. Harrington, P. J.; Hegedus, L. S. *J. Org. Chem.* **1984**, *49*, 2657–62.
21. Hegedus, L. S.; Sestrick, M. R.; Michaelson, E. T.; Harrington, P. J. *J. Org. Chem.* **1989**, *54*, 4141–6.
22. Sundberg, R. J. "The Chemistry of Indoles," Academic Press: New York, 1970.
23. Sundberg, R. J. "Indoles," Academic Press: New York, 1996.
24. For reviews of indole ring synthesis, see (a) Gribble, G. W. *Cont. Org. Syn.* **1994**, *1*, 145–172. (b) Gribble, G. W. *J. Chem. Soc., Perkin Trans. 1* **2000**, 1045–75.
25. Murase, M.; Watanabe, K.; Kurihara, T.; Tobinaga, S. *Chem. Pharm. Bull.* **1998**, *46*, 889–92.
26. Somei, M.; Tsuchiya, M. *Chem. Pharm. Bull.* **1981**, *29*, 3145–57.

27. (a) Thesing, J.; Semler, G.; Mohr, G. *Chem. Ber.* **1962**, *95*, 2205–11. (b) Russell, H. F.; Harris, B. J.; Hood, D. B.; Thompson, E. G.; Watkins, A. D.; Williams, R. D. *Org. Prep. Proc. Int.* **1985**, *17*, 391–9.
28. Miyake, Y.; Kikugawa, Y. *J. Heterocycl. Chem.* **1983**, *20*, 349–52.
29. (a) Carrera, Jr., G. M.; Sheppard, G. S. *Synlett* **1994**, 93–4. (b) Moyer, M. P.; Shiurba, J. F.; Rapoport, H. *J. Org. Chem.* **1986**, *51*, 5106–10.
30. Harrowven, D. C.; Lai, D.; Lucas, M. C. *Synthesis* **1999**, 1300–2.
31. Iwao, M.; Kuraishi, T. *Heterocycles* **1992**, *34*, 1031–8.
32. Meyers, A. I.; Hutchings, R. H. *Tetrahedron Lett.* **1993**, *34*, 6185–8.
33. Hutchings, R. H.; Meyers, A. I. *J. Org. Chem.* **1996**, *61*, 1004–13.
34. Kuehne, M. E.; Hall, T. C. *J. Org. Chem.* **1976**, *41*, 2742–6.
35. Chen, S.; Vasquez, L.; Noll, B. C.; Rakowski DuBois, M. *Organometallics* **1997**, *16*, 1757–64.
36. Sakamoto, T.; Yasuhara, A.; Kondo, Y.; Yamanaka, H. *Heterocycles* **1993**, *36*, 2597–600.
37. Beswick, P. J.; Greenwood, C. S.; Mowlem, T. J.; Nechvatal, G.; Widdowson, D. A. *Tetrahedron* **1988**, *44*, 7325–34.
38. Hollins, R. A.; Colnago, L. A.; Salim, V. M.; Seidl, M. C. *J. Heterocycl. Chem.* **1979**, *16*, 993–6.
39. Somei, M.; Yamada, F.; Kunimoto, M.; Kaneko, C. *Heterocycles* **1984**, *22*, 797–801.
40. Somei, M.; Saida, Y. *Heterocycles* **1985**, *23*, 3113–4.
41. Somei, M.; Saida, Y.; Funamoto, T.; Ohta, T. *Chem. Pharm. Bull.* **1987**, *35*, 3146–54.
42. Ritter, K. *Synthesis* **1993**, 735–62.
43. Conway, S. C.; Gribble, G. W. *Synth. Commun.* **1992**, *22*, 2987–95.
44. Bourlot, A. S.; Desarbre, E.; Mérour, J. Y. *Synthesis* **1994**, 411–6.
45. *Åkermark*, B.; Eberson, L.; Jonsson, E.; Pettersson, E. *J. Org. Chem.* **1975**, *40*, 1365–7.
46. *Åkermark*, B.; Oslob, J. D.; Heuschert, U. *Tetrahedron Lett.* **1995**, *36*, 1325–6.
47. (a) Furukawa, H.; Yogo, M.; Ito, C.; Wu, T.-S.; Kuoh, C.-S. *Chem. Pharm. Bull.* **1985**, *33*, 1320–2. (b) Furukawa, H.; Ito, C.; Yogo, M.; Wu, T. *Chem. Pharm. Bull.* **1986**, *34*, 2672–5. (c) Yogo, M.; Ito, C.; Furukawa, H. *Chem. Pharm. Bull.* **1991**, *39*, 328–34. (d) Bittner, S.; Krief, P.; Massil, T. *Synthesis* **1991**, 215–6.
48. (a) Miller, R. B.; Moock, T. *Tetrahedron Lett.* **1980**, *21*, 3319–22. (b) Oliveira-Campos, A. M. F.; Queiroz, M.-J. R. P.; Rapso, M. M. M.; Shannon, P. V. R. *Tetrahedron Lett.* **1995**, *36*, 133–4.
49. Morel, S.; Boyer, G.; Coullet, F.; Galy, J.-P. *Synth. Commun.* **1996**, *26*, 2443–7.
50. Mandal, A. B.; Delgado, F.; Tamariz, J. *Synlett* **1998**, 87–9.
51. Knölker, H.-J.; O'Sullivan, N. *Tetrahedron Lett.* **1994**, *35*, 1695–8.
52. Knölker, H.-J.; O'Sullivan, N. *Tetrahedron* **1994**, *50*, 10893–908

53. Knölker, H.-J.; Fröhner, W. *J. Chem. Soc., Perkin Trans. 1* **1998**, 173–5.
54. Knölker, H.-J.; Reddy, K. R.; Wagner, A. *Tetrahedron Lett.* **1998**, *39*, 8267–70.
55. Knölker, H.-J.; Reddy, K. R. *Synlett* **1999**, 596–8.
56. Trost, B. M.; Genet, J. P. *J. Am. Chem. Soc.* **1976**, *98*, 8516–7.
57. Trost, B. M.; Godleski, S. A.; Genet, J. P. *J. Am. Chem. Soc.* **1978**, *100*, 3930–1.
58. Trost, B. M.; Godleski, S. A.; Belletire, J. L. *J. Org. Chem.* **1979**, *44*, 2052–4.
59. Trost, B. M.; Fortunak, J. M. D. *Organometallics* **1982**, *1*, 7–13.
60. Cushing, T. D.; Sanz-Cervera, J. F.; Williams, R. M. *J. Am. Chem. Soc.* **1996**, *118*, 557–79.
61. Itahara, T.; Sakakibara, T. *Synthesis* **1978**, 607–8.
62. Itahara, T. *Synthesis* **1979**, 151–2.
63. Itahara, T. *Heterocycles* **1986**, *24*, 2557–62.
64. Itahara, T. *J. Chem. Soc., Chem. Commun.* **1981**, 254–5.
65. Itahara, T. *J. Org. Chem.* **1985**, *50*, 5272–5.
66. Black, D. St C.; Keller, P. A.; Kumar, N. *Tetrahedron Lett.* **1989**, *30*, 5807–8.
67. Black, D. St C.; Keller, P. A.; Kumar, N. *Tetrahedron* **1993**, *49*, 151–64.
68. Black, D. St C.; Kumar, N.; Wong, L. C. H. *J. Chem. Soc., Chem. Commun.* **1985**, 1174–5.
69. Harris, W.; Hill, C. H.; Keech, E.; Malsher, P. *Tetrahedron Lett.* **1993**, *34*, 8361–4.
70. Gribble, G. W.; Berthel, S. J. *Stud. Nat. Prod. Chem.* Atta-ur-Rahman, Ed.; Elsevier: New York, 1993; Vol. 12, 365.
71. Ohkubo, M.; Nishimura, T.; Jona, H.; Honma, T.; Morishima, H. *Tetrahedron* **1996**, *52*, 8099–112.
72. Faul, M. M.; Winneroski, L. L.; Krumrich, C. A. *J. Org. Chem.* **1998**, *63*, 6053–8.
73. Jeevanandam, A.; Srinivasan, P. C. *Synth. Commun.* **1995**, *25*, 3427–34.
74. Itahara, T. *J. Org. Chem.* **1985**, *50*, 5546–50.
75. Billups, W. E.; Erkes, R. S.; Reed, L. E. *Synth. Commun.* **1980**, *10*, 147–54.
76. Fujiwara, Y.; Maruyama, O.; Yoshidomi, M.; Taniguchi, H. *J. Org. Chem.* **1981**, *46*, 851–5.
77. Itahara, T.; Ikeda, M.; Sakakibara, T. *J. Chem. Soc., Perkin Trans. 1* **1983**, 1361–3.
78. Itahara, T.; Kawasaki, K.; Ouseto, F. *Synthesis* **1984**, 236–7.
79. (a) Yokoyama, Y.; Matsumoto, T.; Murakami, Y. *J. Org. Chem.* **1995**, *60*, 1486–7. (b) Osanai, K.; Yokoyama, Y.; Kondo, K.; Murakami, Y. *Chem. Pharm. Bull.* **1999**, *47*, 1587–90.
80. Yokoyama, Y.; Takahashi, M.; Kohno, Y.; Kataoka, K.; Fujikawa, Y.; Murakami, Y. *Heterocycles* **1990**, *31*, 803–4.
81. Yokoyama, Y.; Takahashi, M.; Takashima, M.; Kohno, Y.; Kobayashi, H.; Kataoka, K.; Shidori, K.; Murakami, Y. *Chem. Pharm. Bull.* **1994**, *42*, 832–8.
82. Murakami, Y.; Yokoyama, Y.; Aoki, T. *Heterocycles* **1984**, *22*, 1493–6.

83. Yokoyama, Y.; Takashima, M.; Higaki, C.; Shidori, K.; Moriguchi, S.; Ando, C.; Murakami, Y. *Heterocycles* **1993**, *36*, 1739–42.
84. Yokoyama, Y.; Suzuki, H.; Matsumoto, S.; Sunaga, Y.; Tani, M.; Murakami, Y. *Chem. Pharm. Bull.* **1991**, *39*, 2830–6.
85. Pindur, U.; Adam, R. *Helv. Chim. Acta* **1990**, *73*, 827–38.
86. Jia, C.; Lu, W.; Kitamura, T.; Fujiwara, Y. *Org. Lett.* **1999**, *1*, 2097–100.
87. Abdrakhmanov, I. B.; Mustafin, A. G.; Tolstikov, G. A.; Fakhretdinov, R. N.; Dzhemilev, U. M. *Chem. Heterocycl. Cpds.* **1986**, 262–4.
88. Fukuda, Y.; Furuta, H.; Shiga, F.; Asahina, Y.; Terashima, S. *Heterocycles* **1997**, *45*, 2303–8.
89. Minato, A.; Tamao, K.; Hayashi, T.; Suzuki, K.; Kumada, M. *Tetrahedron Lett.* **1981**, *22*, 5319–22.
90. Minato, A.; Suzuki, K.; Tamao, K.; Kumada, M. *Tetrahedron Lett.* **1984**, *25*, 83–6.
91. Minato, A.; Suzuki, K.; Tamao, K.; Kumada, M. *J. Chem. Soc., Chem. Commun.* **1984**, 511–3.
92. Kondo, Y.; Yoshida, A.; Sato, S.; Sakamoto, T. *Heterocycles* **1996**, *42*, 105–8.
93. Widdowson, D. A.; Zhang, Y.-Z. *Tetrahedron* **1986**, *42*, 2111–6.
94. Vincent, P.; Beaucourt, J. P.; Pichart, L. *Tetrahedron Lett.* **1984**, *25*, 201–2.
95. Sakamoto, T.; Kondo, Y.; Takazawa, N.; Yamanaka, H. *Heterocycles* **1993**, *36*, 941–2.
96. Sakamoto, T.; Kondo, Y.; Takazawa, N.; Yamanaka, H. *Tetrahedron Lett.* **1993**, *34*, 5955–6.
97. Sakamoto, T.; Kondo, Y.; Takazawa, N.; Yamada, H. *J. Chem. Soc., Perkin Trans. 1* **1996**, 1927–34.
98. Kondo, Y.; Takazawa, N.; Yoshida, A.; Sakamoto, T. *J. Chem. Soc., Perkin Trans. 1* **1995**, 1207–8.
99. Amat, M.; Hadida, S.; Bosch, J. *Tetrahedron Lett.* **1993**, *34*, 5005–6.
100. Amat, M.; Hadida, S.; Bosch, J. *Tetrahedron Lett.* **1994**, *35*, 793–6.
101. Amat, M.; Hadida, S.; Pshenichnyi, G.; Bosch, J. *J. Org. Chem.* **1997**, *62*, 3158–75.
102. Danieli, B.; Lesma, G.; Martinelli, M.; Passarella, D.; Peretto, I.; Silvani, A. *Tetrahedron* **1998**, *54*, 14081–8.
103. Fisher, L. E.; Labadie, S. S.; Reuter, D. C.; Clark, R. D. *J. Org. Chem.* **1995**, *60*, 6224–5.
104. Cheng, K.-F.; Cheung, M.-K. *J. Chem. Soc., Perkin Trans. 1* **1996**, 1213–8.
105. Herz, H.-G.; Queiroz, M. J. R. P.; Maas, G. *Synthesis* **1999**, 1013–6.
106. Pimm, A.; Kocienski, P.; Street, S. D. A. *Synlett* **1992**, 886–8.
107. Faul, M. M.; Winneroski, L. L. *Tetrahedron Lett.* **1997**, *38*, 4749–52.
108. Bergman, J.; Venemalm, L. *Tetrahedron* **1990**, *46*, 6061–6.
109. Davidsen, S. K.; Summers, J. B.; Albert, D. H.; Holms, J. H.; Heyman, H. R.; Magoc, T. J.; Conway, R. G.; Rhein, D. A.; Carter, G. W. *J. Med. Chem.* **1994**, *26*, 4423–9.

110. Griffen, E. J.; Roe, D. G.; Snieckus, V. *J. Org. Chem.* **1995**, *60*, 1484–5.
111. Bracher, F.; Hildebrand, D.; *Tetrahedron* **1994**, *50*, 12329–36.
112. Burns, B.; Grigg, R.; Sridharan, V.; Stevenson, P.; Sukirthalingam, S.; Worakun, T. *Tetrahedron Lett.* **1989**, *30*, 1135–8.
113. Luo, F.-T.; Wang, R.-T. *Heterocycles* **1991**, *32*, 2365–72.
114. Claridge, T. D. W.; Long, J. M.; Brown, J. M.; Hibbs, D.; Hursthouse, M. B.; *Tetrahedron* **1997**, *53*, 4035–50.
115. Zheng, Q.; Yang, Y.; Martin, A. R. *Tetrahedron Lett.* **1993**, *34*, 2235–8.
116. Zheng, Q.; Yang, Y.; Martin A. R. *Heterocycles* **1994**, *37*, 1761–72.
117. Kawasaki, I.; Yamashita, M.; Ohta, S. *Chem. Pharm. Bull.* **1996**, *44*, 1831–9.
118. Hoerrner, R. S.; Askin, D.; Volante, R. P.; Reider, P. J. *Tetrahedron Lett.* **1998**, *39*, 3455–8.
119. Neel, D. A.; Jirousek, M. R.; McDonald III, J. H. *Bioorg. Med. Chem. Lett.* **1998**, *8*, 47–50.
120. Wallow, T. I., Novak, B. M. *J. Org. Chem.* **1994**, *59*, 5034–7.
121. Wright, S. W.; Hageman, D. L.; McClure, L. D. *J. Org. Chem.* **1994**, *59*, 6095–7.
122. Merlic, C. A.; McInnes, D. M.; You, Y. *Tetrahedron Lett.* **1997**, *38*, 6787–90.
123. Merlic, C. A.; McInnes, D. M. *Tetrahedron Lett.* **1997**, *38*, 7661–4.
124. Johnson, C. N.; Stemp, G.; Anand, N.; Stephen, S. C.; Gallagher, T. *Synlett* **1998**, 1025–7.
125. Yang, Y.; Martin, A. R.; Nelson, D. L.; Regan, J. *Heterocycles* **1992**, *34*, 1169–75.
126. Yang, Y.; Martin, A. R. *Heterocycles* **1992**, *34*, 1395–8.
127. Carbonnelle, A.-C.; González-Zamora, E.; Beugelmans, R.; Roussi, G. *Tetrahedron Lett.* **1998**, *39*, 4467–70.
128. Carbonnelle, A.-C.; González-Zamora, E.; Beugelmans, R.; Roussi, G. *Tetrahedron Lett.* **1998**, *39*, 4471–2.
129. Zembower, D. E.; Ames, M. M. *Synthesis* **1994**, 1433–6.
130. Siddiqui, M. A.; Snieckus, V. *Tetrahedron Lett.* **1990**, *31*, 1523–6.
131. Yang, Y.; Martin, A. R. *Synth. Commun.* **1992**, *22*, 1757–62.
132. Banwell, M. G.; Bissett, B. D.; Busato, S.; Cowden, C. J.; Hockless, D. C. R.; Holman, J. W.; Read, R. W.; Wu, A. W. *J. Chem. Soc., Chem. Commun.* **1995**, 2551–2.
133. Moody, C. J.; Doyle, K. J.; Elliott, M. C.; Mowlem, T. J. *J. Chem. Soc., Perkin Trans.1* **1997**, 2413–9.
134. Ishikura, M.; Kamada, M.; Terashima, M. *Synthesis* **1984**, 936–8.
135. Somei, M.; Amari, H.; Makita, Y. *Chem. Pharm. Bull.* **1986**, *34*, 3971–3.
136. Tidwell, J. H.; Peat, A. J.; Buchwald, S. L. *J. Org. Chem.* **1994**, *59*, 7164–8.
137. Chu, L.; Fisher, M. H.; Goulet, M. T.; Wyvratt, M. J. *Tetrahedron Lett.* **1997**, *38*, 3871–4.
138. Joseph, B.; Malapel, B.; Mérour, J.-Y. *Synth. Commun.* **1996**, *26*, 3289–95.

139. Malapel-Andrieu, B.; Mérour, J.-Y. *Tetrahedron* **1998**, *54*, 11079–94.
140. Doi, K.; Mori, M. *Heterocycles* **1996**, *42*, 113–6.
141. Choshi, T.; Sada, T.; Fujimoto, H.; Nagayama, C.; Sugino, E.; Hibino, S. *Tetrahedron Lett.* **1996**, *37*, 2593–6.
142. (a) Choshi, T.; Sada, T.; Fujimoto, H.; Nagayama, C.; Sugino, E.; Hibino, S. *J. Org. Chem.* **1997**, *62*, 2535–43. (b) Choshi, T.; Kuwada, T.; Fukui, M.; Matsuya, Y.; Sugino, E.; Hibino, S. *Chem. Pharm. Bull.* **2000**, *48*, 108–13.
143. McCort, G.; Duclos, O.; Cadilhac, C.; Guilpain, E. *Tetrahedron Lett.* **1999**, *40*, 6211–5.
144. Kubota, H.; Rothman, R. B.; Dersch, C.; McCullough, K.; Pinto, J.; Rice, K. C. *Bioorg. Med. Chem. Lett.* **1998**, *8*, 799–804.
145. Bracher, F.; Hildebrand, D. *Liebigs Ann. Chem.* **1992**, 1315–9.
146. Joseph, B.; Alagille, D.; Rousseau, C.; Mérour, J.-Y. *Tetrahedron* **1999**, *55*, 4341–52.
147. (a) Ishikura, M.; Terashima, M. *J. Chem. Soc., Chem. Commun.* **1989**, 135–6. (b) For a review, see Ishikura, M.; Agata, I. *Recent Res. Devel. Org. Chem.* **1997**, *1*, 145–57.
148. Ishikura, M.; Terashima, M.; Okamura, K.; Date, T. *J. Chem. Soc., Chem. Commun.* **1991**, 1219–21.
149. Ishikura, M.; Terashima, M. *Tetrahedron Lett.* **1992**, *33*, 6849–52.
150. Ishikura, M.; Terashima, M. *J. Org. Chem.* **1994**, *59*, 2634–7.
151. Ishikura, M. *J. Chem. Soc., Chem. Commun.* **1995**, 409–10.
152. Ishikura, M.; Agata, I. *Heterocycles* **1996**, *43*, 1591–5.
153. Ishikura, M.; Matsuzaki, Y.; Agata, I. *Chem. Commun.* **1996**, 2409–10.
154. (a) Ishikura, M.; Yaginuma, T.; Agata, I.; Miwa, Y. Yanada, R.; Taga, T. *Synlett* **1997**, 214–6. (b) Ishikura, M.; Hino, A.; Yaginuma, T.; Agata, I.; Katagiri, N. *Tetrahedron* **2000**, *56*, 193–207.
155. Ishikura, M.; Matsuzaki, Y.; Agata, I.; Katagiri, N. *Tetrahedron* **1998**, *54*, 13929–42.
156. Ishikura, M.; Agata, I.; Katagiri, N. *J. Heterocycl. Chem.* **1999**, *36*, 873–9.
157. Ishikura, M.; Matsuzaki, Y.; Agata, I. *Heterocycles* **1997**, *45*, 2309–12.
158. Satoh, M.; Miyaura, N.; Suzuki, A. *Synthesis* **1987**, 373–7.
159. Kumar, V.; Dority, J. A.; Bacon, E. R.; Singh, B.; Lesher, G. Y. *J. Org. Chem.* **1992**, *57*, 6995–8.
160. Gridnev, I. D.; Miyaura, N.; Suzuki, A. *J. Org. Chem.* **1993**, *58*, 5351–4.
161. Sun, L.; Tran, N.; Liang, C.; Tang, F.; Rice, A.; Schreck, R.; Waltz, K.; Shawver, L. K.; McMahon, G.; Tang, C. *J. Med. Chem.* **1999**, *42*, 5120–30.
162. Todd, M. H.; Oliver, S. F.; Abell, C. *Org. Lett.* **1999**, *1*, 1149–51.
163. (a) Rocca, P.; Marsais, F.; Godard, A.; Quéguiner, G.; Adams, L.; Alo, B. *J. Heterocycl. Chem.* **1995**, *32*, 1171–5. (b) Trécourt, F.; Mongin, F.; Mallet, M.; Quéguiner, G. *Synth. Commun.* **1995**, *25*, 4011–24. (c) Arzel, E.; Rocca, P.; Marsais, F.; Godard, A.; Quéguiner, G. *J. Heterocycl. Chem.* **1997**, *34*, 1205–10.

164. Grigg, R.; Sansano, J. M.; Santhakumar, V.; Sridharan, V.; Thangavelanthum, R.; Thornton-Pett, M.; Wilson, D. *Tetrahedron* **1997**, *53*, 11803–26.
165. Amat, M.; Hadida, S.; Sathyanarayana, S.; Bosch, J. *J. Org. Chem.* **1994**, *59*, 10–1.
166. Beak, P.; Lee, W. K. *J. Org. Chem.* **1993**, *58*, 1109–17.
167. Labadie, S. S.; Teng, E. *J. Org. Chem.* **1994**, *59*, 4250–4.
168. Caddick, S.; Joshi, S. *Synlett* **1992**, 805–6.
169. Arnswald, M.; Neumann, W. P. *J. Org. Chem.* **1993**, *58*, 7022–8.
170. Palmisano, G.; Santagostino, M. *Helv. Chim. Acta* **1993**, *76*, 2356–66.
171. Palmisano, G.; Santagostino, M. *Synlett* **1993**, 771–3.
172. Hudkins, R. L.; Diebold, J. L.; Marsh, F. D. *J. Org. Chem.* **1995**, *60*, 6218–20.
173. Brenner, M.; Mayer, G.; Terpin, A.; Steglich, W. *Chem. Eur. J.* **1997**, *3*, 70–4.
174. Murray, P. E.; Mills, K.; Joule, J. A. *J. Chem. Res. (S)* **1998**, 377.
175. Fukuyama, T.; Chen, X.; Peng, G. *J. Am. Chem. Soc.* **1994**, *116*, 3127–8.
176. (a) Kobayashi, Y.; Fukuyama, T. *J. Heterocycl. Chem.* **1998**, *35*, 1043–55. (b) Kobayashi, Y.; Peng, G.; Fukuyama, T. *Tetrahedron Lett.* **1999**, *40*, 1519–22.
177. Kraxner, J.; Arlt, M.; Gmeiner, P. *Synlett* **2000**, 125–7.
178. Ciattini, P. G.; Morera, E.; Ortar, G. *Tetrahedron Lett.* **1994**, *35*, 2405–8.
179. (a) Yokoyama, Y.; Ito, S.; Takahashi, Y.; Murakama, Y. *Tetrahedron Lett.* **1985**, *26*, 6457–60. (b) Yokoyama, Y.; Ikeda, M.; Saito, M.; Yoda, T.; Suzuki, H.; Murakami, Y. *Heterocycles* **1990**, *31*, 1505–11.
180. Yoshida, S.; Kubo, H.; Saika, T.; Katsumura, S. *Chem. Lett.* **1996**, 139–40.
181. Somei, M.; Sayama, S.; Naka, K.; Yamada, F. *Heterocycles* **1988**, *27*, 1585–7.
182. Palmisano, G.; Santagostino, M. *Tetrahedron Lett.* **1993**, *49*, 2533–42.
183. Gribble, G. W.; Conway, S. C. *Synth. Commun.* **1992**, *22*, 2129–41.
184. Sakamoto, T.; Kondo, Y.; Yasuhara, A.; Yamanaka, H. *Heterocycles* **1990**, *31*, 219–21.
185. Sakamoto, T.; Kondo, Y.; Yasuhara, A.; Yamanaka, H. *Tetrahedron* **1991**, *47*, 1877–86.
186. Sakamoto, T.; Yasuhara, A.; Kondo, Y.; Yamanaka, H. *Chem. Pharm. Bull.* **1994**, *42*, 2032–5.
187. Vaillancourt, V.; Albizati, K. F. *J. Am. Chem. Soc.* **1993**, *115*, 3499–02.
188. Choshi, T.; Yamada, S.; Sugino, E.; Kuwada, T.; Hibino, S. *Synlett* **1995**, 147–8.
189. Choshi, T.; Yamada, S.; Sugino, E.; Kuwada, T.; Hibino, S. *J. Org. Chem.* **1995**, *60*, 5899–04.
190. Choshi, Y.; Yamada, S.; Nobuhiro, J.; Mihara, Y.; Sugino, E.; Hibino, S. *Heterocycles* **1998**, *48*, 11–4.
191. Achab, S.; Guyot, M.; Potier, P. *Tetrahedron Lett.* **1995**, *36*, 2615–8.
192. Choshi, T.; Fujimoto, H.; Sugino, E.; Hibino, S. *Heterocycles* **1996**, *43*, 1847–54.

193. Krolski, M. E.; Renaldo, A. F.; Rudisill, D. E.; Stille, J. K. *J. Org. Chem.* **1988**, *53*, 1170–6.
194. Dickens, M. J.; Mowlem, T. J.; Widdowson, D. A.; Slawin, A. M. Z.; Williams, D. J. *J. Chem. Soc., Perkin Trans. 1* **1992**, 323–5.
195. Somei, M.; Yamada, F.; Naka, K. *Chem. Pharm. Bull.* **1987**, *35*, 1322–5.
196. (a) Pearce, B. C. *Synth. Commun.* **1992**, *22*, 1627–43. (b) Benhida, R.; Lecubin, F.; Fourrey, J.-L.; Rivas Castellanos, L.; Quintero, L. *Tetrahedron Lett.* **1999**, *40*, 5701–3.
197. Miki, Y.; Matsushita, K.; Hibino, H.; Shirokoshi, H. *Heterocycles* **1999**, *51*, 1585–91.
198. Grigg, R.; Teasdale, A.; Sridharan, V. *Tetrahedron Lett.* **1991**, *32*, 3859–62.
199. Somei, M.; Kawasaki, T.; Ohta, T. *Heterocycles* **1988**, *27*, 2363–5.
200. Iwao, M.; Takehara, H.; Obata, S. Watanabe, M. *Heterocycles* **1994**, *38*, 1717–20.
201. Bracher, F.; Hildebrand, D. *Liebigs Ann. Chem.* **1993**, 837–9.
202. Lavilla, R.; Gullón, F.; Bosch, J. *J. Chem. Soc., Chem. Commun.* **1995**, 1675–6.
203. Rudisill, D. E.; Stille, J. K. *J. Org. Chem.* **1989**, *54*, 5856–66.
204. Sakamoto, T.; Satoh, C.; Kondo, Y.; Yamanaka, H. *Heterocycles* **1992**, *34*, 2379–84.
205. Iwao, M.; Takehara, H.; Furukawa, S.; Watanabe, M. *Heterocycles* **1993**, *36*, 1483–8.
206. Grigg, R.; Sridharan, V.; Zhang, J. *Tetrahedron Lett.* **1999**, *40*, 8277–80.
207. Sonogashira, K.; Tohda, Y.; Hagihara, N. *Tetrahedron Lett.* **1975**, 4467–70.
208. (a) Castro, C. E.; Stephens, R. D. *J. Org. Chem.* **1963**, *28*, 2163. (b) Stephens, R. D.; Castro, C. E. *J. Org. Chem.* **1963**, *28*, 3313–5. (c) Castro, C. E.; Gaughan, E. J.; Owsley, D. C. *J. Org. Chem.* **1966**, *31*, 4071–8. (d) Castro, C. E.; Havlin, R.; Honwad, V. K.; Malte, A.; Mojé, S. *J. Am. Chem. Soc.* **1969**, *91*, 6464–70.
209. Sakamoto, T.; Kondo, Y.; Yamanaka, H. *Heterocycles* **1984**, *22*, 1347–50.
210. Sakamoto, T.; Kondo, Y.; Yamanaka, H. *Chem. Pharm. Bull.* **1986**, *34*, 2362–8.
211. For a review of the use of palladium catalysis in heterocycle synthesis, with a good summary of the authors' work, see Sakamoto, T.; Kondo, Y.; Yamanaka, H. *Heterocycles* **1988**, *27*, 2225–49.
212. Taylor, E. C.; Katz, A. H.; Salgado-Zamora, H.; McKillop, A. *Tetrahedron Lett.* **1985**, *26*, 5963–6.
213. Tischler, A. N.; Lanza, T. J. *Tetrahedron Lett.* **1986**, *27*, 1653–6.
214. Sakamoto, T.; Kondo, Y.; Yamanaka, H. *Heterocycles* **1986**, *24*, 31–2.
215. (a) Sakamoto, T.; Kondo, Y.; Iwashita, S.; Yamanaka, H. *Chem. Pharm. Bull.* **1987**, *35*, 1823–8. (b) Sakamoto, T.; Kondo, Y.; Iwashita, S.; Nagano, T.; Yamanaka, H. *Chem. Pharm. Bull.* **1988**, *36*, 1305–8.
216. (a) Sakamoto, T.; Nagano, T.; Kondo, Y.; Yamanaka, H. *Chem. Pharm. Bull.* **1988**, *36*, 2248–52. (b) Sakamoto, T.; Numata, A.; Saitoh, H.; Kondo, Y. *Chem. Pharm. Bull.* **1999**, *47*, 1740–3.
217. Zhu, Z.; Moore, J. S. *J. Org. Chem.* **2000**, *65*, 116–23.
218. Arcadi, A.; Cacchi, S.; Marinelli, F. *Tetrahedron Lett.* **1989**, *30*, 2581–4.

219. Zhong, W.; Gallivan, J. P.; Zhang, Y.; Li, L.; Lester, H. A.; Dougherty, D. A. *Proc. Natl. Acad. Sci. USA* **1998**, *95*, 12088–93.
220. Kondo, Y.; Kojima, S.; Sakamoto, T. *J. Org. Chem.* **1997**, *62*, 6507–11.
221. Yu, M. S.; Lopez de Leon, L.; McGuire, M. A.; Botha, G. *Tetrahedron Lett.* **1998**, *39*, 9347–50.
222. Xu, L.; Lewis, I. R.; Davidsen, S. K.; Summers, J. B. *Tetrahedron Lett.* **1998**, *39*, 5159–62.
223. Shin, K.; Ogasawara, K. *Synlett* **1995**, 859–60.
224. Shin, K.; Ogasawara, K. *Chem. Lett.* **1995**, 289–90.
225. Shin, K.; Ogasawara, K. *Synlett* **1996**, 922–3.
226. Ezquerra, J.; Pedregal, C.; Lamas, C.; Barluenga, J.; Perez, M.; García-Martín, M. A.; González, J. M. *J. Org. Chem.* **1996**, *61*, 5804–12.
227. Soloducho, J. *Tetrahedron Lett.* **1999**, *40*, 2429–30.
228. Blurton, P.; Brickwood, A.; Dhanak, D. *Heterocycles* **1997**, *45*, 2395–403.
229. Botta, M.; Summa, V.; Corelli, F.; Di Pietro, G.; Lombardi, P. *Tetrahedron: Asymmetry* **1996**, *7*, 1263–6.
230. Kondo, Y.; Kojima, S.; Sakamoto, T. *Heterocycles* **1996**, *43*, 2741–6.
231. Yasuhara, A.; Kanamori, Y.; Kaneko, M.; Numata, A.; Kondo, Y.; Sakamoto, T. *J. Chem. Soc., Perkin Trans. 1* **1999**, 529–34.
232. Amatore, C.; Blart, E.; Genet, J. P.; Jutand, A.; Lemaire-Audoire, S.; Savignac, M. *J. Org. Chem.* **1995**, *60*, 6829–39.
233. Fagnola, M. C.; Candiani, I.; Visentin, G.; Cabri, W.; Zarini, F.; Mongelli, N.; Bedeschi, A. *Tetrahedron Lett.* **1997**, *38*, 2307–10.
234. Collini, M. D.; Ellingboe, J. W. *Tetrahedron Lett.* **1997**, *38*, 7963–6.
235. Zhang, H.-C.; Brumfield, K. K.; Jaroskova, L.; Maryanoff, B. E. *Tetrahedron Lett.* **1998**, *39*, 4449–52.
236. Zhang, H.-C.; Ye, H.; Moretto, A. F.; Brumfield, K. K.; Maryanoff, B. E. *Org. Lett.* **2000**, *2*, 89–92.
237. Rainier, J. D.; Kennedy, A. R.; Chase, E. *Tetrahedron Lett.* **1999**, *40*, 6325–7.
238. Galambos, G.; Szantay Jr., C.; Tamás, J.; Szántay, C. *Heterocycles* **1993**, *36*, 2241–5.
239. Frank, W. C.; Kim, Y. C.; Heck, R. F. *J. Org. Chem.* **1978**, *43*, 2947–9.
240. Yamada, F.; Makita, Y.; Suzuki, T.; Somei, M. *Chem. Pharm. Bull.* **1985**, *33*, 2162–3.
241. Somei, M.; Hasegawa, T.; Kaneko, C. *Heterocycles* **1983**, *20*, 1983–5.
242. Somei, M.; Yamada, F. *Chem. Pharm. Bull.* **1984**, *32*, 5064–5.
243. Yamada, F.; Hasegawa, T.; Wakita, M.; Sugiyama, M.; Somei, M. *Heterocycles* **1986**, *24*, 1223–6.
244. Somei, M.; Ohnishi, H.; Shoken, Y. *Chem. Pharm. Bull.* **1986**, *34*, 677–81.
245. Somei, M.; Makita, Y.; Yamada, F. *Chem. Pharm. Bull.* **1986**, *34*, 948–9.
246. Somei, M.; Saida, Y.; Komura, N. *Chem. Pharm. Bull.* **1986**, *34*, 4116–25.

247. Somei, M.; Funamoto, T.; Ohta, T. *Heterocycles* **1987**, *26*, 1783–4.
248. Somei, M.; Yamada, F.; Ohnishi, H.; Makita, Y.; Kuriki, M. *Heterocycles* **1987**, *26*, 2823–8.
249. Ohta, T.; Shinoda, J.; Somei, M. *Chem. Lett.* **1993**, 797–8.
250. Harrington, P. J.; Hegedus, L. S. *J. Org. Chem.* **1984**, *49*, 2657–62.
251. Harrington, P. J.; Hegedus, L. S.; McDaniel, K. F. *J. Am. Chem. Soc.* **1987**, *109*, 4335–8.
252. Hegedus, L. S.; Toro, J. L.; Miles, W. H.; Harrington, P. J. *J. Org. Chem.* **1987**, *52*, 3319–22.
253. Merlic, C. A.; Semmelhack, M. F. *J. Organomet. Chem.* **1990**, *391*, C23–7.
254. Busacca, C. A.; Johnson, R. E.; Swestock, J. *J. Org. Chem.* **1993**, *58*, 3299–303.
255. Ralbovsky, J. L.; Scola, P. M.; Sugino, E.; Burgos-Garcia, C.; Weinreb, S. M.; Parvez, M. *Heterocycles* **1996**, *43*, 1497–512.
256. Nakagawa, Y.; Irie, K.; Nakamura, Y.; Ohigashi, H.; Hayashi, H. *Biosci. Biotechnol. Biochem.* **1998**, *62*, 1568–73.
257. Sundberg, R. J.; Cherney, R. J. *J. Org. Chem.* **1990**, *55*, 6028–37.
258. Malapel-Andrieu, B.; Mérour, J.-Y. *Tetrahedron Lett.* **1998**, *39*, 39–42.
259. Cacchi, S.; Ciattini, P. G.; Morera, E.; Ortar, G. *Tetrahedron Lett.* **1988**, *29*, 3117–20.
260. Arcadi, A.; Cacchi, S.; Marinelli, F.; Morera, E.; Ortar, G. *Tetrahedron* **1990**, *46*, 7151–64.
261. (a) Akita, Y.; Inoue, A.; Yamamoto, K.; Ohta, A. *Heterocycles* **1985**, *23*, 2327–33. (b) Akita, Y.; Itagaki, Y.; Takizawa, S.; Ohta, A. *Chem. Pharm. Bull.* **1989**, *37*, 1477–80.
262. Rajeswaran, W. G.; Srinivasan, P. C. *Synthesis* **1992**, 835–6.
263. Yokoyama, Y.; Matsushima, H.; Takashima, M.; Suzuki, T.; Murakami, Y. *Heterocycles* **1997**, *46*, 133–6.
264. (a) Horwell, D. C.; Nichols, P. D.; Roberts, E. *Tetrahedron Lett.* **1994**, *35*, 939–40. (b) Horwell, D. C.; Nichols, P. D.; Ratcliffe, G. S.; Roberts, E. *J. Org. Chem.* **1994**, *59*, 4418–23.
265. Wiedenau, P.; Monse, B.; Blechert, S. *Tetrahedron* **1995**, *51*, 1167–76.
266. Yokoyama, Y.; Kondo, K.; Mitsuhashi, M.; Murakami, Y. *Tetrahedron Lett.* **1996**, *37*, 9309–12.
267. Black, D. St. C.; Keller, P. A.; Kumar, N.; *Tetrahedron* **1992**, *48*, 7601–8.
268. Germain, A. L.; Gilchrist, T. L.; Kemmitt, P. D. *Heterocycles* **1994**, *37*, 697–700.
269. Gilchrist, T. L.; Kemmitt, P. D.; Germain, A. L. *Tetrahedron* **1997**, *53*, 4447–56.
270. Grigg, R.; Sridharan, V.; Stevenson, P.; Sukirthalingam, S.; Worakun, T. *Tetrahedron* **1990**, *46*, 4003–18.
271. Kozikowski, A. P.; Ma, D. *Tetrahedron Lett.* **1991**, *32*, 3317–20.

272. Kozikowski, A. P.; Ma, D.; Brewer, J.; Sun, S.; Costa, E.; Romeo, E.; Guidotti, A. *J. Med. Chem.* **1993**, *36*, 2908–20.
273. Kraus, G. A.; Kim, H. *Synth. Commun.* **1993**, *23*, 55–64.
274. Desarbre, E.; Mérour, J.-Y. *Heterocycles* **1995**, *41*, 1987–98.
275. Shao, H. W.; Cai, J. C. *Chin. Chem. Lett.* **1996**, *7*, 13–4.
276. Miki, Y.; Shirokishi, H.; Matsushita, K. *Tetrahedron Lett.* **1999**, *40*, 4347–8.
277. Rigby, J. H.; Hughes, R. C.; Heeg, M. J. *J. Am. Chem. Soc.* **1995**, *117*, 7834–5.
278. Rigby, J. H.; Mateo, M. E. *Tetrahedron* **1996**, 52, 10569–82.
279. Melnyk, P.; Gasche, J.; Thal, C. *Tetrahedron Lett.* **1993**, *34*, 5449–50.
280. Melnyk, P.; Legrand, B.; Gasche, J.; Ducrot, P.; Thal, C. *Tetrahedron* **1995**, *51*, 1941–52.
281. Kelly, T. R.; Xu, W.; Sundaresan, J. *Tetrahedron Lett.* **1993**, *34*, 6173–6.
282. Grigg, R.; Sridharan, V.; Stevenson, P.; Worakun, T. *J. Chem. Soc., Chem. Commun.* **1986**, 1697–9.
283. Cornec, O.; Joseph, B.; Mérour, J.-Y. *Tetrahedron Lett.* **1995**, *36*, 8587–90.
284. Joseph B.; Cornec, O.; Mérour, J.-Y. *Tetrahedron* **1998**, *54*, 7765–76.
285. Yoneda, R.; Kimura, T.; Kinomoto, J.; Harusawa, S.; Kurihara, T. *J. Heterocycl. Chem.* **1996**, *33*, 1909–13.
286. Birman, V. B.; Rawal, V. H. *Tetrahedron Lett.* **1998**, *39*, 7219–22.
287. Rawal, V. H.; Michoud, C.; Monestel, R. F. *J. Am. Chem. Soc.* **1993**, *115*, 3030–1.
288. Rawal, V. H.; Michoud, C. *J. Org. Chem.* **1993**, *58*, 5583–4.
289. Rawal, V. H.; Michoud, C. *Tetrahedron Lett.* **1991**, *32*, 1695–8.
290. Rawal, V. H.; Iwasa, S. *J. Org. Chem.* **1994**, *59*, 2685–6.
291. Hegedus, L. S.; Allen, G. F.; Waterman, E. L. *J. Am. Chem. Soc.* **1976**, *98*, 2674–6.
292. Hegedus, L. S.; Allen, G. F.; Bozell, J. J.; Waterman, E. L. *J. Am. Chem. Soc.* **1978**, *100*, 5800–7.
293. Hegedus, L. S.; Allen, G. F.; Olsen, D. J. *J. Am. Chem. Soc.* **1980**, *102*, 3583–7.
294. Odle, R.; Blevins, B.; Ratcliffe, M.; Hegedus, L. S. *J. Org. Chem.* **1980**, *45*, 2709–10.
295. Hegedus, L. S.; Weider, P. R.; Mulhern, T. A.; Asada, H.; D'Andrea, S. *Gazz. Chim. Ital.* **1986**, *116*, 213–9.
296. Mori, M.; Chiba, K.; Ban, Y. *Tetrahedron Lett.* **1977**, 1037–40.
297. Mori, M.; Ban, Y. *Tetrahedron Lett.* **1979**, 1133–6.
298. Mori, M.; Kanda, N.; Oda, I.; Ban, Y. *Tetrahedron* **1985**, *41*, 5465–74.
299. Terpko, M. O.; Heck, R. F. *J. Am. Chem. Soc.* **1979**, *101*, 5281–3.
300. For reviews of early work, see (a) Hegedus, L. S. *Angew. Chem., Int. Ed. Engl.* **1988**, *27*, 1113–26. (b) Reference 211.
301. Larock, R. C.; Babu, S. *Tetrahedron Lett.* **1987**, *28*, 5291–4.
302. Larock, R. C.; Hightower, T. R.; Hasvold, L. A.; Peterson, K. P. *J. Org. Chem.* **1996**, *61*, 3584–5.

303. Larock, R. C.; Yang, H.; Pace, P.; Cacchi, S.; Fabrizi, G. *Tetrahedron Lett.* **1998**, *39*, 1885–8.
304. Larock, R. C.; Pace, P.; Yang, H. *Tetrahedron Lett.* **1998**, *39*, 2515–8.
305. Larock, R. C.; Pace, P.; Yang, H.; Russell, C. E.; Cacchi, S.; Fabrizi, G. *Tetrahedron* **1998**, *54*, 9961–80.
306. Yamaguchi, M.; Arisawa, M.; Hirama, M. *Chem. Commun.* **1998**, 1399–400.
307. Kasahara, A.; Izumi, T.; Murakami, S.; Miyamoto, K.; Hino, T. *J. Heterocycl. Chem.* **1989**, *26*, 1405–13.
308. Mahanty, J. S.; De, M.; Das, P.; Kundu, N. G. *Tetrahedron* **1997**, *53*, 13397–418.
309. Trost, B. M.; Pedregal, C. *J. Am. Chem. Soc.* **1992**, *114*, 7292–4.
310. Iritani, K.; Matsubara, S.; Utimoto, K. *Tetrahedron Lett.* **1988**, *29*, 1799–02.
311. Arcadi, A.; Cacchi, S.; Marinelli, F. *Tetrahedron Lett.* **1992**, *33*, 3915–8.
312. Cacchi, S.; Carnicelli, V.; Marinelli, F. *J. Organomet. Chem.* **1994**, *475*, 289–96.
313. Cacchi, S.; Fabrizi, G.; Marinelli, F.; Moro, L.; Pace, P. *Synlett* **1997**, 1363–6.
314. Cacchi, S.; Fabrizi, G.; Pace, P. *J. Org. Chem.* **1998**, *63*, 1001–11.
315. Saulnier, M. G.; Frennesson, D. B.; Deshpande, M. S.; Vyas, D. M. *Tetrahedron Lett.* **1995**, *36*, 7841–4.
316. Yasuhara, A.; Kaneko, M.; Sakamoto, T. *Heterocycles* **1998**, *48*, 1793–9.
317. Genet, J. P.; Blart, E.; Savignac, M. *Synlett* **1992**, 715–7.
318. Carroll, M. A.; Holmes, A. B. *Chem. Commun.* **1998**, 1395–6.
319. Hegedus, L. S.; Mulhern, T. A.; Mori, A. *J. Org. Chem.* **1985**, *50*, 4282–8.
320. Sundberg, R. J.; Pitts, W. J. *J. Org. Chem.* **1991**, *56*, 3048–54.
321. Sakamoto, T.; Kondo, Y.; Uchiyama, M.; Yamanaka, H. *J. Chem. Soc., Perkin Trans. 1* **1993**, 1941–2.
322. Tietze, L. F.; Grote, T. *J. Org. Chem.* **1994**, *59*, 192–6.
323. Tietze, L. F.; Buhr, W. *Angew. Chem., Int. Ed. Engl.* **1995**, *34*, 1366–8.
324. Tietze, L. F.; Hannemann, R.; Buhr, W.; Lögers, M.; Menningen, P.; Lieb, M.; Starck, D.; Grote, T.; Döring, A.; Schuberth, I. *Angew. Chem., Int. Ed. Engl.* **1996**, *35*, 2674–7.
325. Hoffmann, H. M. R.; Schmidt, B.; Wolff, S. *Tetrahedron* **1989**, *45*, 6113–26.
326. Macor, J. E.; Blank, D. H.; Post, R. J.; Ryan, K. *Tetrahedron Lett.* **1992**, *33*, 8011–4.
327. Macor, J. E.; Ogilvie, R. J.; Wythes, M. J. *Tetrahedron Lett.* **1996**, *37*, 4289–92.
328. Wensbo, D.; Annby, U.; Gronowitz, S. *Tetrahedron* **1995**, *51*, 10323–42.
329. Wensbo, D.; Gronowitz, S. *Tetrahedron* **1996**, *52*, 14975–88.
330. Hennings, D. D.; Iwasa, S.; Rawal, V. H. *Tetrahedron Lett.* **1997**, *38*, 6379–82.
331. Yang, C.-C.; Sun, P.-J.; Fang, J.-M. *J. Chem. Soc., Chem. Commun.* **1994**, 2629–30.
332. Yun, W.; Mohan, R. *Tetrahedron Lett.* **1996**, *37*, 7189–92.
333. Zhang, H.-C.; Maryanoff, B. E. *J. Org. Chem.* **1997**, *62*, 1804–9.

334. Arumugam, V.; Routledge, A.; Abell, C.; Balasubramanian, S. *Tetrahedron Lett.* **1997**, *38*, 6473–6.
335. Hegedus, L. S.; Allen, G. F.; Olsen, D. J. *J. Am. Chem. Soc.* **1980**, *102*, 3583–7.
336. Danishefsky, S.; Taniyama, E. *Tetrahedron Lett.* **1983**, *24*, 15–8.
337. For a summary of the author's work, see Grigg, R. *J. Heterocycl. Chem.* **1994**, *31*, 631–9.
338. Grigg, R.; Dorrity, M. J.; Malone, J. F.; Sridharan, V.; Sukirthalingam, S. *Tetrahedron Lett.* **1990**, *31*, 1343–6.
339. Burns, B.; Grigg, R.; Santhakumar, V.; Sridharan, V.; Stevenson, P.; Workun, T. *Tetrahedron Lett.* **1992**, *48*, 7297–320.
340. Grigg, R.; Fretwell, P.; Meerholtz, C.; Sridharan, V. *Tetrahedron* **1994**, *50*, 359–70.
341. Grigg, R.; Sansano, J. M. *Tetrahedron* **1996**, *52*, 13441–54.
342. Grigg, R.; Brown, S.; Sridharan, V.; Uttley, M. D. *Tetrahedron Lett.* **1998**, *39*, 3247–50.
343. Grigg, R.; Sridharan, V.; York, M. *Tetrahedron Lett.* **1998**, *39*, 4139–42.
344. Evans, P.; Grigg, R.; Ramzan, M. I.; Sridharan, V.; York, M. *Tetrahedron Lett.* **1999**, *40*, 3021–4.
345. Grigg, R.; Major, J. P.; Martin, F. M.; Whittaker, M. *Tetrahedron Lett.* **1999**, *40*, 7709–11.
346. Burns, B.; Grigg, R.; Sridharan, V.; Worakun, T. *Tetrahedron Lett.* **1988**, *29*, 4325–8.
347. Burns, B.; Grigg, R.; Ratananukul, P.; Sridharan, V.; Stevenson, P.; Sukirthalingam, S.; Worakun, T. *Tetrahedron Lett.* **1988**, *29*, 5565–8.
348. Grigg, R.; Loganathan, V.; Sridharan, V. *Tetrahedron Lett.* **1996**, *37*, 3399–402.
349. Brown, D.; Grigg, R.; Sridharan, V.; Tambyrajah, V.; Thornton-Pett, M. *Tetrahedron* **1998**, *54*, 2595–606.
350. Grigg, R.; Loganathan, V.; Sukirthalingam, S.; Sridharan, V. *Tetrahedron Lett.* **1990**, *31*, 6573–6.
351. Abelman, M. M.; Oh, T.; Overman, L. E. *J. Org. Chem.* **1987**, *52*, 4130–3.
352. Earley, W. G.; Oh, T.; Overman, L. E. *Tetrahedron Lett.* **1988**, *29*, 3785–8.
353. Madin, A.; Overman, L. E. *Tetrahedron Lett.* **1992**, *33*, 4859–62.
354. Ashimori, A.; Overman, L. E. *J. Org. Chem.* **1992**, *57*, 4571–2.
355. Ashimori, A.; Matsuura, T.; Overman, L. E.; Poon, D. J. *J. Org. Chem.* **1993**, *58*, 6949–51.
356. Overman, L. E.; Poon, D. J. *Angew. Chem., Int. Ed. Engl.* **1997**, *36*, 518–21.
357. Ashimori, A.; Bachand, B.; Overman, L. E.; Poon, D. J. *J. Am. Chem. Soc.* **1998**, *120*, 6477–87.
358. Ashimori, A.; Bachand, B.; Calter, M. A.; Govek, S. P.; Overman, L. E.; Poon, D. J. *J. Am. Chem. Soc.* **1998**, *120*, 6488–99.
359. Matsuura, T.; Overman, L. E.; Poon, D. J. *J. Am. Chem. Soc.* **1998**, *120*, 6500–3.

360. Overman, L. E.; Paone, D. V.; Stearns, B. A. *J. Am. Chem. Soc.* **1999**, *121*, 7702–3.
361. Newcombe, N. J.; Ya, F.; Vijn, R. J.; Hiemstra, H.; Speckamp, W. N. *J. Chem. Soc., Chem. Commun.* **1994**, 767–8.
362. Iida, H.; Yuasa, Y.; Kibayashi, C. *J. Org. Chem.* **1980**, *45*, 2938–42.
363. Luly, J. R.; Rapoport, H. *J. Org. Chem.* **1984**, *49*, 1671–2.
364. Michael, J. P.; Chang, S.-F.; Wilson, C. *Tetrahedron Lett.* **1993**, *34*, 8365–8.
365. Kasahara, A.; Izumi, T.; Murakami, S.; Yanai, H.; Takatori, M. *Bull. Chem. Soc. Jpn.* **1986**, *59*, 927–8.
366. Sakamoto, T.; Nagano, T.; Kondo, Y.; Yamanaka, H. *Synthesis* **1990**, 215–8.
367. Latham, E. J.; Stanforth, S. P. *Chem. Commun.* **1996**, 2253–4.
368. Latham, E. J.; Stanforth, S. P. *J. Chem. Soc., Perkin Trans. 1* **1997**, 2059–63.
369. Koerber-Plé, K.; Massiot, G. *Synlett* **1994**, 759–60.
370. Chen, C.; Liebermann, D. R.; Larsen, R. D.; Verhoeven, T. R.; Reider, P. J. *J. Org. Chem.* **1997**, *62*, 2676–7.
371. (a) Chen, L.-C.; Yang, S.-C.; Wang, H.-M. *Synthesis* **1995**, 385–6. (b) Wang, H.-M.; Chou, H.-L.; Chen, L.-C. *J. Chin. Chem. Soc.* **1995**, *42*, 593–5.
372. Blache, Y.; Sinibaldi-Troin, M.-E.; Voldoire, A.; Chavignon, O.; Gramain, J.-C.; Teulade, J.-C.; Chapat, J.-P. *J. Org. Chem.* **1997**, *62*, 8553–6.
373. Iwaki, T.; Yasuhara, A.; Sakamoto, T. *J. Chem. Soc., Perkin Trans. 1* **1999**, 1505–10.
374. Ames, D. E.; Opalko, A. *Tetrahedron* **1984**, *40*, 1919–25.
375. (a) Tremont, S. J.; Rahman, H. U. *J. Am. Chem. Soc.* **1984**, *106*, 5759–60. (b) Abdrakhmanov, I. B.; Mustafin, A. G.; Tolstikov, G. A.; Dzhemilev, U. M. *Chem. Heterocycl. Cpds.* **1987**, 420–2.
376. Gowan, M.; Caillé, A. S.; Lau, C. K. *Synlett* **1997**, 1312–4.
377. Sakamoto, T.; Kondo, Y.; Yamanaka, H. *Heterocycles* **1988**, *27*, 453–6.
378. Izumi, T.; Sugano, M.; Konno, T. *J. Heterocycl. Chem.* **1992**, *29*, 899–904.
379. Kasahara, A.; Izumi, T.; Xiao-ping, L. *Chem. Ind.* **1988**, 50–1.
380. Kasahara, A.; Izumi, T.; Kikuchi, T.; Xiao-ping, L. *J. Heterocycl. Chem.* **1987**, *24*, 1555–6.
381. Samizu, K.; Ogasawara, K. *Synlett* **1994**, 499–500.
382. Samizu, K.; Ogasawara, K. *Heterocycles* **1995**, *41*, 1627–9.
383. Sakagami, H.; Ogasawara, K. *Heterocycles* **1999**, *51*, 1131–5.
384. Carlström, A.-S.; Frejd, T. *Acta Chem. Scand.* **1992**, *46*, 163–71.
385. Larock, R. C.; Yum, E. K. *J. Am. Chem. Soc.* **1991**, *113*, 6689–90.
386. Larock, R. C.; Yum, E. K.; Refvik, M. D. *J. Org. Chem.* **1998**, *63*, 7652–62.
387. Jeschke, T.; Wensbo, D.; Annby, U.; Gronowitz, S.; Cohen, L. A. *Tetrahedron Lett.* **1993**, *34*, 6471–4.
388. Ujjainwalla, F.; Warner, D. *Tetrahedron Lett.* **1998**, *39*, 5355–8.
389. Park, S. S.; Choi, J.-K.; Yum, E. K.; Ha, D.-C. *Tetrahedron Lett.* **1998**, *39*, 627–30.

390. Kang, S. K.; Park, S. S.; Kim, S. S.; Choi, J.-K.; Yum, E. K. *Tetrahedron Lett.* **1999**, *40*, 4379–82.
391. Chen, C.; Lieberman, D. R.; Larsen, R. D.; Reamer, R. A.; Verhoeven, T. R.; Reider, P. J.; Cottrell, I. F.; Houghton, P. G. *Tetrahedron Lett.* **1994**, *35*, 6981–4.
392. Chen, C.; Lieberman, D. R.; Street, L. J.; Guiblin, A. R.; Larsen, R. D.; Verhoeven, T. R. *Synth. Commun.* **1996**, *26*, 1977–84.
393. Larock, R. C.; Doty, M. J.; Han, X. *Tetrahedron Lett.* **1998**, *39*, 5143–6.
394. Maassarani, F.; Pfeffer, M.; Spencer, J.; Wehman, E. *J. Organomet. Chem.* **1994**, *466*, 265–71.
395. Rosso, V. W.; Lust, D. A.; Bernot, P. J.; Grosso, J. A.; Modi, S. P.; Rusowicz, A.; Sedergran, T. C.; Simpson, J. H.; Srivastava, S. K.; Humora, M. J.; Anderson, N. G. *Org. Proc. Res. Dev.* **1997**, *1*, 311–4.
396. Zhang, H.-C.; Brumfield, K. K.; Maryanoff, B. E. *Tetrahedron Lett.* **1997**, *38*, 2439–42.
397. Smith, A. L.; Stevenson, G. I.; Swain, C. J.; Castro, J. L. *Tetrahedron Lett.* **1998**, *39*, 8317–20.
398. O'Connor, J. M.; Stallman, B. J.; Clark, W. G.; Shu, A. Y. L.; Spada, R. E.; Stevenson, T. M.; Dieck, H. A. *J. Org. Chem.* **1983**, *48*, 807–9.
399. Larock, R. C.; Berrios-Peña, N.; Narayanan, K. *J. Org. Chem.* **1990**, *55*, 3447–50.
400. Larock, R. C.; Guo, L. *Synlett* **1995**, 465–6.
401. Back, T. G.; Bethell, R. J. *Tetrahedron Lett.* **1998**, *39*, 5463–4.
402. Wang, Y.; Huang, T.-N. *Tetrahedron Lett.* **1998**, *39*, 9605–8.
403. Larock, R. C.; Yum, E. K. *Synlett* **1990**, 529–30.
404. Larock, R. C.; Yum, E. K. *Tetrahedron* **1996**, *52*, 2743–58.
405. Larock, R. C.; Berrios-Peña, N. G.; Fried, C. A. *J. Org. Chem.* **1991**, *56*, 2615–7.
406. Larock, R. C.; Zenner, J. M. *J. Org. Chem.* **1995**, *60*, 482–3.
407. Zenner, J. M.; Larock, R. C. *J. Org. Chem.* **1999**, *64*, 7312–22.
408. Desarbre, E.; Mérour, J.-Y. *Tetrahedron Lett.* **1996**, *37*, 43–6.
409. Larock, R. C.; Liu, C.-L.; Lau, H. H.; Varaprath, S. *Tetrahedron Lett.* **1984**, *25*, 4459–62.
410. Roesch, K. R.; Larock, R. C. *Org. Lett.* **1999**, *1*, 1551–3.
411. Hegedus, L. S.; Winton, P. M.; Varaprath, S. *J. Org. Chem.* **1981**, *46*, 2215–21.
412. Pandey, G. D.; Tiwari, K. P. *Synth. Commun.* **1980**, *10*, 523–7.
413. Mori, M.; Kobayashi, H.; Kimura, M.; Ban, Y. *Heterocycles* **1985**, *23*, 2803–6.
414. Edstrom, E. D.; Yu, T. *J. Org. Chem.* **1995**, *60*, 5382–3.
415. Edstrom, E. D.; Yu, T. *Tetrahedron Lett.* **1994**, *35*, 6985–8.
416. Itahara, T. *Chem. Lett.* **1982**, 1151–2.
417. Kondo, Y.; Sakamoto, T.; Yamanaka, H. *Heterocycles* **1989**, *29*, 1013–6.

418. Kondo, Y.; Shiga, F.; Murata, N.; Sakamoto, T.; Yamanaka, H. *Tetrahedron* **1994**, *50*, 11803–12.
419. Arcadi, A.; Cacchi, S.; Carnicelli, V.; Marinelli, F. *Tetrahedron* **1994**, *50*, 437–52.
420. Herbert, J. M.; McNeil, A. H. *Tetrahedron Lett.* **1998**, *39*, 2421–4.
421. Tollari, S.; Demartin, F.; Cenini, S.; Palmisano, G.; Raimondi, P. *J. Organomet. Chem.* **1997**, *527*, 93–102.
422. Tollari, S.; Cenini, S.; Tunice, C.; Palmisano, G. *Inorg. Chim. Acta* **1998**, *272*, 18–23.
423. Dong, Y.; Busacca, C. A. *J. Org. Chem.* **1997**, *62*, 6464–5.
424. Hirao, K.; Morii, N.; Joh, T.; Takahashi, S. *Tetrahedron Lett.* **1995**, *36*, 6243–6.
425. Iwasaki, M.; Kobayashi, Y.; Li, J.-P.; Matsuzaka, H.; Ishii, Y.; Hidai, M. *J. Org. Chem.* **1991**, *56*, 1922–7.
426. Grigg, R.; Monteith, M.; Sridharan, V.; Terrier, C. *Tetrahedron* **1998**, *54*, 3885–94.
427. Grigg, R.; Kennewell, P.; Teasdale, A. J. *Tetrahedron Lett.* **1992**, *33*, 7789–92.
428. Grigg. R.; Putnikovic, B.; Urch, C. J. *Tetrahedron Lett.* **1996**, *37*, 695–8.
429. Ishikura, M. *Heterocycles* **1995**, *41*, 1385–8.
430. *Åkermark*, B.; Bäckvall, J. E.; Hegedus, L. S.; Zetterberg, K.; Siirala-Hansén, K.; Sjöberg, K. *J. Organomet. Chem.* **1974**, *72*, 127–38.
431. Hegedus, L. S.; Mulhern, T. A.; Asada, H. *J. Am. Chem. Soc.* **1986**, *108*, 6224–8.
432. Martin, P. *Helv. Chim. Acta* **1984**, *67*, 1647–9.
433. Martin, P. *Tetrahedron Lett.* **1987**, *28*, 1645–6.
434. Martin, P. *Helv. Chim. Acta* **1988**, *71*, 344–7.
435. Martin, P. *Helv. Chim. Acta* **1989**, *72*, 1554–82.
436. Boger, D. L.; Panek, J. S. *Tetrahedron Lett.* **1984**, *25*, 3175–8.
437. Boger, D. L.; Duff, S. R.; Panek, J. S.; Yasuda, M. *J. Org. Chem.* **1985**, *50*. 5782–9.
438. Boger, D. L.; Duff, S. R.; Panek, J. S.; Yasuda, M. *J. Org. Chem.* **1985**, *50*, 5790–5.
439. (a) Wolfe, J. P.; Wagaw, S.; Buchwald, S. L. *J. Am. Chem. Soc.* **1996**, *118*, 7215–6, and references cited therein. (b) Louie, J.; Driver, M. S.; Hamann, B. C.; Hartwig, J. F. *J. Org. Chem.* **1997**, *62*, 1268–73.
440. Driver, M. S.; Hartwig, J. F. *J. Am. Chem. Soc.* **1996**, *118*, 7217–8, and references cited therein.
441. Hartwig, J. F. *Synlett* **1997**, 329–40.
442. (a) Marcoux, J.-F.; Wagaw, S.; Buchwald, S. L. *J. Org. Chem.* **1997**, *62*, 1568–9. (b) Sadighi, J. P.; Harris, M. C.; Buchwald, S. L. *Tetrahedron Lett.* **1998**, *39*, 5327–30. (c) Wolfe, J. P.; Buchwald, S. L. *J. Org. Chem.* **1997**, *62*, 1264–7.
443. Hartwig, J. F. *Angew. Chem., Int. Ed.* **1998**, *37*, 2046–67.
444. (a) Wolfe, J. P.; Wagaw, S.; Marcoux, J.-F.; Buchwald, S. L. *Acc. Chem. Res.* **1998**, *31*, 805–18. (b) Yang, B. H.; Buchwald, S. L. *J. Organomet. Chem.* **1999**, *576*, 125–46.
445. Hartwig, J. F. *Angew. Chem., Int. Ed.* **1998**, *37*, 2090–2.

446. Belfield, A. J.; Brown, G. R.; Foubister, A. J. *Tetrahedron* **1999**, *55*, 11399–428.
447. Huang, J.; Grasa, G.; Nolan, S. P. *Org. Lett.* **1999**, *1*, 1307–9.
448. Peat, A. J.; Buchwald, S. L. *J. Am. Chem. Soc.* **1996**, *118*, 1028–30.
449. Wolfe, J. P.; Rennels, R. A.; Buchwald, S. L. *Tetrahedron* **1996**, *52*, 7525–46.
450. Wagaw, S.; Rennels, R. A.; Buchwald, S. L. *J. Am. Chem. Soc.* **1997**, *119*, 8451–8.
451. Aoki, K.; Peat, A. J.; Buchwald, S. L. *J. Am. Chem. Soc.* **1998**, *120*, 3068–73.
452. Yang, B. H.; Buchwald, S. L. *Org. Lett.* **1999**, *1*, 35–7.
453. Wagaw, S.; Yang, B. H.; Buchwald, S. L. *J. Am. Chem. Soc.* **1998**, *120*, 6621–2.
454. Wagaw, S.; Yang, B. H.; Buchwald, S. L. *J. Am. Chem. Soc.* **1999**, *121*, 10251–63.
455. MacNeil, S. L.; Gray, M.; Briggs, L. E.; Li, J. J.; Snieckus, V. *Synlett* **1998**, 419–21.
456. Abouabdellah, A.; Dodd, R. H. *Tetrahedron Lett.* **1998**, *39*, 2119–22.
457. He, F.; Foxman, B. H.; Snider, B. B. *J. Am. Chem. Soc.* **1998**, *120*, 6417–8.
458. Akazome, M.; Kondo, T.; Watanabe, Y. *Chem. Lett.* **1992**, 769–72.
459. Akazome, M.; Kondo, T.; Watanabe, Y. *J. Org. Chem.* **1994**, *59*, 3375–80.
460. Söderberg, B. C.; Shriver, J. A. *J. Org. Chem.* **1997**, *62*, 5838–45.
461. (a) Söderberg, B. C.; Rector, S. R.; O'Neil, S. N. *Tetrahedron Lett.* **1999**, *40*, 3657–60. (b) Söderberg, B. C.; Chisnell, A. C.; O'Neil, S. N.; Shriver, J. A. *J. Org. Chem.* **1999**, *64*, 9731–4.
462. Tollari, S.; Cenini, S.; Crotti, C.; Gianella, E. *J. Mol. Cat.* **1994**, *87*, 203–14.
463. Tollari, S.; Cenini, S.; Rossi, A.; Palmisano, G. *J. Mol. Cat.*, **1998**, *135*, 241–8.
464. Aoyagi, Y.; Mizusaki, T.; Ohta, A. *Tetrahedron Lett.* **1996**, *37*, 9203–6.
465. Izumi, T.; Soutome, M.; Miura, T. *J. Heterocycl. Chem.* **1992**, *29*, 1625–9.
466. (a) Uozumi, Y.; Mori, M.; Shibasaki, M. *J. Chem. Soc., Chem. Commun.* **1991**, 81–3. (b) Mori, M.; Uozumi, Y.; Shibasaki, M. *Heterocycles* **1992**, *33*, 819–30.
467. Malapel-Andrieu, B.; Mérour, J.-Y. *Tetrahedron* **1998**, *54*, 11095–110.
468. Trost, B. M.; Molander, G. A. *J. Am. Chem. Soc.* **1981**, *103*, 5969–72.
469. (a) Mann, G.; Hartwig, J. F.; Driver, M. S.; Fernández-Rivas, C. *J. Am. Chem. Soc.* **1998**, *120*, 827–8. (b) Hartwig, J. F.; Kawatsura, M.; Hauck, S. I.; Shaughnessy, K. H.; Alcazar-Roman, L. M. *J. Org. Chem.* **1999**, *64*, 5575–80. (c) Watanabe, M.; Nishiyama, M.; Yamamoto, T.; Koie, Y. *Tetrahedron Lett.* **2000**, *41*, 481–3.
470. Kato, K.; Ono, M.; Akita, H. *Tetrahedron Lett.* **1997**, *38*, 1805–8.
471. Dickens, M. J.; Gilday, J. P.; Mowlem, T. J.; Widdowson, D. A. *Tetrahedron* **1991**, *47*, 8621–34.
472. Takayama, H.; Sakai, S. *Chem. Pharm. Bull.* **1989**, *37*, 2256–7.
473. Takayama, H.; Kitajima, M.; Sakai, S. *Heterocycles* **1990**, *30*, 325–7.
474. Anderson, B. A.; Bell, E. C.; Ginah, F. O.; Harn, N. K.; Pagh, L. M.; Wepsiec, J. P. *J. Org. Chem.* **1998**, *63*, 8224–8.
475. Bracher, F.; Hildebrand, D.; Ernst, L. *Arch. Pharm.* **1994**, *327*, 121–2.
476. Zhang, D.; Liebeskind, L. S. *J. Org. Chem.* **1996**, *61*, 2594–5.

477. Bailey, W. F.; Jiang, X.-L. *J. Org. Chem.* **1996**, *61*, 2596–7.
478. Yokum, T. S.; Tungaturthi, P. K.; McLaughlin, M. L. *Tetrahedron Lett.* **1997**, *38*, 5111–4.
479. Lemaire-Audoire, S.; Savignac, M.; Genet, J. P.; Bernard, J.-M. *Tetrahedron Lett.* **1995**, *36*, 1267–70.
480. Genet, J. P.; Blart, E.; Savignac, M.; Lemeune, S.; Paris, J.-M. *Tetrahedron Lett.* **1993**, *34*, 4189–92.
481. Takacs, J. M.; Zhu, J. *Tetrahedron Lett.* **1990**, *31*, 1117–20.
482. Reding, M. T.; Fukuyama, T. *Org. Lett.* **1999**, *1*, 973–6.
483. Godleski, S. A.; Villhauer, E. B. *J. Org. Chem.* **1986**, *51*, 486–91.
484. Vangveravong, S.; Nichols, D. E. *J. Org. Chem.* **1995**, *60*, 3409–13.
485. Matsumoto, M.; Watanabe, N.; Kobayashi, H. *Heterocycles* **1987**, *26*, 1479–82.
486. Isomura, K.; Uto, K.; Taniguchi, H. *J. Chem. Soc., Chem. Commun.* **1977**, 664–5.
487. Alper, H.; Mahatantila, C. P. *Heterocycles* **1983**, *20*, 2025–8.
488. Yang, C.-C.; Tai, H.-M.; Sun, P.-J. *J. Chem. Soc., Perkin Trans. 1* **1997**, 2843–50.
489. Genet, J. P.; Grisoni, S. *Tetrahedron Lett.* **1986**, *27*, 4165–8.
490. Genet, J. P.; Grisoni, S. *Tetrahedron Lett.* **1988**, *29*, 4543–6.
491. Kardos, N.; Genet, J.-P. *Tetrahedron: Asym.* **1994**, *5*, 1525–33.
492. Sakai, S.; Yamanaka, E.; Kitajima, M.; Yokota, M.; Aimi, N.; Wongseripatana, S.; Ponglux, D. *Tetrahedron Lett.* **1986**, *27*, 4585–8.

# CHAPTER 4

# Pyridines

Many pyridine-containing molecules are important for their biological and pharmacological properties. A salient example is (+)-epibatidine, a naturally occurring pyridine alkaloid isolated from the skin of an Ecuadorian poisonous frog. It exhibits strong non-opioid analgesic activity and is 200 times more potent than morphine without the addictive effects. Nicotine is another well-known pyridine-containing alkaloid with significant pharmacological activity. In addition to the naturally occurring pyridines, there are numerous synthetic drugs, fungicides, and herbicides possessing the pyridine moiety. Moreover, appropriately substituted pyridines are widely used as polydentate ligands.

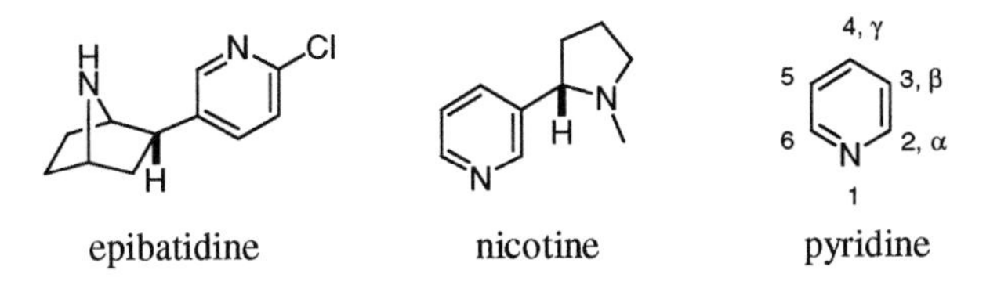


Pyridine is a π-electron-deficient heterocycle. Due to the electronegativity of the nitrogen atom, the α and γ positions bear a partial positive charge, making the C(2), C(4), and C(6) positions prone to nucleophilic attacks. A similar trend occurs in the context of palladium chemistry. The α and γ positions of halopyridines are more susceptible to the oxidative addition to Pd(0) relative to simple carbocyclic aryl halides. Even α- and γ-chloropyridines are viable electrophilic substrates for Pd-catalyzed reactions under standard conditions.

## 4.1 Synthesis of halopyridines

### 4.1.1 Direct metalation followed by quenching with halogens

Several methods exist for the synthesis of intricate, commercially unavailable halopyridines. By taking advantage of the different kinetic acidity at each site [C(4) > C(3) > C(2)], Gribble and Saulnier deprotonated 3-halopyridine regioselectively at C(4) and quenched the resulting 4-lithio-3-halopyridine with iodine to give 4-iodo-3-chloropyridine [1].

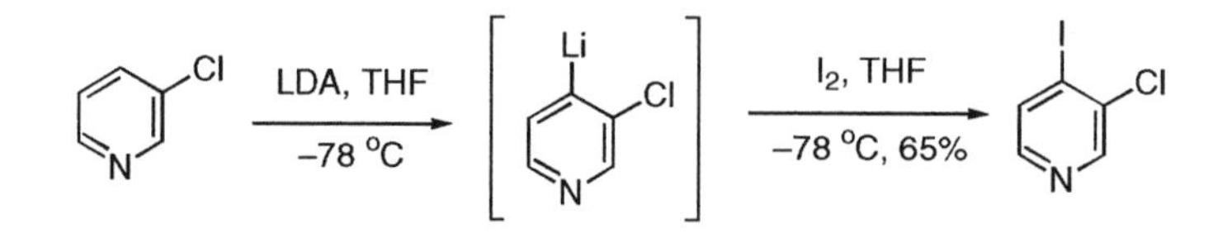


Direct α-metalation of π-deficient heteroaromatics such as pyridine is a challenging task due to the instability of the metalated species. Nonetheless, Kondo *et al.* successfully accomplished the direct α-metalation of pyridine utilizing TMP–zincate (**1**), a bulky base, to afford 2-pyridinylzincate, which upon treatment with $I_2$ provided access to 2-iodopyridine [2]. In another case, a directed *ortho*-metalation of 2-methoxy-3-pivaloylaminopyridine (**2**) was carried out regioselectively at C(4) using a combination of *n*-butyllithium and TMEDA. The resulting lithiopyridine intermediate was quenched with iodine, leading to 4-iodo-2-methoxy-3-pivaloylaminopyridine (**3**). However, a similar reaction of 2-chloro-3-pivaloylaminopyridine utilizing *t*-BuLi gave a low yield (18%) [3].

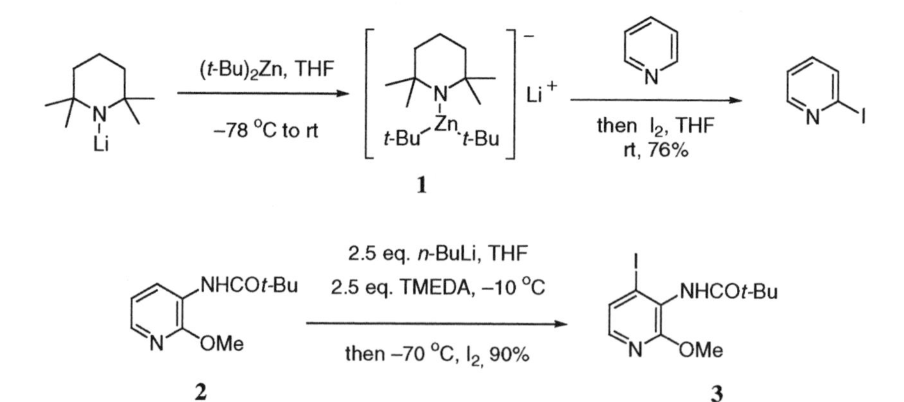


Regiospecific C(4) lithiation of nicotinaldehyde was achieved using *n*-BuLi with the assistance of *N,N,N'*-trimethylethylenediamine. Subsequent treatment of the resulting dianion with 1,2-dibromotetrafluoroethane furnished 4-bromonicotinaldehyde [4].

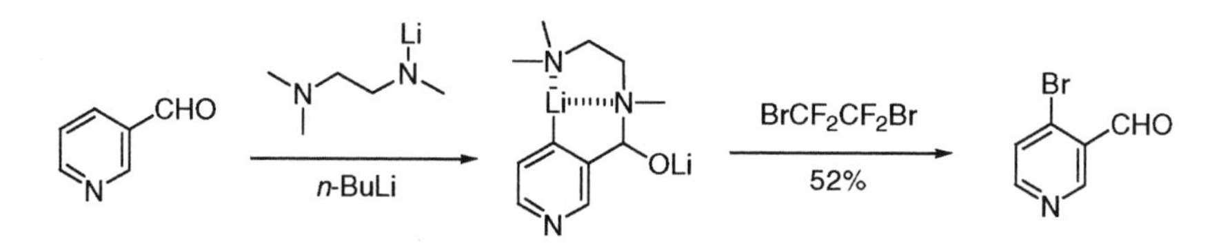

### 4.1.2 Dehydroxy-halogenation

Appropriately substituted pyridines can be directly halogenated. For example, direct bromination of 2-methylamino-3-nitropyridine (**4**) was realized employing pyridinium bromide perbromide to furnish the 5-bromo derivative **5** [5]. Dehydroxy-bromination of a six-membered lactam, oxo-β-carboline **6**, with $POBr_3$ in anisole gave rise to monobromocarboline **7** [6]. The addition of anisole prevented undesired ring bromination by trapping bromine formed from decomposition of $POBr_3$.

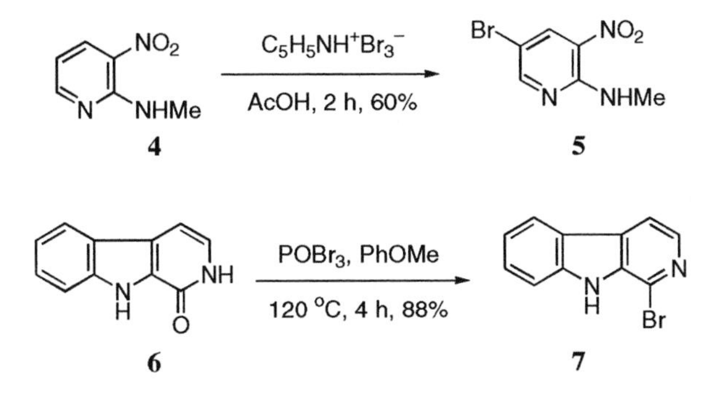


Another synthesis of halopyridines, unique to pyridine and other *N*-containing heteroarenes, involves transformation of pyridine *N*-oxide into the corresponding pyridone followed by halogenation. In one case, treatment of 3-chloro-2,4'-bipyridine-1'-oxide (**8**) with acetic anhydride produced the pyridone, which was then converted to dichloride **9** with $POCl_3$/DMF [7].

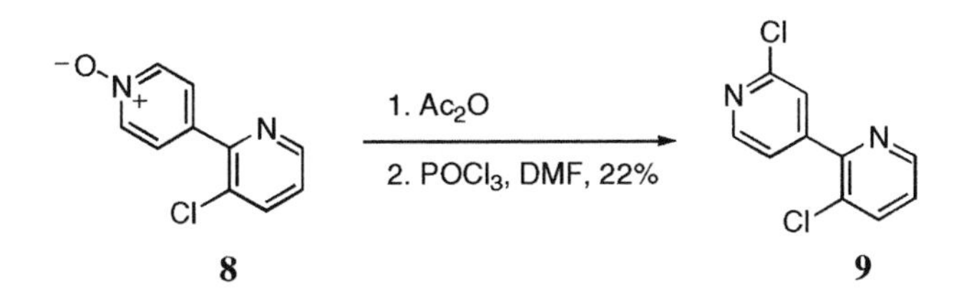


Although the aforementioned dehydroxy-halogenations using phosphorus oxyhalides produces the desired halopyridines, both $POCl_3$ and $POBr_3$ are moisture-sensitive and $POBr_3$ is relatively expensive. Extending the well-known alcohol to halide conversion employing triphenylphosphine and *N*-halosuccinimide, Sugimoto *et al.* halogenated π-deficient hydroxyheterocycles such as 2-hydroxypyridine to their respective halides [8].

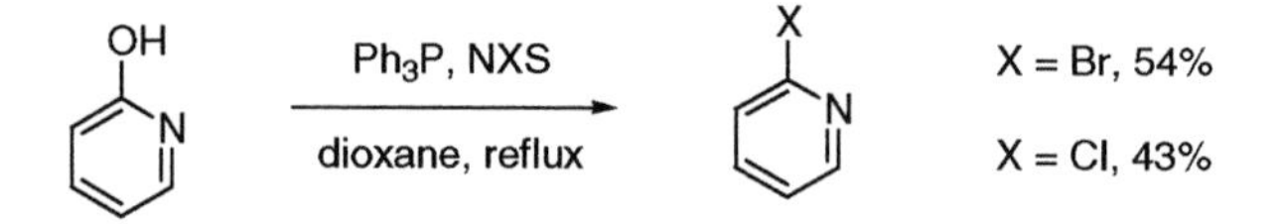


## 4.2 Coupling reactions with organometallic reagents

### 4.2.1 Organomanganese reagents

Pd-catalyzed reactions are particularly effective for the construction of unsymmetrical biaryls. Regularly-used nucleophilic partners for these heterocoupling reactions include Mg, Si, Sn, Zn and B organometallic reagents. A Pd-catalyzed reaction of organomanganese halides with functionalized aryl halides and triflates has been unveiled [9]. The reaction of phenylmanganese chloride with 2-bromopyridine gave 2-phenylpyridine in 96% yield, whereas the same reaction with pyridyl-2-triflate also led to 2-phenylpyridine in 77% yield. The reaction works for substrates bearing electron-donating groups as well, although at a slower rate and lower yields.

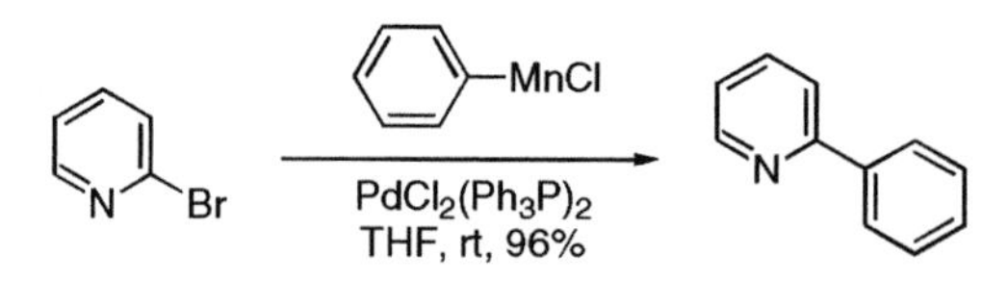


### 4.2.2 Kumada coupling

The Kumada coupling has broad applicability since a wide variety of Grignard reagents are commercially available and others can be readily prepared from the corresponding organic halides. Another advantage of the Kumada coupling is that it often can be run at room temperature, whereas its pitfall is that many functional groups are not compatible with Grignard reagents. In a simple case, the Kumada cross-coupling of 5-(2,2'-bithienyl)magnesium bromide (**10**) with 4-bromopyridine was carried out in refluxing ether to give 4-[5-(2-bithien-2'-yl)]pyridine (**11**) [10]. Under the Kumada coupling conditions, 2,5-dibromopyridine was mono-heteroarylated with one equivalent of indolyl Grignard reagent regioselectively at C(2), affording indolylpyridine **12**. Subsequent reaction of **12** with the second Grignard reagent under more rigorous conditions gave trimer **13** [11, 12]. A similar reaction of 2,6-dichloropyridine and 2-thienylmagnesium bromide led to mono-heteroarylation product **14** [12].

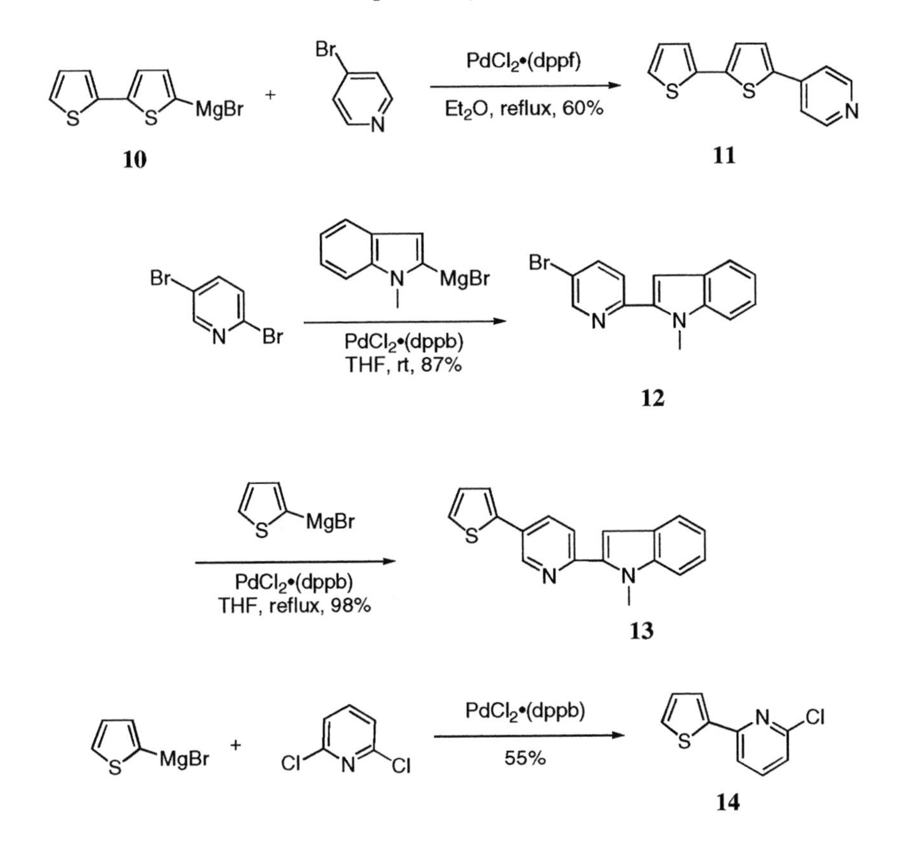


### 4.2.3 Negishi coupling

Organozinc reagents exhibit greater functional group compatibility than organolithium and Grignard reagents. As a consequence, the Negishi coupling has found wide application in organic synthesis. While all chloro- [10] bromo- [13] and iodopyridine [14, 15] derivatives are applicable as electrophiles in Negishi couplings, the choice of which halide to use is substrate-dependent and a spectrum of reactivities has been observed. At one extreme, 2,5-dichloropyridine readily underwent the Negishi reaction with benzylzinc bromide to afford a 56% yield of the monobenzylated adduct **15** and 13% of the dibenzylated adduct **16** [10].

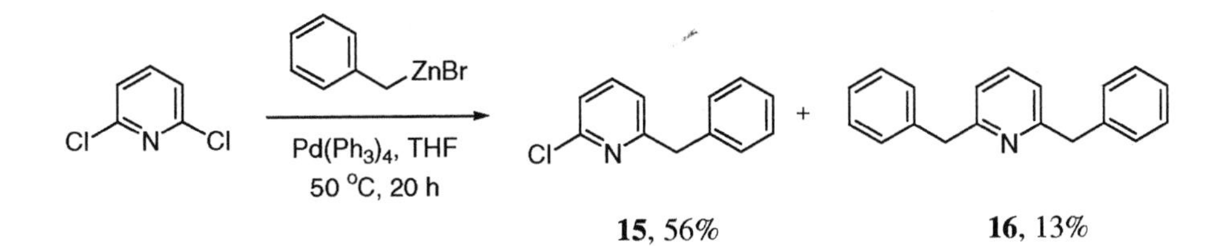

At the other extreme of reactivity, 2-bromopyridine had to be transformed into the corresponding iodide to achieve good efficiency for the coupling reaction [14]. For example, *D-altro*-2-(2,4:3,5-di-*O*-benzylidenepentitol-1-yl)-6-bromopyridine (**17**) was first converted to the corresponding iodide **18**, which was then allowed to react with the Reformatsky reagent to give adduct **19**.

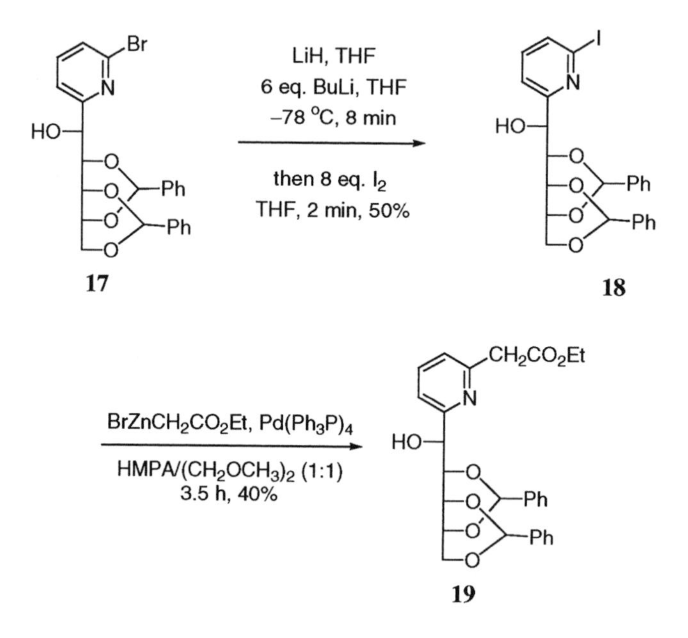


Many aryl- and heteroarylzinc reagents have been coupled with halopyridines. Examples include arylzincate **20** derived from ethyl benzoate [3], thienylzinc reagent **21** [16], 2-imidazolylzinc chloride **22** [17], pyrazolylzinc reagent **23** [18], and 2-pyridylzinc reagent **24** [19].

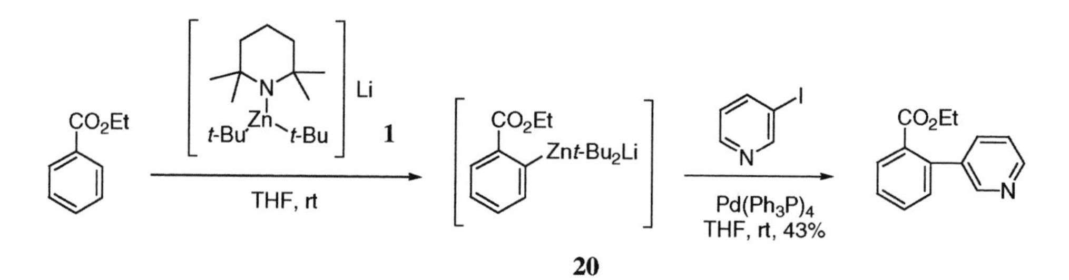

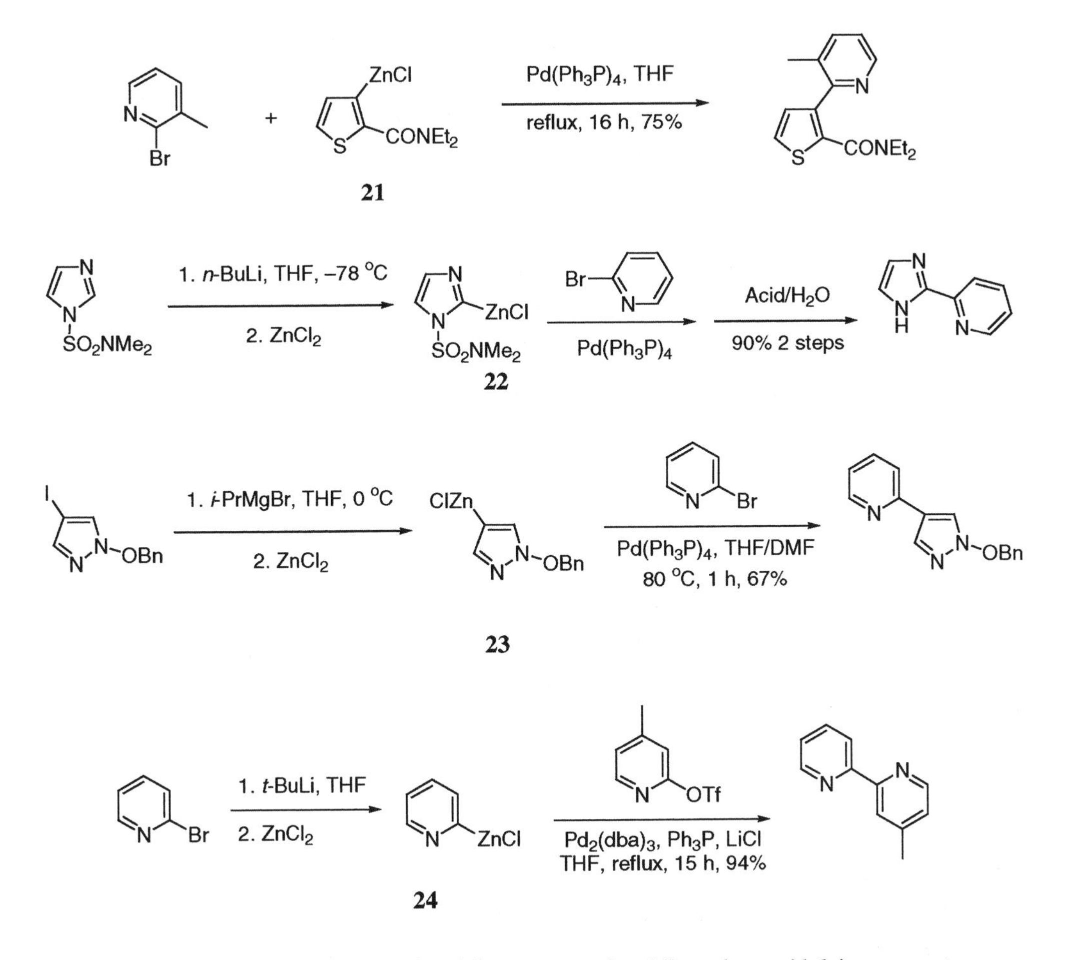


For Negishi reactions in which the pyridines are nucleophiles, the pyridylzinc reagents are usually prepared from the corresponding halopyridines [6, 20, 21]. An excess of 2-chlorozincpyridine *N*-oxide (**26**), arising from 2-bromopyridine *N*-oxide hydrochloride (**25**), was coupled with vinyl triflate **27** in the presence of $Pd(Ph_3P)_4$ to furnish adduct **28** [20]. Recently, an efficient Pd-catalyzed cyanation of 2-amino-5-bromo-6-methylpyridine (**29**) using zinc cyanide has been reported to afford pyridyl nitrile **30** [22].

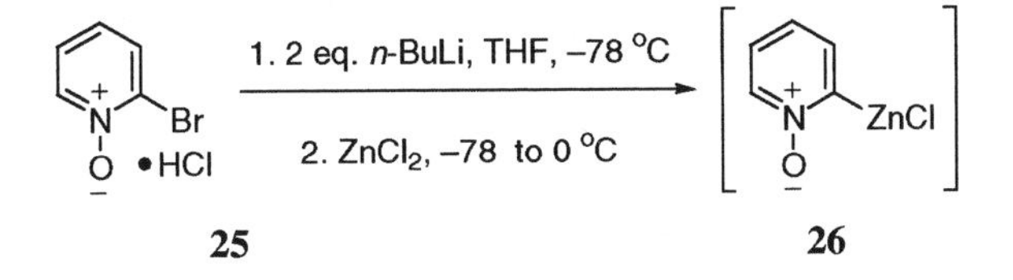

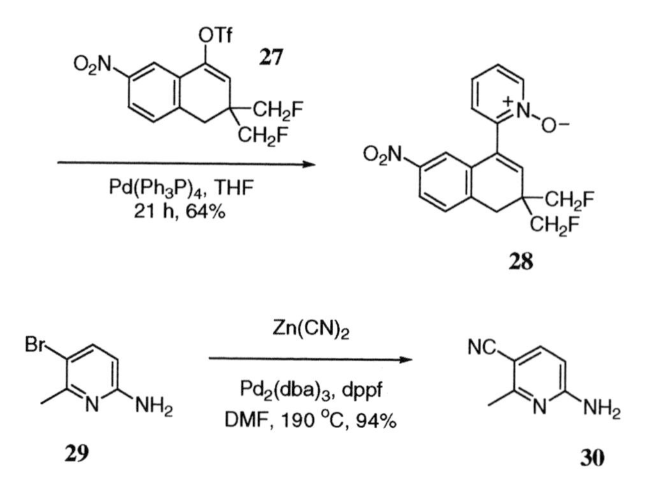


Bosch's group cross-coupled both 2- and 3-indolylzinc derivatives with diversely substituted 2-halopyridines to assemble 2- and 3-(2-pyridyl)indoles, which have become important intermediates in indole alkaloid synthesis [23–28]. Thus, 1-(phenylsulfonyl)indole (**30**) was converted to 2-indolylzinc reagent **32**, which was then coupled with 2-halopyridine **33** to secure 2-(2-pyridyl)indole **34**.

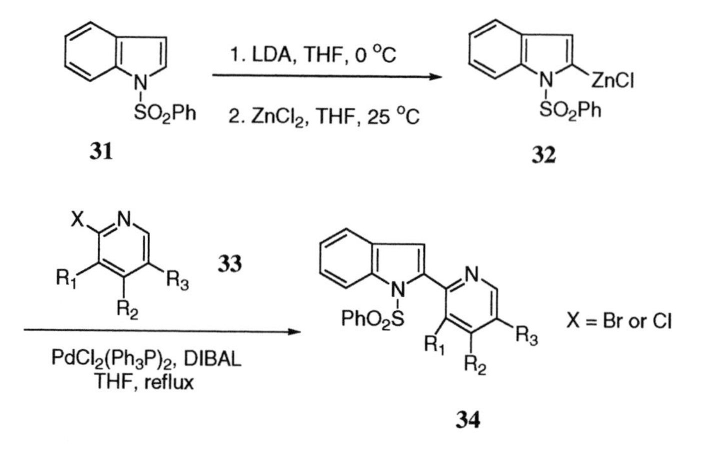


Recognizing that 1-(phenylsulfonyl)-3-lithioindole tends to isomerize to the corresponding 2-lithioindole derivative, Bosch *et al.* used a silyl ether protection to solve the problem. They prepared 3-indolylzinc reagent **36** from 3-bromo-1-(*tert*-butyldimethylsilyl)indole (**35**) and then coupled **36** with 2-halopyridine **33** to form 3-(2-pyridyl)indole **37**. Finally, the Negishi adduct **37** was further manipulated to a naturally occurring indole alkaloid, (±)-nordasycarpidone (**38**) [23, 27].

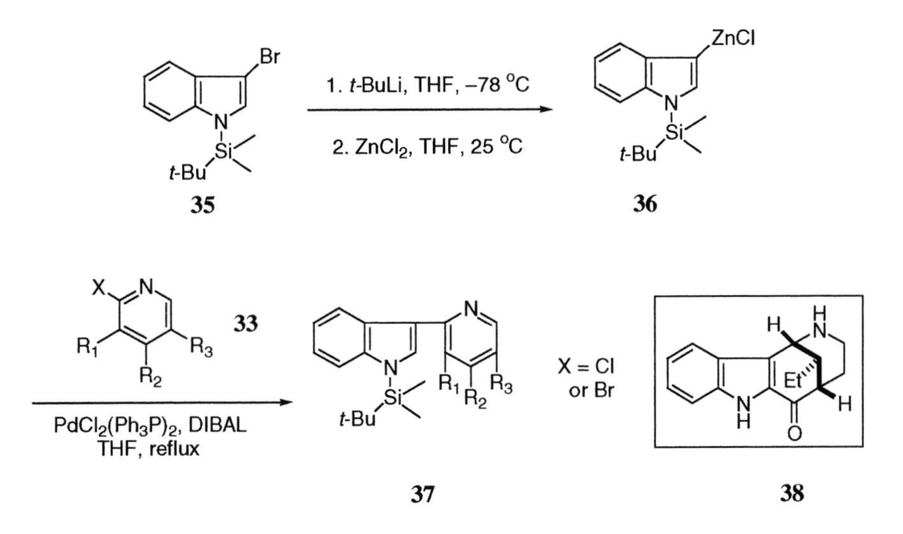


### 4.2.4 Suzuki coupling

There are many examples of the Suzuki coupling in which the pyridine motif is employed as a nucleophile in the form of a pyridylboron reagent. With the growing number of commercially available pyridylboronic acids, more and more applications using this approach are expected. Both diethyl (3-pyridyl)borane and diethyl (4-pyridyl)borane are readily accessible from 3-bromopyridine and 4-bromopyridine, respectively, *via* halogen-metal exchange and reaction with triethylborane [29–32]. The two pyridylboranes have been coupled with a variety of aryl and heteroaryl halides, as exemplified by the coupling of diethyl (3-pyridyl)borane and 2-nitrophenylbromide to form phenylpyridine **39** [30, 31].

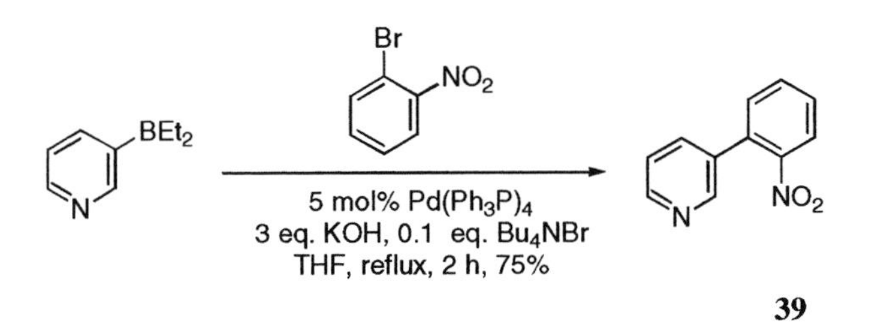


Two additional examples of the Suzuki coupling using diethyl (3-pyridyl)borane include the coupling with 4-chloro-2,3'-bipyridine 1'-oxide (**40**) to give terpyridine **41** [7] and the coupling with 1,8-dibromonaphthalene to afford a mixture of two atropisomers **42** and **43** of 1,8-di(3'-pyridyl)naphthalene [33, 34]. Interestingly, the N atoms in **42** and **43** provide enough steric and spectral differentiation to make the detection of atropisomers possible at room temperature.

Therefore, chirality is achieved without an exocyclic barrier to rotation as is required in the corresponding carbocyclic aryl substrates.

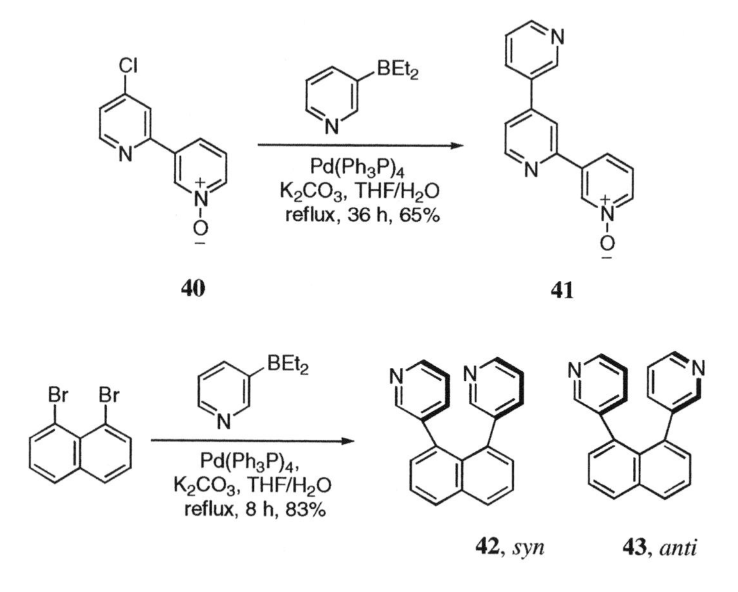


Li and Yue observed an ethyl transmetalation during the Suzuki coupling of diethyl (3-pyridyl)borane and 3-bromoquinoxaline **44** [35]. In the presence of a strong base (NaOH), the Suzuki reaction of diethyl (3-pyridyl)borane with 3-bromo-6,7-dichloroquinoxalin-2-ylamine (**44**) proceeded to give 52% of the 3-pyridylquinoxaline **45** accompanied by 21% of 3-ethylquinoxaline **46**.

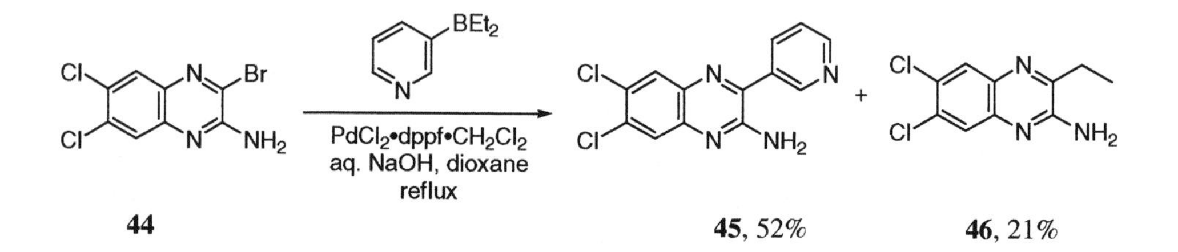


In contrast to 3- and (4-pyridyl)boranes, 2-pyridylborane is considered an unsuitable Suzuki coupling partner because it forms an unusually stable cyclic dimer resembling a dihydroanthracene. This obstacle can be circumvented by using 2-pyridylboronic ester in place of 2-pyridylborane. For example, Diederich's group synthesized 2-(2-pyridyl)-8-methylquinoline (**49**) *via* the Suzuki coupling of 2-pyridylboronic ester **47** and 2-bromo-8-methylquinoline (**48**) [36].

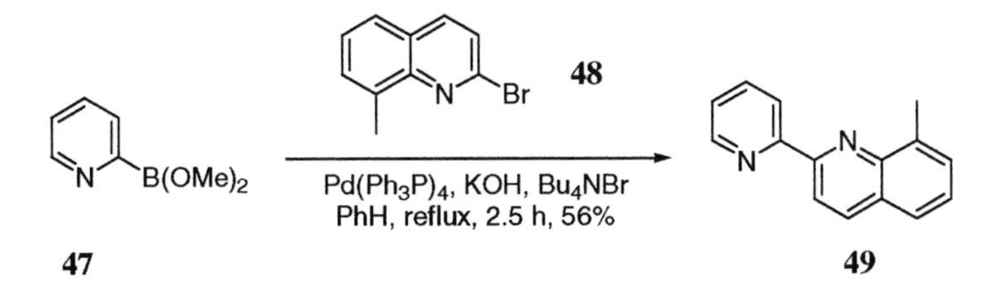


By taking advantage of the kinetic acidity at the β-position of 2-halopyridines, the Achad group prepared pyridylboronic acid **51** from *ortho*-lithiation of 2,6-dichloropyridine and quenching the resulting 3-lithio-2,6-dichloropyridine **50** with trimethylborate [37]. Obviously, pyridylboronic acid **51** is advantageous over diethyl pyridylborane because no ethyl transmetalation can occur. In another case, 4-pyridylboronic acid was used to construct a fully-substituted 2-(4'-pyridyl)pyridine **53** from 3-bromopyridine **52** [38].

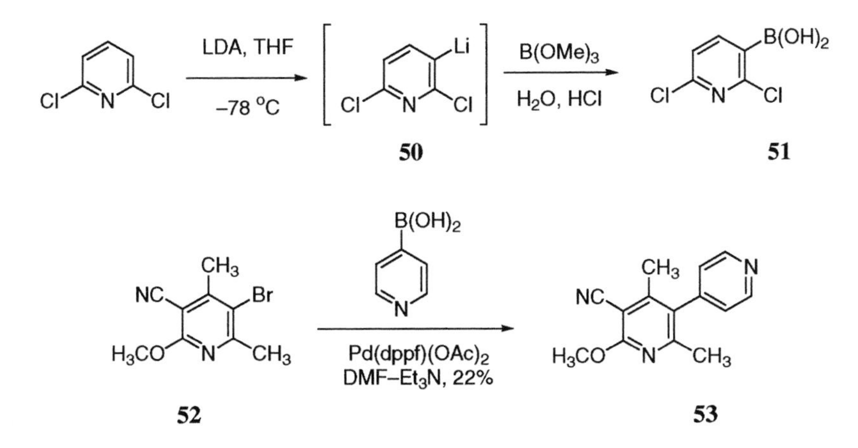


Due to the abundance of halopyridines, Suzuki reactions in which they serve as electrophiles have been used to synthesize a plethora of arylpyridines and heteroarylpyridines [39, 40]. 6-Phenyl-2-nitro-3-methylpyridine (**55**) was obtained *via* the reaction of 6-bromo-2-nitro-3-methylpyridine (**54**) and phenylboronic acid [39]. Not surprisingly, 3-chloropyridine was virtually inert under such conditions in contrast to 2-chloro- and 4-chloropyridine. The Suzuki coupling of 2-hydroxy-6-bromopyridine failed as well, possibly because 2-hydoxypyridines exist almost entirely in the keto form.

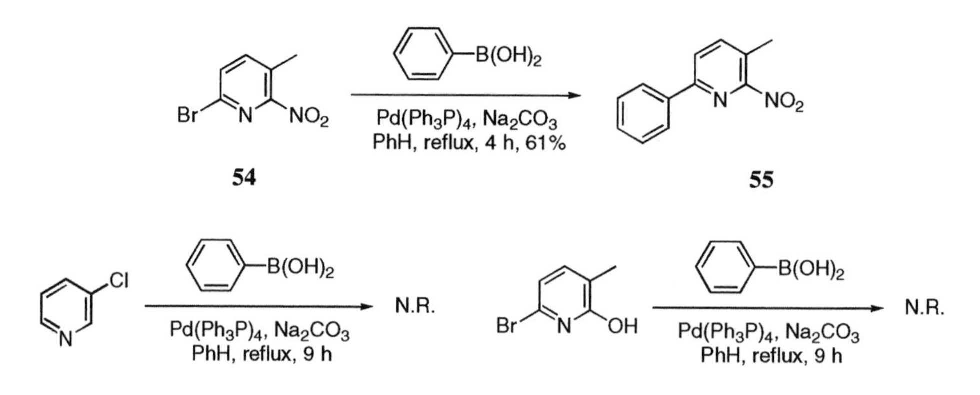


Because of their π-electron-deficient nature, halopyridines normally act to retard the insertion of carbon monoxide into Ar—Pd—X intermediates, and decrease the rate of transmetalation for generating the Ar—Pd—Ar' species. As a result, Pd-catalyzed three-component cross-coupling reactions between halopyridines, arylmetal reagents, and carbon monoxide generally require relatively high pressure to suppress side-reactions. Nevertheless, the Pd-catalyzed carbonylative cross-coupling reaction of 2-iodopyridine and phenylboronic acid resulted in unsymmetrical biaryl ketone **56** under mild conditions [40]. In another case, the Suzuki coupling of 2-bromo-3-methylpyridine and the very bulky phenylboronic acid **57** was feasible using potassium *tert*-butoxide as the base to furnish phenylpyridine **58** [41].

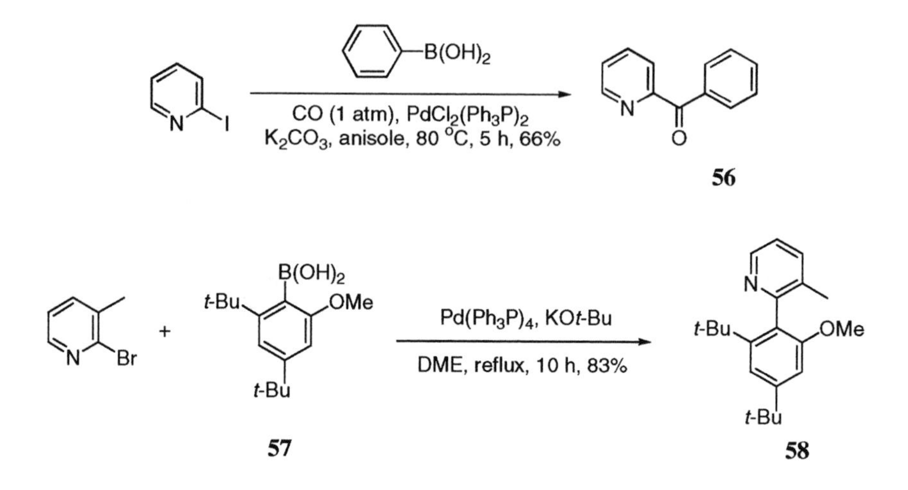


Halopyridines have been also cross-coupled with phenyl- [42], quinolinyl- [43], indolyl- [44], pyrimidyl- [45, 46], and a great many other heteroarylboron reagents, providing access to a wide variety of heterobiaryls.

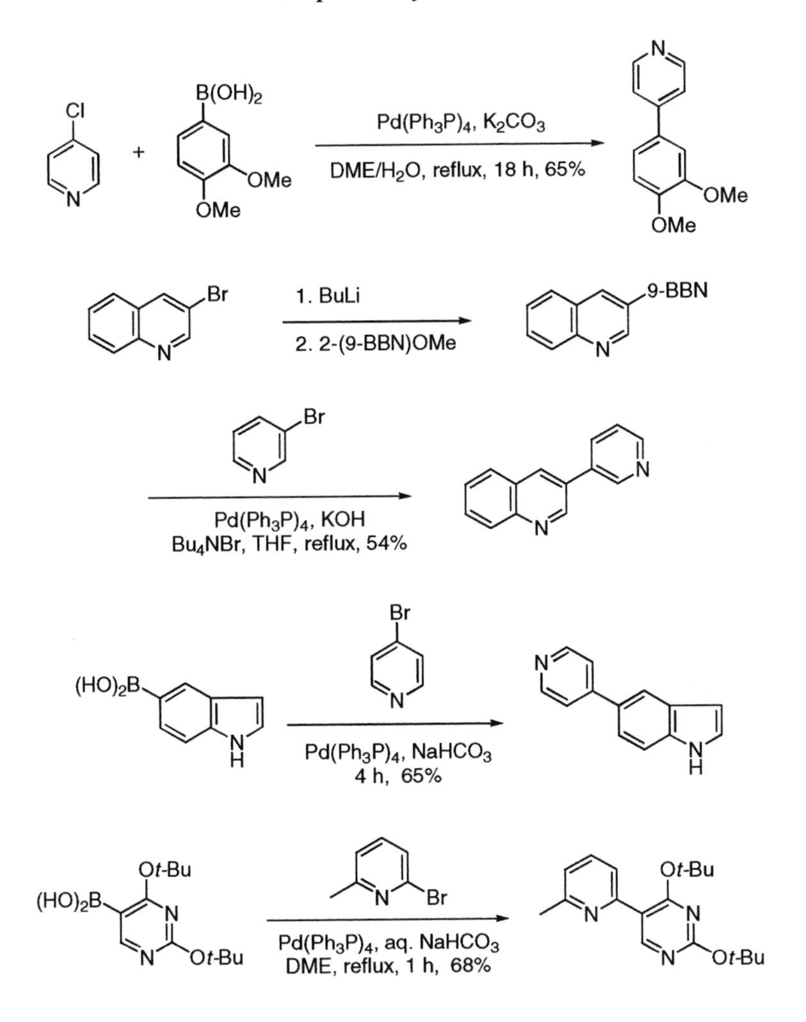


Pyridine-containing tricyclic compounds have been produced *via* a sequence consisting of a Suzuki reaction and a subsequent annulation. Gronowitz *et al.* coupled 2-formylthienyl-3-boronic acid with 3-amino-4-iodopyridine. The resulting adduct spontaneously condensed to yield thieno[2,3-*c*]-1,7-naphthyridine **59** [47]. They also synthesized thieno[3,4-*c*]-1,5-naphthyridine-9-oxide (**60**) in a similar fashion [48]. Neither the amino nor the *N*-oxide functional group was detrimental to the Suzuki reactions.

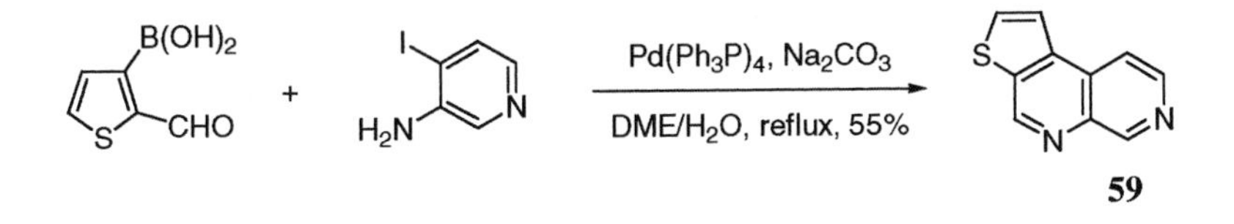

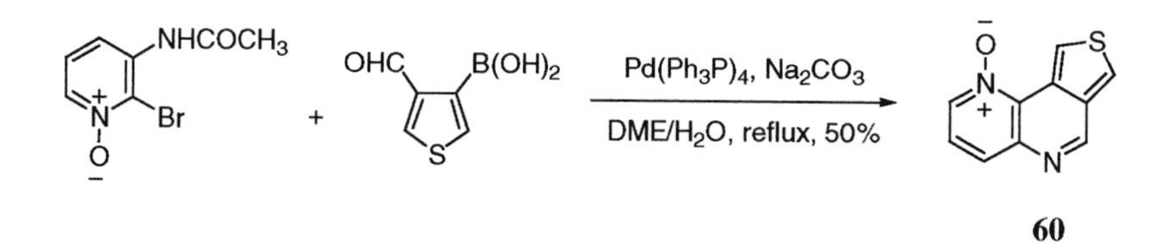


Quéguiner's group enlisted a combination of directed metalation and a Pd-catalyzed cross-coupling reaction for the construction of heteroaryl natural products [49]. One example was the total synthesis of bauerine B (**64**), a β-carboline natural product [50]. *Ortho*-lithiation of 2,3-dichloro-*N*-pivaloylaminobenzene (**61**) was followed by reaction with trimethylborate to provide boronic acid **62** after hydrolysis. The Suzuki reaction between **62** and 3-fluoro-4-iodopyridine led to the desired biaryl product **63** contaminated with the primary amine (ca. 30%), both of which were utilized in the total synthesis of bauerine B (**64**). Another β-carboline natural product, the antibiotic eudistomin T (**65**), and a few other hydroxy β-carbolines have also been synthesized in the same fashion [3, 51].

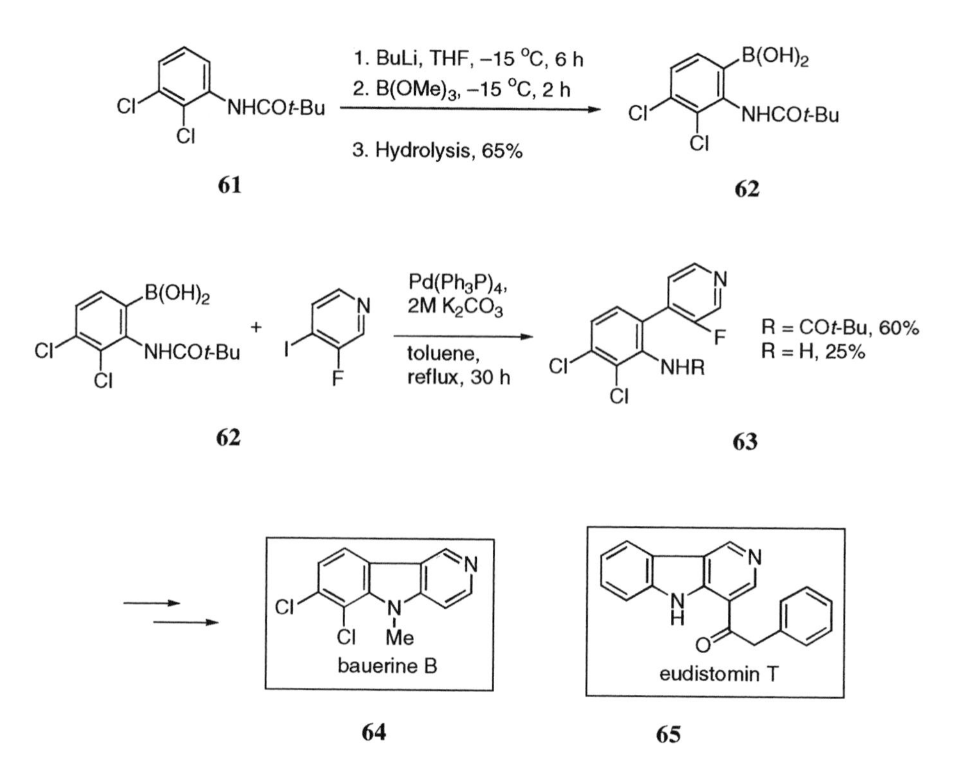

In Baldwin's formal total synthesis of haliclamines A and B, a Suzuki coupling of 3-bromopyridine was the central operation [52]. Chemoselective hydroboration of diene **66** employing 9-BBN occurred at the less hindered terminal olefin. Suzuki coupling of the resulting alkylborane with 3-bromopyridine then furnished alkylpyridine **67** as a common intermediate for the synthesis of haliclamines A and B.

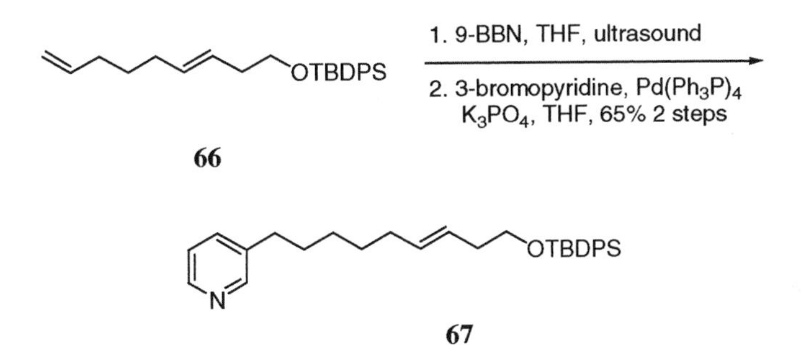


### 4.2.5 Stille coupling

Stille coupling is regarded as the most versatile method among all Pd-catalyzed cross-coupling reactions with organometallic reagents. For a Stille coupling involving a pyridine moiety, the pyridine fragment may serve as either a nucleophilic or an electrophilic coupling partner. Therefore, a choice can be judiciously made when designing coupling partners.

#### *4.2.5.a The pyridine motif as a nucleophile*

Several different approaches have been employed to prepare pyridylstannanes, one of which involves direct metalation followed by transmetalation with $R_3SnCl$. A regioselective metalation of 2-methoxypyridine was achieved using a mixture of bases comprised of BuLi–$LiO(CH_2)_2NMe_2$ [53]. Subsequent transmetalation with $Bu_3SnCl$ led to 2-methoxy-6-(tributylstannyl)pyridine (**68**), which was then coupled with pyridyl-, pyrimidyl-, and quinolinyl bromides in 31–56% yield.

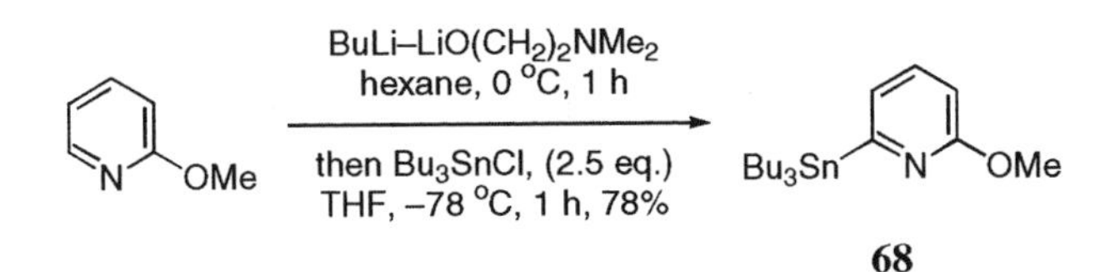

In addition, *ortho*-lithiation of halopyridines with bulky bases such as LDA was followed by reaction with $R_3SnCl$ to produce the desired halostannylpyridines as bifunctional building blocks [54, 55].

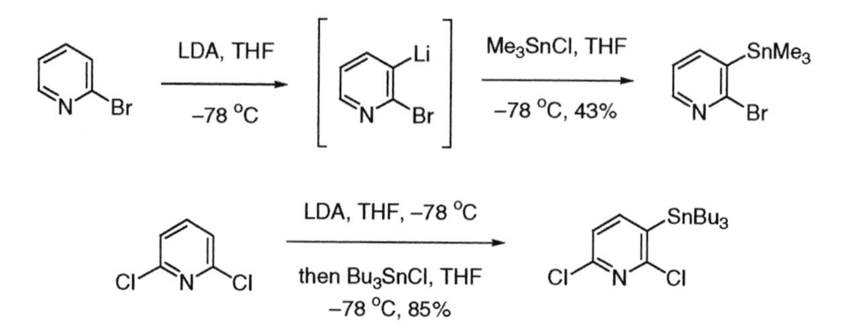


The second procedure for preparing pyridylstannanes is halogen-metal exchange of a halopyridine using BuLi followed by quenching the resulting lithiopyridine with $Bu_3SnCl$. Interestingly, halogen-metal exchange of 2,5-dibromopyridine occurred *regiospecifically at C(5)*, giving rise to 2-bromo-5-lithiopyridine, which upon treatment with $Bu_3SnCl$ afforded 2-bromo-5-(tributylstannyl)pyridine (**69**) [56].

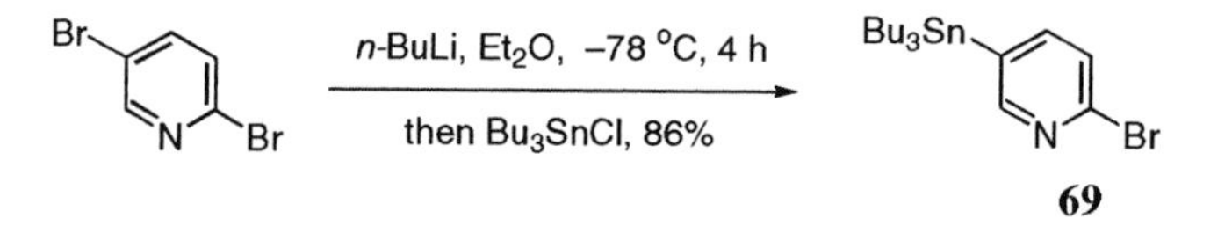


The third methodology in pyridylstannane synthesis is unique to 2,6-dihalopyridines whose synthesis using the conventional lithiation method is low-yielding (17% yield) [57]. Schubert *et al.* prepared 2,6-bis(trimethyltin)pyridine (**70**) *via* an $S_NAr$ displacement of 2,6-dichloropyridine with sodium trimethylstannane, derived *in situ* from trimethyltin chloride [58–60].

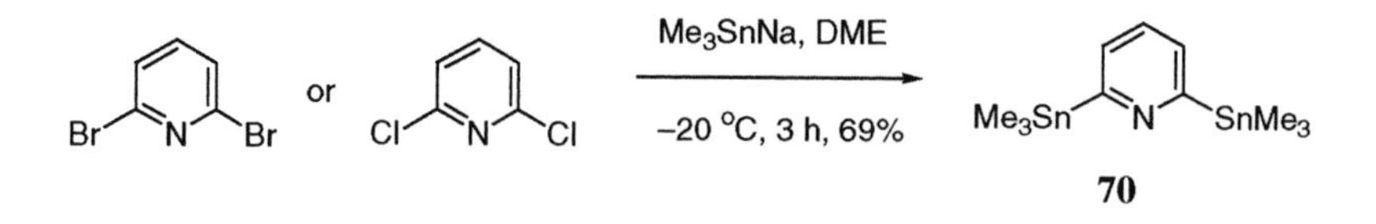


The fourth approach involves the Pd-catalyzed reaction between a halopyridine (or a triflate) with hexaalkylditin. This protocol is especially applicable to substrates with base-sensitive moieties. Pyridylstannane **72** was synthesized from the Pd-catalyzed cross-coupling of bromopyridine **71** and hexamethyldistannane [61].

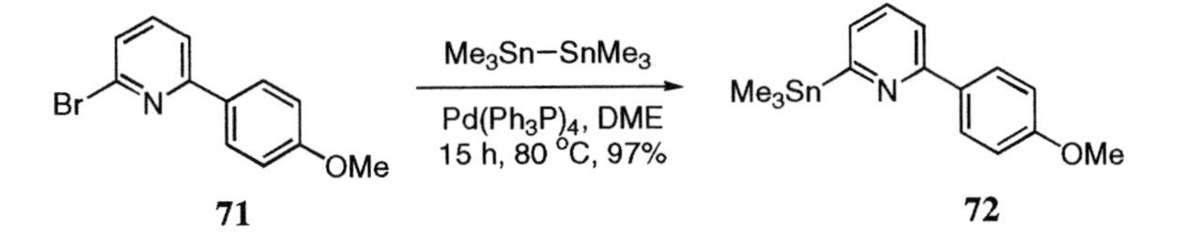


Pyridylstannanes have been cross-coupled with numerous aryl- and heteroaryl halides as well as various other electrophiles. For example, an extension of Stille's original methodology to 3-trimethylstannylpyridine and acid chloride **73** gave the corresponding ketone **74**, which was then converted to 2*S*-(+)-nicotinylalanine **75**, a neuroprotection agent [62].

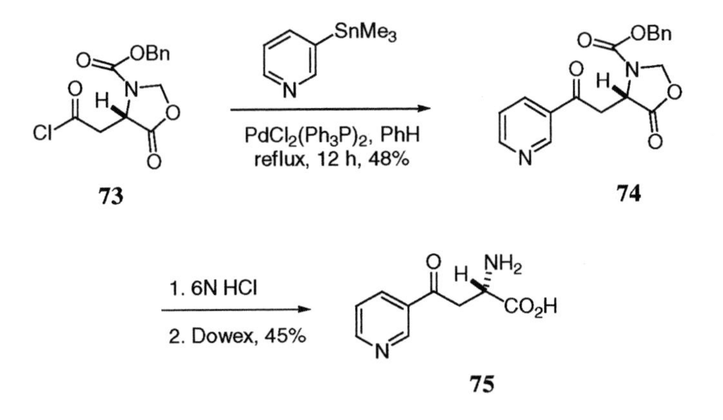


The Stille coupling of α-iodo enones is sluggish under standard conditions. Significant rate enhancement was observed for the Stille reaction of 2-chloro-5-tributylstannylpyridine and α-iodo enone **76** using triphenylarsine as the soft palladium ligand and CuI as the co-catalyst [63]. Oxygenated functionalities did not affect the efficiency of the reaction provided both $Ph_3As$ and CuI were added. Additional manipulations of **77** resulted in the synthesis of (+)-epibatidine (**78**).

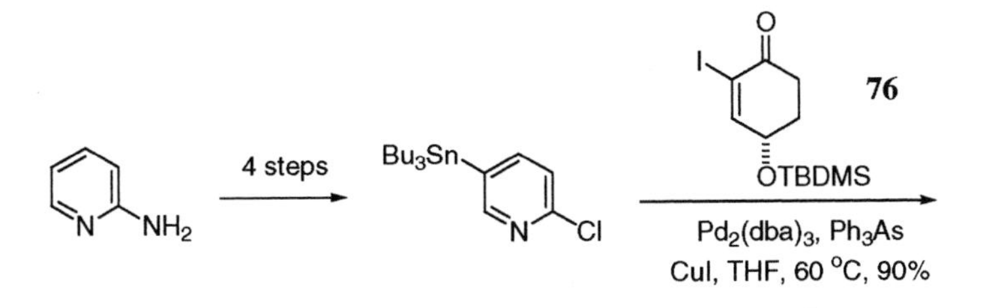

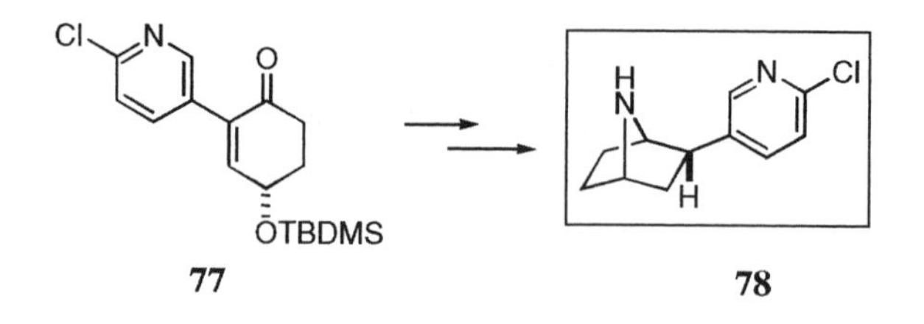


Vinyl- [64] and aryl triflates [65] as pseudo organic halides readily couple with stannylpyridines as long as more than one equivalent of LiCl is present in the reaction mixture. Presumably, the transmetalation is facilitated by replacing triflate with $Cl^-$ at the palladium intermediate generated from oxidative addition. For example, 2-(benzopyran-4-yl)pyridine **80** was obtained from treatment of vinyl triflate **79** with 2-trimethylstannylpyridine in the presence of $Pd_2(dba)_3(CHCl_3)$–$Ph_3P$ and LiCl [64].

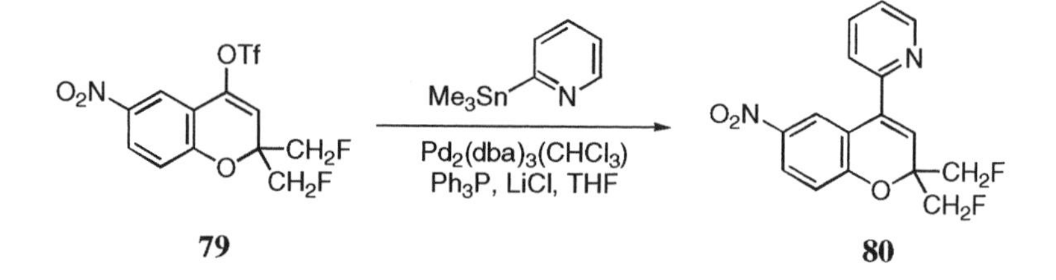


The parent 2-tributylstannylpyridine has been used in cross-coupling reactions with aryl iodides [66], bromobenzyl phosphonates [67], and bromophosphinine **81**, although the lifetime of mono-coupled pyridylphosphinine **82** was rather short [68]. Moreover, ample examples have been found for the cross-coupling of pyridylstannanes with halothiophenes [69–71], halothiazoles [72], halopyrimidines [45], halofuran, and halopyrroles [73].

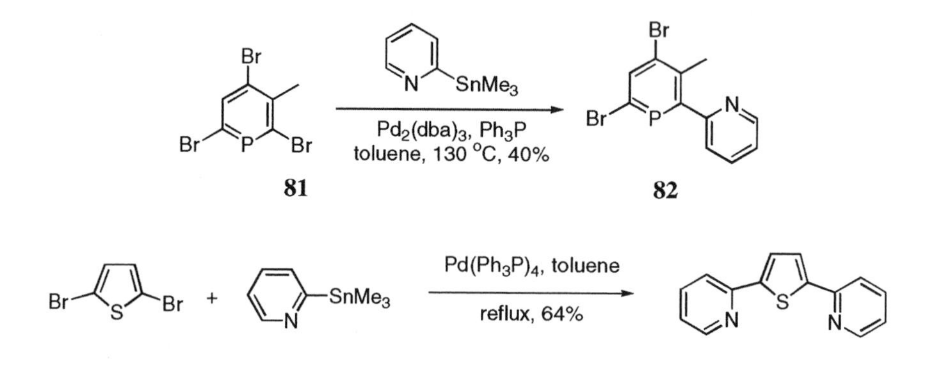

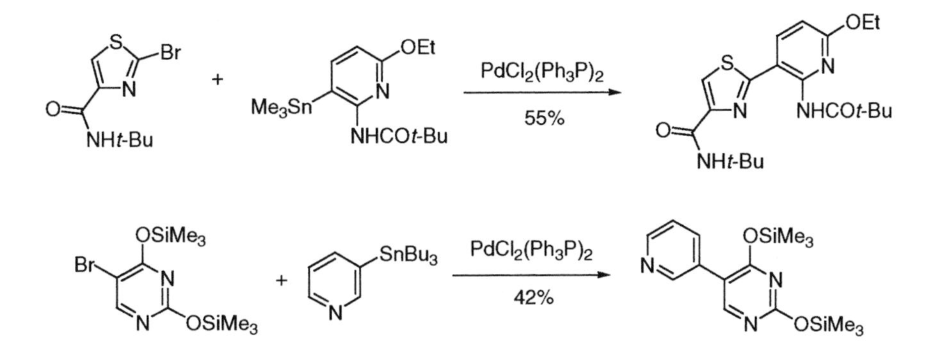


Pyridine-pyridine connections *via* a Stille reaction have been well precedented for the syntheses of bipyridine, terpyridine, tetrapyridine and other oligomers of pyridine [73–75]. These reactions are exemplified by the synthesis of tetrapyridine **84** [75] from terpyridyl chloride **83** and 2-tributylstannylpyridine.

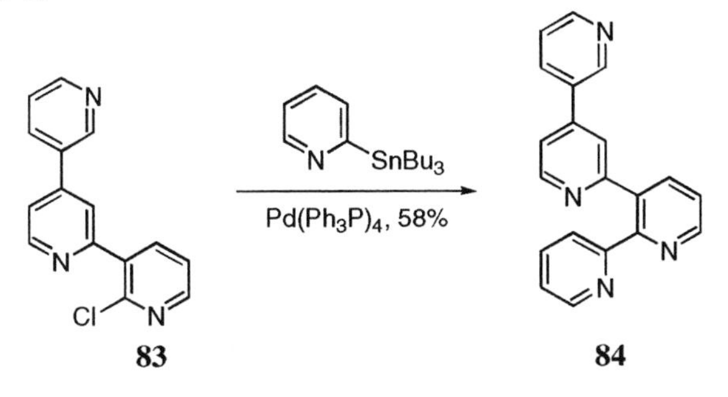


The foregoing observations have been extended to their respective *N*-quaternized pyridinium derivatives, either in the nucleophile fragment [75] or the electrophile fragment [75–78]. In one case [76], the cross-coupling of the quaternary pyridylstannane **85**, readily prepared by refluxing 3-(tributylstannyl)pyridine with methyl tosylate, with 2-chloropyrazine in the presence of Pd(dppe)$Cl_2$ proceeded smoothly to give adduct **86** in 87% yield. In contrast, only a 29% yield of the coupling adduct was isolated from the Stille reaction of 3-(tributylstannyl)pyridine *N*-oxide and 2-chloropyrazine. In another case, the cross-coupling of 3-bromopyridine *N*-oxide (**87**) and 2-(trimethylstannyl)pyridine produced pyridine-pyridine *N*-oxide **88** [77].

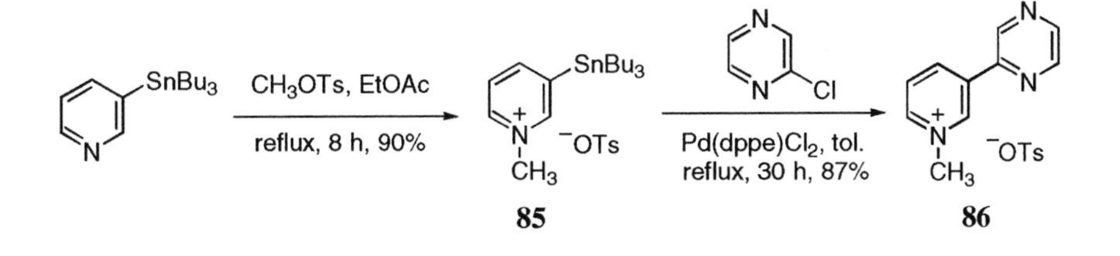

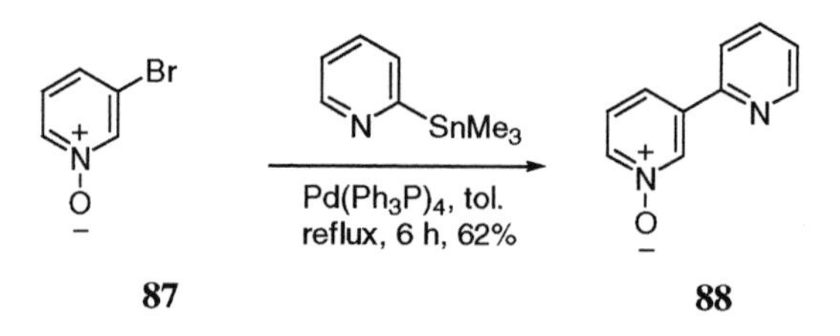


Interestingly, adduct **89**, resulting from the Stille coupling of a pyridylstannane and a bromofuran, may be further manipulated into triflate **90**, a precursor in a novel phenol-forming reaction [79]. In the presence of an appropriate base, Pd-catalyzed cyclization of **90** gave rise to phenol **91**, which was trapped *in situ* with chloroformate to give methylcarbonate **92**. The cyclization mechanism presumably involved a nucleophilic attack of the enolate on the palladium-aryl group, resulting in a cyclic palladium species.

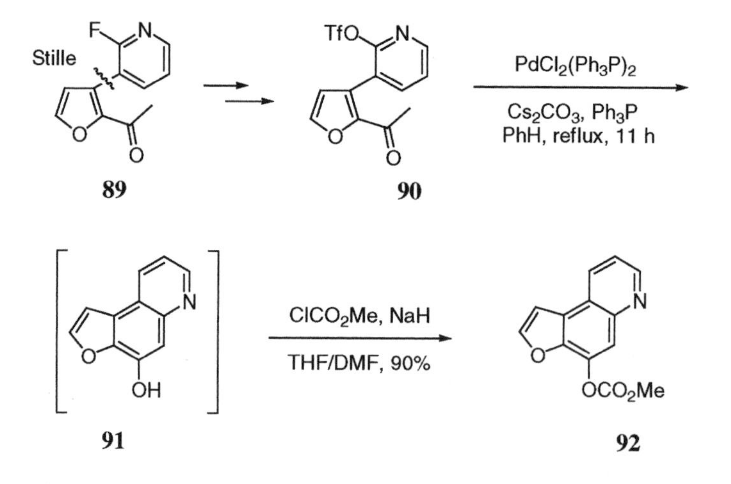


In a sequence analogous to the Suzuki reaction and annulation described in section 4.2.4, pyridine-containing tricyclic compounds have also been prepared *via* the Stille reaction and a subsequent annulation [80, 81]. For instance, benzo[c]-2,7-naphthyridine **94** was assembled from the adduct of 3-formyl-4-(trimethylstannyl)pyridine (**93**) and 2-bromo-4-methoxyacetanilide. The reaction was facilitated by addition of CuO as the co-catalyst.

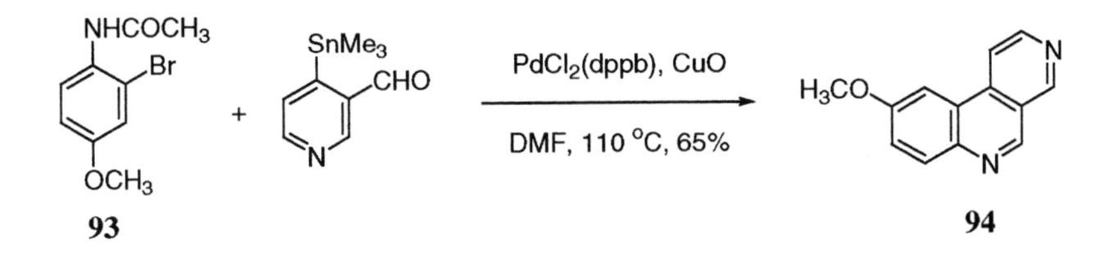


### *4.2.5.b The pyridine motif as an electrophile*

Numerous stannanes have been coupled with halopyridines as electrophiles in the Stille coupling. One of the simplest of these is vinylstannane [82–84]. The Stille reaction of bromopyridine **95** with tributylvinyltin gave angustine (**96**) [84], an indolopyridine alkaloid. Bromopyridine **95** also took part in a three-component carbonylative-Stille coupling sequence to provide an entry to another indolopyridine alkaloid, naucletine (**97**) [84].

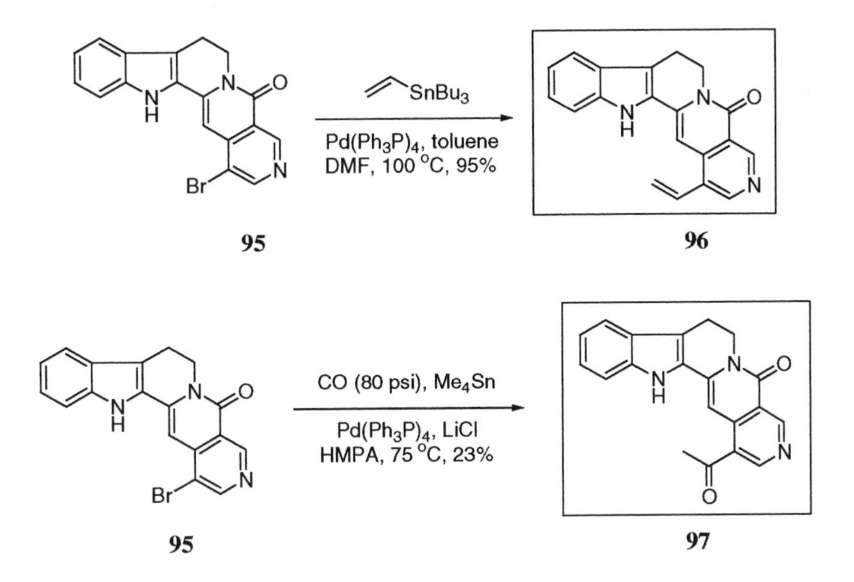


Both 1-ethoxy-2-tributylstannylethene [85, 86] and 1-ethoxy-1-tributylstannylethene [6, 87] are versatile building blocks utilized in Stille couplings. Their Stille adducts can then be hydrolyzed to the corresponding aldehyde and methyl ketone, respectively, as exemplified by the transformation of bromopyridine **98** into methyl ketone **99** [6].

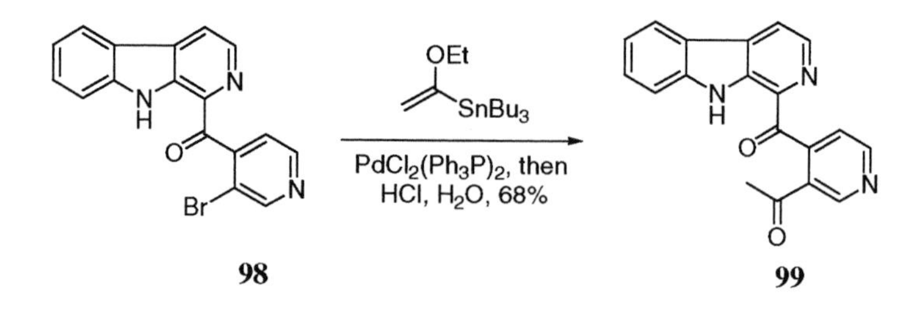


Other tin reagents have found use in Pd-catalyzed cross-couplings with halopyridines as well. The Stille coupling of 3-iodopyridine with ethoxy(tributylstannyl)acetylene gave rise to 3-ethoxyethynylpyridine (**100**), which was then hydrolyzed to the corresponding ethyl 3-pyridylacetate (**101**) [88]. Carbamoylstannane **102** was prepared by sequential treatment of lithiated piperidine with carbon monoxide and trimethyltin chloride. Stille coupling of carbamoylstannane **102** and 3-bromopyridine provided a unique entry to amide **103** [89].

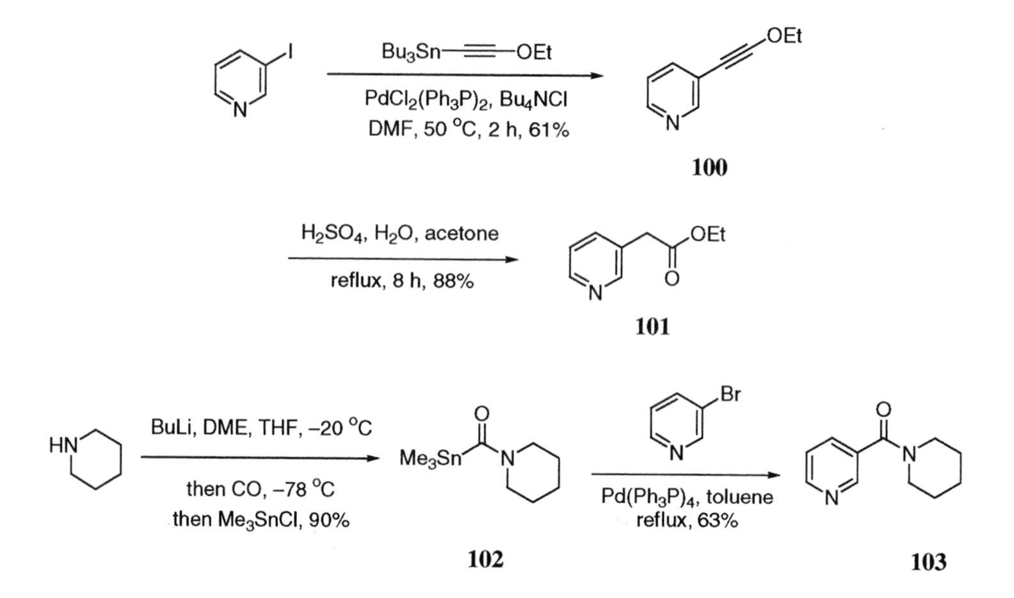


Innumerable aryl- and heteroarylstannanes take part in Stille couplings with halopyridines. In one such example, the union of arylstannane **106** and 4-bromopyridine furnished arylpyridine **107** [90]. Arylstannane **106** was prepared from the Pd-catalyzed reaction of hexabutylditin with iodoarene **105**, which arose from aminobenzolactam **104** *via* a Sandmeyer reaction.

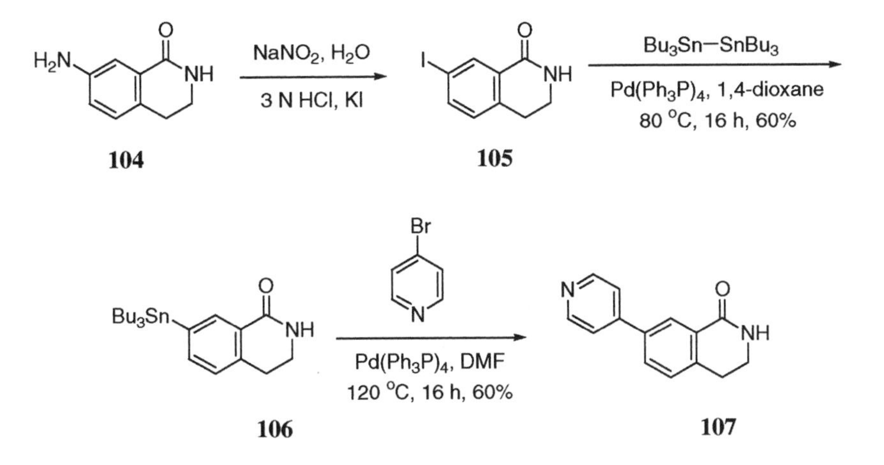


Many pyridine-indole compounds are biologically active. A growing number of methods for the preparation of indolylstannanes have been developed. 2-Trialkylstannylindoles, for example, have been synthesized *via* directed metalation followed by reaction with tin chloride [91–93]. The latest indolylstannane syntheses include Fukuyama's free radical approach to 2-trialkylstannylindoles from novel isonitrile-alkenes [94], and its extension to an isonitrile-alkyne cascade [95]. Assisted by the chelating effect of the SEM group oxygen atom, direct metalation of 1-SEM-indole and transmetalation with $Bu_3SnCl$ afforded 2-(tributylstannyl)-1*H*-indole **108**, which was then coupled with 2,6-dibromopyridine to give adduct **109**.

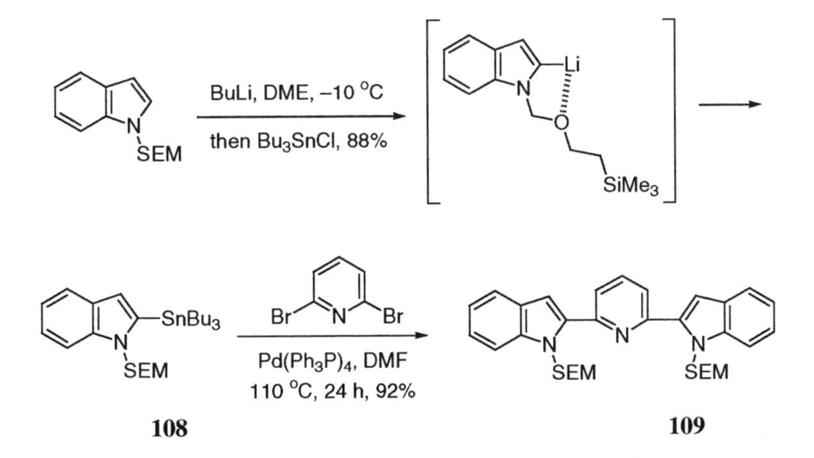


In comparison to 2-trialkylstannylindoles, the synthesis of 3-trialkylstannylindoles is more complicated since the halogen-metal exchange process usually leads to rapid isomerization to the corresponding 2-lithioindole species. In contrast, the Pd-catalyzed reaction of 3-iodoindole and hexabutylditin has proven to be a reliable preparative approach for making 3-tributylstannylindole (**110**) [96]. Thus, using the ditin chemistry, 1-tosyl-3-iodoindole was converted to 1-tosyl-3-tributylstannylindole, which was then coupled with 2-bromopyridine to furnish 1-tosyl-3-(2'-pyridyl)indole (**111**) [97].

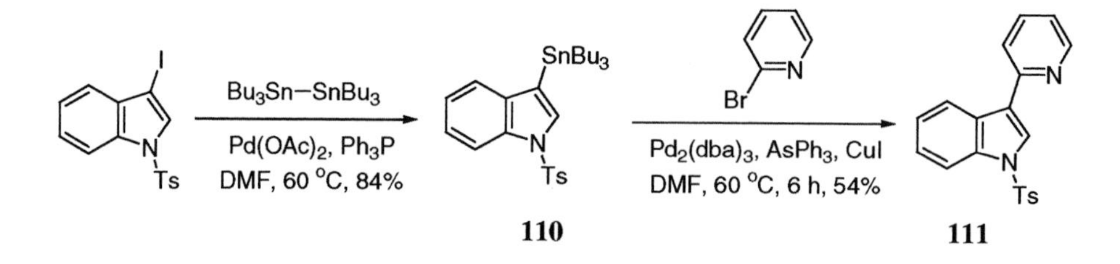


Quéguiner's group conducted a model study towards a convergent synthesis of the streptonigrin and lavendamycin alkaloid skeleton using the Stille coupling strategy. They coupled (2-quinolyl)trimethylstannane with 2-(4-phenyl-3-pivaloylamino)pyridyl triflate (**113**), derived from lactam **112**, to prepare the expected 2-[2-(4-phenyl-3-pivaloylamino)pyridyl]quinoline (**114**) [3, 98].

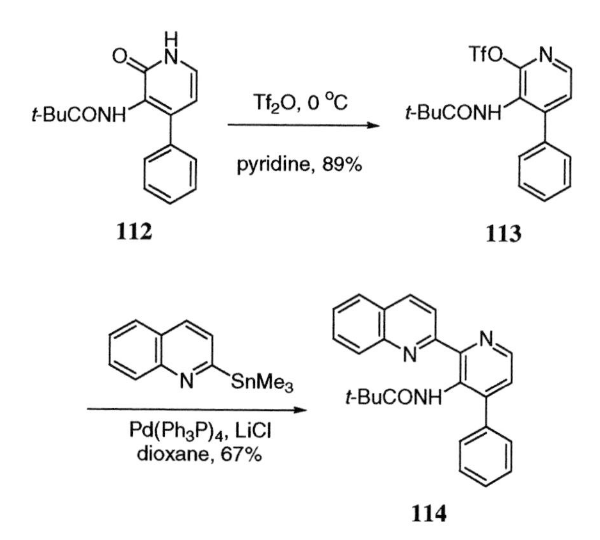


5-Tributylstannyl-2-methylthiopyrimidine (**115**) was prepared from 5-bromo-2-methylthiopyrimidine using either a Pd-catalyzed reaction with hexabutyldistannane in the presence of fluoride or a substitution reaction using a tributylstannyl anion. The subsequent coupling of **115** with 4-bromopyridine delivered the expected pyridylpyrimidine **116** [99].

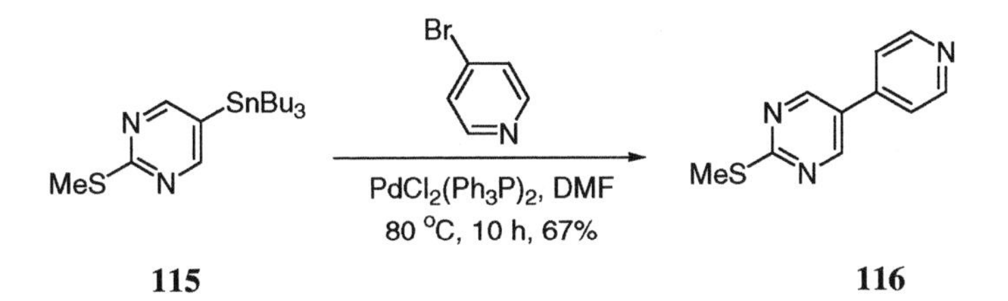


In a series of papers, the Gronowitz group described the synthesis of pyridine-substituted hydroxythiophenes *via* the Stille strategy [100]. 2-*tert*-Butoxythiophene was lithiated and quenched with $Me_3SnCl$ to give 5-tri-*n*-methylstannyl-2-butoxythiophene (**117**), which was coupled with 3-bromopyridine to produce thienylpyridine **118**. The Gronowitz group also synthesized many pyridine-containing condensed heteroaromatics. See section 4.2.4. for references and details.

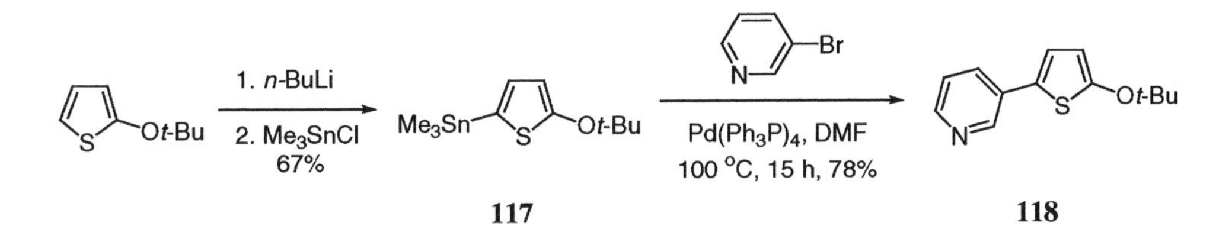


A Stille coupling of a bromopyridine on solid support was described by Snieckus' group [101]. Merrifield resin **119** was esterified with 3-bromopyridine-5-carboxylic acid to afford ester **120**. The Stille coupling of ester **120** on a solid support led to the expected hetero phenylpyridine **121**, which was then cleaved *via* basic hydrolysis to produce **122**. Snieckus' work has the potential for application to combinatorial chemistry and high throughput screening.

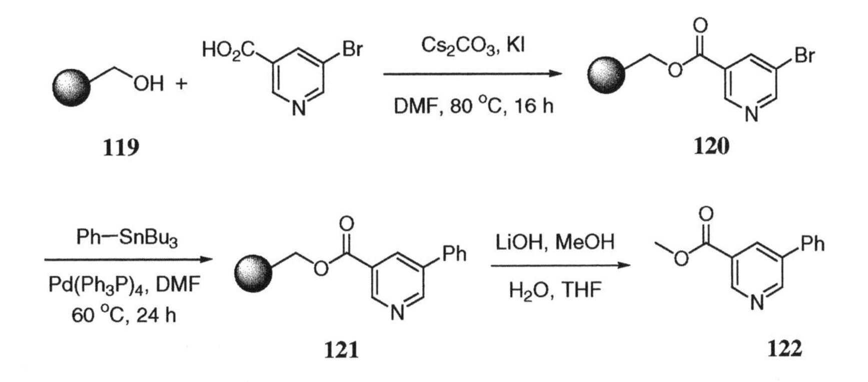

### 4.2.6 Hiyama coupling

There have been no reports on Pd-catalyzed cross-coupling involving a pyridylsilicon reagent. However, several publications have appeared describing Hiyama couplings in which halopyridines served as the electrophilic partners. Treating chlorosilane **123** with KF resulted in a pentacoordinate species, which was coupled with 3-bromopyridine to give the unsymmetrical arylpyridine **124**. The same reaction furnished an isomer of **124** in only 23% yield when 2-bromopyridine was used in place of 3-bromopyridine [102].

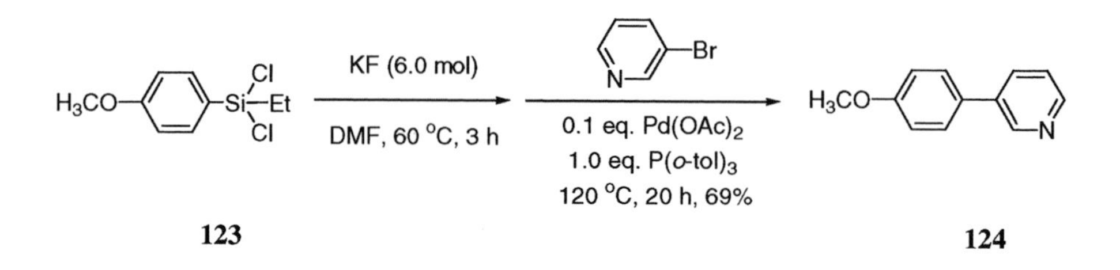


The C—Si bond can be activated by NaOH as well as a fluoride source, whereas a series of other inorganic bases such as LiOH, KOH, $K_2CO_3$, and $Na_2CO_3$ are less effective. Thus, (*E*)-hexenylchlorosilane **125** was coupled with 2-chloropyridine, giving rise to 2-(*E*)-hexenyl-pyridine (**126**) [103].

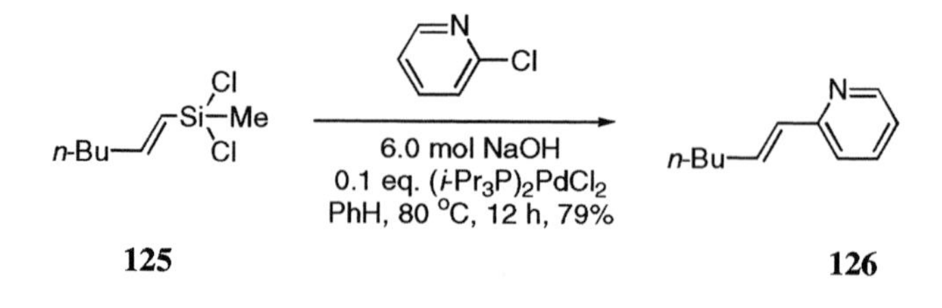


While chlorosilanes are susceptible to hydrolysis, aryltrialkoxysilanes are not. Treating the Grignard reagent generated from aryl bromide **127** with $SiCl_4$ then with a solution of pyridine in methanol furnished aryltrimethoxysilane **128** [104]. Subsequent Pd-catalyzed coupling between **128** and 3-bromopyridine assembled biaryl **129** in 72% yield with the aid of TBAF.

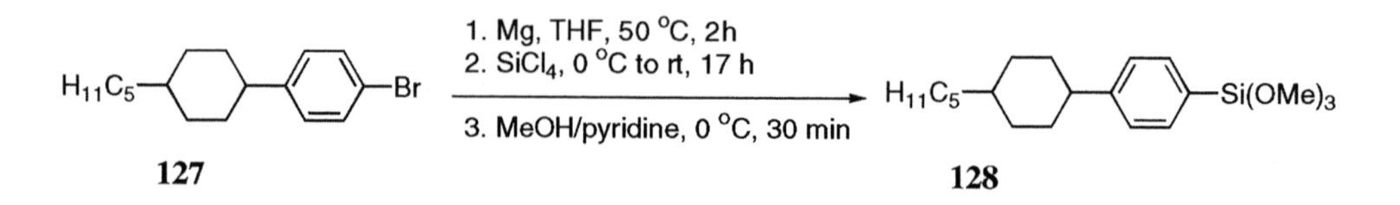

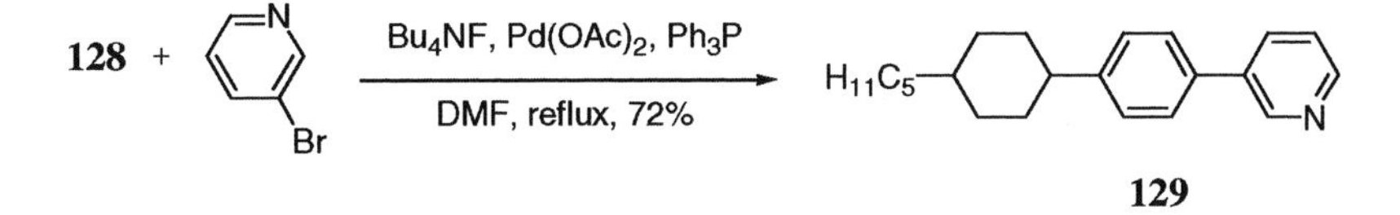


In addition, commercially available phenyltrimethoxysilane has been successfully cross-coupled with heteroaryl bromides including both bromopyridines and bromothiophenes [105]. Both 2-bromopyridine and 3-bromopyridine were coupled with phenyltrimethoxysilane to give the corresponding phenylpyridines in 76% and 62% yield, respectively. Employing phenyl tris(trifluoroethoxy)silane, a siloxane with electron-withdrawing groups, did not improve the yields.

## 4.3 Sonogashira reaction

Halopyridines, like simple carbocyclic aryl halides, are viable substrates for Pd-catalyzed cross-coupling reactions with terminal acetylenes in the presence of Pd/Cu catalyst. The Sonogashira reaction of 2,6-dibromopyridine with trimethylsilylacetylene afforded 2,6-bis(trimethylsilyl-ethynyl)pyridine (**130**), which was subsequently hydrolyzed with dilute alkali to provide an efficient access to 2,6-diethynylpyridine (**131**) [106]. Extensions of the reactions to 2-chloropyridine, 2-bromopyridine, and 3-bromopyridine were also successful albeit at elevated temperatures [107].

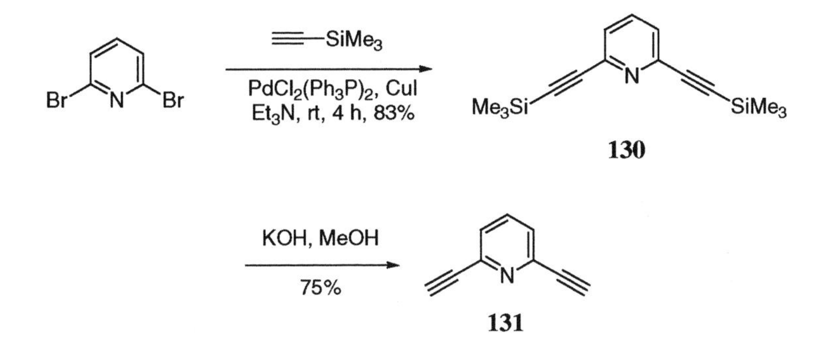


Regioselective mono-acetylenation of 2,5-dibromopyridine with trimethylsilylacetylene afforded 2-trimethylsilylethynyl-5-bromopyridine (**132**) [108, 109]. The regioselectivity was in contrast to the lithiation of 2,5-dibromopyridine in which the 5-position was more reactive.

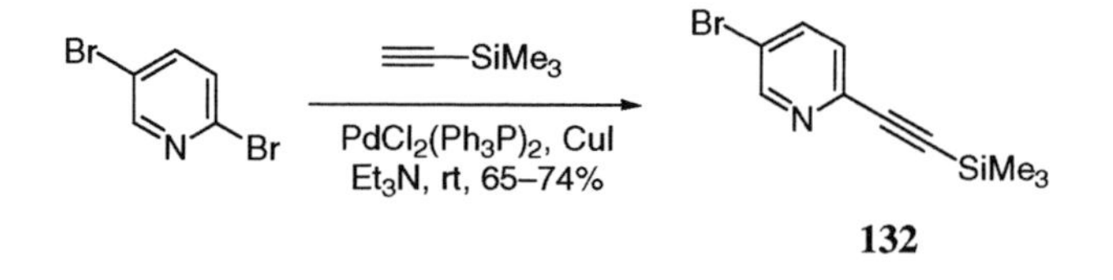


Due to the ease of desilylation either *via* alkaline hydrolysis or using fluoride sources [110], the aforementioned Sonogashira/desilylation sequence has been routinely employed to synthesize ethynylpyridines [111]. An interesting application of such a process was the synthesis of the 7-azaindole derivative **135** [112]. Thus, the Sonogashira adduct **134** was generated from the coupling of 3-iodopyridine **133** and trimethylsilylacetylene. Upon refluxing **134** with CuI in DMF, the cyclization product, 7-azaindole derivative **135**, was isolated in 40% yield, along with 35% of ethynylpyridine **136** resulting from proteodesilylation of **134** during either the reaction or workup.

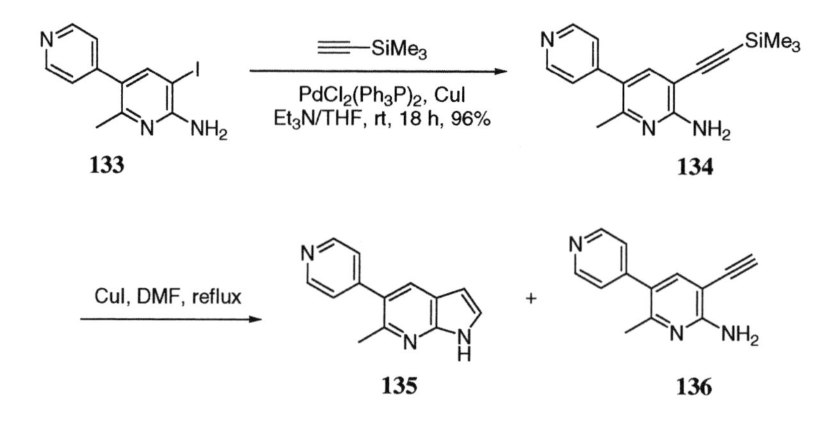


Preparation of 2,3-diethynylpyridine (**140**) proved to be a challenging task. When 2,3-dichloropyridine was used as the substrate, using a combination of $Pd(Ph_3P)_4$ and diisopropylamine at elevated temperature, only monoacetylenation was achieved at the C(2) position, giving rise to 3-chloro-2-trimethylsilylethynylpyridine [113]. The outcome was not completely unexpected considering that the C(3) position was not as activated. An alternative route began with 2-bromo-3-pyridinol, which was converted to its corresponding triflate (**138**). The subsequent coupling reaction of **138** with trimethylsilylacetylene resulted in 2,3-bis(trimethylsilylethynyl)pyridine (**139**), which was desilylated to secure the desired 2,3-diethynylpyridine (**140**).

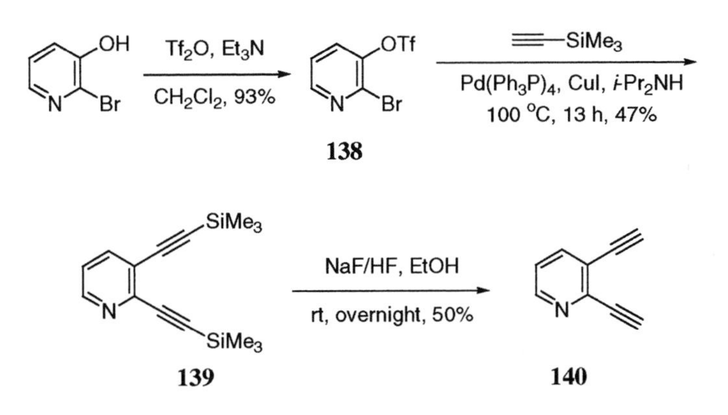


Additional examples of the Sonogashira reactions of pyridine triflates include coupling of 2-pyridyltriflate and 3-hydroxy-3-methylbut-1-yne to afford alkyne **141** [114]. The carbinol adduct could be readily unmasked to give 2-ethynylpyridine *via* a basic-catalyzed "retro-Favorsky" elimination of acetone. Due to the volatility of 2-ethynylpyridine, use of a high boiling liquid such as paraffin oil for the basic hydrolysis made the distillation more convenient [115].

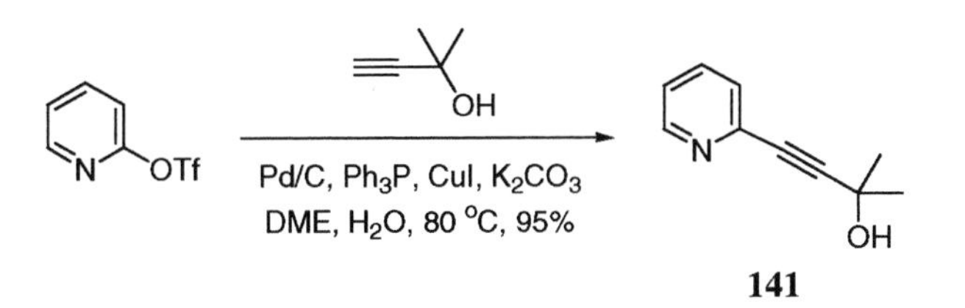


An unusual Sonogashira coupling between an *alkynyl bromide* and a 4-ethynylpyridine **142** was reported to produce a new complexing agent **143** containing a highly-conjugated pyridine subunit [116]. In addition, alkynylpyridines, as important intermediates in medicinal chemistry,

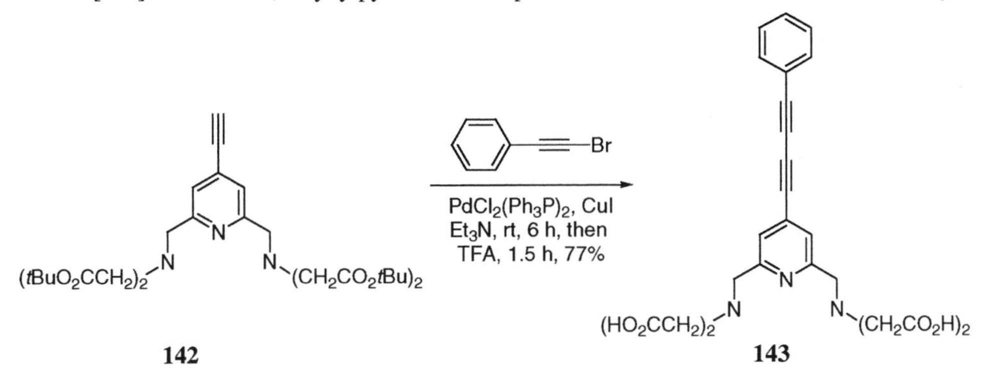

have been prepared utilizing the Sonogashira approach. 3-Pyridylhexylnitrile (**144**), derived from 3-bromopyridine and 5-hexynylnitrile, was transformed to *N*-(pyridylhexyl)pyrido[2,1-*b*]quinazoline-8-carboxamide (**145**), an orally active antiallergy agent [117]. Ethynylpyridine **147**, arising from 3-bromopyridine derivative **146** and 3-hydroxy-undec-5-(*Z*)-en-1-yne, was one of the later intermediates for the synthesis of novel antagonists of leukotriene $B_4$ ($LTB_4$) [118].

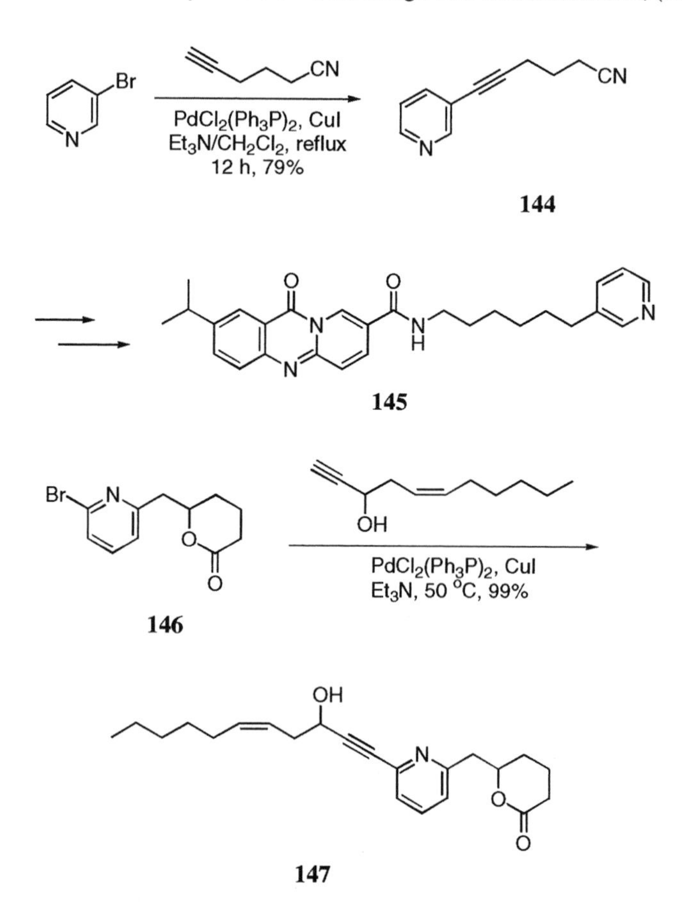


In the presence of a pendant hydroxyl group, the Sonogashira alkynylpyridine adduct can cyclize *in situ* to produce 2-substituted furopyridines [119–121]. Thus, alkynylpyridinol **148**, from the coupling of 2-iodopyridin-3-ol and phenylacetylene, underwent a spontaneous 5-*endo-dig* cyclization promoted by either Pd or Cu to give 2-phenylfuropyridine **149** [118].

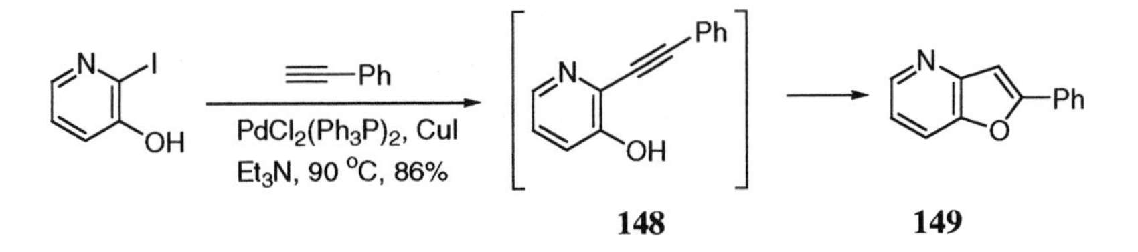


An unexpected isomerization was observed for several Sonogashira adducts of certain heteroaromatics [122]. Although the reaction between 3-bromopyridine and phenyl propargyl alcohol resulted in the normal adduct **150** [122, 123], 2-iodopyridine **151** produced the isomerized chalcone **152**, presumably due to promotion by triethylamine, under the same conditions [122].

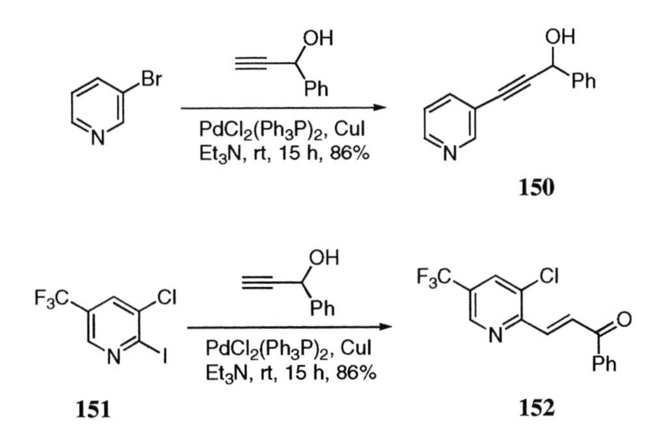


## 4.4 Heck and intramolecular Heck reactions

The intermolecular Heck reaction of halopyridines provides an alternative route to functionalized pyridines, circumventing the functional group compatibility problems encountered in other methods. 3-Bromopyridine has often been used as a substrate for the Heck reaction [124–126]. For example, ketone **155** was obtained from the Heck reaction of 3-bromo-2-methoxy-5-chloropyridine (**153**) with allylic alcohol **154** [125]. The mechanism for such a synthetically useful coupling warrants additional comments: oxidative addition of 3-bromopyridine **153** to Pd(0) proceeds as usual to give the palladium intermediate **156**. Subsequent insertion of allylic alcohol **154** to **156** gives intermediate **157**. Reductive elimination of **157** gives enol **158**, which then isomerizes to afford ketone **155** as the ultimate product. This tactic is frequently used in the synthesis of ketones from allylic alcohols.

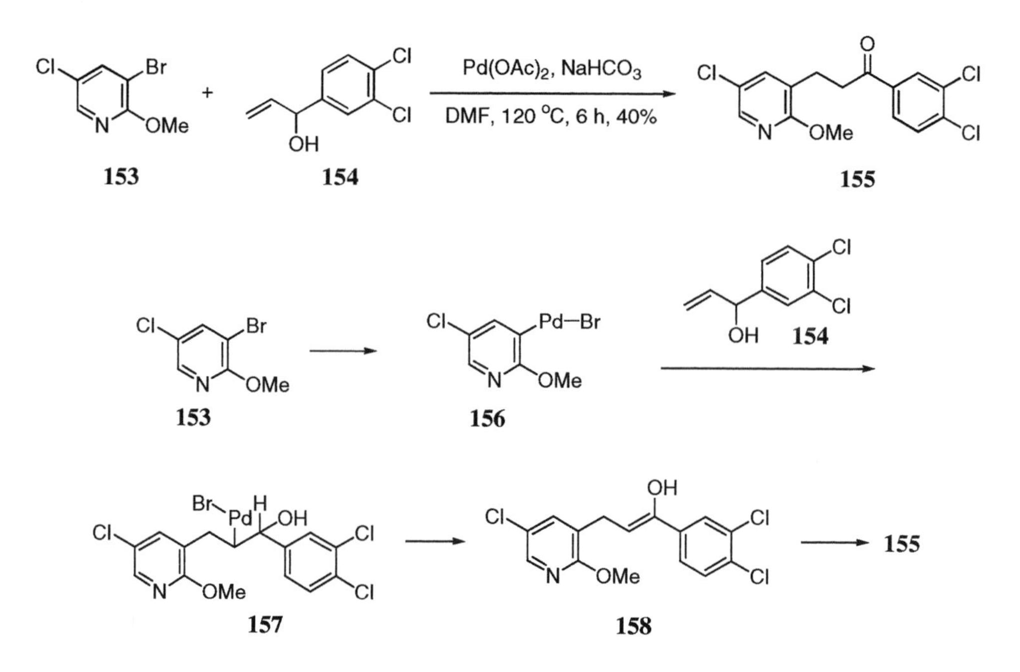


In one case, the intermolecular Heck reaction of 3-pyridyltriflate with ethyl acrylate was accelerated by LiCl to give **159** [127, 128]. Here, both electronic and steric effects all favored β-substitution. In another case, however, electronic effects prevailed and complete α-substitution was observed. In the presence of an electron-donating substituent (i.e., a protected amine), 3-bromopyridine **160** was coupled with *t*-butoxyethylene to give 3-pyridyl methyl ketone **162** [126]. The regiochemistry of the Heck reaction was governed by inductive effects, leading to intermediate **161**.

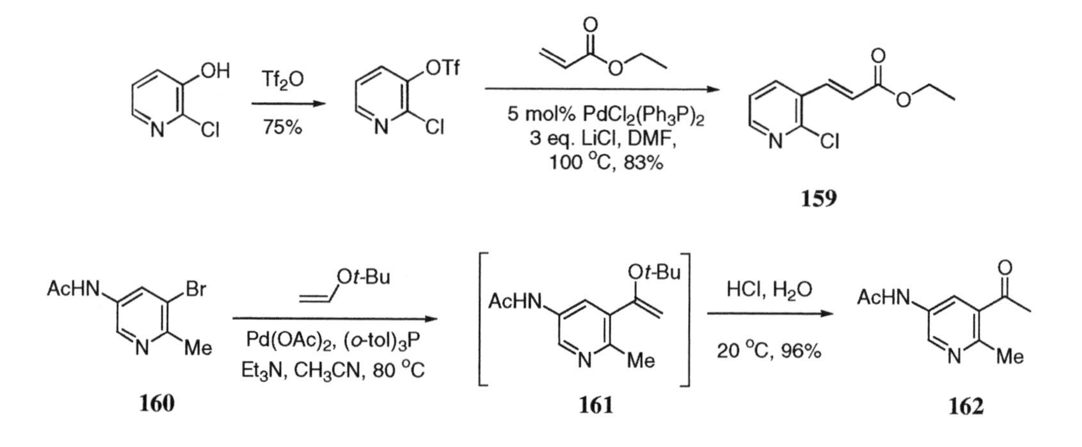

Without the alkoxy substituent on the olefin, steric effects prevail. The Heck adduct **164** was isolated when 2-chloro-3-pivaloylamidopyridine (**163**) was allowed to react with an excess of styrene [129]. Steric effects can play a major role in the regioselectivity of these reactions.

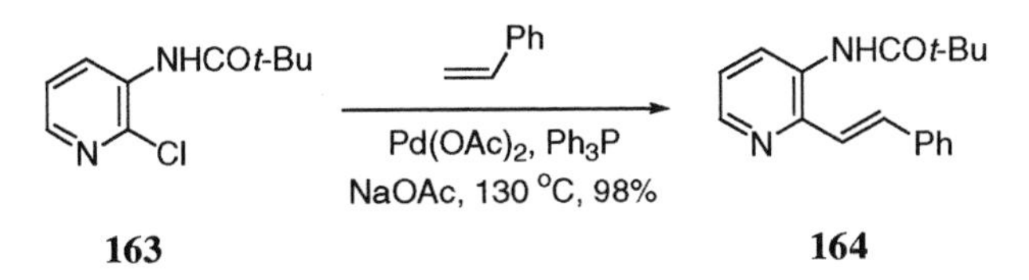


While the aforementioned reaction works well for aminopyridines and alkoxypyridines, it is not operative for most electron-deficient pyridines as well as 2- and 4-bromopyridines. One of the possible reasons for its failure with 2-halopyridines is the formation of an unreactive dimer complex from the oxidative addition intermediate [130].

A reductive intermolecular Heck heteroarylation (hydroheteroarylation) of *N*-protected azabicyclo[2,2,1]heptene **165** has been used to construct 7-azabicyclo[2.2.1]heptane **166** in moderate yield [131, 132]. An asymmetric version of such a transformation to provide enantiomerically-enriched *N*-protected epibatidine has also been described [128, 133]. It was found that introduction of Noyori's BINAP ligand resulted in the best enantioselectivities with 72–81% *ee* and a 53% yield. By using either the (*R*)- or (*S*)-BINAP ligand, either enantiomer was easily accessible.

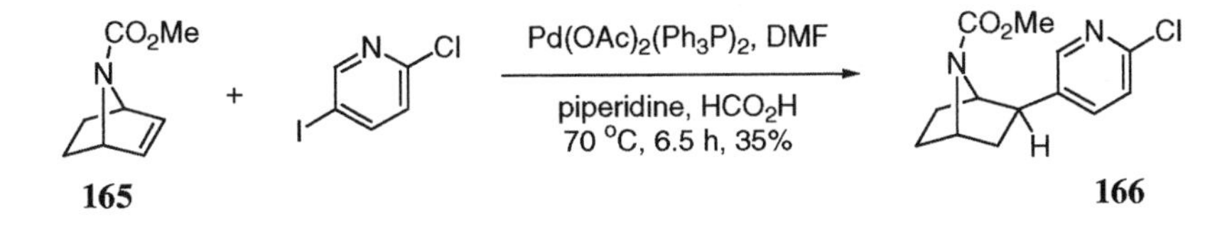


Recently, Malpass *et al.* reported a synthesis of epibatidine isomers also under reductive Heck conditions [134, 135]. 2-Azabicyclo[2.2.1]heptene **167** was assembled by cycloaddition of an iminium salt with cyclopentadiene. Treatment of **167** with 2-chloro-5-iodopyridine provided a mixture of *exo*-5-(6-chloro-3-pyridyl) derivative **168** and *exo*-6-(6-chloro-3-pyridyl) derivative **169**.

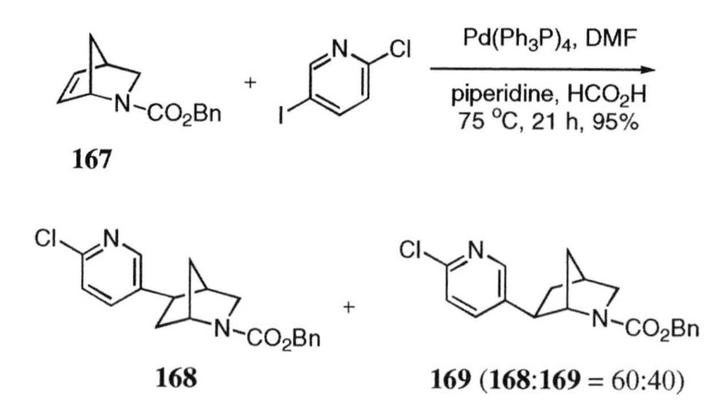


An intramolecular 7-*exo* Heck cyclization was the key to Kelly's synthesis of maxonine (**172**), originally isolated from the root of *Simira maxonii*, a plant endemic to the Costa Rican tropical forest. The migratory insertion step of the intramolecular cyclization of substrate **170** took place at the pendant olefin, giving rise to the seven-membered product **171** [136]. Oxidative cleavage of the stilbene double bond in **171** then produced maxonine (**172**). Kelly's synthesis of maxonine (**172**) resulted in revision of the original structural assignment of the naturally occurring alkaloid.

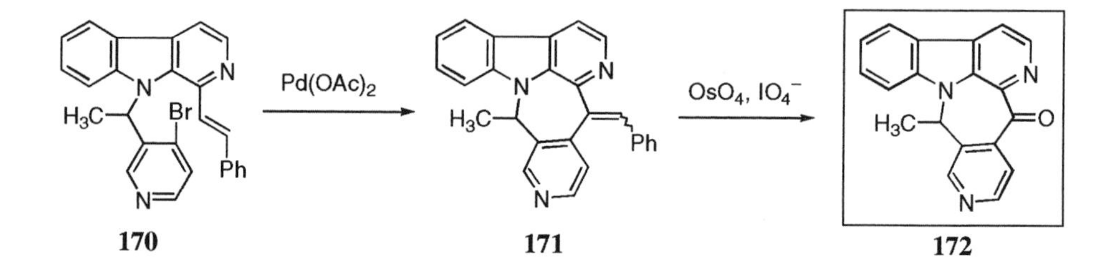


By taking advantage of the C(2) activation, 2-allyloxy-3-iodopyridine (**173**) was prepared by an $S_NAr$ displacement of 2-chloro-3-iodopyridine with sodium allyloxide [137]. 2-Chloro-3-iodopyridine was prepared by *ortho*-lithiation of 2-chloropyridine followed by iodine quench. The intramolecular Heck reaction of allyl ether **173** under Jeffery's ligand-free conditions resulted in 3-methylfuro[2,3-*b*]pyridine (**174**).

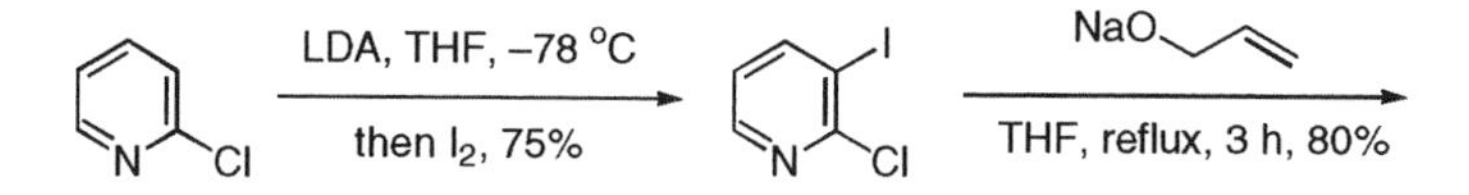

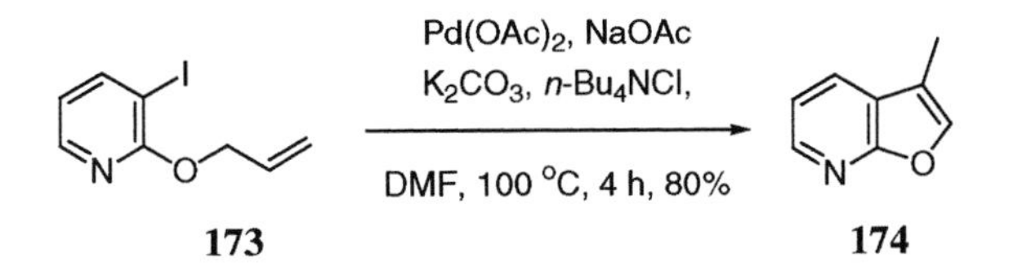


The intramolecular Heck cyclization of 4-iodopyridine **175** with a pendant crotyl ether led to a mixture of bicyclic ethers **176** and **177** in which the endocyclic isomer **176** was predominant [138].

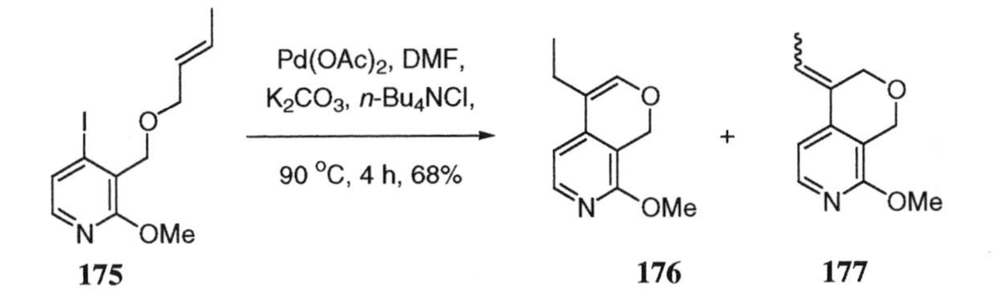


An intramolecular Heck cyclization of substrate **178** was the central feature in Overman's synthesis of pyridinomorphinans [139]. Octahydroisoquinoline **178**, derived from an allylsilane–iminium ion cyclization as a single stereoisomer, was cyclized under forcing conditions to afford enantiomerically pure **179**, an intermediate for a (–)-morphine (**180**) analog.

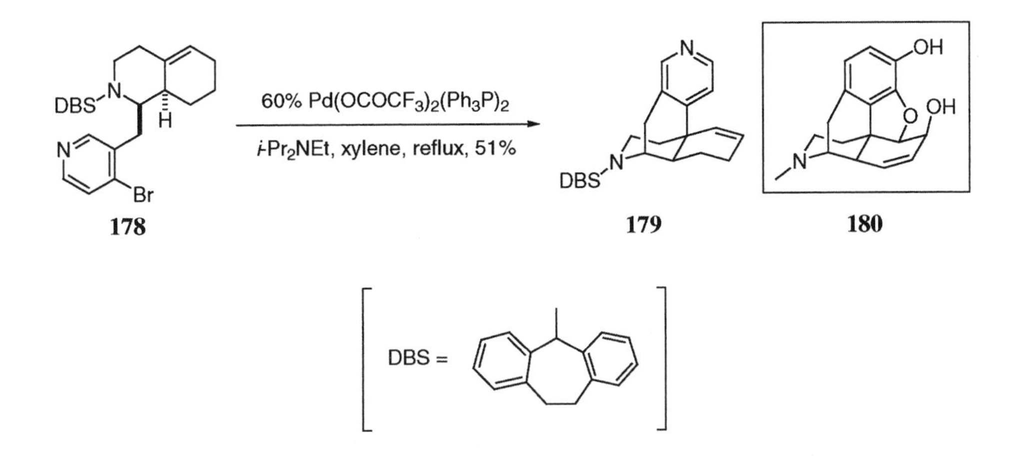


In a synthetic route to the *E*-azaeburnane series, an intramolecular heteroaryl Heck reaction was the major cyclization strategy [140]. Under Jeffery's ligand-free conditions, *E*-azaeburnane skeleton **182** was prepared from bromopyridine **181**. The migratory insertion occurred at C(2) of the indole ring.

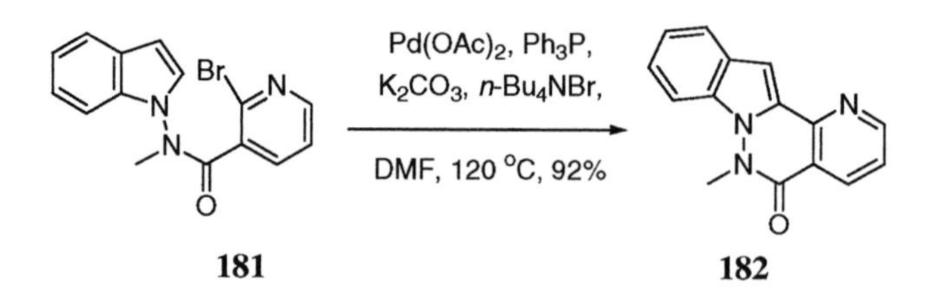


## 4.5 C—N bond formation

An early example of Pd-catalyzed C—N bond formation involving an aminopyridine is found in Boger's synthesis of lavendamycin [141–143]. Boger's initial attempts to construct the aryl-nitrogen bond of **183** to make the desired β-carboline **184** using copper reagents failed. In contrast, treatment of **183** with 1.5 equivalents of $Pd(Ph_3P)_4$ under conditions conducive to oxidative insertion delivered the desired β-carboline **184**. The requirement for stoichiometric catalyst for effective transformation potentially arose from a competing oxidative addition of liberated HBr to $Pd(Ph_3P)_4$.

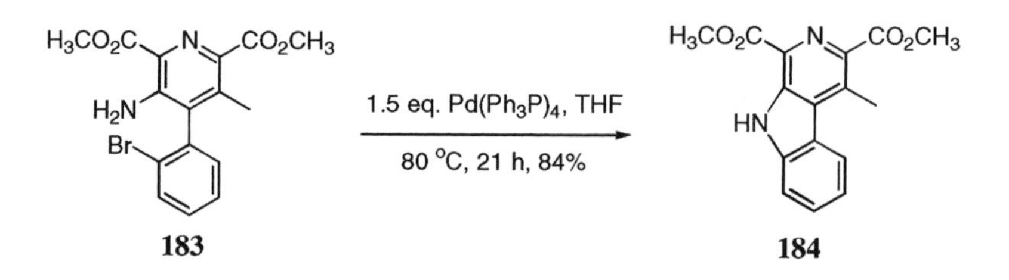


Since 1994, the successful development of the amination of aryl halides with amines using *catalytic* palladium has uncovered an efficient and practical method of C—N bond formation. Buchwald and Wagaw described an efficient and relatively general catalytic amination of halopyridines using chelating bis(phosphine) ligands [144]. This is a significant improvement over existing methods in which both activated substrates and harsh reaction conditions were required. As the three following examples entail, Buchwald's protocol worked for both chloro- and bromopyridine, activated or non-activated substrates. Primary amines and anilines were efficiently arylated. The major limitation of the Pd/(±)-BINAP or Pd/dppp-catalyzed amination protocol is the inability of this system to cross-couple halopyridines with acyclic dialkylamines.

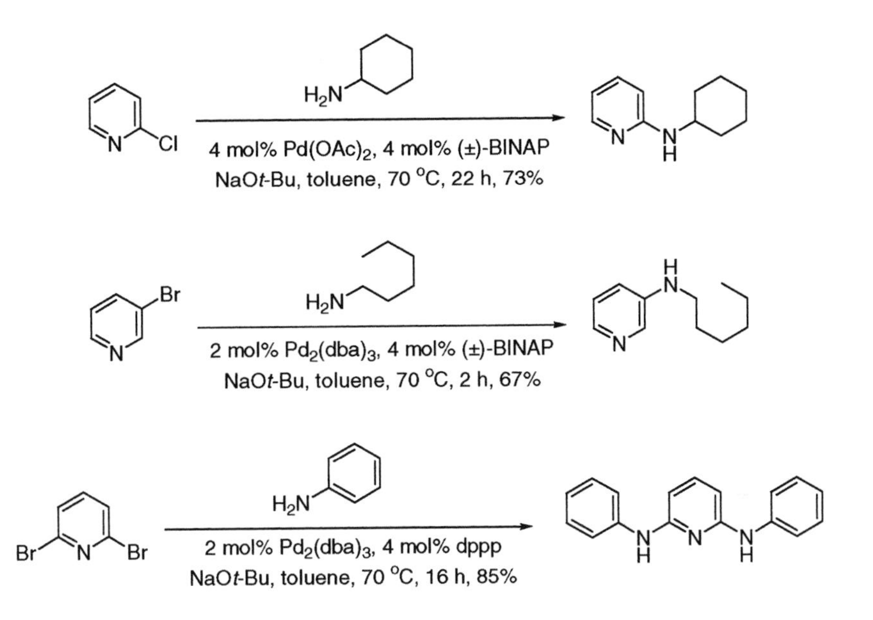


The efficiency observed with the bis(phosphine) ligand system is presumably due to the inhibition of side reactions by its *bidentate* chelating phosphine ligand. The major side reactions are β-hydride elimination from the amidopalladium intermediate and formation of bis-amine complexes, leading to products of hydrodebromination. In cases where pyridine substrates are involved, chelating ligands also prevent the formation of bispyridyl complexes that terminate the catalytic cycle. In another case, however, a *monodentate* phosphine ligand was found to facilitate the Pd-catalyzed amination of bromopyridines [145]. Bis-*N*-pyridyl-aza-crown ether **185** was synthesized from the reaction of 2,5-dibromopyridine and aza-18-crown-6 under the original conditions developed by the Buchwald and Hartwig groups. Some of the mono-amination product was also generated from the Pd-imido intermediate *via* a β-hydride elimination-reduction sequence.

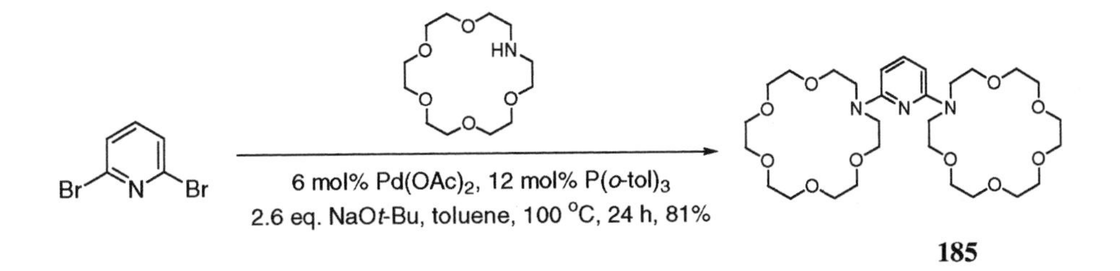


Uemura and coworkers utilized an arene–chromium complex ligand to form the C—N bond on a pyridine ring [146]. In the presence of 3 mol% of the monophosphine(dicarbonyl)chromium complexed-arene ligand **186**, the Pd-catalyzed amination of 2-bromopyridine with acyclic secondary amines such as $Et_2NH$ was effected. This method complements Buchwald's amination protocol, which failed for acyclic secondary amines.

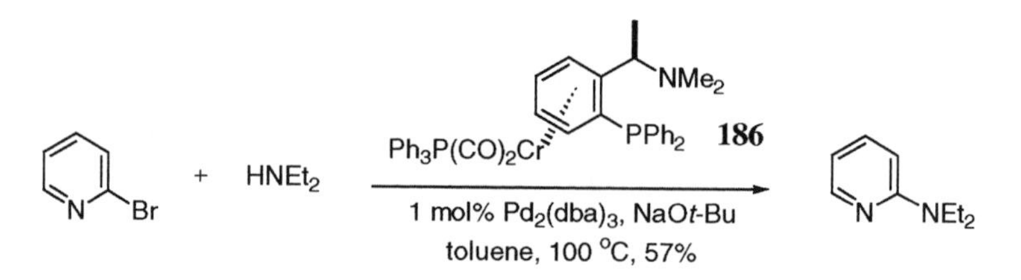


A synergistic combination of Pd-catalyzed amination and arylation was the central operation of Sakamoto's synthesis of carbolines [147]. Diarylamine **187** was first installed *via* the Buchwald-Hartwig amination protocol. Subsequent intramolecular Heck-like arylation of **187** provided a novel route to α-carboline **188**.

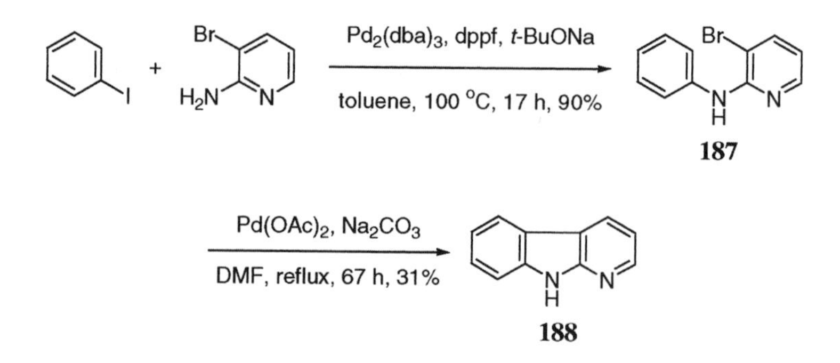


## 4.6 Carbonylation of halopyridines

### 4.6.1 Alkoxycarbonylation

The Pd-catalyzed alkoxycarbonylation of bromopyridines readily takes place regardless of the position occupied by the bromine atom on the ring. 3,5-Diethylnicotinate (**189**) was obtained by carbonylation of 3,5-dibromopyridine using $PdCl_2(Ph_3P)_2$ as the catalyst [148]. The role of triethylamine was to scavenge HBr generated during the reaction.

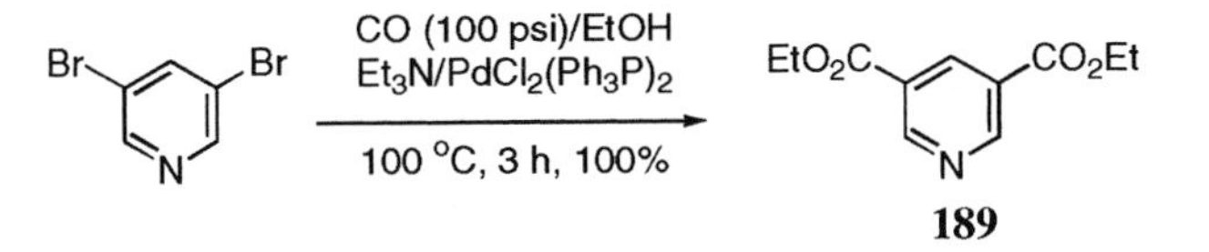


Although 2,5-dibromopyridine may be regiospecifically lithiated at C(5) [149], Pd-catalyzed couplings, including the Sonogashira, Negishi [150], and alkoxycarbonylation reactions [151] all preferentially occur at the C(2) position, as illustrated by the formation of **190**.

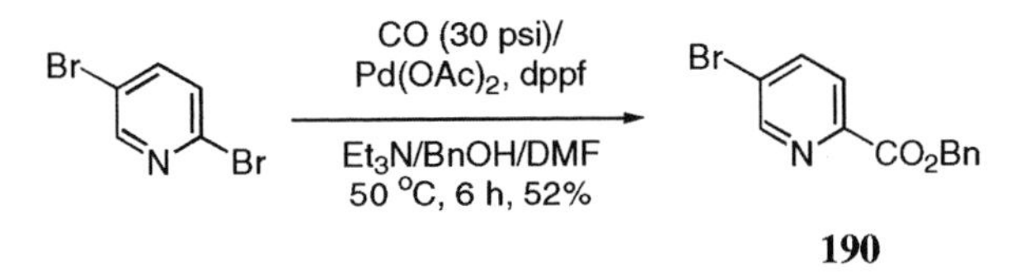


Carbonylation of chloropyridines is less straightforward than that of bromo- and iodopyridines. The $\alpha$ and $\gamma$ positions are sufficiently activated to undergo facile oxidative addition to Pd(0), whereas the reactivity is greatly diminished at $\beta$ positions. Thus, 2,6-dichloropyridine was converted to the corresponding dimethyl ester in good yield under normal Pd-catalyzed carbonylation conditions [152], and 2-chloropyridine was readily carbonylated to furnish methyl 2-pyridinecarboxylate (**191**), but 3-chloropyridine gave no carbonylated products under the same conditions [153, 154].

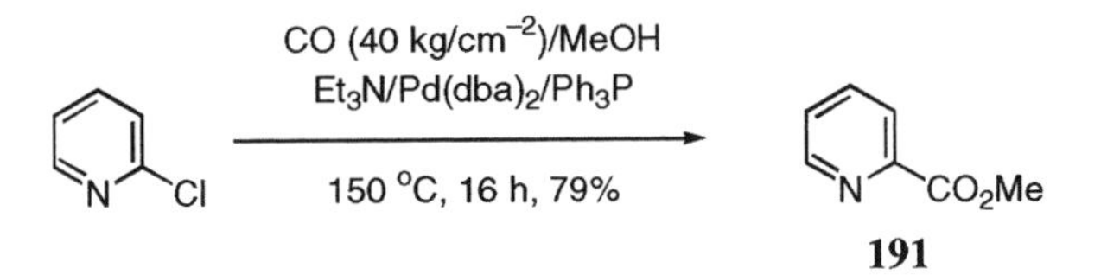


Fortunately, there is a solution to this problem. Since electron-withdrawing groups on an aryl halide facilitate oxidative addition to Pd(0) and pyridine is already electron-deficient, an additional electron-withdrawing group can provide sufficient activation of a 3-chloropyridine for the Pd-catalyzed alkoxycarbonylation to occur. An example is the alkoxycarbonylation of 2,3-dichloro-5-(methoxymethyl)pyridine (**192**). Initially, the 2-position of **192** was adequately activated and monocarboxylate **193** was obtained [155]. The newly introduced ester group, in turn, activated C(3) enough for another alkoxycarbonylation to take place in the formation of bis-ester **194** under conditions slightly more rigorous.

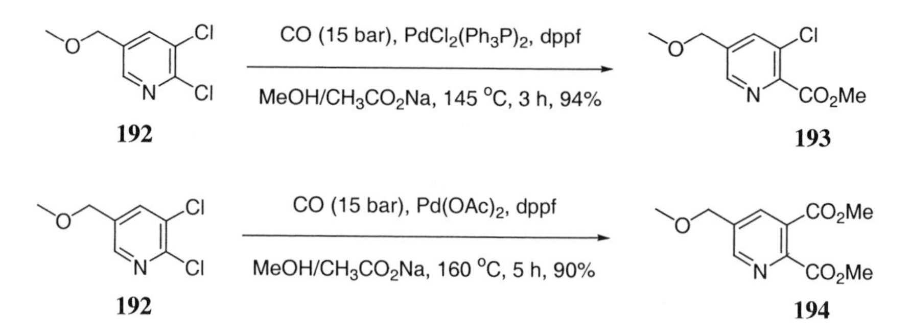


### 4.6.2 Aminocarbonylation

The Pd-catalyzed carbonylation of 2,6-dibromopyridine in the presence of 2-pyridylamine gave *N,N'*-di(2-pyridyl)pyridine-2,6-dicarboxamide under relatively high pressure and prolonged reaction time [156]. Partial aminocarbonylation of 2,6-dibromopyridine was more cumbersome—only 55% of the monocarboxamide **195** was isolated in a shorter reaction time accompanied by the corresponding 2,6-dicarboxamide in 32% yield.

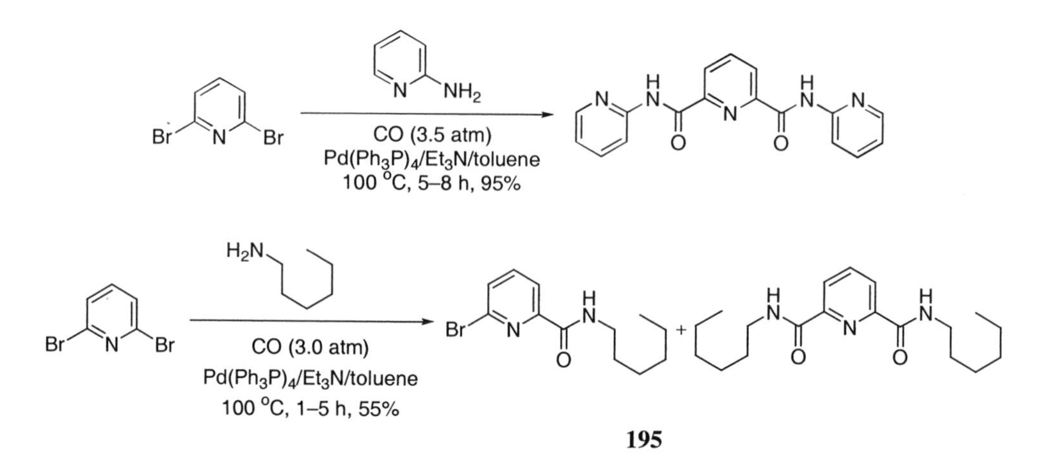


As previously discussed, the Pd-catalyzed reactions of 2,5-dibromopyridine including Sonogashira, Negishi, and alkoxycarbonylation all take place at C(2). It has also been demonstrated that aminocarbonylation of 2,5-dibromo-3-methylpyridine occurred regioselectively at C(2) to afford **196** despite the steric hindrance of the 3-methyl group [157].

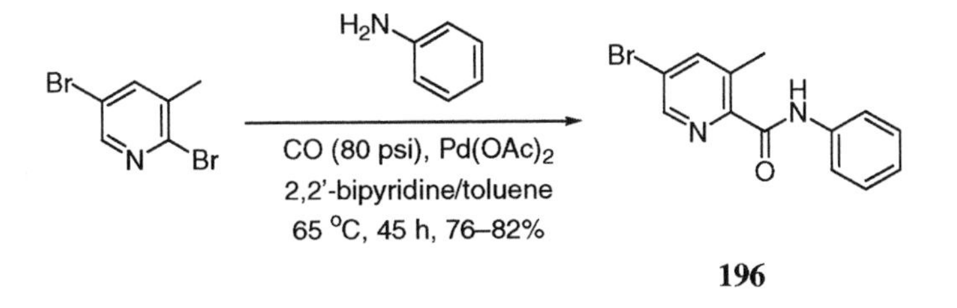


A unique Pd-catalyzed carbonyl insertion reaction of **197** furnished pyrido[2,1-*b*]quinazoline **198**, an antiallergy agent [158]. This particular outcome may be substrate-specific.

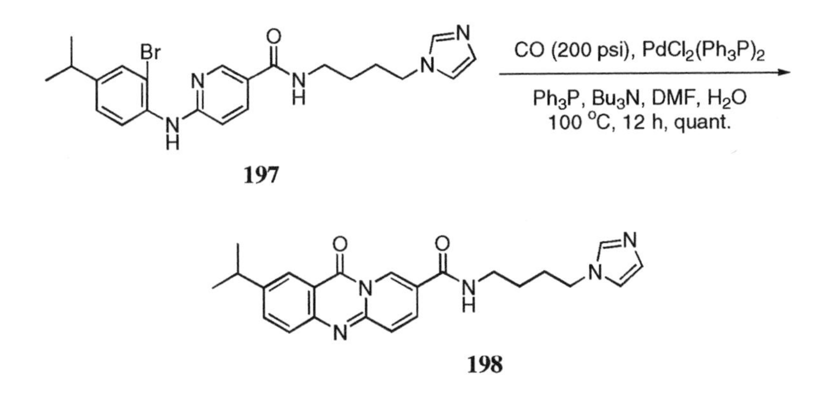


Pd-catalyzed double carbonylation reactions are rare due to the inefficiency of formation of bis-carbonylation products. Addition of bulky tricyclohexylphosphine to the normal carbonylation system greatly facilitated the formation of the double carbonylation product. Subjecting 4-iodopyridine **199** to such conditions afforded primarily 4-pyridylglyoxylic acid derivative **200**, which was not easily attainable *via* classical synthetic methods [159]. The monoamide **201** was isolated as a minor by-product.

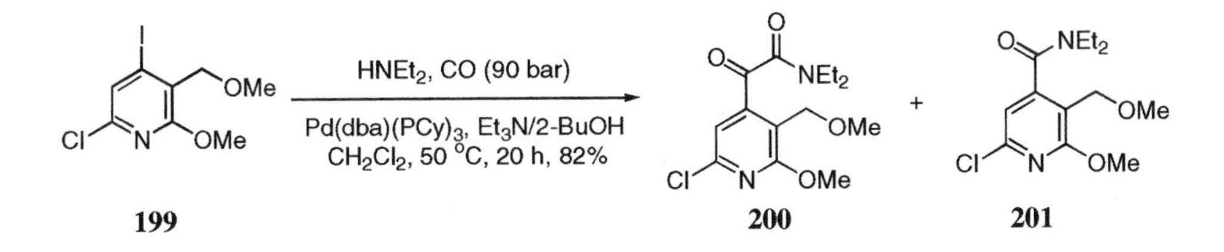

## 4.7 Phosphonation of halopyridines

The Pd-catalyzed reaction is a useful approach for the $C_{sp2}$—P bond formation since the Michaelis-Arbuzov reaction is not suitable for such substrates. Phosphonation of 3-bromopyridine was carried out using diethyl phosphite and catalytic $Pd(Ph_3P)_4$ to give diethyl pyridylphosphonate (**202**) [161, 161]. The Pd-catalyzed phosphonation of 3-bromopyridine may be rationalized by the following mechanism. At first, oxidative addition of 3-bromopyridine to Pd(0) gives rise to intermediate **203**. The attack of dialkyl phosphite on this pyridylpalladium complex leads to diethyl pyridylphosphonate (**202**), along with Pd(II) hydride **204**. **204** is subsequently reduced by triethylamine to regenerate the Pd(0) species with the deposition of $Et_3N{\bullet}HBr$. The regenerated Pd(0) species is then available for another reaction cycle.

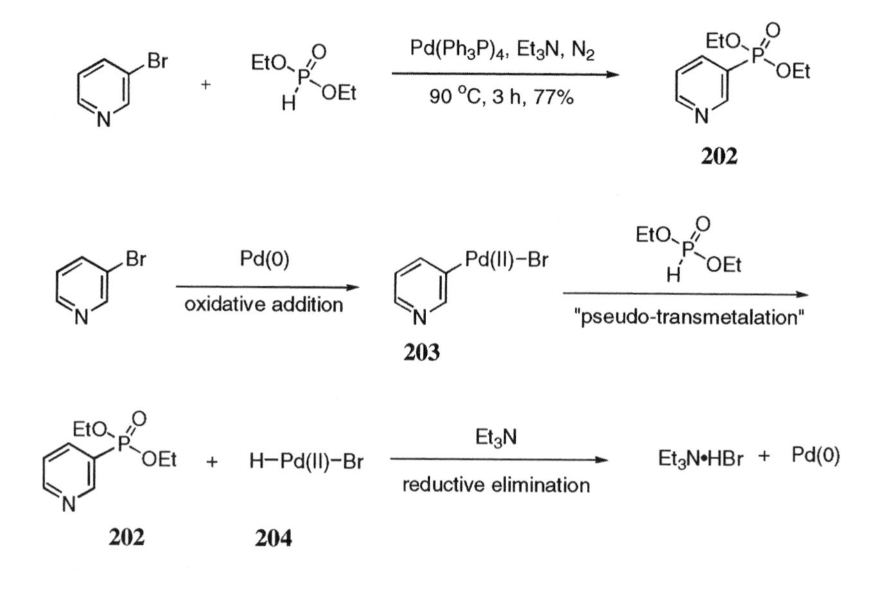


## 4.8 Heteroannulation

An attempt to extend Larock's indole synthesis [162, 163] to azaindole gave poor yields using Larock's original conditions [164]. When the chloride source was tetrabutylammonium chloride, reaction of propargyl alcohol **206** on 3-iodo-4-amidopyridine **205** gave 5-azaindole **207** in only 40% yield. It was later discovered that simply changing the chloride source from tetraalkylammonium chloride to lithium chloride doubled the yield [165], as exemplified by the transformation of **208** to **209**. The silyl group at C(2) of **209** allowed further elaboration to provide a variety of 2-substituted-3-methylpyrrolo[2,3-*b*]pyridines.

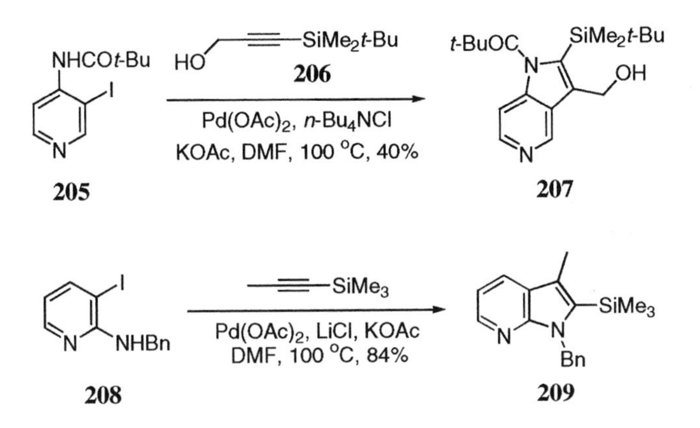


The acyclic version of Larock's heteroannulation was successfully applied to the synthesis of highly substituted pyridines [166]. The annulation of *tert*-butylimine **210** with phenyl propargyl alcohol produced pyridine **211** regioselectively in excellent yield. The regiochemistry obtained was governed by steric effects. Furthermore, the choice of imines was crucial to the success of the heteroannulations. *tert*-Butylimine was the substrate of choice, since all other imines including methyl, isopropyl, allyl and benzyl imines failed completely to produce the desired heterocyclic products.

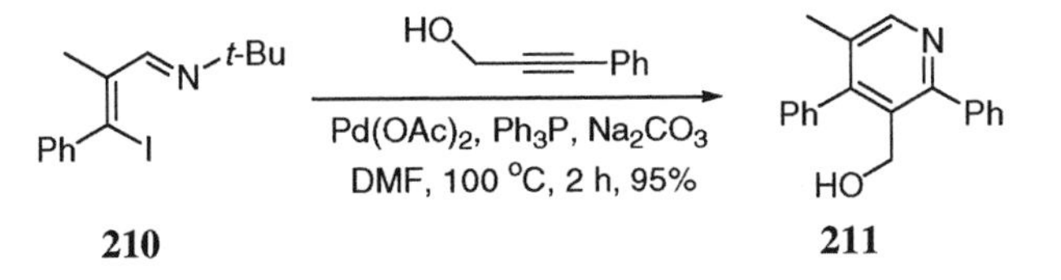


The mechanism is analogous to that of the Larock indole synthesis (see Section 1.10, pages 27–8). Specifically, oxidative addition of vinylic iodide **210** to Pd(0) produces the organopalladium intermediate **212**. Coordination of **212** with phenyl propargyl alcohol brings the two components together to form complex **213**. The alkyne inserts to afford the vinylic palladium intermediate **214**. The regiochemistry here is governed by steric effects as opposed to the coordination effect of the hydroxyl group. This stands in contrast to the indole formation example shown in Section 1.10. Intermediate **214** then reacts with the neighboring imine substituent to form a seven-membered palladacyclic immonium salt **215**. Subsequent reductive elimination produces a *tert*-butylpyridinium salt **216** and regenerates Pd(0). As previously suggested by Heck [167, 168], the *tert*-butyl group apparently fragments to relieve the strain resulting from interaction with the substituent present in the 3-position, giving rise to pyridine **211**.

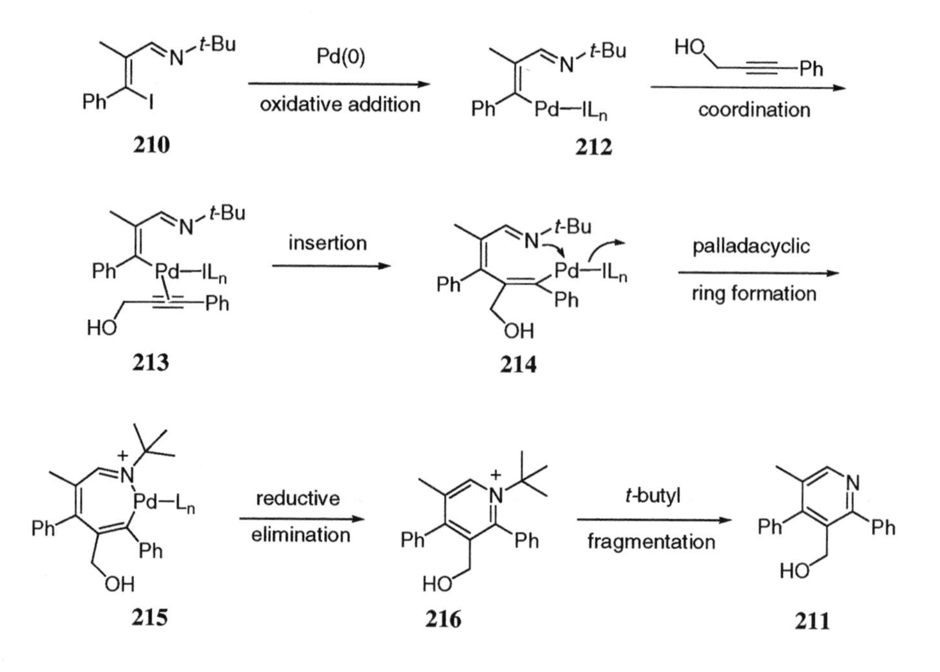


The 5-, 6-, and 7-azaindoles were synthesized *via* the Pd-catalyzed heteroannulation of internal alkynes using *ortho*-aminopyridine derivatives in an extention of Larock's indole synthesis [169]. LiCl was found to be an essential component in order to obtain regioselectivity, reproducibility and improved yields.

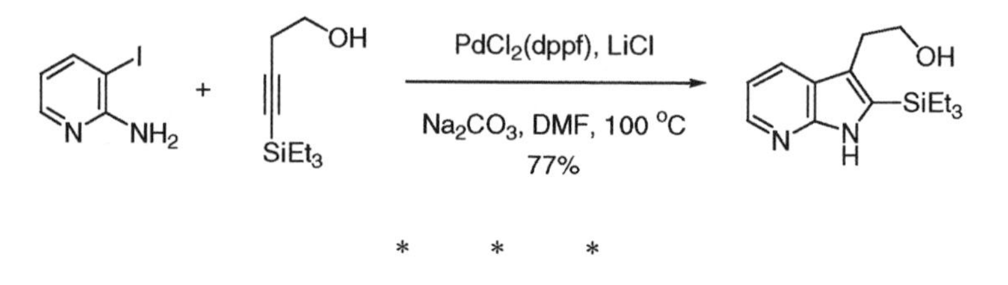


* * *

In summary, activated halopyridines take part in nucleophilic displacement more readily than their unactivated counterparts. A similar trend has been observed for their Pd-catalyzed reactions. Even 2-, 4- and 6-chloropyridines are viable substrates. Among all cross-coupling reactions, the Stille coupling seems to be the most suitable choice for substrates with base-sensitive moieties. Nonetheless, the Kumada, Negishi, and Suzuki versions are also good choices provided the substrates can survive the reaction conditions. The latter three choices have the benefit of avoiding toxic tin reagents. In addition, although most halopyridines behave analogously to simple carbocyclic aryl halides, 2-halopyridines are unique. Their Heck reactions tend to fail due to the formation of a less reactive dimer complex from the oxidative addition intermediate. With regard to 2,5-dibromopyridine, Pd-catalyzed reactions take place regioselectively at the C(2) position, whereas halogen-metal exchange reactions take place at C(5) predominantly.

Finally, the Pd-catalyzed amination of halopyridines may be achieved using chelating bis-phosphine ligands. Akin to the cross-coupling reactions, regioselective carbonylations at C(2) have been demonstrated for both 2,3-dichloro- and 2,5-dibromopyridines. In addition, 3-chloropyridine may be sufficiently activated by an adjacent electron-withdrawing group to also undergo carbonylation.

## 4.9 References


1. (a) Gribble, G. W.; Saulnier, M. G. *Tetrahedron Lett.* **1980**, *21*, 4137–40. For more extensive investigations of the same reaction, see, (b) Gribble, G. W.; Saulnier, M. G. *Heterocycles* **1993**, *35*, 151–69. (c) Mallet, M.; Quéguiner, G. *Tetrahedron* **1982**, *38*, 3035–42. (d) Mallet, M.; Quéguiner, G. *Tetrahedron* **1986**, *42*, 2253–62.
2. Kondo, Y.; Shilai, M.; Uchiyama, M.; Sakamoto, T. *J. Am. Chem. Soc.* **1999**, *121*, 3539–40.
3. Godard, A.; Rovera, J. C.; Marsais, F.; Plé, N.; Quéguiner, G. *Tetrahedron* **1992**, *48*, 4123–34.
4. (a) Kelly, T. R.; Xu, W.; Sundaresan, J. *Tetrahedron Lett.* **1993**, *34*, 6173–6. (b) Comins, D. L.; Killpack, M. O. *J. Org. Chem.* **1990**, *55*, 69–73.
5. Lindströem, S.; Eriksson, M.; Grivas, S. *Acta Chem. Scand.* **1993**, *47*, 805–12.
6. Bracher, F.; Hildebrand, D. *Tetrahedron* **1994**, *50*, 12329–36.
7. Zoltewicz, J. A.; Cruskie, M. P., Jr. *Tetrahedron* **1995**, *51*, 11393–400.
8. Sugimoto, O.; Mori, M.; Tanji, K.-i. *Tetrahedron Lett.* **1999**, *40*, 7477–8.
9. Riguet, E.; Alami, M.; Cahiez, G. *Tetrahedron Lett.* **1997**, *38*, 4397–400.
10. Abbotto, A.; Bradamante, S.; Facchetti, A.; Pagani, G. A. *J. Org. Chem.* **1997**, *62*, 5755–65.
11. Meth-Cohn, O.; Jiang, H. *J. Chem. Soc., Perkin Trans. 1* **1998**, 3737–45.
12. Minato, A.; Suzuki, K.; Tamao, K.; Kumada, M. *J. Chem. Soc., Chem. Commun.* **1984**, 511–3.
13. Turner, R. M.; Ley, S. V.; Lindell, S. D. *Synlett* **1993**, 748–50.
14. De Vos, E.; Esmans, E. L.; Alderweireldt, F. C.; Balzarini, J.; De Clercq, E. *J. Heterocycl. Chem.* **1993**, *30*, 1245–52.
15. Yamanaka, H.; An-Naka, M.; Kondo, Y.; Sakamoto, T. *Chem. Pharm. Bull.* **1985**, *33*, 4309–13.

16. Brandão, M. A.; de Oliveira, A. B.; Snieckus, V. *Tetrahedron Lett.* **1993**, *34*, 2437–40.
17. Bell, A. S.; Ruddock, K. S. *Tetrahedron Lett.* **1988**, *29*, 5013–6.
18. Felding, J.; Kristensen, J.; Bjerregaaed, T.; Sander, L.; Vesdø, P.; Begtrup, M. *J. Org. Chem.* **1999**, *64*, 4196–8.
19. Savage, S. A.; Smith, A. P.; Fraser, C. L. *J. Org. Chem.* **1998**, *63*, 10048–51.
20. Takahashi, T.; Koga, H.; Sato, H.; Ishizawa, T.; Taka, N. *Heterocycles* **1995**, *41*, 2405–8.
21. Bolm, C.; Ewald, M.; Felder, M.; Schlingloff, G. *Chem. Ber.* **1992**, *125*, 1169–90.
22. Maligres, P. E.; Waters, M. S.; Fleitz, F.; Askin, D. *Tetrahedron Lett.* **1999**, *40*, 8193–5.
23. Amat, M.; Hadida, S.; Bosch, J. *Tetrahedron Lett.* **1993**, *34*, 5005–6.
24. Amat, M.; Hadida, S.; Bosch, J. *Tetrahedron Lett.* **1994**, *35*, 793–6.
25. Amat, M.; Sathyanarayana, S.; Hadida, S.; Bosch, J. *Tetrahedron Lett.* **1994**, *35*, 7123–6.
26. Amat, M.; Hadida, S.; Sathyanarayana, S.; Bosch, J. *Tetrahedron Lett.* **1996**, *37*, 3071–4.
27. Amat, M.; Hadida, S.; Bosch, J. *An. Quím. Int. Ed.* **1996**, *92*, 62–3.
28. Amat, M.; Hadida, S.; Pshenichnyi, G.; Bosch, J. *J. Org. Chem.* **1997**, *62*, 3158–75.
29. Ishikura, M.; Ohta, T.; Terashima, M. *Chem. Pharm. Bull.* **1985**, *33*, 4755–63.
30. Ishikura, M.; Kamada, M.; Terashima, M. *Heterocycles* **1984**, *22*, 265–8.
31. Ishikura, M.; Kamada, M.; Terashima, M. *Synthesis* **1984**, 936–8.
32. Ishikura, M.; Kamada, M.; Ohta, T.; Terashima, M. *Heterocycles* **1984**, *22*, 2475–8.
33. Zoltewicz, J. A.; Maier, N. M.; Fabian, W. M. F. *J. Org. Chem.* **1997**, *62*, 3215–9.
34. Zoltewicz, J. A.; Maier, N. M.; Fabian, W. M. F. *Tetrahedron* **1996**, *52*, 8703–6.
35. Li, J. J.; Yue, W. S. *Tetrahedron Lett.* **1999**, *40*, 4507–10, footnote 8.
36. Deshayes, K.; Broene, R. D.; Chao, I.; Knobler, C. B.; Diederich, F. *J. Org. Chem.* **1991**, *56*, 6787–95.
37. Achad, S.; Guyot, M.; Potier, P. *Tetrahedron Lett.* **1993**, *34*, 2127–30.
38. Thompson, W. J.; Jones, J. H.; Lyle, P. A.; Thies, J. E. *J. Org. Chem.* **1988**, *53*, 2052–5.
39. Stavenuiter, J.; Hamzink, M.; Van der Hulst, R.; Zomer, G.; Westra, G.; Kriek, E. *Heterocycles* **1987**, *26*, 2711–6.
40. Ishiyama, T.; Kizaki, H.; Miyaura, N.; Suzuki, A. *Tetrahedron Lett.* **1993**, *34*, 7595–8.
41. Zhang, H.; Kwong, F. Y.; Tian, Y.; Chan, K. S. *J. Org. Chem.* **1998**, *63*, 6886–90.
42. Lohse, O.; Thevenin, P.; Waldvogel, E. *Synlett* **1999**, 45–8.
43. Ishikura, M.; Oda, I.; Terashima, M. *Heterocycles* **1985**, *23*, 2375–86.
44. Yang, Y,; Martin, A. R. *Heterocycles* **1992**, *34*, 1395–8.
45. Peters, D.; Hörnfeldt, A.-B.; Gronowitz, S. *J. Heterocycl. Chem.*, **1990**, *27*, 2165–73.
46. Wellmar, U.; Hörnfeldt, A.-B.; Gronowitz, S. *J. Heterocycl. Chem.*, **1995**, *32*, 1159–63.
47. Malm, J.; Rehn, B.; Hörnfeldt, A.-B.; Gronowitz, S. *J. Heterocycl. Chem.*, **1994**, *31*, 11–5.
48. Malm, J.; Hörnfeldt, A.-B.; Gronowitz, S. *Heterocycles* **1993**, *35*, 245–62.
49. Godard, A.; Marsais, F.; Plé, N.; Trecourt, F.; Turck, A.; Quéguiner, G. *Heterocycles* **1995**, *40*, 1055–91.

50. Rocca, P.; Rovera, J. C.; Marsais, F.; Quéguiner, G. *Synth. Commun.* **1995**, *25*, 3901–8.
51. Rocca, P.; Rovera, J. C.; Marsais, F.; Quéguiner, G. *Synth. Commun.* **1995**, *25*, 3373–9.
52. Baldwin, J. E.; James, D. A.; Lee, V. *Tetrahedron Lett.* **2000**, *41*, 733–6.
53. Gros, P.; Fort, Y. *Synthesis* **1999**, 754–6.
54. Muratake, H.; Tanegawa, M.; Natsume, M. *Chem. Pharm. Bull.* **1998**, *46*, 400–12.
55. Achad, S.; Guyot, M.; Potier, P. *Tetrahedron Lett.* **1993**, *34*, 2127–30.
56. Henze, O.; Lehmann, U.; Schlüter, A. D. *Synthesis* **1999**, 683–7.
57. Hanan, G. S.; Schubert, U. S.; Vilkmer, D.; Riviere, E.; Lehn, J.-M.; Kyritskas, N.; Fischer, J. *Can. J. Chem.* **1997**, *75*, 169–82.
58. Yamamoto, Y.; Yanagi, A. *Chem. Pharm. Bull.* **1982**, *30*, 2003–10.
59. Schubert, U.; Eschbaumer, C.; Weidl, C. H. *Synlett* **1999**, 342–4.
60. Schubert, U.; Eschbaumer, C. *Org. Lett.* **1999**, *1,* 1027–9.
61. Benaglia, M.; Toyota, S.; Woods, C. R.; Siegel, J. S. *Tetrahedron Lett.* **1997**, *38*, 4737–40.
62. Pellicciari, R.; Gallo-Mezo, M. A.; Natalini, B.; Amer, A. M. *Tetrahedron Lett.* **1992**, *33*, 3003–4.
63. Barros, M. T.; Maycock, C. D.; Ventura, M. R. *Tetrahedron Lett.* **1999**, *40*, 557–60.
64. Takahashi, T.; Koga, H.; Sato, H.; Ishizawa, T.; Taka, N. *Heterocycles* **1995**, *41*, 2405–8.
65. Hanefeld, W.; Jung, M. *Tetrahedron* **1994**, *50*, 2459–68.
66. Bailey, T. R. *Tetrahedron Lett.* **1986**, *27*, 4407–10.
67. Kennedy, G.; Perboni, A. D. *Tetrahedron Lett.* **1996**, *37*, 7611–4.
68. Floch, P. L.; Carmichael, D.; Ricard, L.; Mathey, F. *J. Am. Chem. Soc.* **1993**, *115*, 10665–70.
69. Dondoni, A.; Fantin, G.; Fogagnolo, M.; Medici, A.; Pedrini, P. *Synthesis* **1989**, 693–6.
70. Kitamura, C.; Tanaka, S.; Yamashita, Y. *J. Chem. Soc., Chem. Commun.* **1994**, 1585–6.
71. Gronowitz, S.; Björk, P.; Malm, J.; Hörnfeldt, A-B. *J. Organomet. Chem.* **1993**, *460*, 127–9.
72. Kelly, T. R.; Jagoe, C. T.; Gu, Z. *Tetrahedron Lett.* **1991**, *32*, 4263–6.
73. Yamamoto, Y.; Azuma, Y.; Mitoh, H. *Synthesis* **1986**, 564–5.
74. Bolm, C.; Ewald, M.; Felder, M.; Schlingloff, G. *Chem. Ber.* **1992**, *125*, 1169–90.
75. Zoltewicz, J. A.; Cruskie, M. P. *Tetrahedron* **1995**, *51*, 11393–400.
76. Zoltewicz, J. A.; Cruskie, M. P. *J. Org. Chem.* **1995**, *60*, 3487–93.
77. Yamamoto, Y.; Tanaka, T.; Yagi, M.; Inamoto, M. *Heterocycles* **1996**, *42*, 189–94.
78. Barchín, B. M.; Valenciano, J.; Cuadro, A. M.; Alvarez-Builla, J.; Vaquero, J. *Org. Lett.* **1999**, *1*, 545–7.
79. Muratake, H.; Hayakawa, A.; Natsume, M. *Tetrahedron Lett.* **1997**, *38*, 7577–80.
80. Malm, J.; Björk, P.; Gronowitz, S.; Hörnfeldt, A.-B. *Tetrahedron Lett.* **1994**, *35*, 3195–6.
81. Malm, J.; Björk, P.; Gronowitz, S.; Hörnfeldt, A.-B. *Tetrahedron Lett.* **1992**, *33*, 2199–202.

82. Bracher, F.; Papke, T. *Monatsh. Chem.* **1995**, *126*, 805–9.
83. Lebarbier, C.; Carreaux, F.; Carboni, B. *Tetrahedron Lett.* **1999**, *40*, 6233–5.
84. Lavilla, R.; Gullón, F.; Bosch, J. *J. Chem. Soc., Chem. Commun.* **1995**, *16,* 1675–6.
85. Sakamoto, T.; Kondo, Y.; Yasuhara, A.; Yamanaka, H. *Heterocycles* **1990**, *31*, 219–21.
86. Sakamoto, T.; Satoh, C.; Kondo, Y.; Yamanaka, H. *Heterocycles* **1992**, *34*, 2379–84.
87. Sakamoto, T.; Kondo, Y.; Yasuhara, A.; Yamanaka, H. *Tetrahedron* **1991**, *47*, 1877–86.
88. Sakamoto, T.; Yasuhara, A.; Kondo, Y.; Yamanaka, H. *Chem. Pharm. Bull.* **1994**, *42*, 2032–5.
89. Lindsay, C. M.; Widdowson, D. A. *J. Chem. Soc., Perkin Trans. 1* **1988**, 569–73.
90. Hutchinson, J. H.; Cook, J. J.; Brashear, K. M.; Breslin, M. J.; Glass, J. D.; Gould, R. J.; Halczenko, W.; Holahan, M. A.; Lynch, R. J.; Sitko, G. R.; Stranieri, M.; Hartman, G. D. *J. Med. Chem.* **1996**, *39*, 4583–91.
91. Palmisano, G.; Santagostino, M. *Helv. Chim. Acta* **1993**, *76*, 2356–66.
92. Palmisano, G.; Santagostino, M. *Synlett* **1993**, 771–3.
93. Hudkins, R. L.; Diebold, J. L.; Marsh, F. D. *J. Org. Chem.* **1995**, *60*, 6218–20.
94. Tokuyama, H.; Yamashita, T.; Reding, M. T.; Kaburagi, Y.; Fukuyama, T. *J. Am. Chem. Soc.* **1999**, *121*, 3791–2. And references cited therein.
95. Rainier, J. D.; Kennedy, A. R.; Chase, E. *Tetrahedron Lett.* **1999**, *40*, 6325–7.
96. Azizian, H.; Eaborn, C.; Pidcock, A. *J. Organomet. Chem.* **1981**, *215*, 48–58.
97. Ciattini, P. G.; Morera, E.; Ortar, G. *Tetrahedron Lett.* **1994**, *35*, 2405–8.
98. Godard, A.; Rocca, P.; Fourquez, J. M.; Rovera, J. C.; Marsais, F.; Quéguiner, G. *Tetrahedron Lett.* **1993**, *34*, 7919–22.
99. Sandosham, J.; Undhelm, K. *Acta Chem. Scand.* **1989**, *43*, 684–9.
100. Zhang, Y.; Hörnfeldt, A.-B.; Gronowitz, S. *J. Heterocycl. Chem.*, **1995**, *32*, 771–7.
101. Chamoin, S.; Houldsworth, S.; Snieckus, V. *Tetrahedron Lett.* **1998**, *39*, 4175–8.
102. Hatanaka, Y.; Goda, K.; Okahara, Y.; Hiyama, T. *Tetrahedron* **1994**, *50,* 8301–16.
103. Hagiwara, E.; Gouda, K-I.; Hatanaka, Y.; Hiyama, T. *Tetrahedron Lett.* **1997**, *38,* 439–42.
104. Shibata, K.; Miyazawa, K.; Goto, Y. *Chem. Commun.* **1997**, 1309–10.
105. Mowery, M. E.; DeShong, P. *Org. Lett.* **1999**, *1*, 2137–40.
106. Takahashi, S.; Kuroyama, Y.; Sonogashira, K.; Higihara, N. *Synthesis* **1980**, 627–30.
107. Sakamoto, T.; Shiraiwa, M.; Kondo, Y.; Yamanaka, H. *Synthesis* **1983**, 312–4.
108. Ernst, A.; Gobbi, L.; Vasella, A. *Tetrahedron Lett.* **1996**, *37*, 7959–62.
109. Tilley, J. W.; Zawoiski, S. *J. Org. Chem.* **1988**, *53*, 386–90.
110. Marsais, F.; Pineau, P.; Nivolliers, F.; Mallet, M.; Turck, A.; Godard, A.; Quéguiner, G. *J. Org. Chem.* **1992**, *57*, 565–73.
111. Houpis, I. N.; Choi, W. B.; Reider, P. J.; Molina, A.; Churchill, H.; Lynch, J.; Volante, R. P. *Tetrahedron Lett.* **1994**, *35,* 9355–8.

112. Kumar, V.; Dority, J. A.; Bacon, E. R.; Singh, B.; Lesher, G. Y. *J. Org. Chem.* **1992**, *57*, 6995–8.
113. Kim, C.-S.; Russell, K. C. *J. Org. Chem.* **1998**, *63*, 8229–34.
114. Bleicher, L. S.; Cosford, N. D.; Herbaut, A.; McMillum, J. S.; McDonald, I. A. *J. Org. Chem.* **1998**, *63*, 1109–18.
115. Mal'kina, A. G.; Brandsma, L.; Vasilevsku, S. F.; Trofimov, B. A. *Synthesis* **1996**, 589–10.
116. Hänninen, E.; Takalo, H.; Kankare, J. *Acta Chem. Scand., Ser. B* **1988**, *B42*, 614–9.
117. Tilley, J. W.; Levitan, P.; Lind, J.; Welton, A. F.; Crowley, H. J.; Tobias, L. D.; O'Donnell, M. *J. Med. Chem.* **1987**, *30*, 185–93.
118. Morris, J.; Wishka, D. G. *Tetrahedron Lett.* **1988**, *29*, 143–6.
119. Sakamoto, T.; Kondo, Y.; Watanabe, R.; Yamanaka, H. *Chem. Pharm. Bull.* **1986**, *34*, 2719–24.
120. Arcadi, A.; Marinelli, F.; Cacchi, S. *Synthesis* **1986**, 749–51.
121. Torii, S.; Hu, L. H.; Okumoto, H. *Synlett* **1992**, 515–6.
122. Minn, K. *Synlett* **1991**, 115–6.
123. Ames, D. E.; Bull, D.; Takunda, C. *Synthesis* **1981**, 364–5.
124. Main, A. J.; Bhagwat, S.; *et al. J. Med. Chem.* **1992**, *35*, 4366–72.
125. Bargar, T. M.; Wilson, T.; Daniel, J. K. *J. Heterocycl. Chem.*, **1985**, *22*, 1583–92.
126. Wright, S. W.; Hageman, D. L.; McClure, L. D. *J. Heterocycl. Chem.* **1998**, *35*, 719–23.
127. Namyslo, J.; C.; Kaufman, D. E. *Synlett* **1999**, 804–6.
128. Draper, T. L.; Bailey, T. R. *Synlett* **1995**, 157–8.
129. Niu, C.; Li, J.; Doyle, T. W.; Chen, S.-H. *Tetrahedron* **1998**, *54*, 6311–8.
130. Basu, B.; Freijd, T. *Acta Chem. Scand.* **1996**, *50*, 316–22.
131. Clayton, S. C.; Regan, A. C. *Tetrahedron Lett.* **1993**, *34*, 7493–6.
132. Kasyan, A.; Wagner, C.; Maier, M. E. *Tetrahedron* **1998**, *54*, 8047–54.
133. Namyslo, J. C.; Kaufman, D. E. *Synlett* **1999**, 114–6.
134. Malpass, J. R.; Cox, C. D. *Tetrahedron Lett.* **1999**, *40*, 1419–22.
135. Cox, C. D.; Malpass, J. R. *Tetrahedron* **1999**, *55*, 11879–88.
136. Kelly, T. R.; Xu, W.; Sundaresan, J. *Tetrahedron Lett.* **1993**, *34*, 6173–6.
137. Cho, S. Y.; Kim, S. S.; Park, K.-H.; Kang, S. K.; Choi, J.-K.; Hwang, K.-J.; Yum, E. K. *Heterocycles* **1996**, *43*, 1641–52.
138. Bankston, D.; Fang, F.; Huie, E.; Xie, S. *J. Org. Chem.* **1999**, *64*, 3461–6.
139. Hong, C. Y.; Overman, L. E.; Romero, A. *Tetrahedron Lett.* **1997**, *38*, 8439–42.
140. Melnyk, P.; Legrand, L.; Gasche, J.; Ducrot, P.; Thai, C. *Tetrahedron Lett.* **1995**, *51*, 1941–52.
141. Boger, D. L.; Panek, J. S. *Tetrahedron Lett.* **1984**, *25*, 3175–8.
142. Boger, D. L.; Duff, S. R.; Panek, J. S.; Yasuda, M. *J. Org. Chem.* **1985**, *50*, 5782–9.
143. Boger, D. L.; Duff, S. R.; Panek, J. S.; Yasuda, M. *J. Org. Chem.* **1985**, *50*, 5790–5.

144. Wagaw, S.; Buchwald, S. L. *J. Org. Chem.* **1996**, *61*, 7240–41.
145. Kamikawa, K.; Sugimoto, S.; Uemura, M. *J. Org. Chem.* **1998**, *63*, 8407–10.
146. Witulski, B. *Synlett* **1999**, 1223–6.
147. Iwaki, T.; Yasuhara, A.; Sakamoto, T. *J. Chem. Soc., Perkin Trans. 1* **1999**, 1505–10.
148. Head, R. A.; Ibbotson, A. *Tetrahedron Lett.* **1984**, *25*, 5939–42.
149. Parham, W. E.; Piccirilli, R. M. *J. Org. Chem.* **1977**, *42*, 257–60.
150. Tilley, J. W.; Zawoiski, S. *J. Org. Chem.* **1988**, *53*, 386–90.
151. Najiba, D.; Carpentier, J.-F.; Castanet, Y.; Biot, C.; Brocard, J.; Mortreux, A. *Tetrahedron Lett.* **1999**, *40*, 3719–22.
152. Chambers, R. J.; Marfat, A. *Synth. Commun.* **1997**, *27*, 515–20.
153. Takeuchi, R.; Suzuki, K.; Sato, N. *Synthesis* **1990**, 923–4.
154. Takeuchi, R.; Suzuki, K.; Sato, N. *J. Mol. Cat.* **1991**, *66*, 277–8.
155. Bessard, Y.; Roduit, J. P. *Tetrahedron* **1999**, *55*, 393–404.
156. Horino, H.; Sakaba, H.; Arai, M. *Synthesis* **1989**, 715–18.
157. Wu, G. G.; Wong, Y.; Poirier, M. *Org. Lett.* **1999**, *1*, 745–7.
158. (a) Tilley, J. W.; Zawoiski, S. *J. Org. Chem.* **1987**, *52*, 2469–74. (b) Tilley, J. W.; Burghardt, B.; Burghardt, C.; Mowles, T. F.; Leinweber, F. J.; Klevans, L.; Young, R.; Hirkaler, G.; Fahrenholtz, K.; *et al. J. Med. Chem.* **1988**, *31*, 466–72. (c) Tilley, J. W.; Levitan, P.; Lind, J.; Welton, A. F.; Crowley, H. J.; Tobias, L. D.; O'Donnell, M. *J. Med. Chem.* **1987**, *30*, 185–93.
159. Couve-Bonnaire, S.; Carpentier, J.-F.; Castanet, Y.; Mortreux, A. *Tetrahedron Lett.* **1999**, *40*, 3717–18.
160. Hirao, T.; Masunaga, T.; Ohshiro, Y.; Agawa, T. *Synthesis* **1981**, 56–7.
161. Hirao, T.; Masunaga, T.; Yamada, Y.; Ohshiro, Y.; Agawa, T. *Bull. Chem. Soc. Jpn.* **1982**, *55,* 909–13.
162. Larock, R. C.; Yum, E. K. *J. Am. Chem. Soc.* **1991**, *113,* 6689–90.
163. Larock, R. C.; Yum, E. K.; Refvik, M. D. *J. Org. Chem.* **1998,** *63,* 7652–62.
164. Wensbo, D.; Eriksson, A.; Jeschke, T.; Annby, U.; Gronowitz, S.; Cohen, L. *Tetrahedron Lett.* **1993**, *34*, 2823–6.
165. Park, S. S.; Choi, J.-K.; Yum, E. K.; Ha, D.-C. *Tetrahedron Lett.* **1998**, *39*, 627–30.
166. Roesch, K. R.; Larock, R. C. *J. Org. Chem.* **1998**, *63*, 5306–7.
167. Wu, G.; Rheingold, A. L.; Geib, S. J.; Heck, R. F. *Organometallics* **1987**, *6*, 1941–6.
168. Wu, G.; Geib, S. J.; Rheingold, A. L.; Heck, R. F. *J. Org. Chem.* **1988**, *53*, 3238–41.
169. Ujjainwalla, F.; Warner, D. *Tetrahedron Lett.* **1998**, *39*, 5355–8.

# CHAPTER 5

## Thiophenes and Benzo[*b*]thiophenes

Thiophene-containing molecules can be found in both natural products and synthetic chemotherapeutics. Bithiophene **1**, a naturally occurring nematocide, is isolated from the roots of *Echinops spaerocephalus*, whereas tiaprofenic acid, an anti-inflammatory agent, is a synthetic thiophene derivative. Moreover, thiophene is a useful template for four-carbon homologation *via* reduction [1], as well as a bioisostere of the benzene ring and other heterocycles in medicinal chemistry.

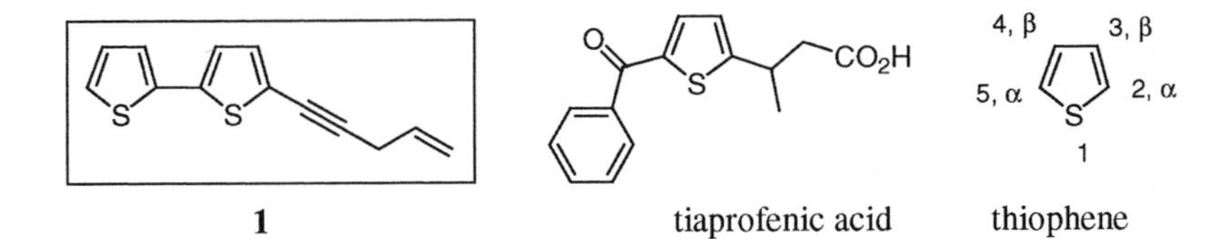


**1** tiaprofenic acid thiophene

Thiophene is a π-electron-excessive heterocycle. It favors electrophilic substitution, which, similar to metalation, takes place preferably at the α-positions due to the electronegativity of the sulfur atom. In comparison to the oxygen atom in furan, the sulfur atom in thiophene has lower electronegativity, so its lone pair electrons are more effectively incorporated into the aromatic system. The aromaticity of thiophene is in between that of benzene and furan. As a consequence, the difference in reactivity of α-halothiophenes and β-halothiophenes is not as pronounced as that of the corresponding halofurans. In the context of palladium chemistry, while regioselective reactions are routinely performed for α,β-dihalofurans, regioselectivity for α,β-dihalothiophenes is not as easily achieved. Furthermore, Pd(II), like many other transition metals, possesses strong thiophilicity. This is reflected in the "poisoning effects" of the sulfur atom on reactions involving thiophenes and benzothiophenes [2].

## 5.1 Preparation of halothiophenes and halobenzothiophenes

### 5.1.1 Direct halogenation

Direct halogenation of thiophenes is undoubtedly the most commonly used method for making halothiophenes. Treatment of 3,4-bis[(methoxycarbonyl)methyl]-thiophene (**2**) with NBS resulted in the monobromination product **3**, whereas treatment of **2** with bromine gave rise to the corresponding bisbromination product **4** [3]. Using a direct halogenation with NBS, Gronowitz *et al.* converted 2-methoxythiophene to 2-methoxy-3,5-dibromothiophene (**5**), which was subsequently transformed to 2-methoxy-3-bromothiophene (**6**) *via* regioselective halogen-metal exchange at C(5) followed by quenching with $H_2O$ [4]. 2-Methyl-3-bromo-5-chlorothiophene (**8**), in turn, was obtained from bromination of 2-methyl-5-chlorothiophene (**7**), derived from regioselective chlorination of 2-methylthiophene at C(5) with NCS [5].

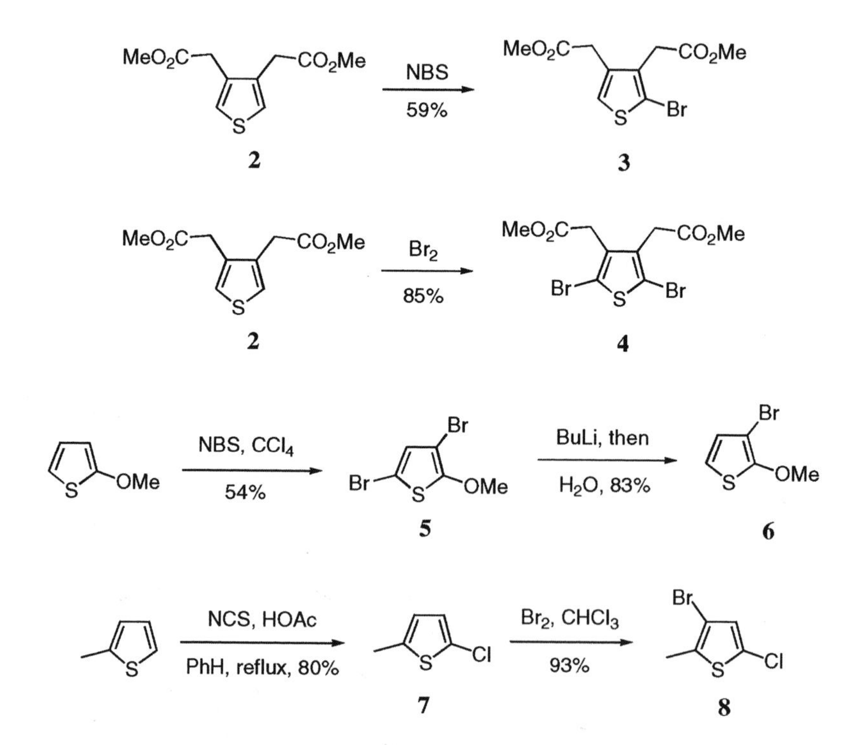


Halogenation can be run under milder conditions using more active quaternary ammonium polyhalides such as pyridinium tribromide. The reaction between thiophene and benzyltrimethylammonium tribromide in acetic acid–$ZnCl_2$ provided 2,5-dibromothiophene [6]. Chloro- and iodo-substituted thiophene derivatives may be prepared in the same manner. In comparison, bromination of thiophene employing 2 equivalents of NBS in chloroform gave 2,5-dibromothiophene in 56% yield [7].

$PhCH_2NMe_3{\bullet}Br_3$

HOAc–$ZnCl_2$, 54%

The aforementioned direct halogenation methods are also applicable to bisthiophenes [8] and terthiophenes [9]. Taking advantage of the α-activation, debromination of tetrabromobisthiophene **10** was achieved to afford the corresponding derivative **11** *via* regioselective reduction [10]. Recognizing that mono-bromination of benzothiophene gave an almost equal amount of 2- and 3-bromothiophene, the preparation of 3-bromo-6-methylbenzothiophene (**13**) began with dibromination of 6-methylbenzothiophene to give dibromide **12**. Selective removal of the C(2)-bromide of **12** using *n*-BuLi followed by treatment with $H_2O$ then provided **13** [11, 12].

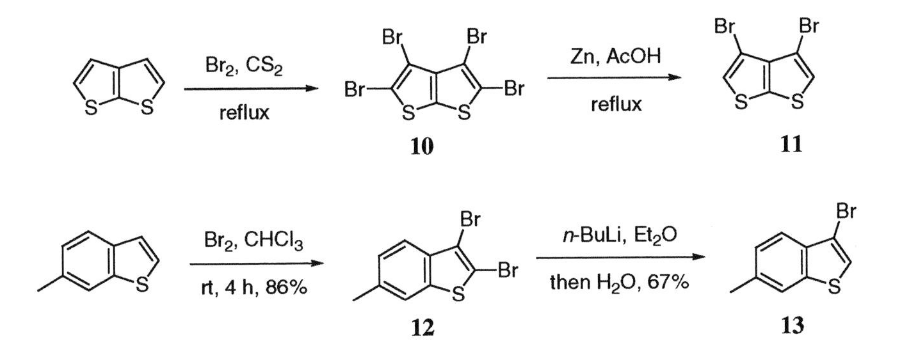


3-Iodobenzothiophene **15** was prepared by iodination of 2-trifluoroacetylamino-benzo[*b*]thiophene (**14**), although the same reaction with the regioisomeric 3-trifluoroacetylaminobenzo[*b*]thiophene gave only unidentified polymeric material [13]. In another case, thienylpyridine **16** could be either brominated [14], or iodinated to give **17** [15, 16].

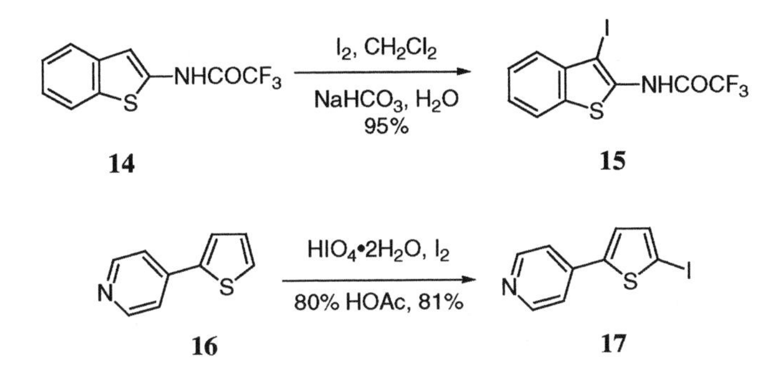


### 5.1.2 Quenching lithiothiophenes with halogens

2-Lithiothiophene, arising from lithiation of thiophene with *n*-BuLi, was treated with iodine to give 2-iodothiophene, which was allowed to react with sodium malononitrile in the presence of catalytic $PdCl_2(Ph_3P)_2$ to afford thienylmalononitrile **18**. Interestingly, α-metalation of ethyl 3-

thiophenecarboxylate was achieved using TMP–zincate **19**, a bulky base. The resulting thienylzincate was treated with $I_2$ to give 2-iodothiophene **20**. The same sequence with ethyl 2-thiophenecarboxylate furnished 5-iodothiophene **21** [17].

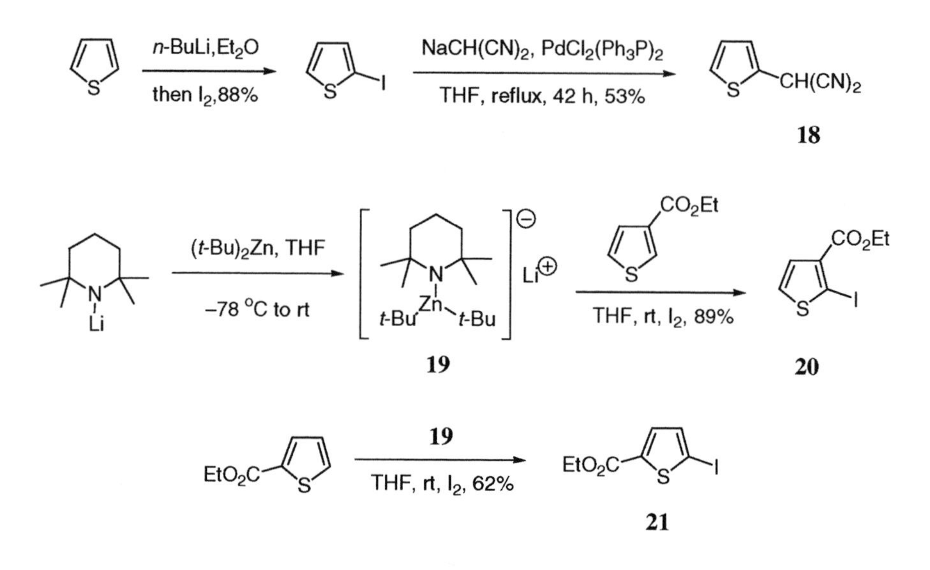


## 5.2 Oxidative and reductive coupling reactions

A small amount of bisthiophene was isolated when thiophene was treated with $Pd(OAc)_2$ [18]. The oxidative couplings of a thiophene with thiophene, furan, or substituted arenes were achieved in poor to moderate yields by using $Pd(OAc)_2$ in HOAc [19–21]. The oxidative couplings of thiophene or benzo[*b*]thiophene with olefins also suffer from inefficiency [22].

2-Bromothiophene-5-carboxaldehyde (**22**) was converted to bisthiophene **23** *via* a Pd-catalyzed reductive homocoupling [23]. The reaction was also applicable to chloro- and iodothiophenes bearing many functional groups. In addition, an Ullmann-type reductive homocoupling of 2-iodothiophene utilizing catalytic Pd/C and three equivalents of Zn provided a practical entry to bisthiophene in 64% yield [24].

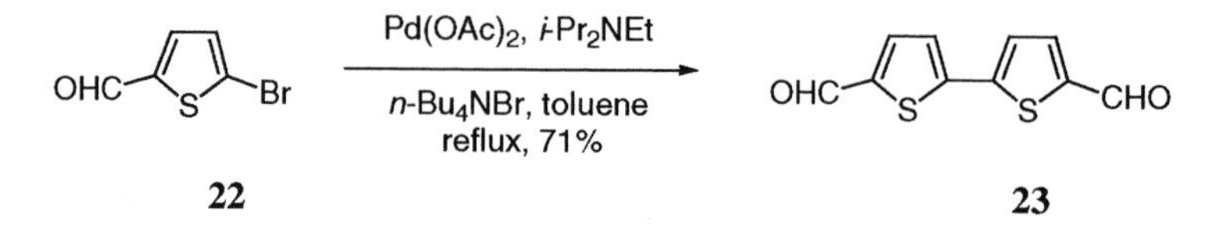

## 5.3 Cross-coupling with organometallic reagents

### 5.3.1 Kumada coupling

Organomagnesium reagents, which can serve as the nucleophiles in the Kumada coupling, are easy to make and many of them are commercially available. Even though some Kumada reactions can be run at room or lower temperature, many functional groups are not tolerant of Grignard reagents. Nonetheless, in the synthesis of thienylbenzoic acid **24**, the carboxylic acid moiety did survive the reaction conditions [25].

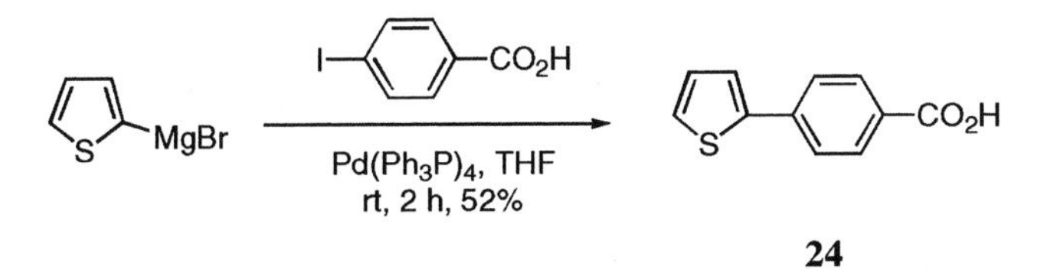


In the synthesis of 2-bromobisthiophene (**25**), mono-arylation was achieved using the Kumada coupling of 2-thienylmagnesium bromide with 2,5-dibromothiophene although nickel(0)-catalyzed coupling failed [26]. Another mono-substitution was realized when one equivalent of 2-thienylmagnesium bromide was allowed to react with 2,5-dichloropyridine, furnishing thienylpyridine **26** [27]. Furthermore, the Kumada coupling of 5-(2,2'-bithienyl)magnesium bromide (**27**) with 4-bromopyridine proceeded in refluxing ether to give 4-[5-(2-bithien-2'-yl)]pyridine (**28**) [28].

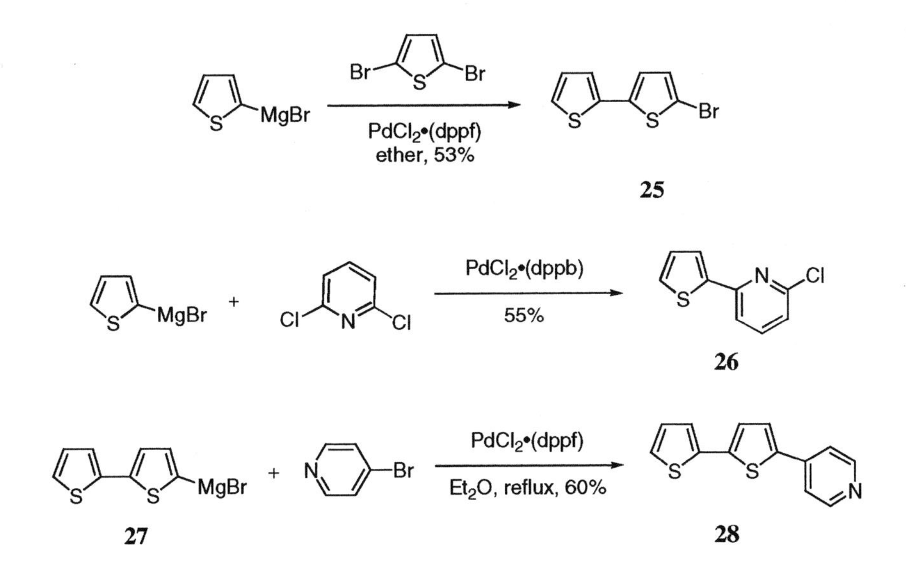

In the direct synthesis of aryl terminal alkynes *via* Pd-catalyzed cross-coupling of aryl halides with ethynylmetals, formation of diarylethynes is one of the potential side reactions. Indeed, the Kumada coupling of 2-iodo-5-methylthiophene (**29**) with ethynylmagnesium chloride gave the desired 2-ethynyl-5-methylthiophene (**30**) in only 35% yield, along with 24% of bis(5-methyl-2-thienyl)ethyne (**31**) [29]. The high propensity for H–Mg exchange reaction to occur was blamed for the diarylethyne formation.

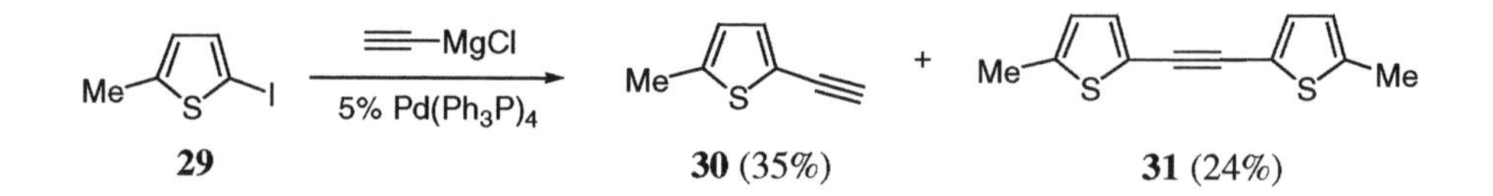


### 5.3.2 Negishi coupling

In comparison to the Kumada coupling, the Negishi coupling tolerates a wider range of functional groups including carbonyl, nitro and amino, avoiding some tedious protection–deprotection sequences in synthesis. The requisite organozinc reagents are most conveniently prepared either by reaction of organolithium (organomagnesiun) reagents with zinc chloride or oxidative addition of organohalides to zinc metal.

While the Kumada reaction of 2-iodo-5-methylthiophene (**29**) with ethynylmagnesium chloride gave a substantial amount of diarylacetylene, the Negishi reaction of **29** with ethynylzinc bromide produced ethynylthiophene **30** in 87% yield. Similarly, the Negishi reaction between 2-iodobenzo[*b*]thiophene and ethynylzinc bromide led to 2-ethynylbenzo[*b*]thiophene (**32**). The limitations of the Negishi reaction for synthesizing arylalkynes reside in the fact that it generally fails with unactivated aryl bromides. However, coupling with many aryl iodides proceeds smoothly [30].

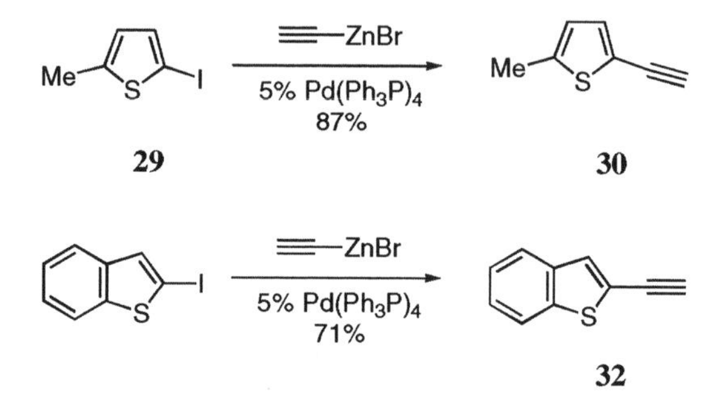


Halothiophenes were coupled with heteroarylzinc reagents including furylzinc chloride [31] and pyrazolylzinc chloride [32] to elaborate heterobiaryls **33** and **34**, respectively.

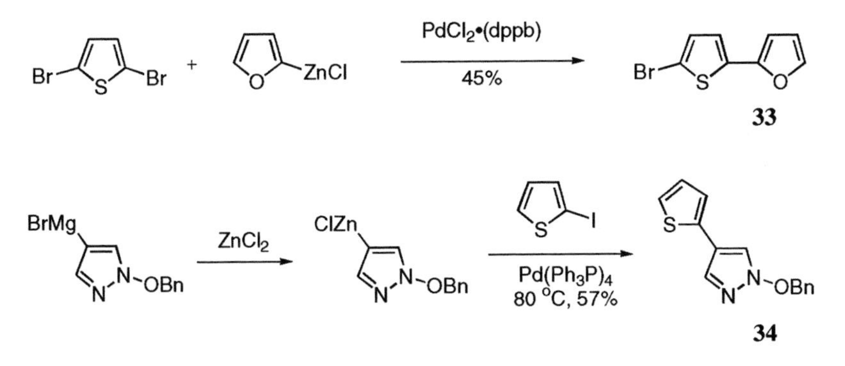


Gilchrist *et al.* prepared **36** *via* the Negishi coupling of bromoaldehyde **35** and 2-thienylzinc bromide without interfering the enal motif [33, 34]. 3-Thienylzinc bromide **38** was derived from a regioselective *ortho*-lithiation at the C(3) position of 4,4-dimethyl-2-(2-thienyl)oxazoline (**37**) followed by treatment with $ZnCl_2$. The subsequent Negishi reaction of **38** with iodobenzene afforded tricycle **39** [35]. Additional examples of Negishi couplings of thienylzinc reagents with iodoarenes include the synthesis of arylthiophenes **40** [36], **41** [37] and **42** [38]. Moreover, thienylpyridine **45** was elaborated *via* a Negishi coupling of 2-bromopyridine **43** and thienylzinc bromide **44** [39].

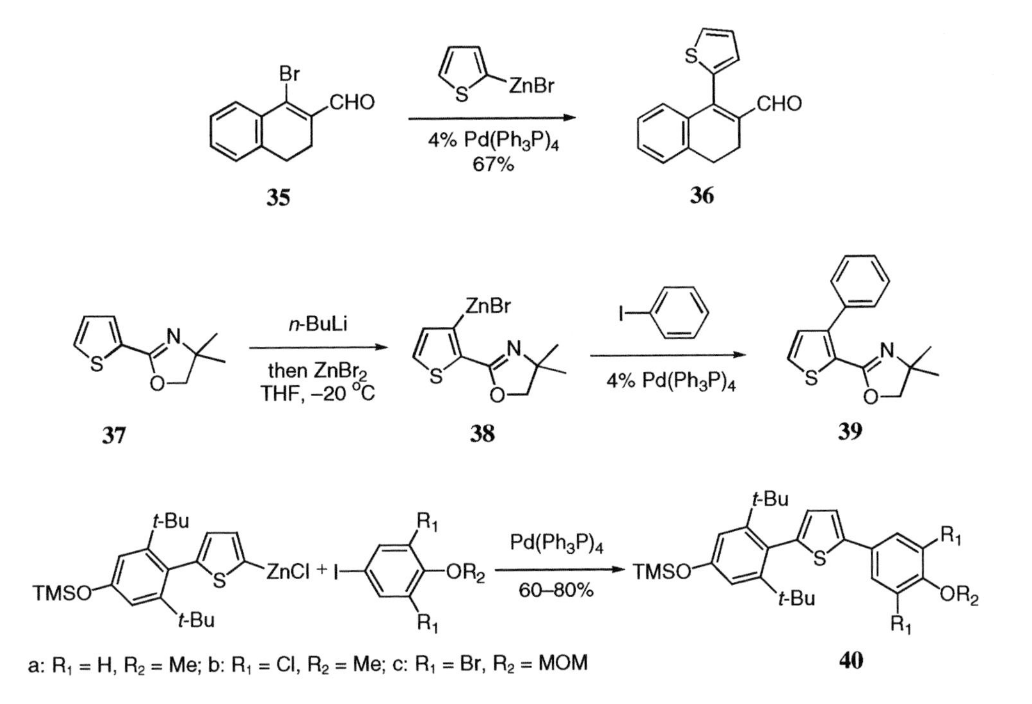

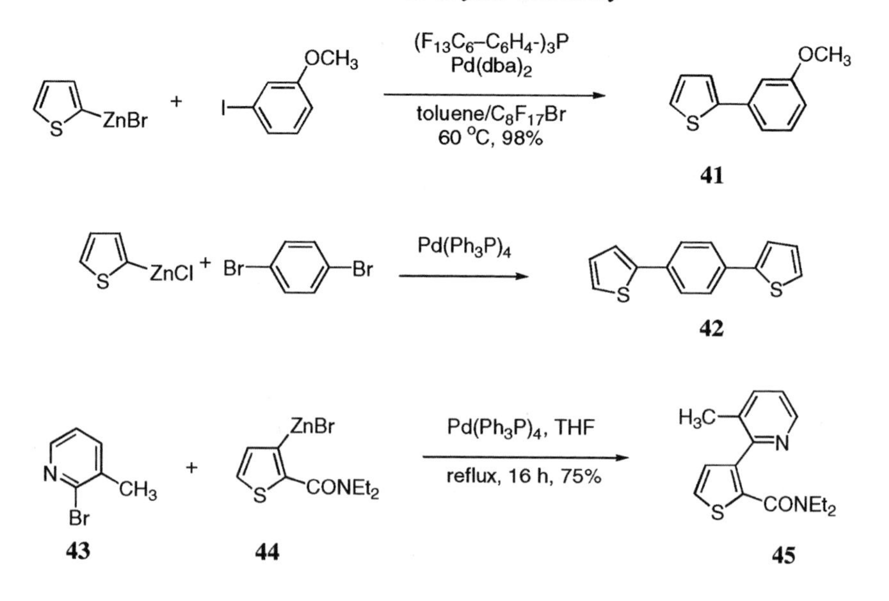


Conventionally, lithiation of thiophene at the C(3) position has been achieved using halogen-metal exchange of 3-bromothiophene with *n*-butyllithium. This method is troublesome because 3-lithiothiophene has a temperature-dependent stability in polar ethereal solvents and slowly undergoes 2- and 3-positional isomerization as well as decomposition at temperatures higher than –25 °C [40, 41]. Rieke metal Zn* was generated from the reduction of $ZnCl_2$ with lithium using naphthalene or biphenyl as an electron carrier in THF. 3-*Iodo*thiophene oxidatively added to Zn* to give the corresponding organozinc reagent **46**, although 3-*bromo*thiophene was inert to Zn*. Subsequently, Negishi reaction of **46** and 1-iodo-4-nitrobenzene then led to arylthiophene **47**.

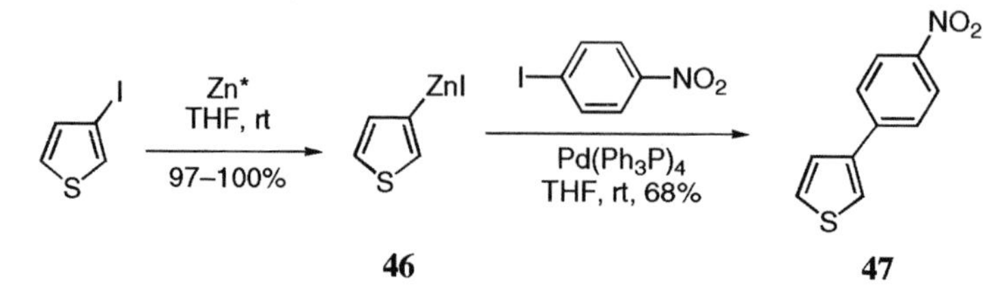


### 5.3.3 Suzuki coupling

The Suzuki reaction has been utilized with increasing frequency as more and more organoboranes become commercially available. Thienylboronic acid is readily prepared by treatment of a thienyl Grignard or a thienyllithium reagent with a trialkylborate followed by acidic hydrolysis, as exemplified by the conversion of bisthiophene **48** to bisthienylboronic acid **49** [42]. Other methods for preparing thienylboronic acid include halogen-metal exchange of a halothiophene followed by quenching with trialkylborate and Pd-catalyzed reaction of a halothiophene with dialkoxyborane [43].

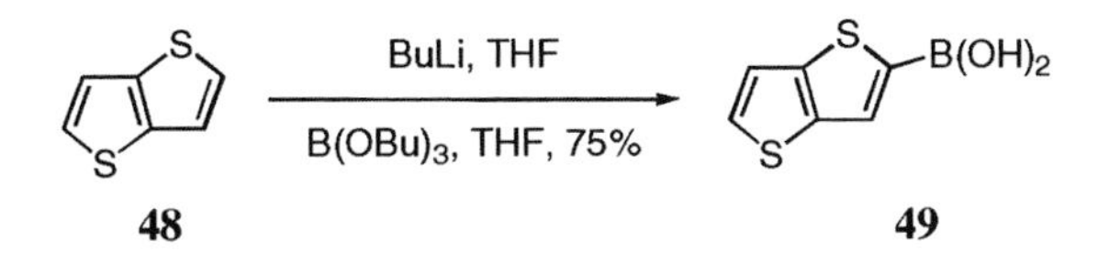


Gronowitz and colleagues prepared various thienylpyrimidines using the Suzuki reaction approach [44, 45]. The union of 5-bromo-2,4-di-*t*-butoxypyrimidine (**50**) and 3-methyl-2-thiopheneboronic acid gave thienylpyrimidine **51**, which was then hydrolyzed to the corresponding uracil **52**, a potential antiviral agent. The adduct **51** was also assembled by switching the coupling partners and coupling 2,4-di-*t*-butoxy-5-pyrimidineboronic acid (**53**) with 2-bromo-3-methylthiophene. In addition, the Gronowitz group synthesized a condensed aromatic compound, thieno[2,3-*c*]-1,7-naphthyridine (**54**), enlisting the Suzuki reaction of 2-formylthiophene-3-boronic acid and 3-amino-4-iodopyridine [46].

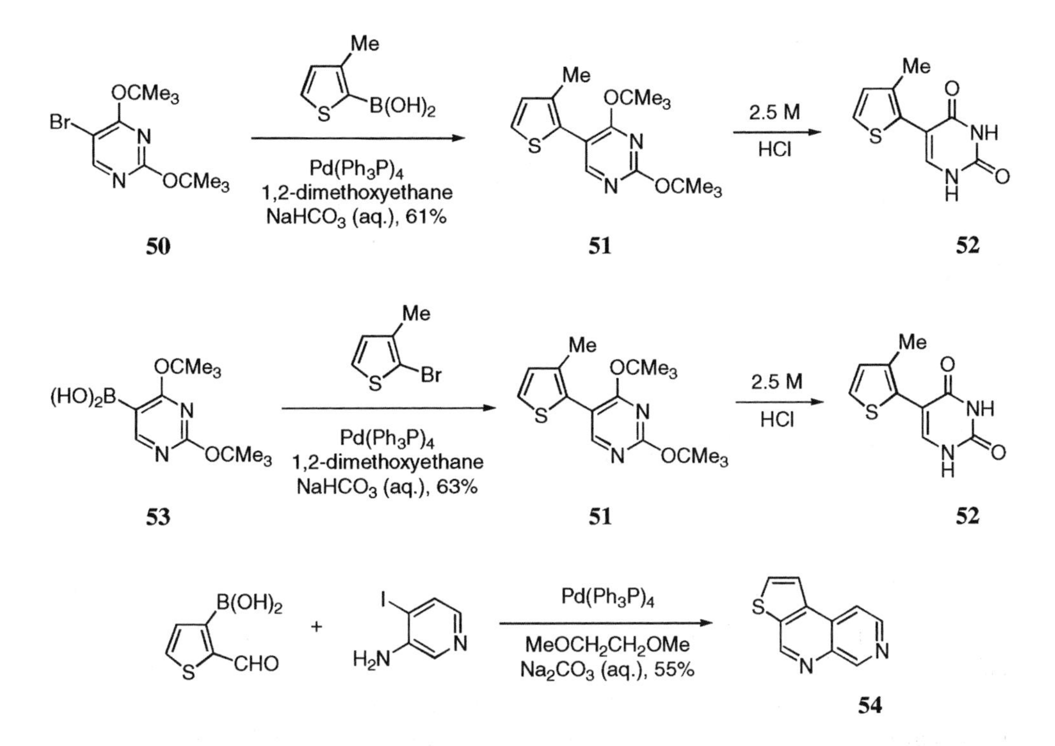


Thienylboronic acids are useful building blocks for preparing biaryls and heterobiaryls employing the Suzuki reaction. In one case, a Suzuki coupling between thiophene-3-boronic acid and iodocyclopropane **55** was promoted by cesium fluoride to furnish the adduct **56** with retention of configuration [47]. In another example, the union between thiophene-3-boronic acid and 5-bromo-2,2-dimethoxy-1,3-indandione (**57**) provided ninhydrin derivative **58** [48].

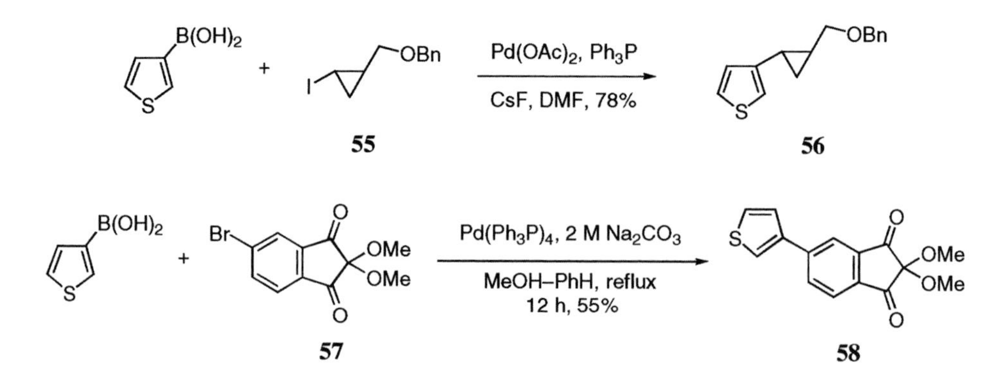


Treatment of thiophene *t*-butyl sulfonamide (**59**) with 2 equivalents of *n*-BuLi formed a dianion in which the second anion resided at the C(5) position. Quenching the resulting dianion with triisopropylborate followed by acidic workup furnished thienylboronic acid **60**, which was then coupled with *p*-bromobenzyl alcohol under basic conditions to afford arylthiophene **61** [49].

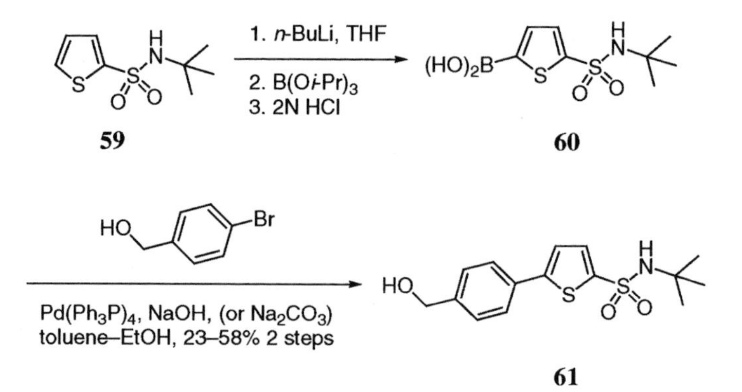


Introduction of a thiophene substituent onto the pyrazine ring was realized by coupling thiophene-2-boronic acid with bromopyrazine **62** to give thienylpyrazine **63** [51]. In the coupling of indole triflate **64** and thiophene-2-boronic acid, an organic base (triethylamine) gave better results than an inorganic base to provide 3-thienylindole **65** [51]. The Suzuki reaction of *p*-chlorobenzoyl chloride and thiophene-2-boronic acid was carried out under anhydrous conditions to furnish ketone **66** [52], providing an alternative synthesis of ketones. Behaving like simple aryl halides, iodothiophenes served as coupling partners with phenylboronic acid [53] and thienylboronic acid [54] to deliver biaryls **67** and **68**, respectively.

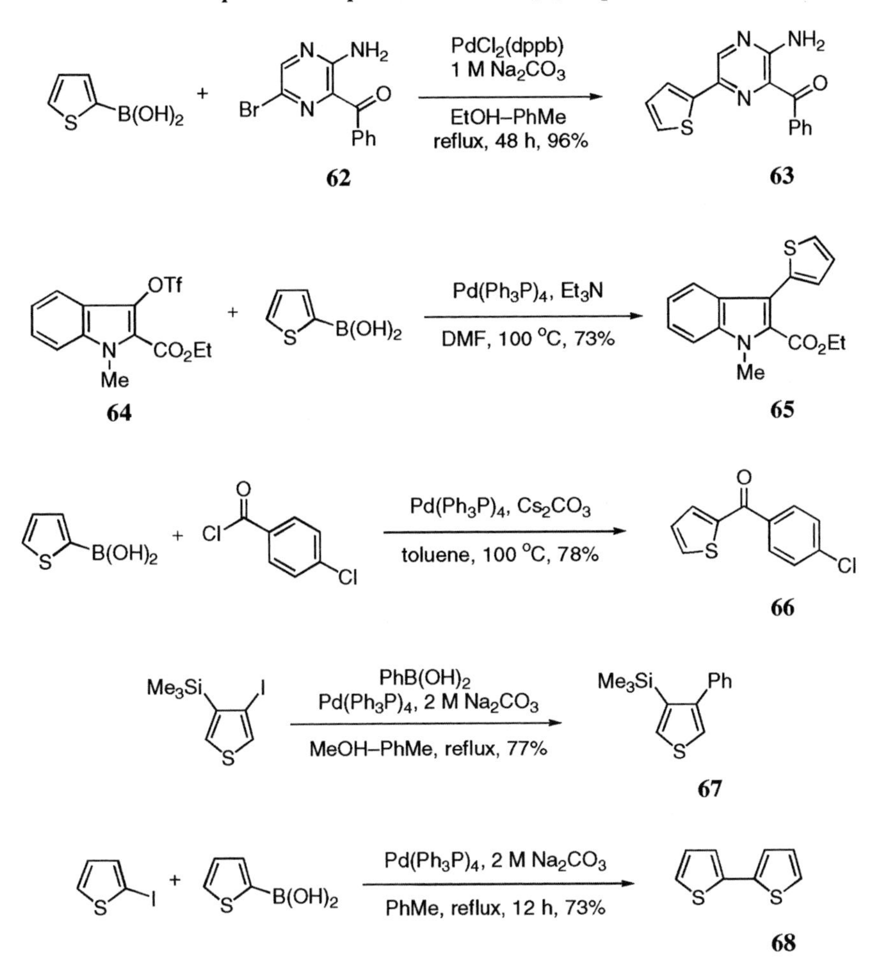


In addition, arylthiophene **70** was obtained by a one-pot Suzuki coupling of *p*-methoxyiodobenzene and 3-bromothiophene *via* an *in situ* boronate formation using one equivalent of the thermally stable diborane **69** [55]. This method avoids the isolation of boronic acids and is advantageous when base-sensitive groups such as aldehyde, nitriles and esters are present. However, the cross-coupling yields are low when both aryl halides are electron-poor because of competitive homocoupling during the reaction.

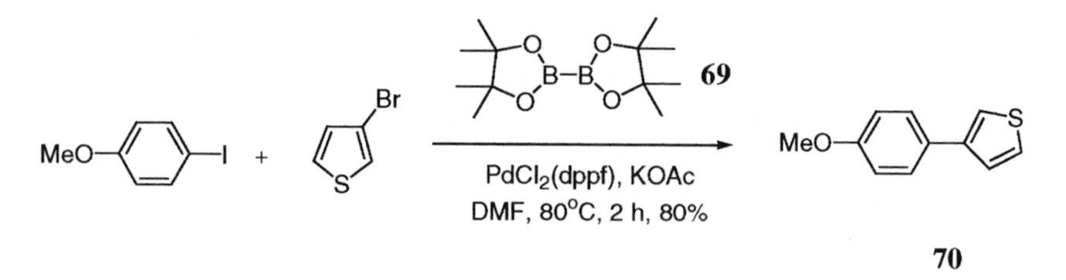


The Suzuki coupling of soluble polyethylene glycol (PEG)-bound bromothiophene **71** and *p*-formylphenylboronic acid provided biaryl **72** [56]. Due to the high solubilizing power of PEG, the reaction was conducted as a liquid-phase synthesis. Treatment of **72** with *o*-pyridinediamine resulted in a two-step-one-pot heterocyclization through an imine intermediate. Nitrobenzene served as an oxidant in the ring closure step. Finally, transesterification with NaOMe in MeOH resulted in 1*H*-imidazole[4,5-*c*]pyridine **73**.

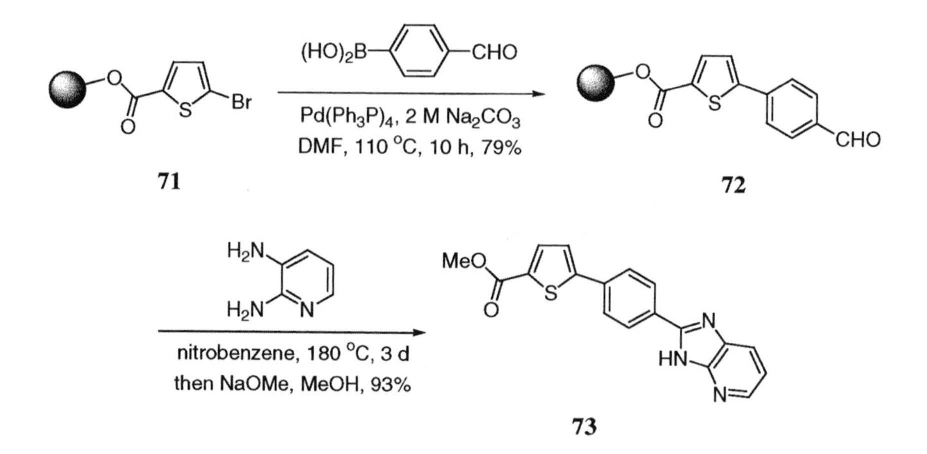


### 5.3.4 Stille coupling

#### *5.3.4.a Preparation of stannylthiophenes*

There are three favored methods for the preparation of stannylthiophenes although other approaches exist [57]: (a) direct metalation of a thiophene followed by quenching with a stannyl electrophile; (b) halogen-metal exchange of a halothiophene followed by reaction with a stannyl electrophile; and (c) Pd-catalyzed reaction between a halothiophene and hexaalkylditin. Direct metalation of thiophenes followed by reaction with a stannyl electrophile is the most frequently utilized approach to prepare stannyl thiophenes. In the absence of *ortho*-directing groups, the metalation occurs at the α positions due to the inductive effect of the $C_{sp2}$—S bond [58–61]. Thienylstannane **75** was prepared by treatment of 2-(3-thienyl)-1,3-dioxolane (**74**) with *n*-butyllithium followed by quenching with $Me_3SnCl$ [60–63]. Acidic hydrolysis of dioxolane **75** then produced 2-stannylthiophene-3-carboxaldehyde (**76**). In some cases, addition of TMEDA facilitated the metalation process [64, 65]. In other cases, employing a bulky base was advantageous if the substrates possessed functional groups susceptible to nucleophilic attack. Thus, direct metalation of (*E*)-3-(2-thienyl)propenoate (**77**) was achieved using LDA and the resulting lithiothiophene was quenched with trimethyltin chloride to afford stereoisomerically pure (*E*)-3-(5-trimethylstannyl-2-thienyl)propenoate (**78**) [66].

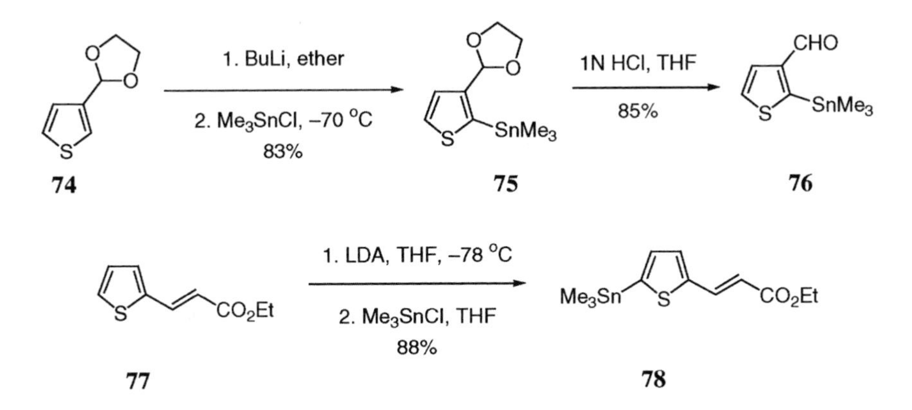


Halogen-metal exchange of a halothiophene followed by quenching with a stannyl electrophile is another approach for preparing stannylthiophenes [67–70]. This protocol is especially useful when regiochemistry is critical because the regiochemistry is often complicated for direct metalation. In the preparation of 4-stannylthiophene (**80**), addition of *t*-butyl *N*-(4-bromo-3-thienyl)carbamate (**79**) to a cooled solution of butyllithium resulted in a cleaner reaction than when the addition was reversed [67]. Stannane **80** was stable and could be purified by flash chromatography, whereas other regioisomers readily decomposed on silica gel to give the C—Sn bond cleaved product. When 2,4-dibromo-5-methylthiophene (**81**) was treated with one equivalent of *n*-BuLi, a regioselective halogen-metal exchange at the C(2) position was achieved, giving 2-silylthiophene **82** after treatment with TMSCl [70]. A second halogen-metal exchange at C(4) of **82** was followed by stannylation to give 4-stannylthiophene **83**.

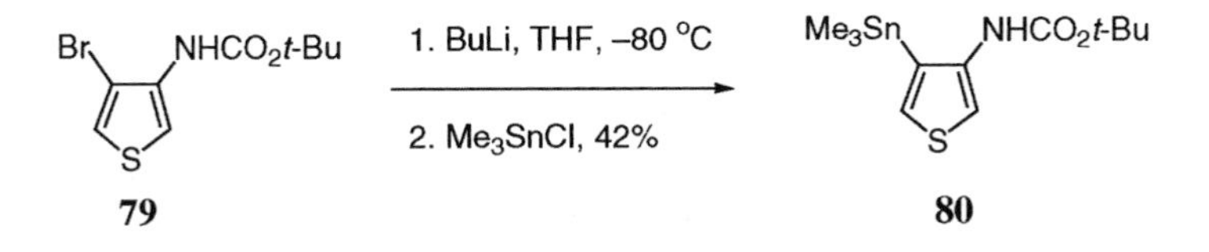

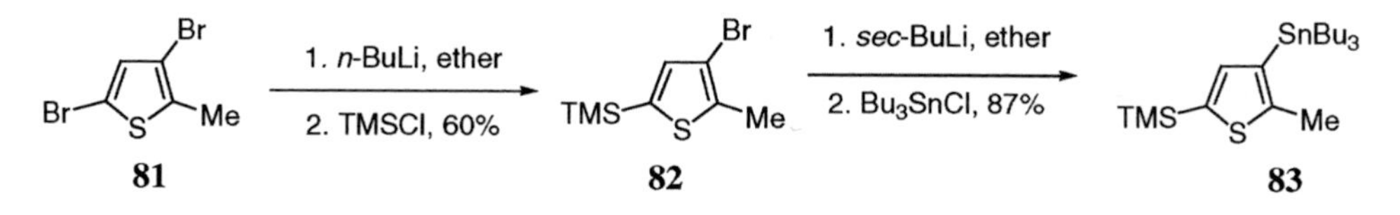


Most functional groups will tolerate Pd-catalyzed reactions between a halothiophene and hexaalkylditin for the preparation of stannylthiophenes [71, 72]. In practice, however, this method often suffers from consumption of large quantities of hexaalkylditin because of its disproportionation reaction. When bromothiophene **84** was refluxed with hexamethylditin under the agency of $Pd(Ph_3P)_4$ catalysis, stannane **85** was obtained in good yield [73].

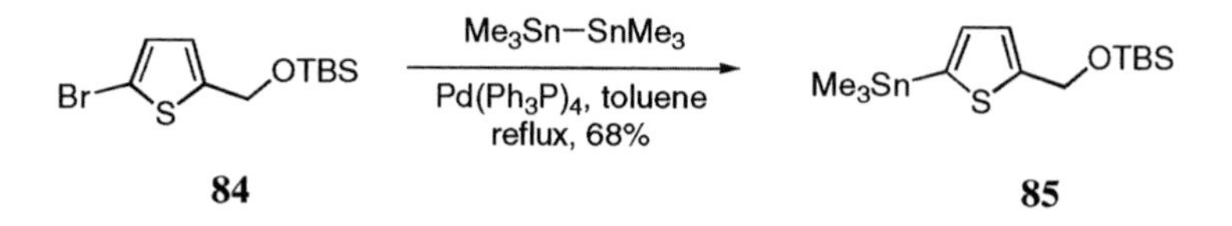


### *5.3.4.b Alkyl-, vinyl- and alkynylthiophenes*

Methylation of halothiophenes **86** and **88** was accomplished *via* the Stille reaction with tetramethyltin to give methylated thieno[3,2-*b*]pyran **87** [74] and thienyldeoxyuridine **89** [75], respectively. Analogously, the coupling of an allyl chloride, chloromethylcephem **90** and 2-tri-*n*-butylstannylthiophene furnished **91**, an intermediate for a C(3) thiophene analog of cephalosporin [76].

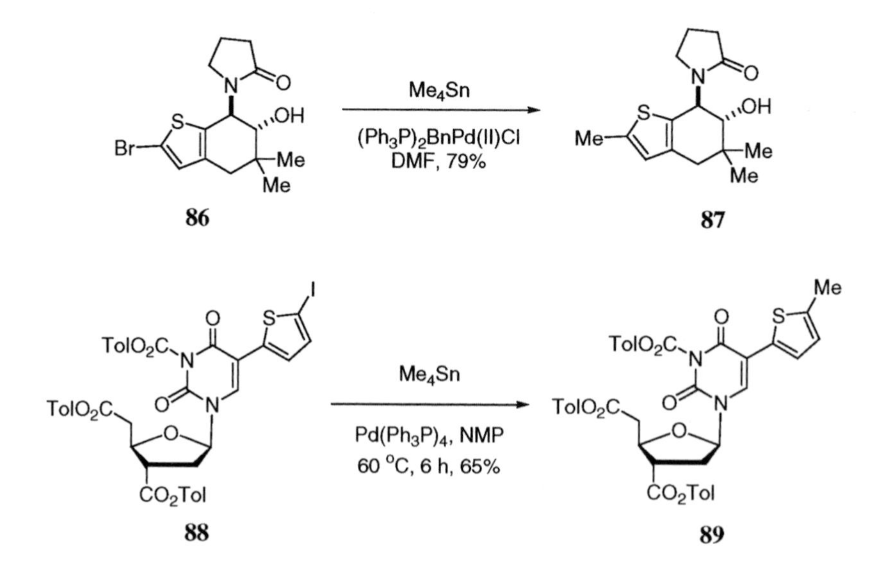

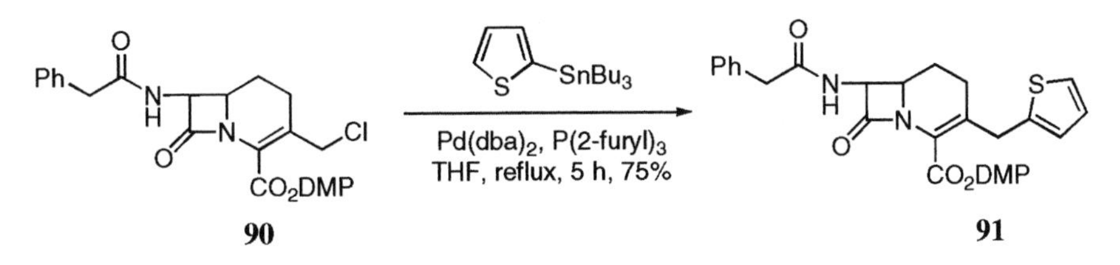


The Stille adduct of 2-bromothiophene and 1-ethoxy-2-tributyl-*n*-stannylethene or 1-ethoxy-1-tri-*n*-butylstannylethene is a masked thienyl aldehyde or a masked ketone, respectively [77–80]. Vinylstannane **93**, derived exclusively as the *E*-isomer from hydrostannation of bis(trimethylsilyl)propargyl amine (**92**), was coupled with 2-bromothiophene to form (*E*)-cinnamyl amine **94** upon acidic hydrolysis [81, 82]. In another case, stereoisomerically pure phenyl (*E*)-2-tributylstannyl-2-alkenoate **95**, arising from Pd-mediated hydrostannation of phenyl (*E*)-2-alkynoate, was joined with 2-iodothiophene to deliver the stereodefined trisubstituted α,β-unsaturated ester **96** [83, 84].

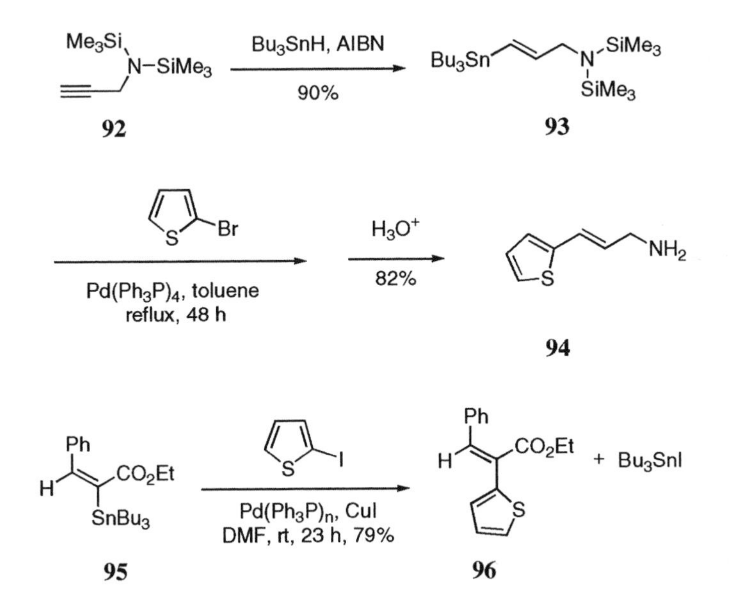


Liebeskind and associates converted 3,4-diisopropyl squarate (**97**) to stannylcyclobutenedione **98** *via* a 1,4-addition-elimination sequence. **98** was then coupled with 2-iodothiophene to afford substituted cyclobutenedione **99** [85]. In another case, 3-lithioquinuclidin-2-ene, generated from the Shapiro reaction of 3-quinuclidinone (**100**), was quenched with $Bu_3SnCl$ to afford a unique "enamine stannane" **101**. The Stille reaction of stannane **101** and 5-bromo-2-formyl-thiophene then furnished 3-thienylquinuclidine **102** [86].

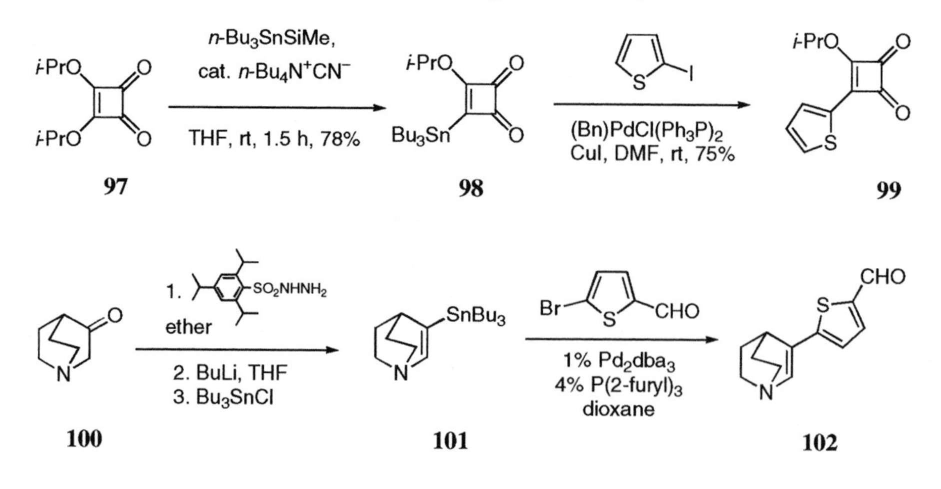


Vinylphosphates and some vinyl chlorides have been utilized as electrophiles in couplings with stannylthiophenes. The union between vinylphosphate **103** and 2-tributylstannylthiophene afforded 3-thienyl-4*H*-1,4-benzoxazine **104** [87], whereas the coupling of 6-chloropyrone **105** and 2-tributylstannylthiophene gave rise to 6-thienyl-4-hydroxy-2*H*-pyan-2-one **106**, offering an opportunity for introducing substitution at the C(6) position of the pyrone nucleus [88]. Interestingly, allenylstannane, readily prepared from propargyl tosylate by copper-mediated addition of lithium tributylstannane, was joined with 2-iodothiophene to provide a rapid entry to 2-thienylallene (**107**) [89].

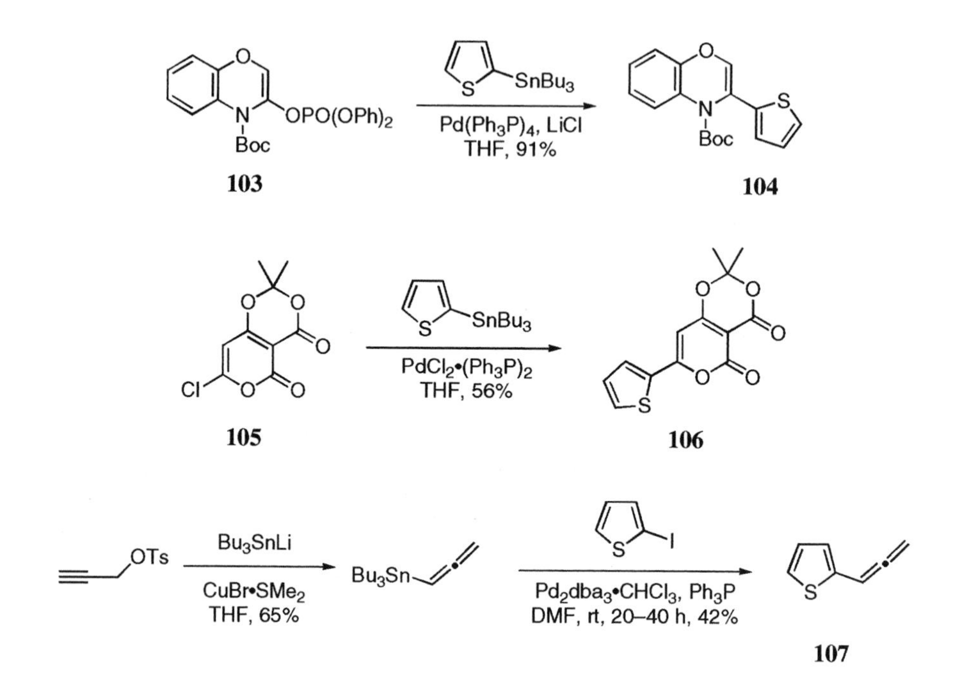

Although the Sonogashira reaction (*vide infra*) is the mildest and most common approach employed to introduce alkyne functionality, it is of limited utility for acetylenes having either electron-withdrawing or electron-donating groups. The Stille reaction of 2-iodothiophene and ethoxy(trimethylstannyl)acetylene, however, allowed formation of ethoxyethynylthiophene (**108**) [90], which was hydrated to ester **109**. Other advantages of using stannanes to prepare arylacetylene include tolerance of more functional groups and better maintenance of stereochemical integrity [91, 92]. In Stille's total synthesis of the naturally occurring thienyl dienynol **113** [92], alkynylstannane **110**, derived from the reaction of ethynylthiophene and (diethylamino)trimethylstannane, was coupled with vinyl iodide **111** to provide isomerically pure dienyne **112**. Subsequent hydrolysis of **112** then gave stereochemically pure (3*E*, 5*E*)-8-(2-thienyl)-3,5-octadien-7-yn-1-ol (**113**), a possible natural insecticide isolated from *Crysanthemum macrotum* (Dur.) Ball.

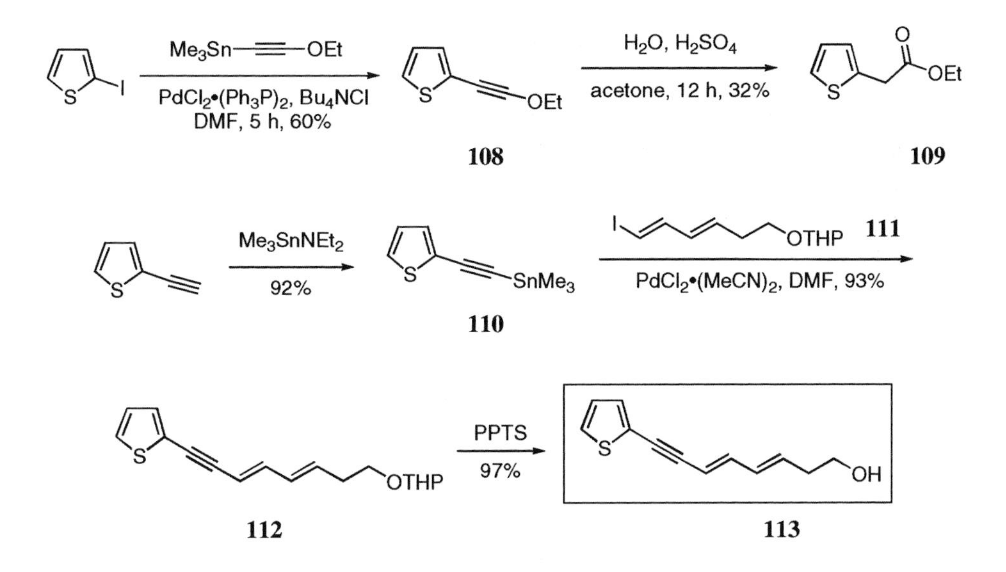


### *5.3.4.c Arylthiophenes*

A tremendous amount of work has been reported on the synthesis of arylthiophenes and heteroarylthiophenes utilizing the Stille reaction approach. In one case, 2-tributylstannylthiophene was coupled with *p*-acetoxyphenyl iodide to give thienylphenol **114** after hydrolysis [93]. In another, the union of 2-tributylstannylbenzo[*b*]thiophene and *p*-acetyliodobenzene provided arylbenzothiophene **115** using inexpensive Pd/C as a heterogeneous catalyst, CuI as a co-catalyst, and $AsPh_3$ as a soft ligand [94]. Moreover, Kennedy *et al.* coupled 2-tributylstannylthiophene and 2-chloro-4-bromobenzylphosphonate (**116**) to make heterobiaryl phosphonate **117** [95]. During this reaction, portion-wise addition of the catalyst every two hours was found to be necessary owing to its continuous slow decomposition.

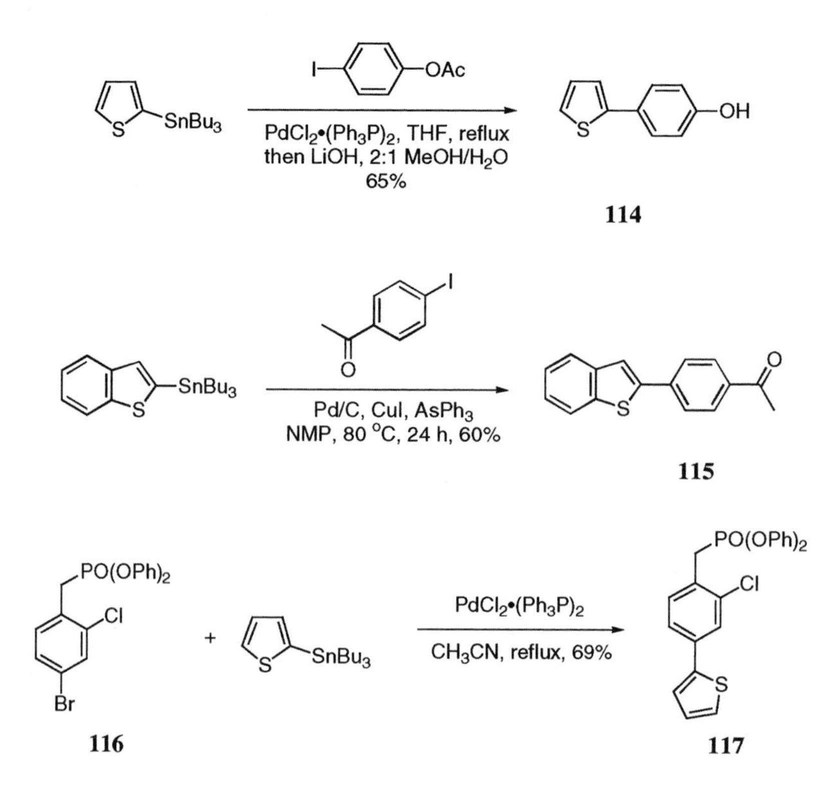


Furthermore, arylthiophenes have been prepared using the Stille coupling of hypervalent iodonium salts [96] or organolead compounds [97, 98] as electrophiles in place of aryl or vinyl halides and triflates. Hypervalent iodonium salts are sufficiently reactive to undergo coupling at room temperature.

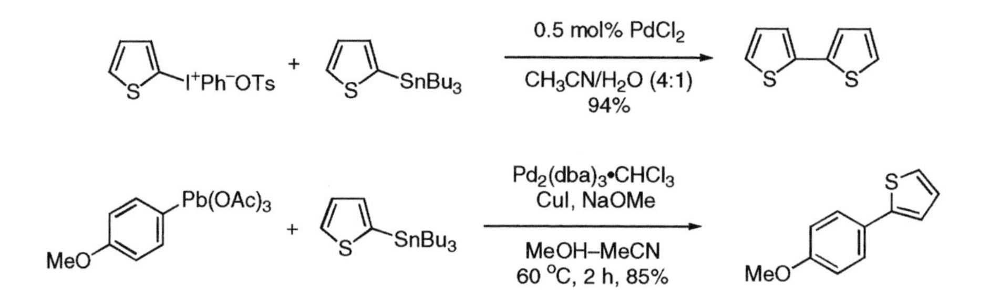

### *5.3.4.d Heteroarylthiophenes*

The 2-oxazolinyl group can serve as a masked aldehyde or carboxylic acid and as a director of metalation for both nucleophilic and electrophilic aromatic substitutions. The cross-coupling of aryl Grignard reagents and methylthiooxazoline has been reported with limited examples [99]. However, the Stille reaction of 3-bromothiophene and 2-stannyloxazoline **118**, derived from 4,4-dimethyl-2-oxazoline, did indeed give 2-thienyloxazoline **119**. An asymmetric version of the Stille coupling of chiral 2-bromooxazolines and thiophenestannanes was also described [101] (albeit in only 20–28% yield), as was the coupling of oxazol-2-yl and 2-oxazolin-2-yltrimethylstannanes with aryl halides [102].

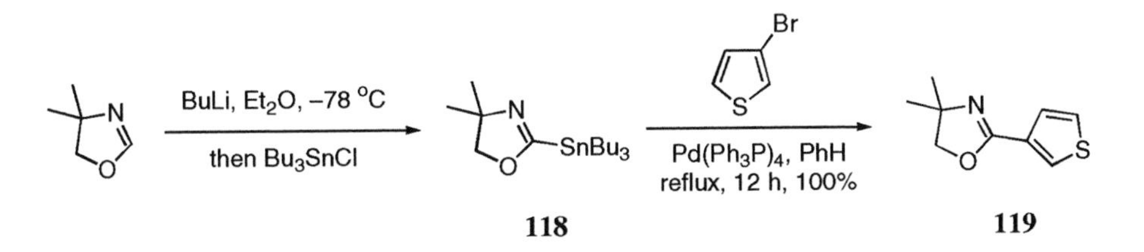


Some bithienyl and terthienyl derivatives display biological activities including antifungal, nematocidal and inhibiting seed germination. The Stille reaction is regarded as the method of choice for the preparations of these thiophenes. For example, the union of 5-iodo-2-thiophenecarboxaldehyde and 2-tributylstannylthiophene furnished 5-(2'-thienyl)-2-thiophenecarboxaldehyde (**120**) [58]. In addition, terthienyl and bithienyl derivatives have been synthesized using 3,4-dinitro-2,5-dibromothiophene [103] and 3-methylsulfonyl-2-bromothiophene [104, 105] as electrophiles. Furthermore, tetrathiafulvalene (TTF), a π-electron donor in organic conductors, was transformed into the corresponding trimethylstannyl-tetrathiafulvalene, which was then coupled with 2,5-dibromothiophene to secure 2,5-thienylbistetrathiafulvalene (**121**) [106].

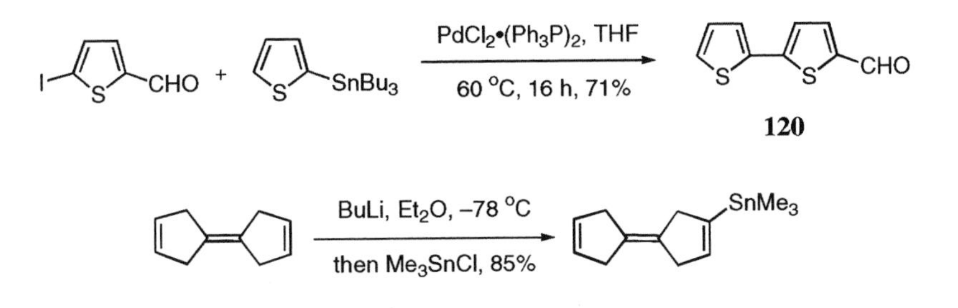

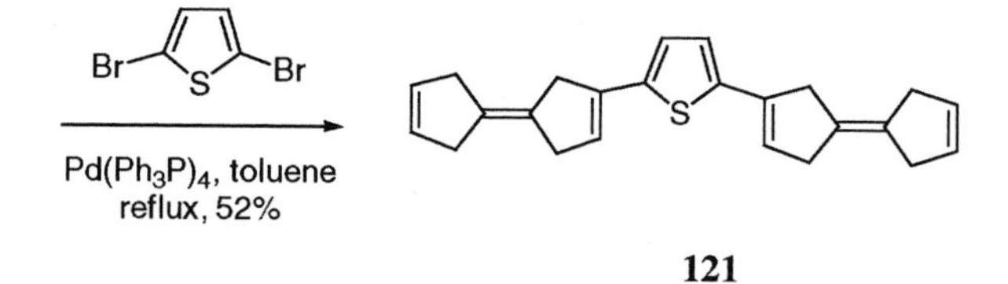


Thienylpyridines are also of great biological interest. Their syntheses *via* the Stille reaction are well precedented. In one case, two equivalents of 2-(trimethylstannyl)pyridine and 2,5-dibromothiophene underwent a Pd-catalyzed cross-coupling reaction to give 2,5-bis(2-pyridyl)thiophene (**122**) [107, 108]. In a series of papers, Gronowitz's group described the synthesis of pyridine-substituted hydroxythiophenes employing the Stille approach [109, 110]. 3-Bromo-2-methoxythiophene was converted to 3-tri-*n*-methylstannyl-2-methoxythiophene (**123**) *via* regiochemical halogen-metal exchange at the C(5) position. The subsequent Stille reaction of **123** with 2-bromopyridine produced the desired thienylpyridine **124**.

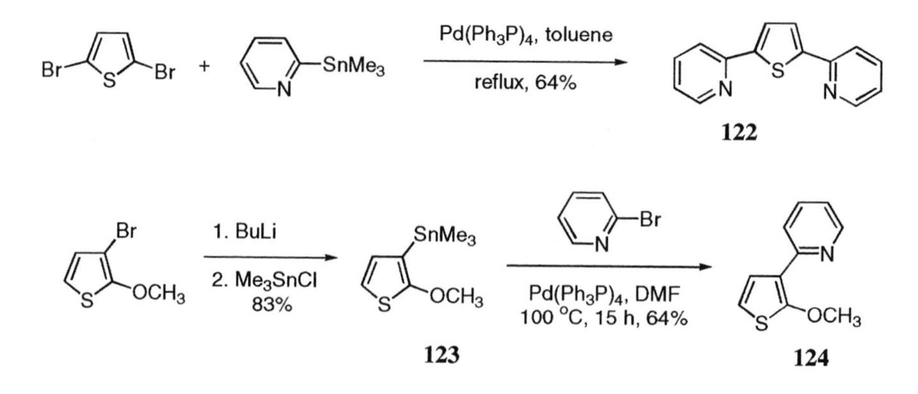


### *5.3.4.e Thiophene-containing condensed heteroaromatics*

The Stille reaction is the key step in some synthesis of thiophene-containing condensed heteroaromatics. Enlisting a Stille-Kelly reaction, Iyoda *et al.* treated dibromide **125** with hexamethylditin in the presence of $Pd(Ph_3P)_4$ to afford dithienothiophene (**126**) [111].

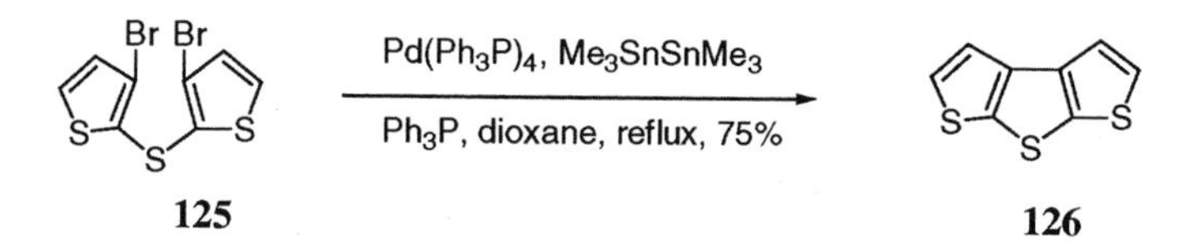


Gronowitz and associates coupled 2-(2-trimethylstannyl-3-thienyl)-1,3-dioxolane (**75**) with *tert*-butyl *N*-(*ortho*-bromothienyl)carbamate (**127**) to give the Stille adduct, which underwent acid-catalyzed deprotection and cyclization to deliver dithienopyridine **128** [112]. The uncharacteristically low yield of the Stille coupling was presumably due to the steric bulk of **75** because it led to a crowded intermediate, decreased the ease of transmetalation. In contrast, *t*-butyl *N*-(2-trimethylstannyl-3-thienyl)carbamate (**129**) was coupled with 3-iodo-2-formylpyridine to produce thieno[3,2-*b*][2,8]naphthyridine (**130**) in good yield [113].

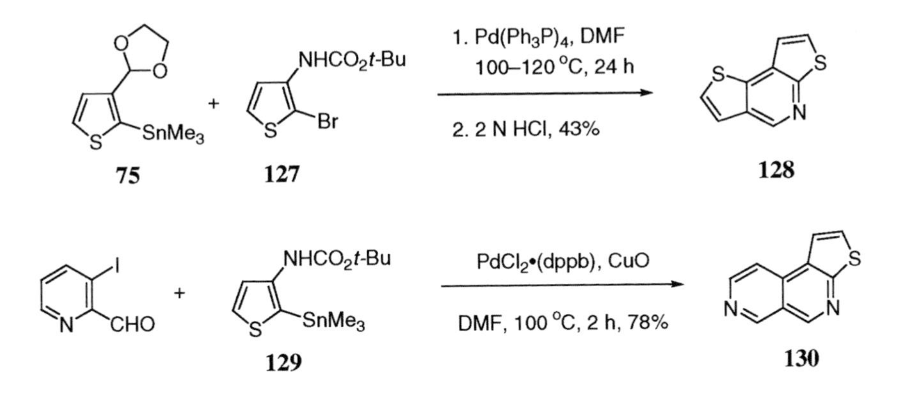


### 5.3.4.f Cu(I) thiophene-2-carboxylate (CuTC)

Since its introduction by Allred and Liebsekind in 1996 [114], copper thiophene-2-carboxylate (CuTC) has emerged as a mild and useful reagent for mediating the cross-coupling of organostannanes with vinyl iodides at room temperature. CuTC is especially effective for substrates that are not stable at high temperature. In Paterson's total synthesis of elaiolide, he enlisted a CuTC-promoted Stille cyclodimerization of vinyl iodide **131** to afford the 16-membered macrocycle **132** under very mild conditions [115].

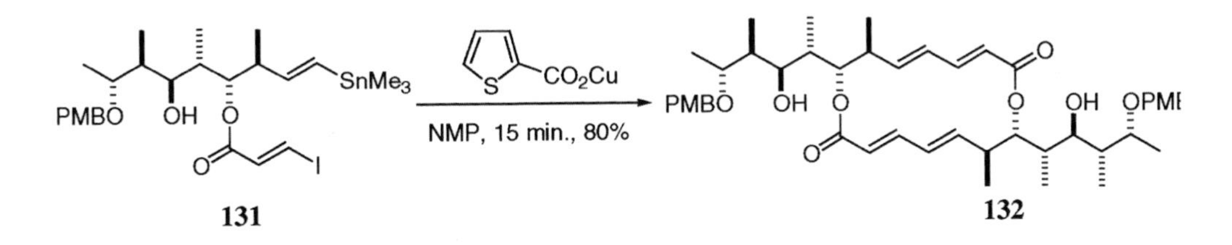


### 5.3.5 Hiyama coupling

Pd-catalyzed cross-coupling reactions of organosilicon compounds and organic halides display higher stereoselectivity, and chemoselectivity than other organometallic reagents, and organosilicon substrates are readily available. However, because the C—Si bond is not as polarized as other carbon-metal bonds, introduction of a fluorine atom into the silyl group of organosilanes is necessary to accelerate Pd-mediated cross-coupling reactions. In the presence of catalytic $\eta^3$-allylpalladium chloride dimer and two equivalents of KF, the cross-coupling of ethyl(2-thienyl)difluorosilane and methyl 3-iodo-2-thiophenecarboxylate gave adduct **133** under relatively forcing conditions [116].

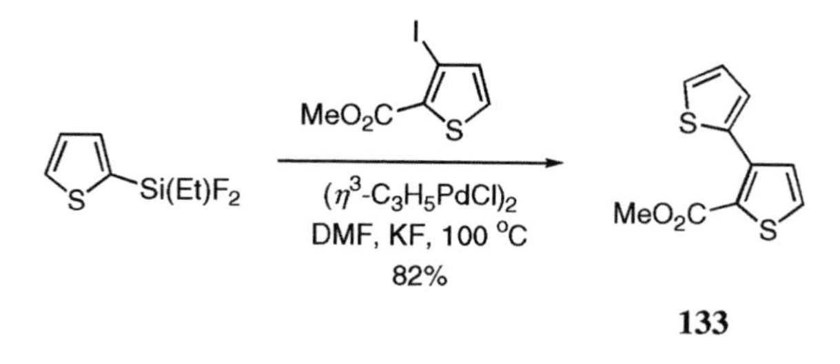


## 5.4 Sonogashira reaction

Sonogashira reactions of both α-halothiophenes [117] and β-halothiophenes [118] proceed smoothly even for fairly complicated molecules as illustrated by the transformation of brotizolam (**134**) to alkyne **135** [119]. Interestingly, 3,4-bis(trimethylsilyl)thiophene (**137**), derived from the intermolecular cyclization of 4-phenylthiazole (**136**) and bis(trimethylsilyl)acetylene, underwent consecutive iodination and Sonogashira reaction to make 3,4-bisalkynylthiophenes [120]. Therefore, a regiospecific mono-*ipso*-iodination of **137** gave iodothiophene **138**, which was coupled with phenylacetylene to afford alkynylthiophene **139**. A second iodination and a Sonogashira reaction then provided the unsymmetrically substituted 3,4-bisalkynylthiophene **140**.

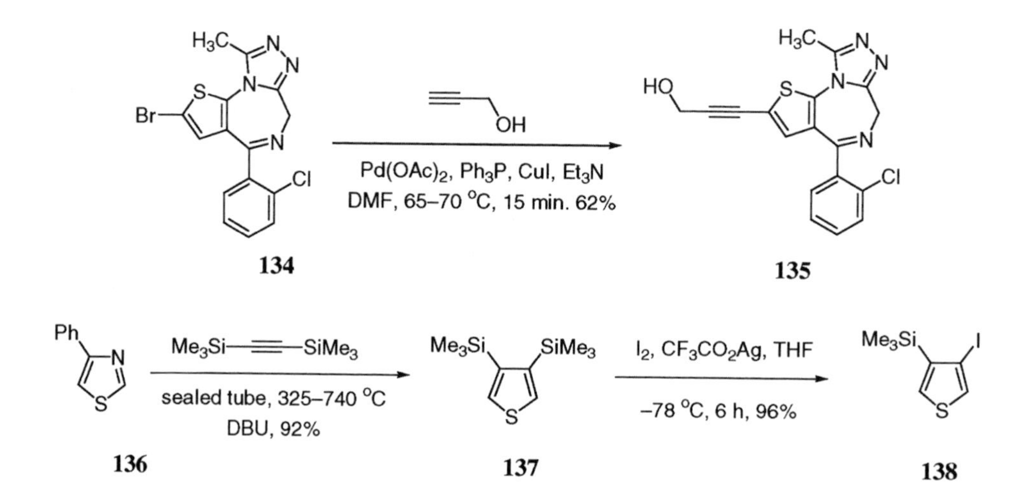

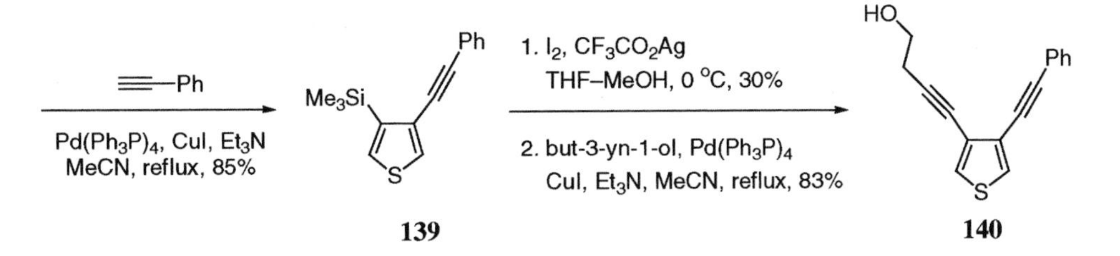


Using NaOH as the base, diarylacetylenes have been synthesized from either 2-methyl-3-butyn-2-ol [121] or trimethylsilylacetylene [122]. In both cases, NaOH unmasked the protections after the first coupling reaction, revealing the additional terminal alkynyl functionality. Therefore, coupling the adduct **141**, derived from 2-iodothiophene and 2-methyl-3-butyn-2-ol, with 2-iodobenzothiophene provided diarylacetylene **142** [121]. Analogously, dithienylacetylene (**143**) was obtained when 2-iodothiophene and trimethylsilylacetylene were subjected to the same conditions [122].

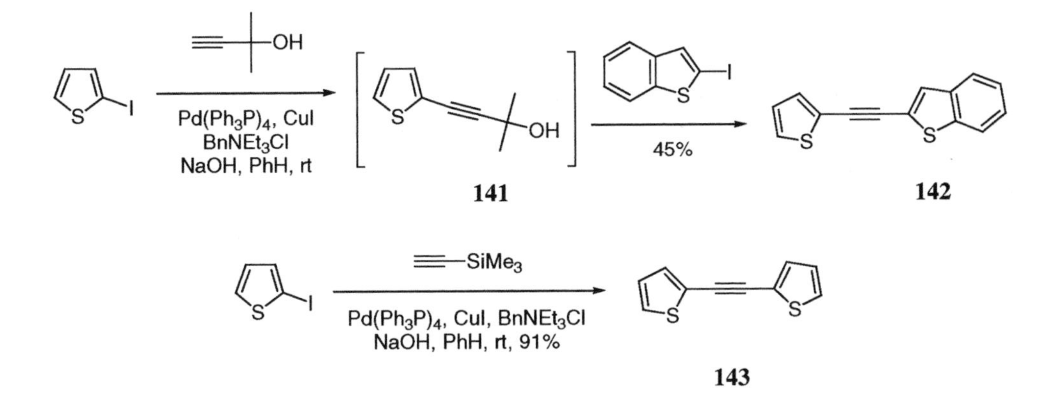


Advantage has been taken of the aforementioned observations in the synthesis of a terthiophene natural product, arctic acid (**147**) [123]. Pd-catalyzed carbonylation of bromobisthiophene **25**, obtained from the Kumada coupling of 2-thienylmagnesium bromide and 2,5-dibromothiophene, gave bithiophene ester **144**, which was converted to iodide **145** by reaction with iodine and yellow mercuric oxide. Subsequent propynylation of **145** was then realized using the Sonogashira reaction with prop-1-yne to give bisthienyl alkyne **146**, which was subsequently hydrolyzed to 5'-(1-propynyl)-2,2'-bithienyl-5-carboxylic acid (**147**), a natural product isolated from the root of *Arctium lappa*.

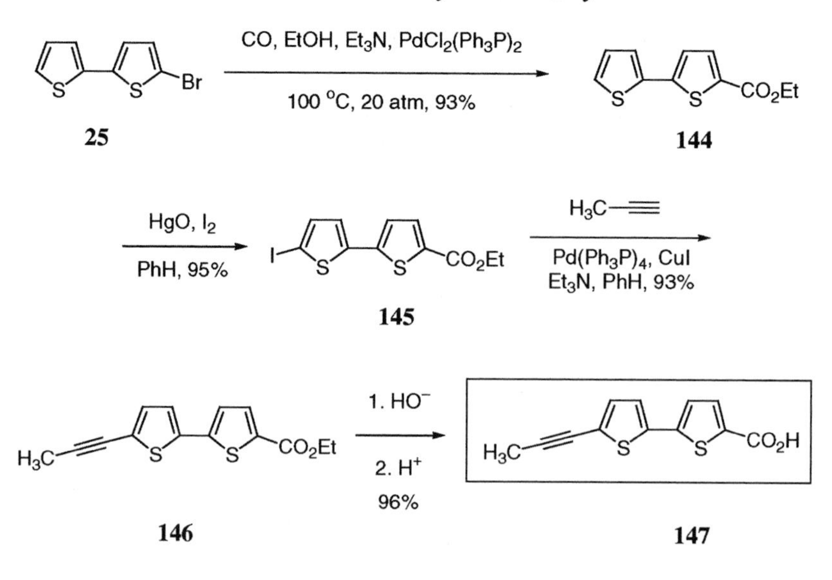


## 5.5 Heck and intramolecular Heck reactions

Both 2-bromothiophene and 3-bromothiophene have been coupled with allyl alcohols to make thienylated α,β-unsaturated ketones [124]. Iodothiophenes were more reactive than the corresponding bromides, whereas the chlorothiophenes were unreactive. As expected, 2-bromothiophene was two to three times more reactive than 3-bromothiophene. In addition to the expected Heck adduct **149**, the reaction of 2-bromothiophene **148** with 1-methylprop-2-en-1-ol also resulted in the regiosmer **150**.

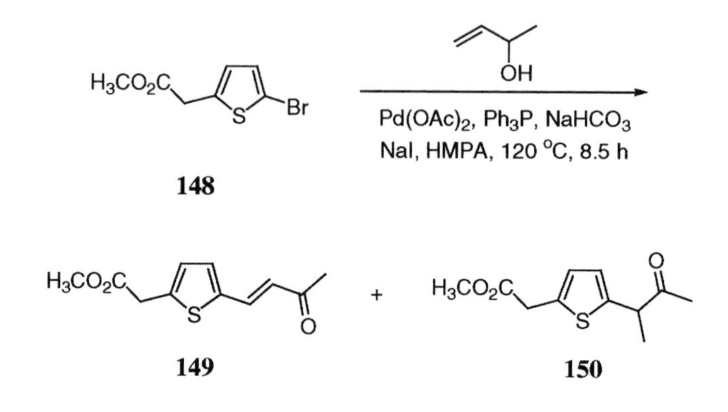


While the intramolecular Heck reaction has been widely used to synthesize indoles and benzofurans, not many applications have been found in the preparation of benzothiophenes because of the thiophilicity of the Pd(II) species. Pleixats and coworkers treated iodophenylsulfide **151**, obtained from *o*-iodoaniline and crotyl bromide in two steps, with

$Pd(Ph_3P)_4$ and $Et_3N$ in refluxing acetonitrile to form the intramolecular Heck cyclization product **152** [125]. The mechanism is akin to that of the Mori-Ban indole synthesis (see page 24). In another case, the intramolecular Heck cyclization of enamidone **153** with a pendant thienylbromide moiety furnished the 6-*trig-endo* product, indolizine **154**, in 63% yield, along with the debrominated enamidone **155** in 37% yield [126].

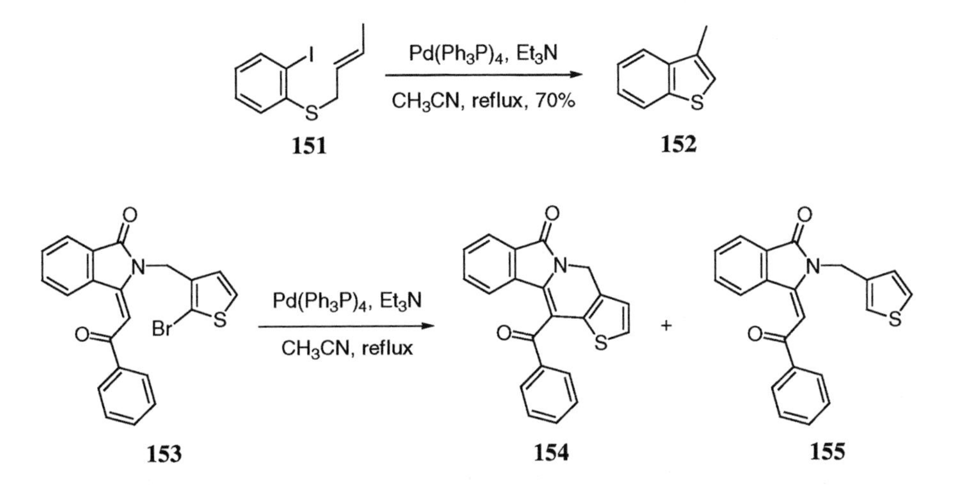


Ohta's group investigated the heteroaryl Heck reaction of thiophenes and benzothiophenes with aryl halides [127] and chloropyrazines [128]. Addition of the electrophiles invariably took place at C(2) as exemplified by the formation of arylbenzothiophene **156** from the reaction of benzothiophene and *p*-bromobenzaldehyde [127]. As expected, the heteroaryl Heck reaction of 2-thienylnitrile, an activated thiophene, with iodobenzene afforded the arylation product **157** [129].

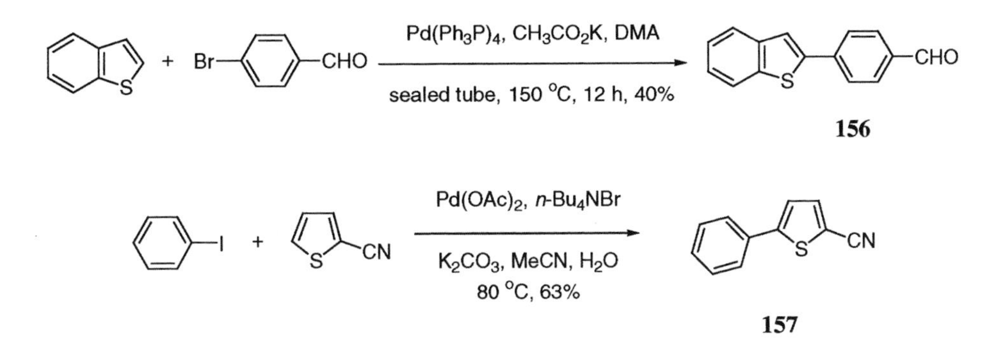

## 5.6 Carbonylation reactions

Halothiophenes take part in Pd-catalyzed alkoxycarbonylations in the presence of CO, alcohol and base. In order to avoid the inconvenience of pressurized carbon monoxide, alkyl formate may be used as a safe surrogate [130]. In one of the many examples, 2-Iodothiophene was carbonylated to the corresponding methyl ester using methylformate in place of CO.

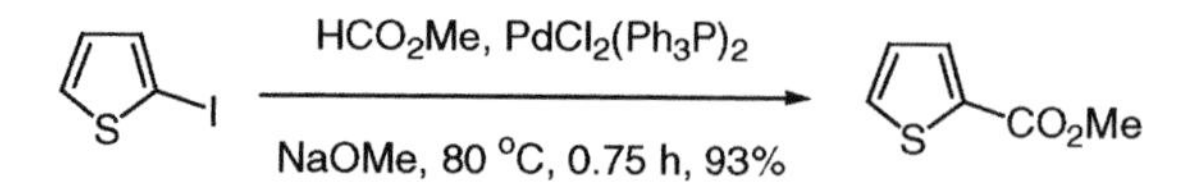


Lin and Yamamoto described a Pd-catalyzed carbonylation of benzyl alcohols [131]. Thus, under the agency of palladium catalysis and promotion by HI, 3-thiophenemethanol was carbonylated to give 3-thiopheneacetic acid as a major product along with methylthiophene as a minor one.

Pd($Ph_3P)_4$, CO (90 atm), HI; acetone/$H_2O$, 90 °C, 42 h

The Hiyama group discovered that transmetalation of pentacoordinate silicate occurs in Pd-catalyzed reactions. They also successfully conducted Pd-catalyzed alkoxycarbonylation of organofluorosilanes with organic halides under the promotion of fluoride ion [132]. One salient feature of such a silicon-based carbonylative reaction is its remarkable functional group accommodation, even allowing a carbonyl group on either coupling partner. For example, a three-component carbonylation of 2-thienyldifluorosilane and *m*-iodobenzaldehyde was carried out in the presence of CO (1 atm), $(\eta^3\text{-}C_3H_5Pd)_2$ and KF in 1,3-dimethyl-2-imidazolidinone (DMI) to form ketone **158**. In another three-component carbonylative cross-coupling reaction, the hypervalent iodonium salt **159,** a halide surrogate, was joined with *p*-methoxyphenylboronic acid at room temperature to afford ketone **160** [133].

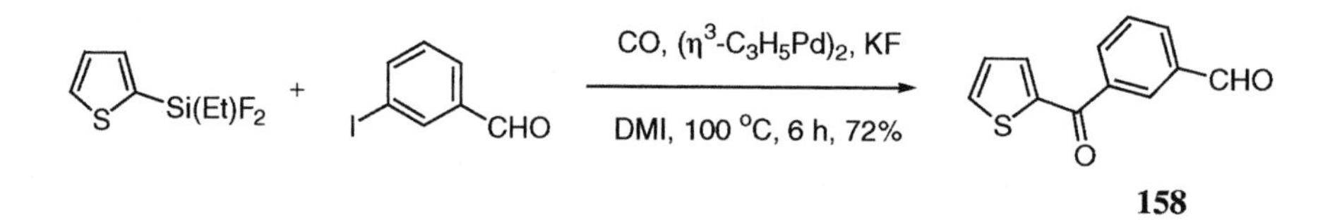

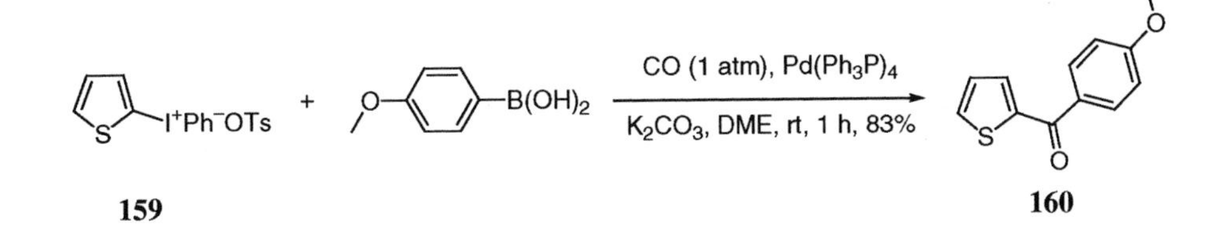


## 5.7 Miscellaneous

### 5.7.1 Pd-catalyzed C—P bond formation

The formation of an $sp^3$-hybridized C—P bond is readily achievable using the Michaelis-Arbuzov reaction. Such an approach is not applicable to form heteroaryl C—P bonds in which the carbon atoms are $sp^2$ hybridized, whereas palladium catalysis does provide a useful method for $Csp^2$—P bond formation. The first report on Pd-catalyzed C—P bond formation was revealed by Hirao *et al.* [134–136]. Xu's group further expanded the scope of these reactions [137, 138]. They coupled 2-bromothiophene with *n*-butyl benzenephosphite to form *n*-butyl arylphosphinate **161** [137]. In addition, the coupling of 2-bromothiophene and an alkylarylphosphinate was also successful [138]. For the mechanism, see page 19–21.

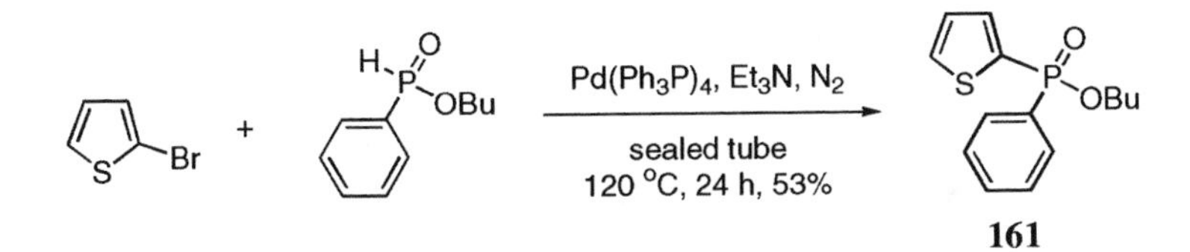


### 5.7.2 Pd-catalyzed C—N bond formation

Historically, α-halothiophenes had been considered poor substrates for the Pd-catalyzed amination because of the strong thiophilicity of Pd(II). However, enlisting the Buchwald-Hartwig arylamination conditions, Watanabe and coworkers successfully aminated both α- and β-halothiophenes [139]. In a strategy employing the sterically hindered, electron-rich phosphine ligand $P(t\text{-}Bu)_3$, NaO*t*-Bu as the base, and $Pd(OAc)_2$ as the catalyst, 2,5-dibromothiophene was bisaminated with diphenylamine to afford 2,5-bis(diphenylamino)-thiophene (**162**). This method is also applicable to 3-bromothiophene (69% yield), whereas the monoamination of 3,4-dibromothiophene was low-yielding (12%).

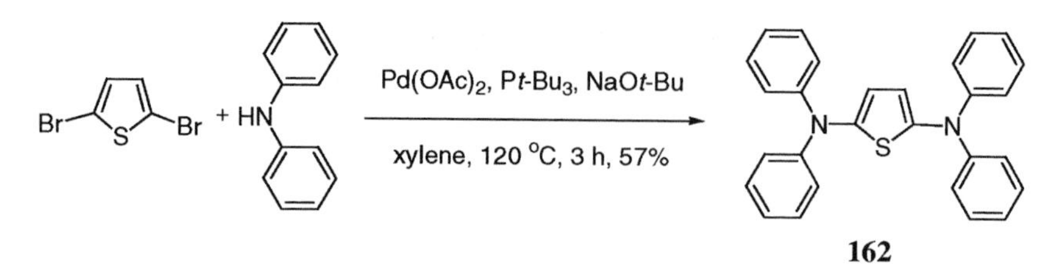


### 5.7.3 Pd-catalyzed cycloisomerization

In contrast to the prevalence of furan preparation *via* Pd-catalyzed heteroannulation methods, a scarcity of literature precedents are found for thiophene syntheses *via* such an approach, again possibly due to the thiophilicity of the Pd(II)-intermediates. Nevertheless, Gabriele's group synthesized substituted thiophenes from (*Z*)-2-en-4-yne-1-thiols utilizing Pd-catalyzed cycloisomerization [140]. For example, using $PdI_2$ as the catalyst and KI as the solubilizing agent, the Pd(II)-catalyzed cycloisomerization of **163** gave rise to thiophene **166**. Presumably, coordination of the triple bond with the Pd(II) species gives **164**, which is followed by nucleophilic attack by the SH group, forming **165**. Protonolysis of **165** is then followed by aromatization to furnish **166**.

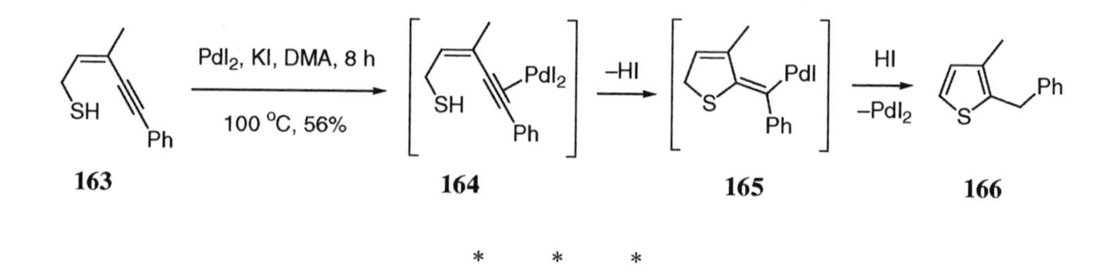


* * *

To summarize, electrophilic substitutions and metalations of thiophenes take place preferably at the α-positions due to the electronegativity of the sulfur atom. This is the consequence of the more effective incorporation of lone pair electrons on the sulfur into the aromatic system. Although regioselective reactions are routinely performed for α,β-dihalofurans, regioselectivity is not as easily achieved in Pd-mediated chemistry with α,β-dihalothiophenes.

Electrophilic palladations of thiophenes and benzothiophenes are not very synthetically useful due to low yields and the consumption of stoichiometric amounts of $Pd(OAc)_2$. For the direct synthesis of aryl terminal alkynes *via* Pd-catalyzed cross-coupling of halothiophenes with ethynylmetals, the Negishi reaction is more advantageous than the Kumada conditions. Formation of dithienylethyne is a serious side reaction due to the high propensity for H–Mg exchange in the latter approach. In addition, although both 2- and 3-thiopheneboronic acid are frequently used in the Suzuki coupling, the Stille coupling employing a halo- or a stannylthiophene is undoubtedly better. The Stille coupling partners span the whole spectrum of organic halides and stannanes. As a result, the thiophene moiety, an important bioisostere of a phenyl ring in medicinal chemistry, can be incorporated into a wide variety of molecules employing the Stille coupling approach.

## 5.8 References

1. Yoshida, Z.-i.; Yamada, Y.; Tamaru, Y. *Chem. Lett.* **1977**, 423–4.
2. Smith, G. V.; Notheisz, F.; Zsigmond, A. G.; Bartok, M. *Stud. Surf. Sci. Catal.* **1993**, *75* (*New Frontiers in Catalysis, Pt. C*), 2463–6.
3. Fazio, A.; Gabriele, B.; Salerno, G.; Destri, S. *Tetrahedron* **1999**, *55*, 485–502.
4. Zhang, Y.; Hörnfeldt, A-B.; Gronowitz, S. *J. Heterocycl. Chem.* **1995**, *32*, 435–44.
5. Lucas, L.; van Esch, J.; Kellogg, R. M.; Feringa, B. L. *Tetrahedron Lett.* **1999**, *40*, 1775–8.
6. Okamoto, T.; Kakinami, T.; Fujimoto, H.; Kajigaeshi, S. *Bull. Chem. Soc. Jpn.* **1991**, *64*, 2566–8.
7. Mitchell, R. H.; Chen, Y.; Zhang, J. *Org. Prep. Proc. Int.* **1997**, *29*, 715–9.
8. Rossi, R.; Carpita, A.; Lezzi, A. *Tetrahedron* **1984**, *40*, 2773–9.
9. Hucke, A.; Cava, M. P. *Tetrahedron* **1998**, *63*, 7413–7.
10. Otsubo, T.; Kono, Y.; Hozo, N.; Miyamoto, H.; Aso, Y.; Ogura, F.; Tanaka, T.; Sawada, M. *Bull. Chem. Soc. Jpn.* **1993**, *66*, 2033–41.
11. Clark, P. D.; Clarke, K.; Scrowston, R. M.; Sutton, T. M. *J. Chem. Res. Synop.* **1978**, 10.
12. Cross, P. E.; Dickinson, R. P.; Parry, M. J.; Randall, M. J. *J. Med. Chem.* **1986**, *29*, 1637–43.
13. Prats, M.; Gálvez, C. *Heterocycles* **1992**, *34*, 149–56.
14. Abbotto, A.; Bradmante, S.; Facchetti, A.; Pagani, G. *J. Org. Chem.* **1997**, *62*, 5755–65.
15. Nakajima, R.; Iida, H.; Hara, T. *Bull. Chem. Soc. Jpn.* **1990**, *63*, 636–7.
16. Takahashi, K.; Tarutani, S. *Heterocycles* **1996**, *43*, 1927–35.
17. Kondo, Y.; Shilai, M.; Uchiyama, M.; Sakamoto, T. *J. Am. Chem. Soc.* **1999**, *121*, 3539–40.
18. Eberson, L.; Gomez-Gozalez, L. *Acta Chem. Scand.* **1973**, *27*, 1249–54.
19. Kozhevnikov, I. V. *React. Kinet. Catal. Lett.* **1976**, *4*, 451–8.
20. Itahara, T.; Hashimoto, M.; Yumisashi, H. *Synthesis* **1984**, 255–6.
21. Itahara, T. *J. Org. Chem.* **1985**, *50*, 5272–5.
22. Fujiwara, Y.; Maruyama, O.; Yoshidomi, M.; Taniguchi, H. *J. Org. Chem.* **1981**, *46*, 851–5.
23. Hassan, J.; Lavenot, L.; Gozzi, C.; Lemaire, M. *Tetrahedron Lett.* **1999**, *40*, 857–8.
24. Venkatraman, S.; Li, C.-J. *Org. Lett.* **1999**, *1*, 1133–5.

25. Amatore, C.; Jutand, A.; Negri, S.; Fauvarque, J. F. *J. Organomet. Chem.* **1990**, *390*, 389–98.
26. Minato, A.; Tamao, K.; Hayashi, T.; Suzuki, K.; Kumada, M. *Tetrahedron Lett.* **1980**, *21*, 845–8.
27. Abbotto, A.; Bradmante, S.; Facchetti, A.; Pagani, G. *J. Org. Chem.* **1997**, *62*, 5755–65.
28. Minato, A.; Suzuki, K.; Tamao, K.; Kumada, M. *J. Chem. Soc., Chem. Commun.* **1984**, 511–3.
29. Gilchrist, T. L.; Healy, M. A. M. *Tetrahedron Lett.* **1990**, *31*, 5807–10.
30. Gilchrist, T. L.; Healy, M. A. M. *Tetrahedron* **1993**, *49*, 2543–56.
31. Minato, A.; Suzuki, K.; Tamao, K.; Kumada, M. *J. Chem. Soc., Chem. Commun.* **1984**, 511–3.
32. Felding, J.; Kristensen, J.; Bjerregaard, T.; Sander, L.; Vedsø, Bergtrup, M. *J. Org. Chem.* **1999**, *64*, 4196–8.
33. Ennis, D. S.; Gilchrist, T. L. *Tetrahedron Lett.* **1990**, *46*, 2623–32.
34. Gronowitz, S. in *Organic Sulphur—Structure, Mechanism, and Synthesis*; Sterling, C. J. M., Ed.; Butterworths: London, **1975**, 203–28.
35. Moses, P.; Gronowitz, S. *Arkiv. Kemi.* **1961**, *18*, 119.
36. Takahashi, K.; Sakai, T. *Chem. Lett.* **1993**, 157–60.
37. Betzemeier, B,; Knochel, P. *Angew. Chem., Int. Ed. Eng.* **1997**, *36*, 2623–4.
38. Ribereau, P.; Pasteur, P. *Bull. Soc. Chim. Fr*, **1969**, 2076–9.
39. Brandão, M. A.; de Oliveira, A. B.; Snieckus, V. *Tetrahedron Lett.* **1993**, *34*, 2437–40.
40. Wu, X.; Rieke, R. D. *J. Org. Chem.* **1995**, *60*, 6658–9.
41. Rieke, R. D.; Kim, S.-H.; Wu, X. *J. Org. Chem.* **1997**, *62*, 6921–7.
42. Prim, D.; Kirsch, G. *J. Chem. Soc., Perkin Trans. 1* **1994**, 2603–6.
43. (a) Murata, M.; Oyama, T.; Watanabe, S.; Masuda, Y. *J. Org. Chem.* **2000**, *65*, 164–8; (b) Murata, M.; Watanabe, S.; Masuda, Y. *J. Org. Chem.* **1997**, *62*, 6458–9.
44. Wellmar, U.; Hörnfeldt, A.-B.; Gronowitz, S. *J. Heterocycl. Chem.* **1995**, *32*, 1159–63.
45. Peters, D.; Hörnfeldt, A.-B.; Gronowitz, S. *J. Heterocycl. Chem.* **1990**, *27*, 2165–73.
46. Malm, J.; Rehn, B.; Hörnfeldt, A.-B.; Gronowitz, S. *J. Heterocycl. Chem.* **1994**, *31*, 11–5.
47. Charette, A. B.; Giroux, A. *J. Org. Chem.* **1996**, *61*, 8718–9.
48. Hark, R. R.; Hauze, D. B.; Petrovskaia, O.; Joullié, M. M.; Jaouhari, R.; McComiskey, P. *Tetrahedron Lett.* **1994**, *35*, 7719–22.
49. Kevin, N. J.; Rivero, R. A.; Greenlee, W. J.; Chang, R. S. L.; Chen, T. B. *Biorg. Med. Chem. Lett.* **1994**, *4*, 189–94.
50. Jones, K.; Keenan, M.; Hibbert, F. *Synlett* **1996**, 509–10.
51. Malapel-Andrieu, B.; Mérour, J.-Y. *Tetrahedron* **1998**, *54*, 11079–94.
52. Haddach, M.; McCathy, J. R. *Tetrahedron Lett.* **1999**, *40*, 3109–12.
53. Ye, X.-S.; Wong, H. N. C. *J. Org. Chem.* **1997**, *62*, 1940–54.
54. Andersen, N. G.; Maddaford, S. P.; Keay, B. A. *J. Org. Chem.* **1996**, *61*, 9556–9.
55. Giroux, A.; Han, Y.; Prasit, P. *Tetrahedron Lett.* **1997**, *38*, 3841–4.

56. Blettner, C.; König, W. A.; Rühter, G.; Stenzel, W.; Schotten, T. *Synlett* **1999**, 307–10.
57. Davies, A. G. *Organotin Chemistry* VCH, Weinheim, FRG, **1997**, 329pp.
58. Crisp, G. T. *Synth. Comm.* **1989**, *19*, 307–16.
59. Zhang, Y.; Hörnfeldt, A-B.; Gronowitz, S. *Synthesis*, **1989**, *2*, 130–1.
60. Prim, D.; Kirsch, G. *J. Chem. Soc., Perkin Trans. 1* **1994**, 2603–6.
61. Zhang, Y.; Hörnfeldt, A-B.; Gronowitz, S. *J. Heterocycl. Chem.* **1995**, *32*, 771–7.
62. Gronowitz, S.; Hörnfeldt, A-B.; Yang, Y. *Chem. Scrip.* **1988**, *28*, 275–9.
63. Malm, J.; Hörnfeldt, A-B.; Gronowitz, S. *Heterocycles* **1993**, *35*, 245–62.
64. Tamao, K.; Yamaguchi, S.; Shiozaki, M.; Nakagawa, Y.; Ito, Y. *J. Am. Chem. Soc.* **1992**, *114*, 5867–9.
65. Hucke, A.; Cava, M. P. *J. Org. Chem.* **1998**, *63*, 7413–7.
66. Rossi, R.; Carpita, A.; Ciofalo, M.; Lippolis, V. *Tetrahedron* **1991**, *47*, 8443–60.
67. Björk, P.; Aekermann, T.; Hörnfeldt, A-B.; Gronowitz, S. *J. Heterocycl. Chem.* **1995**, *32*, 751–4.
68. Otsubo, T.; Kono, Y.; Hozo, N.; Miyamoto, H.; Aso, Y.; Ogura, F.; Tanaka, T.; Sawada, M. *Bull. Chem. Soc. Jpn.* **1993**, *66*, 2033–41.
69. Gronowitz, S.; Yang, Y.; Hörnfeldt, A-B. *Acta Chem. Scand.* **1992**, *46*, 654–60.
70. Yoshida, S.; Kubo, H.; Saika, T.; Katsumura, S. *Chem. Lett.* **1996**, 139–40.
71. Kosugi, M.; Shimizu, K.; Ohtani, A.; Migita, T. *Chem. Lett.* **1981**, 829–30.
72. Kosugi, M.; Ohta, T.; Migita, T. *Bull. Chem. Soc. Jpn.* **1983**, *56*, 3855–6.
73. Rivero R. A.; Kevin, N. J.; Allen, E. E. *Biorg. Med. Chem. Lett.* **1993**, *3*, 1119–24.
74. Sanfilippo, P. J.; McNally, J. J.; Press, J. B.; Fitzpatrick, L. J.; Urbanski, M. J.; Katz, L. B.; Giardino, E.; Falotico, R.; Salata, J.; Moore, Jr., J. B.; Miller, W. *J. Med. Chem.* **1992**, *35*, 4425–33.
75. Wigerinck, P.; Kerremans, L.; Claes, P.; Snoeck, R.; Maudgal, P.; De Clercq, E.; Herdewijn, P. *J. Med. Chem.* **1993**, *36*, 538–43.
76. Park, H.; Lee, J. Y.; Lee, Y. S.; Park, J. O.; Koh, S. B.; Ham, W-H. *J. Antibiotics* **1994**, *47*, 606–8.
77. Sakamoto, T.; Kondo, Y.; Yasuhara, A.; Yamanaka, H. *Heterocycles* **1990**, *31*, 219–21.
78. Sakamoto, T.; Kondo, Y.; Yasuhara, A.; Yamanaka, H. *Tetrahedron* **1991**, *47*, 1877–86.
79. Crisp, G.; Glink, P. T. *Tetrahedron* **1994**, *50*, 3213–34.
80. Johannes, H.-H.; Grahn, W.; Reisner, A.; Jones, P. G. *Tetrahedron Lett.* **1995**, *36*, 7225–8.
81. Corriu, R. J. P.; Bolin, G.; Moreau, J. J. E. *Tetrahedron Lett.* **1991**, *32*, 4121–4.
82. Corriu, R. J. P.; Geng, B.; Moreau, J. J. E. *J. Org. Chem.* **1993**, *58*, 1443–8.
83. Hollingworth, G. J.; Sweeney, J. B. *Tetrahedron Lett.* **1992**, *33*, 7049–52.
84. Bellina, F.; Carpita, A.; De Santis, M.; Rossi, R. *Tetrahedron* **1994**, *50*, 12029–46.
85. Liebeskind, L. S.; Fengl, R. W. *J. Org. Chem.* **1990**, *55*, 5359–64.

86. Nodvall, G.; Sundquist, S.; Nilvebrant, L.; Hacksell, U. *Biorg. Med. Chem. Lett.* **1994**, *4*, 2837–40.
87. Buon, C.; Bouyssou, P.; Coudert, G. *Tetrahedron Lett.* **1999**, *40*, 701–2.
88. May, P. D.; Larsen, S. D. *Synlett* **1997**, 895–6.
89. Aidhen, I. S.; Braslau, R. *Synth. Comm.* **1994**, *24*, 789–97.
90. Sakamoto, T.; Yasuhara, A.; Kondo, Y.; Yamanaka, H. *Chem. Pharm. Bull.* **1994**, *42*, 2032–5.
91. Ye, X.-S.; Wong, H. N. C. *Chem. Commun.* **1996**, 339–40.
92. Stille, J. K.; Simpson, J. H. *J. Am. Chem. Soc.* **1987**, *109*, 2138–52.
93. Bailey, T. R. *Tetrahedron Lett.* **1986**, *27*, 4407–10.
94. Roth, G. P.; Farina, V. *Tetrahedron Lett.* **1995**, *36*, 2191–4.
95. Kennedy, G.; Perboni, A. D. *Tetrahedron Lett.* **1996**, *37*, 7611–4.
96. Kang, S.-K.; Lee, H.-W.; Jang, S.-B.; Kim, T.-H.; Kim, J.-S. *Synth. Commun.* **1996**, *26*, 4311–8.
97. Kang, S.-K.; Ryu, H.-C.; Choi. S.-C. *Chem. Commun.* **1998**, *12*, 1317–8.
98. Kang, S.-K.; Lim, K.-H.; Ho, P.-S.; Yoon, S.-K.; Son, H.-J. *Synth. Commun.* **1998**, *28*, 1481–9.
99. Pridgen, L. N.; Killmer, L. B. *J. Org. Chem.* **1981**, *46*, 5402–4.
100. Dondoni, A.; Fogagnolo, M.; Fantin, G.; Medici, A.; Pedrini, P. *Tetrahedron Lett.* **1986**, *27*, 5269–70.
101. Meyers, A. I.; Novachek, K. A. *Tetrahedron Lett.* **1996**, *37*, 1747–8.
102. Dondoni, A.; Fantin, G.; Fogagnolo, M.; Medici, A.; Pedrini, P. *Synthesis*, **1987**, 693–6.
103. Kitamura, C.; Tanaka, S.; Yamashita, Y. *J. Chem. Soc., Chem. Commun.* **1994**, 1585–6.
104. Folli, U.; Iarossi, D.; Montorsi, M.; Mucci, A.; Schenetti, L. *J. Chem. Soc., Perkin Trans. 1* **1995**, 537–40.
105. Barbarella, G.; Zambianchi, M.; Sotgiu, G.; Bongini, A. *Tetrahedron* **1997**, *53*, 9401–6.
106. Iyoda, M.; Kuwatani, Y.; Ueno, N.; Oda, M. *J. Chem. Soc., Perkin Trans. 1* **1992**, 158–9.
107. Takahashi, K.; Nihira, T. *Bull. Chem. Soc. Jpn.* **1992**, *65*, 1855–9.
108. Takahashi, K.; Nihira, T.; Akiyama, K.; Ikegami, Y.; Fukuyo, E. *J. Chem. Soc., Chem. Commun.* **1992**, 620–2.
109. Gronowitz, S.; Björk, P.; Malm, J.; Hörnfeldt, A-B. *J. Organomet. Chem.* **1993**, *460*, 127–9.
110. Zhang, Y.; Hörnfeldt, A-B.; Gronowitz, S. *J. Heterocycl. Chem.* **1995**, *32*, 435–44.
111. Iyoda, M.; Miura, M.; Sasaki, S.; Kabir, S. M. H.; Kuwatani, Y.; Yoshida, M. *Tetrahedron Lett.* **1997**, *38*, 4581–2.
112. Zhang, Y.; Hörnfeldt, A-B.; Gronowitz, S. *Synthesis*, **1989**, *2*, 130–1.
113. Malm, J.; Hörnfeldt, A-B.; Gronowitz, S. *Tetrahedron Lett.* **1992**, *33*, 2199–202.
114. Allred, G. D.; Liebeskind, L. S. *J. Am. Chem. Soc.* **1996**, *118*, 2748–9.
115. Paterson, I.; Lombart, H.-G.; Allerton, C. *Org. Lett.* **1999**, *1*, 19–22.

116. Hatanaka, Y.; Fukushima, S.; Hiyama, T. *Heterocycles* **1990**, *30*, 303–6.
117. Nguefack, J.-F.; Bolitt, V.; Sinou, D. *Tetrahedron Lett.* **1996**, *37*, 5527–30.
118. John, J. A.; Tour, J. M. *Tetrahedron* **1997**, *53*, 15515–34.
119. Walser, A.; Flynn, T.; Mason, C.; Crowley, H.; Maresca, C.; O'Donnell, M. *J. Med. Chem.* **1991**, *34*, 1440–6.
120. Ye, X.-S.; Wong, H. N. C. *Chem. Commun.* **1996**, 339–40.
121. Rossi, R.; Carpita, A.; Lezzi, A. *Tetrahedron* **1984**, *40*, 2773–9.
122. D'Auria, M. *Synth. Commun.* **1992**, *22*, 2393–9.
123. Carpita, A.; Rossi, R. *Gazz. Chim. Ital.* **1985**, *115*, 575–83.
124. Tamaru, Y.; Yamada, Y.; Yoshida, Z.-I. *Tetrahedron* **1979**, *35*, 329–40.
125. Arnau, N.; Moreno-Manãs, M.; Pleixats, R. *Tetrahedron* **1993**, *49*, 11019–28.
126. Pigeon, P.; Decroix, B. *Tetrahedron Lett.* **1996**, *37*, 7707–10.
127. Ohta, A.; Akita, Y.; Ohkuwa, T.; Chiba, M.; Fukunaka, R.; Miyafuji, A.; Nakata, T.; Tani, N. Aoyagi, Y. *Heterocycles* **1990**, *31*, 1951–7.
128. Aoyagi, Y.; Inoue, A.; Koizumi, I.; Hashimoto, R.; Tokunaga, K.; Gohma, K.; Komatsu, J.; Sekine, K.; Miyafuji, A.; Konoh, J. Honma, R. Akita, Y.; Ohta, A. *Heterocycles* **1992**, *33*, 257–72.
129. Lavenot, L.; Gozzi, C.; Ilg, K.; Orlova, I.; Penalva, V.; Lemaire, M. *J. Organomet. Chem.*, **1998**, *567, 49*–55.
130. Carpentier, J.-F.; Castanet, Y.; Brocard, J.; Mortreux, A.; Petit, F. *Tetrahedron Lett.* **1991**, *32*, 4705–8.
131. Lin, Y.-S.; Yamamoto, A. *Tetrahedron Lett.* **1997**, *38*, 3747–50.
132. Hatanaka, Y.; Fukushima, S.; Hiyama, T. *Tetrahedron* **1992**, *48*, 2113–26.
133. Kang, S.-K.; Lim, K.-H.; Ho, P.-S.; Yoon, S.-K.; Son, H.-J. *Synth. Commun.* **1998**, *28*, 1481–9.
134. Hirao, T.; Masunaga, T.; Ohshiro, Y.; Agawa, T. *Tetrahedron Lett.* **1980**, *21,* 3595–8.
135. Hirao, T.; Masunaga, T.; Ohshiro, Y.; Agawa, T. *Synthesis* **1981**, 56–7.
136. Hirao, T.; Masunaga, T.; Yamada, Y.; Ohshiro, Y.; Agawa, T. *Bull. Chem. Soc. Jpn.* **1982**, *55,* 909–13.
137. Xu, Y.; Li, Z.; Xia, J.; Guo, H.; Huang, Y. *Synthesis* **1983**, 377–8.
138. Xu, Y.; Zhang, J. *Synthesis* **1984**, 778–80.
139. Watanabe, M.; Yamamoto, T.; Nishiyama, M. *Chem. Commun.* **2000**, 133–4.
140. Gabriele, B.; Salerno, G.; Fazio, A. *Org. Lett.* **2000**, *2*, 351–2.

# CHAPTER 6

## Furans and Benzo[*b*]furans

Furan-containing molecules are found in both natural products and pharmaceuticals. At one time, furfural was produced in great quantities from corncobs. Perillene, a secondary plant metabolite, is an example of naturally occurring furans. Furan is frequently used as a bioisostere of a benzene ring in medicinal chemistry. For example, Ranitidine (Zantac) marketed by Glaxo was one of the first "blockbuster drugs" with annual sales over 1 billion dollars.

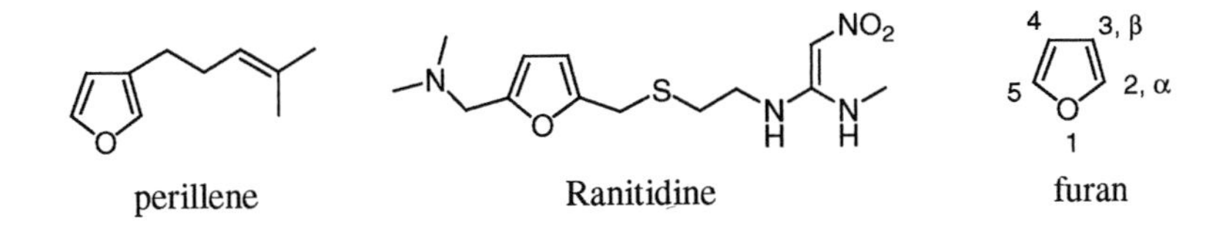


Furan is a π-electron-excessive heteroarene. Electrophilic substitution and metalation take place regioselectively at C(2) due to the electronegativity of the oxygen atom. In comparison to the sulfur atom in thiophene, the oxygen in furan has higher electronegativity. Therefore, the lone pair electrons on the oxygen are less effectively incorporated into the aromatic system, which contributes to the pronounced difference of reactivity of the α and β-positions. In the context of palladium chemistry, furan and indole are the two classes of heterocycles that have attracted most attention. The synthesis of benzofurans and indoles using palladium chemistry has been a very active and prolific field. Contrary to all-carbon aryl halides, regioselective coupling reactions are routinely accomplished at C(2) for a 2,3-dihalofuran derivative. However, unlike 2-chloropyridines (an electron-deficient heterocycle), 2-chlorofurans are not sufficiently activated to add oxidatively to Pd(0).

## 6.1 Synthesis of halofurans and halobenzo[*b*]furans

### 6.1.1 Halofurans

Commercially available 3-bromofuran is prepared from furfural [1, 2]. 3-Iodofuran can be synthesized by sequential deiodination of tetraiodofuran [3].

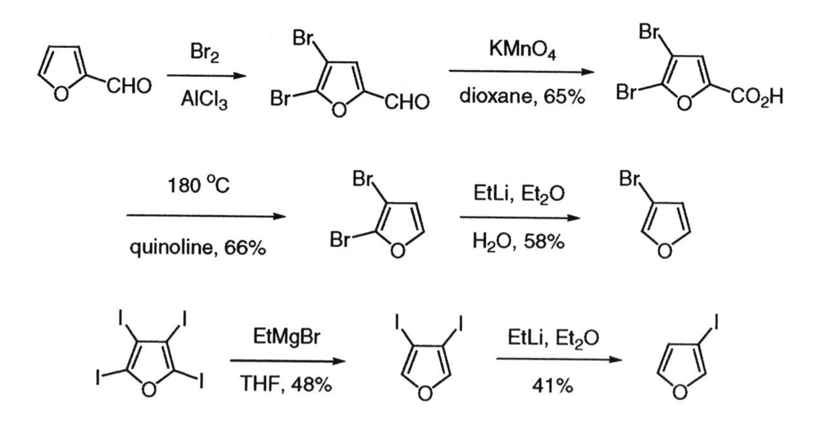

Simply treating furan in DMF with $Br_2$ at 20–30 °C resulted in 70% of 2-bromofuran and 8% of 2,5-dibromofuran. Further treatment of 2-bromofuran with another equivalent of $Br_2$ at 30–40 °C produced 2,5-dibromofuran in 48% yield [4, 5]. Alternatively, deprotonation of furan at the C(2) position using EtLi followed by quenching with $Br_2$ provided 2-bromofuran [6]. Similarly, quenching 3-lithiofuran, derived from halogen-metal exchange between 3-bromofuran and EtLi, with hexachloroethane gave 3-chlorofuran [7].

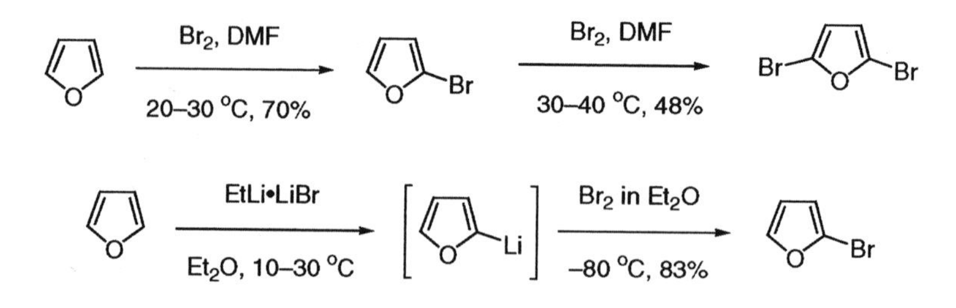

### 6.1.2 Halobenzofurans

2-Iodobenzofuran was readily prepared by lithiation of benzofuran followed by iodine quench [8]. Bromination of benzofuran with bromine gave 2,3-dibromobenzofuran, which, when subjected to a regioselective halogen/metal exchange at the C(2) position with *n*-BuLi followed by quenching with methanol furnished 3-bromobenzofuran [9].

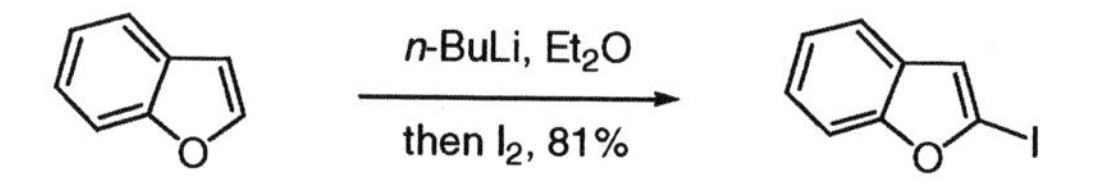

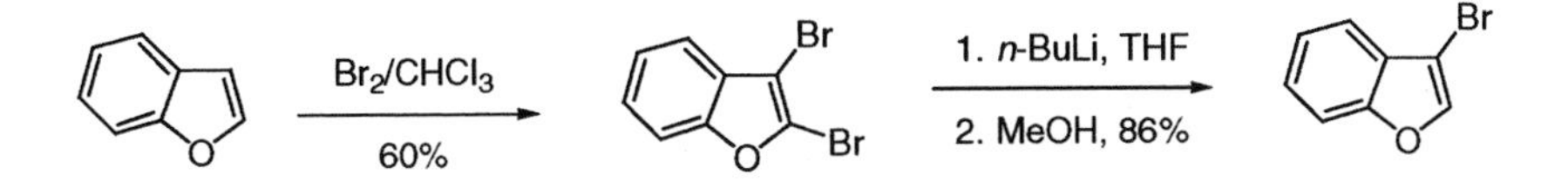


The literature is replete with synthetic methods to prepare 5-bromofurans. One of the more practical syntheses [10, 11] commenced with etherification of 4-bromophenol with bromoacetaldehyde diethyl acetal using either NaH in DMF or KOH in DMSO. Treatment of the resulting aryloxyacetaldehyde acetal with polyphosphoric acid (PPA) afforded 5-bromofuran in good yield *via* intramolecular cyclocondensation. However, cyclization of *m*-aryloxyacetaldehyde acetal **1** resulted in a mixture of two regioisomers, 6-bromofuran (**2**) and 4-bromofuran (**3**). Finally, 7-bromofuran **5** can be prepared similarly using the intramolecular cyclocondensation of aryloxyacetaldehyde acetal **4** generated from etherification of 2-bromophenol with bromoacetaldehyde diethyl acetal.

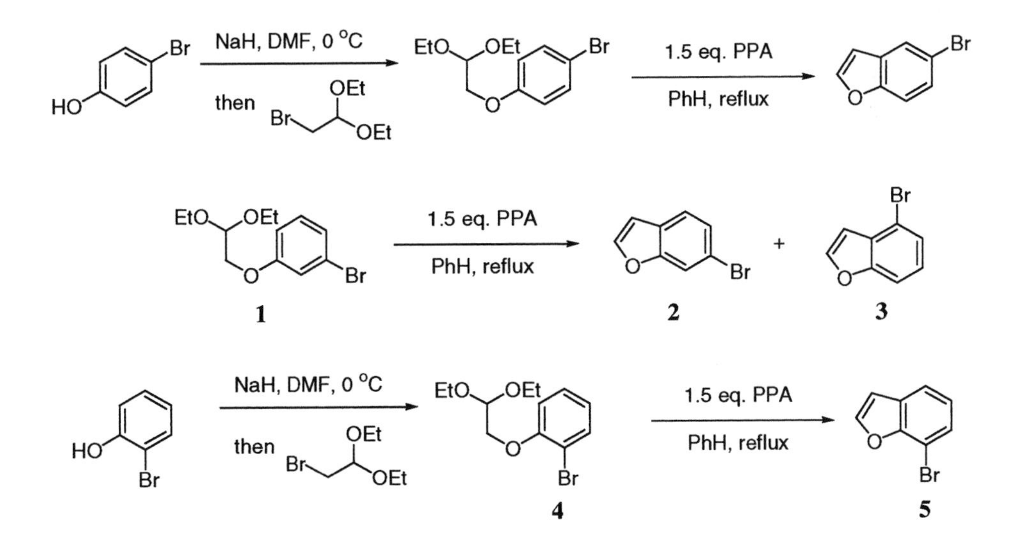


## 6.2 Oxidative coupling/cyclization

When furan or substituted furans were subjected to the classic oxidative coupling conditions [Pd(OAc)$_2$ in refluxing HOAc], 2,2'-bifuran was the major product, whereas 2,3'-bifuran was a minor product [12, 13]. Similar results were observed for the arylation of furans using Pd(OAc)$_2$ [14]. The oxidative couplings of furan or benzo[*b*]furan with olefins also suffered from inefficiency [15]. These reactions consume at least one equivalent of palladium acetate, and therefore have limited synthetic utility.

In contrast, Tsuji's group coupled 2-methylfuran with ethyl acrylate to afford adduct **6** *via* a Pd-*catalyzed* reaction using *tert*-butyl peroxybenzoate to reoxidize Pd(0) to Pd(II) [16]. The palladation of 2-methylfuran took place at the electron-rich C(5) in a fashion akin to electrophilic aromatic substitution. The perbenzoate acted as a hydrogen acceptor.

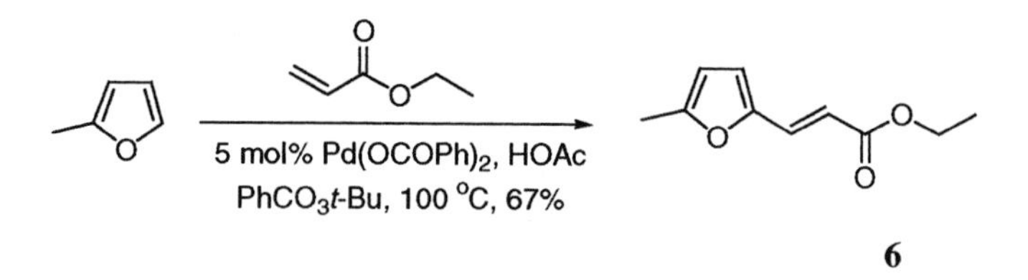


The phenolic oxygen on 2-allyl-4-bromophenol (**7**) readily underwent oxypalladation using a catalytic amount of $PdCl_2$ and three equivalents of $Cu(OAc)_2$, to give the corresponding benzofuran **8**. This process, akin to the Wacker oxidation, was catalytic in terms of palladium, and $Cu(OAc)_2$ served as oxidant [17]. Benzofuran **10**, a key intermediate in Kishi's total synthesis of aklavinone [18], was synthesized *via* the oxidative cyclization of phenol **9** using stoichiometric amounts of a Pd(II) salt.

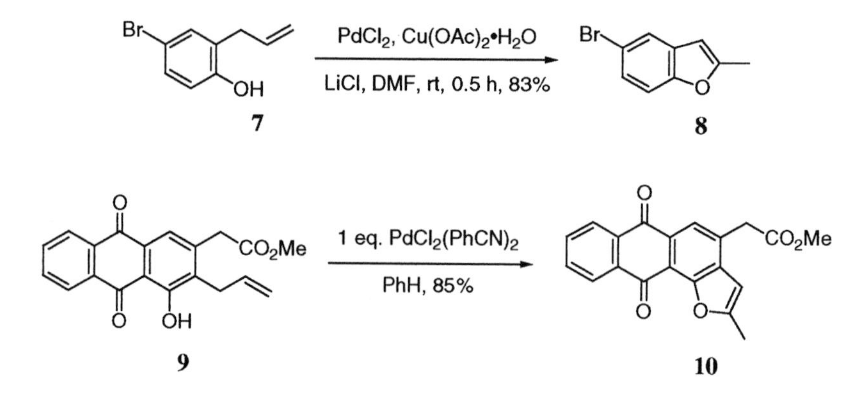


## 6.3 Coupling reactions with organometallic reagents

### 6.3.1 Kumada coupling

Rossi and associates described a Kumada coupling of 2-bromofuran with 2-thienylmagnesium bromide to assemble thienylfuran **11** [19]. The reaction proceeded readily at room temperature.

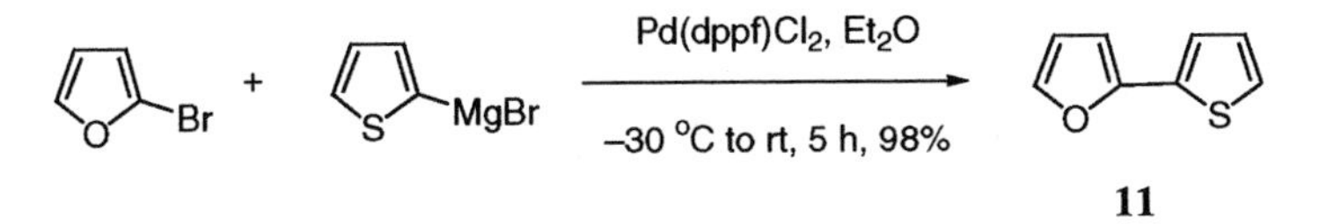


### 6.3.2 Negishi coupling

Treating 2-furyllithium (derived from deprotonation of furan with *n*-BuLi) with $ZnCl_2$ gave 2-furylzinc chloride, which then was coupled with 1,3-dibromobenzene to furnish bis-furylbenzene **12** [20, 21]. In addition, 2-furylzinc chloride was coupled with 4-iodobenzoic acid to give adduct **13**. The Negishi reaction conditions were compatible with the carboxylic acid [22].

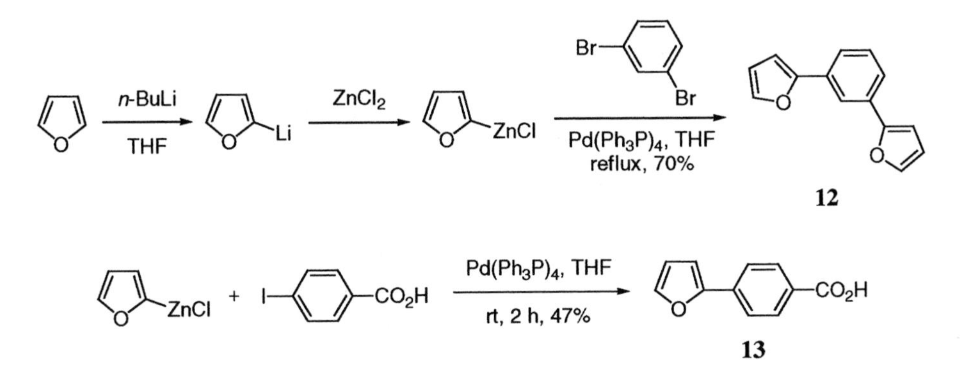


Snieckus *et al.* prepared the furan amide organozinc reagent **16** *via ortho*-lithiation of furyl amide **14** followed by reaction of the resulting 2-lithiofuran species **15** with $ZnBr_2$ [23]. The subsequent Negishi reaction of **16** and 2-bromotoluene generated phenylfuran **17**. Gilchrist's group carried out an *ortho*-lithiation using an oxazoline directing group [24]. Thus, directed *ortho*-lithiation of 2-furyloxazoline **18** with *sec*-BuLi followed by treatment with $ZnBr_2$ provided 3-furylzinc bromide **19**, which was subsequently joined with aryl-, benzyl- and vinyl halides to give the expected adducts such as **20**.

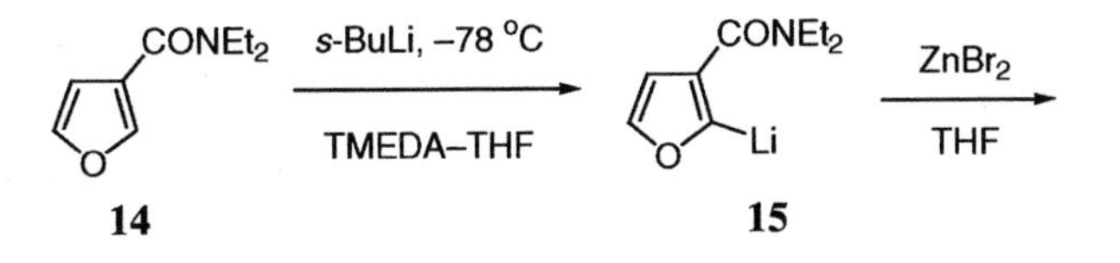

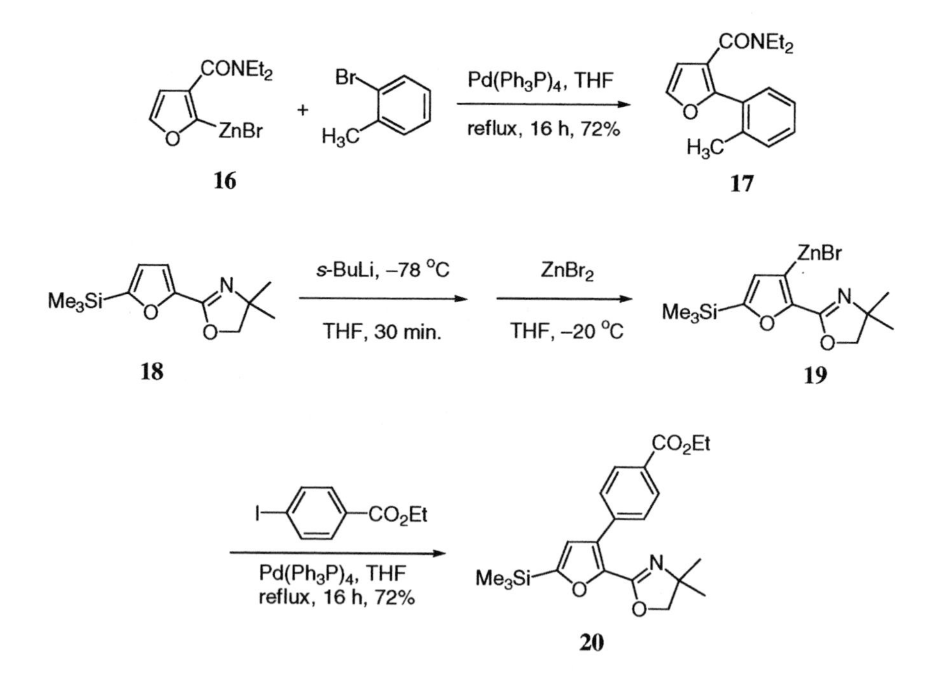


These couplings exemplify the functional group tolerance of the Negishi reaction conditions. Carboxylic acids, esters, amides, and even free anilines are compatible with the reaction conditions, as illustrated by the synthesis of furylaniline **21** [25].

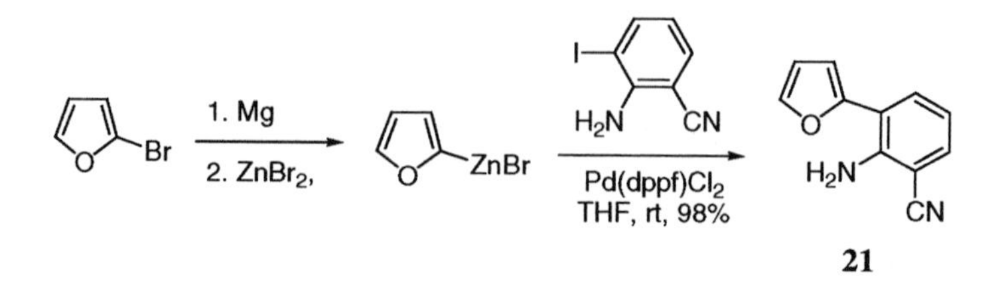


Both vinyl- and aryl triflates have been cross-coupled with 2-furylzinc chloride [26–28]. Since vinyl triflates are easily obtained from the corresponding ketones, they are useful substrates in Pd-catalyzed reactions. In the following example, a Negishi coupling of 2-furylzinc chloride and indol-5-yl triflate (**22**) provided an expeditious entry to 2-(5'-indolyl)furan (**23**). Protection of the NH in the indole ring was not required. A similar reaction was successful with pyridyl- and quinolinyl triflates.

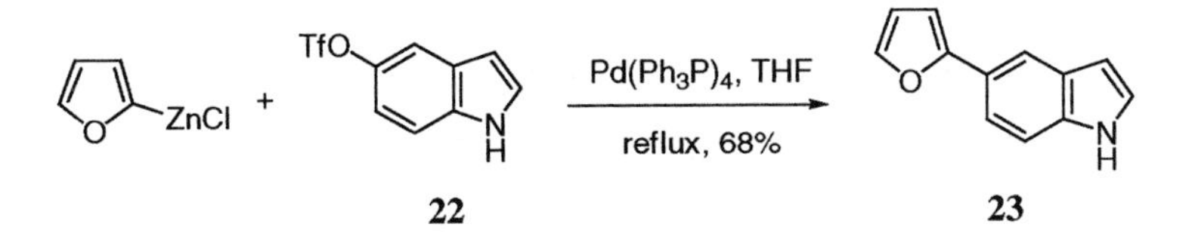


Like ordinary aryl halides, furyl- and benzofuryl halides participate in Negishi couplings as nucleophiles. The electrophilic coupling partners ranged from aryl- [29], alkynyl- [30, 31], vinyl- [32, 33], to alkylzinc reagents [34]. In particular, with the aid of ultrasonic activation, organozinc reagent **25** was generated from the protected β-iodoalanine derivative **24** [34]. Organozinc reagent **25** was then effectively coupled *in situ* with 2-furoyl chloride to give enantiomerically pure protected 4-oxo-α-amino acid **26** with palladium catalysis.

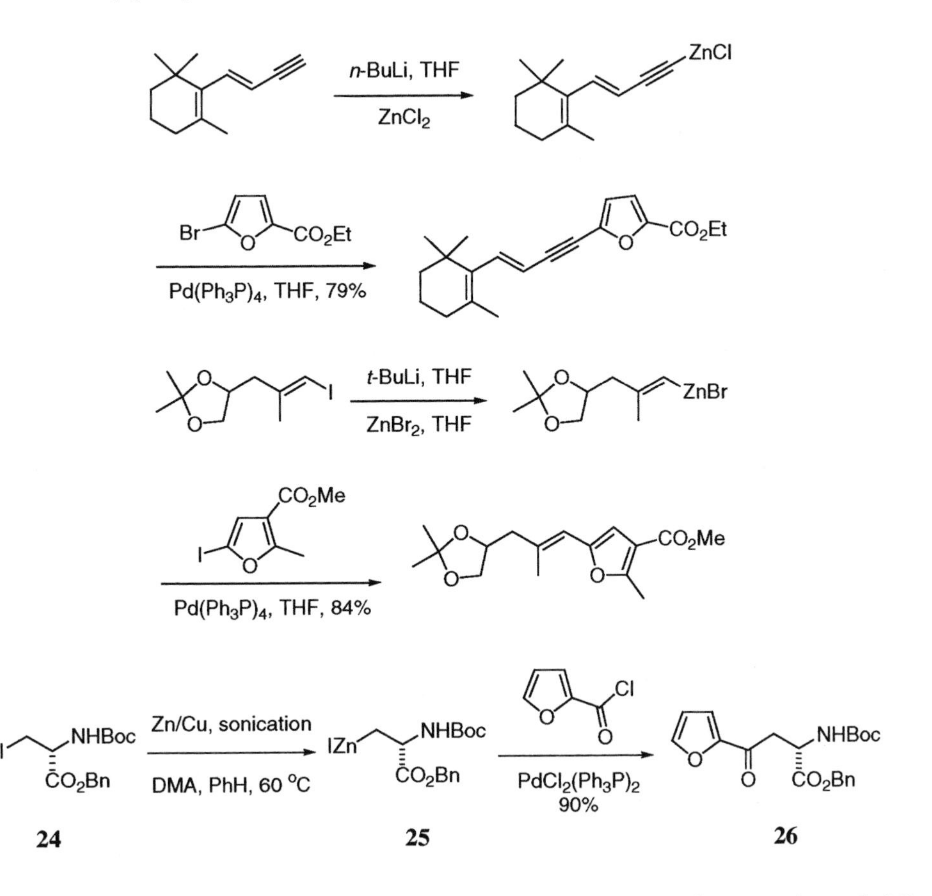


Unlike normal all-carbon aryl halides, regioselective Negishi coupling reaction of 2,3-dibromofuran **27** with 2-furylzinc chloride was achieved at C(2) to afford bisfuran **28** [35].

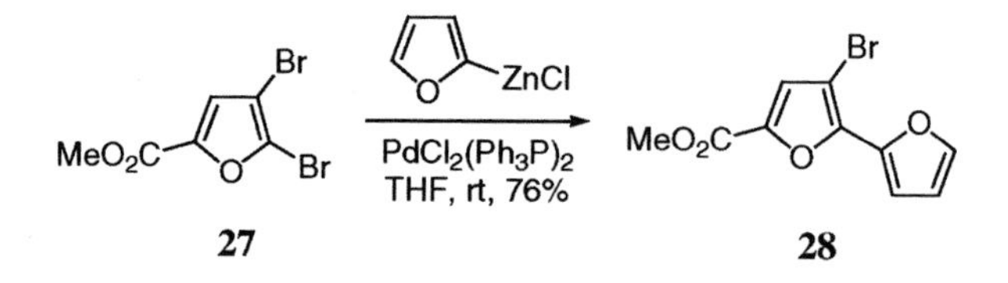


### 6.3.3 Suzuki coupling

#### *6.3.3.a Furylboronic acids*

Furylboronic acid, behaving like a simple arylboronic acid, has been coupled with a myriad of oragnohalides and triflates. The coupling between 2-furylboronic acid and bromopyridine **29** produced furylpyridine **30** [36].

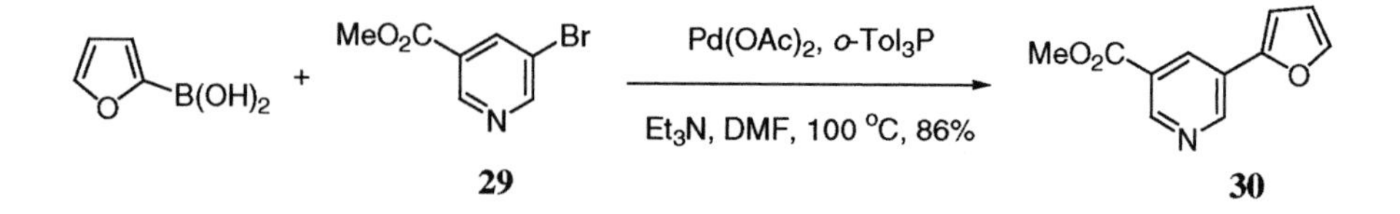


Adapting Gribble's method for synthesizing indol-3-yl triflate [37], Mérour *et al.* converted 2-formyl-1-(phenylsulfonyl)-1*H*-indole (**31**) to indol-2-yl triflate **32** in two steps. **32** was subsequently coupled with benzofuryl-2-boronic acid to furnish 2-benzofurylindole **33** [38, 39]. In another case, 2-bromoacetaniline was coupled with 2-formyl-3-furylboronic acid **35** [40]. The resulting Suzuki coupling adduct underwent a spontaneous cyclization, forming tricyclic furo[2,3-*c*]quinoline **36**.

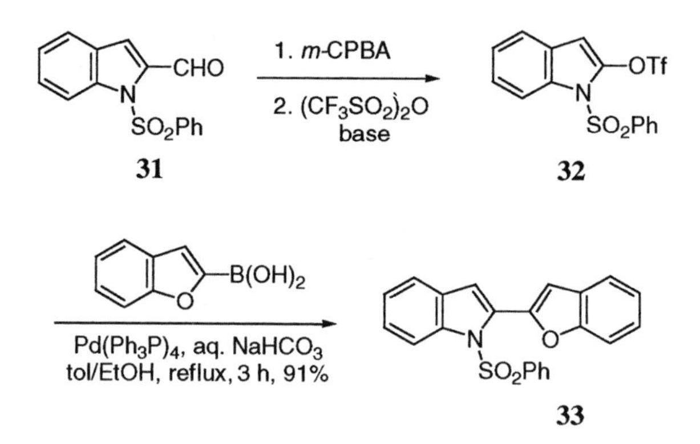

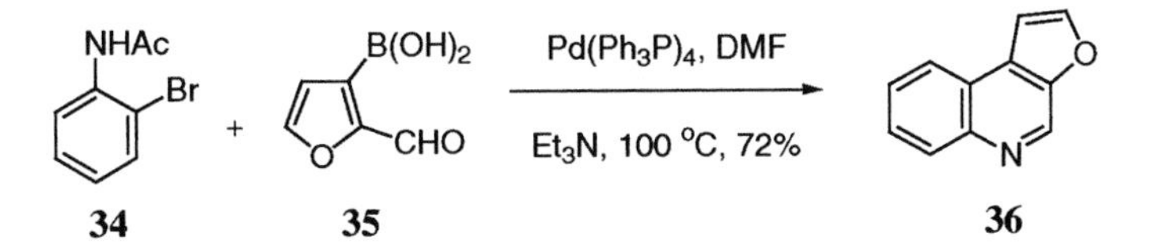


### *6.3.3.b Furans as electrophiles*

Like simple aryl halides, furyl halides take part in Suzuki couplings as electrophiles [41, 42]. Young and Martin coupled 2-bromofuran with 5-indolylboronic acid to prepare 5-substituted indole **37** [43]. Terashima's group cross-coupled 3-bromofuran with diethyl-(4-isoquinolyl)borane **38** to make 4-substituted isoquinoline **39** [44]. Similarly, 2- and 3-substituted isoquinolines were also synthesized in the same fashion [45].

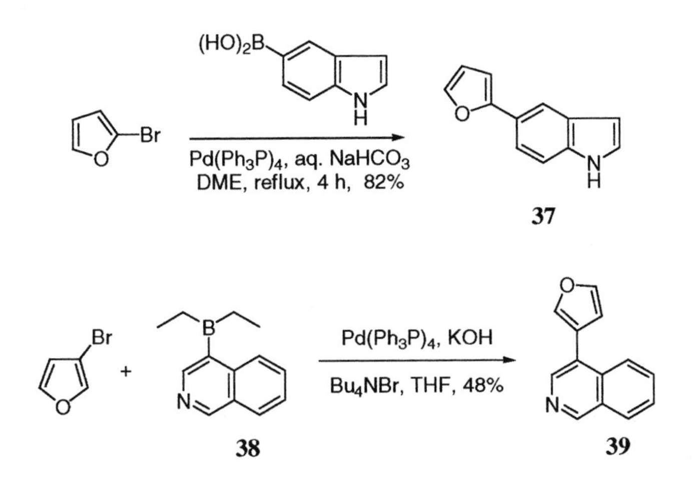


Unlike simple aryl halides, a regioselective Suzuki coupling of 2,4-dibromofuran may be achieved at C(2) [46]. The coupling between 2,4-dibromofuran and pyrimidylboronic acid **40** provided furylpyrimidine **41**, which was then hydrolyzed to 5-substituted uracil **42**.

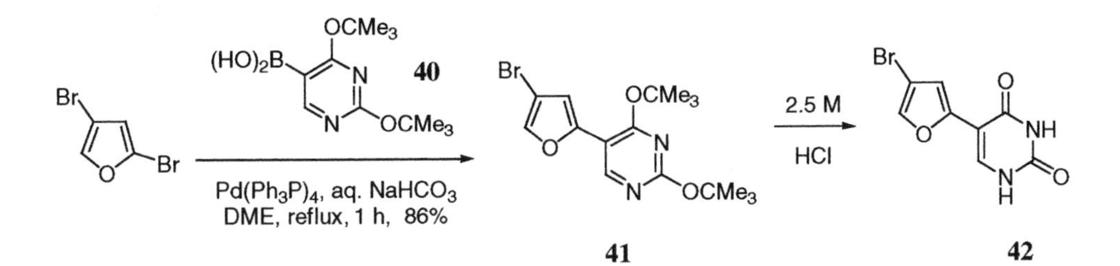

Fuostifoline (**47**), a furo[3,2-*a*]carbazole, was isolated from *Murraya euchrestifolia*. Timári's total synthesis of **47** commenced with alkylation of bromocresol **43** with bromoacetaldehyde diethyl acetal and $P_4O_{10}$-promoted cyclization to furnish 5-bromo-7-methylbenzofuran (**44**) [47]. The Suzuki coupling of boronic acid **45**, derived from **44**, with *o*-bromonitrobenzene yielded biaryl **46**. Nitrene generation, achieved *via* deoxygenation of nitro compound **46** using triethyl phosphite, was followed by cyclization to fuostifoline (**47**).

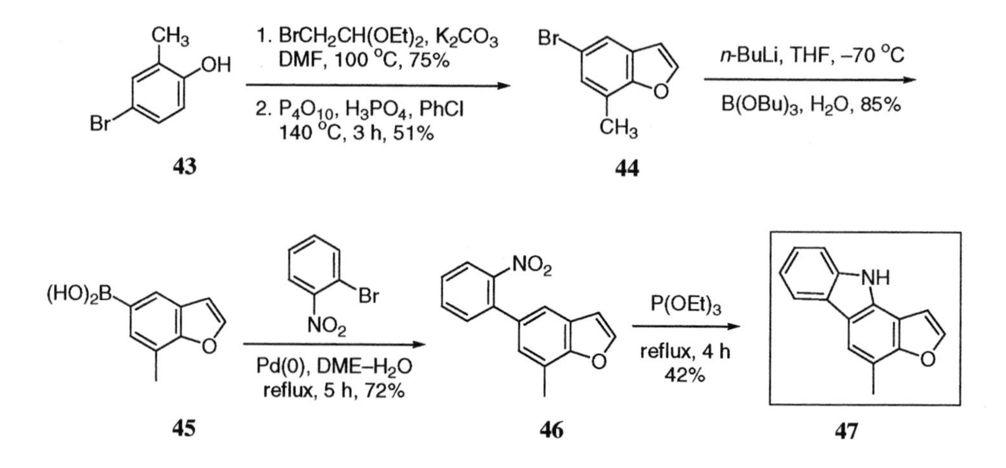


The Pd-catalyzed three-component cross-coupling reaction among aryl metal reagents, carbon monoxide, and aryl electrophiles is a straightforward and convenient route for the synthesis of unsymmetrical biaryl ketones. The reaction of electron-deficient electrophiles generally suffers from a side reaction that gives the direct coupling product without monoxide insertion. Miyaura developed an efficient Pd-catalyzed carbonylative three-component cross-coupling reaction of an arylboronic acid with aryl electrophiles including a bromofuran substrate [48]. Using $Pd(Ph_3P)_4$ as catalyst, the unsymmetrical biaryl ketone **49** was synthesized from 2-bromofuran **48**. Very interestingly, $PdCl_2$(dppf), the catalyst of choice for other aryl halides, gave exclusive direct coupling product without insertion of CO.

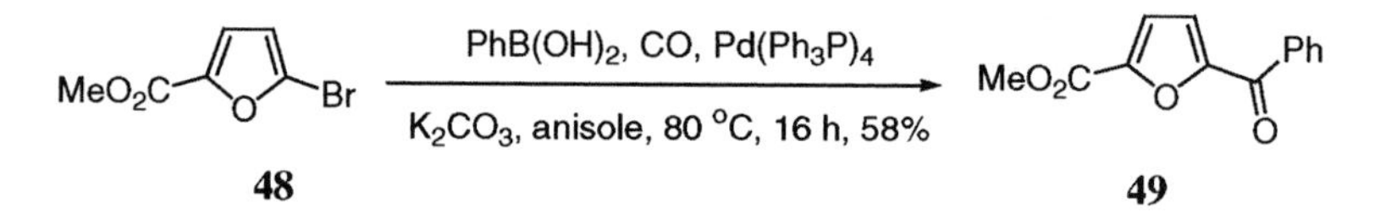

### 6.3.4 Stille coupling

#### *6.3.4.a Furan motif as a nucleophile (stannane)*

Furylstannanes may be prepared by a number of methods, one of which involves direct metalation of a furan followed by quenching with tin chloride [49]. The second method for furylstannane preparation uses halogen-metal exchange to generate the lithiofuran species, which is then quenched by tin chloride [50]. The third method, more suitable to base-sensitive substrates, is Pd-catalyzed coupling of a halofuran or halobenzofuran with hexabutylditin [51].

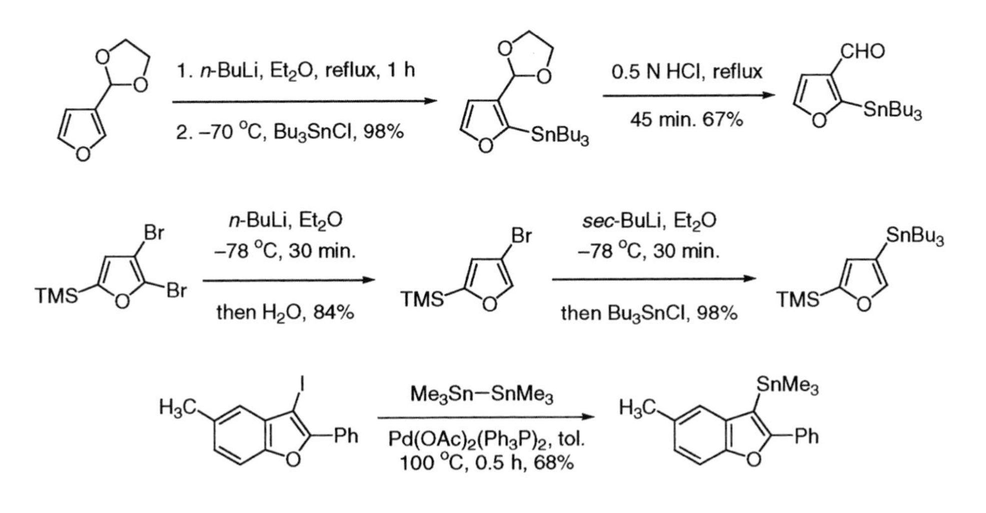


Interestingly, the alkyne-oxazole Diels-Alder cycloaddition strategy provides a unique entry to some furyl stannanes [52]. Thus, thermolysis of bis(tributylstannyl)acetylene (**50**) and 4-phenyloxazole (**51**) led to a separable mixture of 3,4-bis(tributylstannyl)furan (**52**, 19% yield) and 3-tributylstannylfuran (**53**, 23% yield).

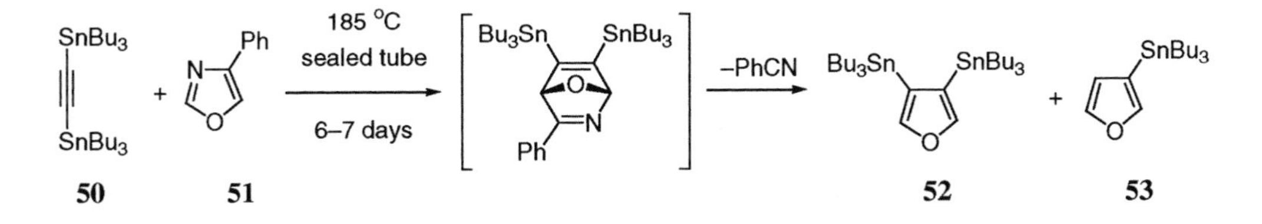


The furyl- and benzofurylstannanes can be coupled with a variety of electrophiles including aryl [53] and heteroaryl halides [54, 55]. In their synthetic studies towards lophotoxin and pukalide, Paterson and coworkers explored both intermolecular and intramolecular Stille coupling reactions [56]. The intermolecular approach between vinyl iodide **54** and furylstannane **55** was more successful, giving adduct **56** in 67% yield. The intramolecular version provided the macrocyclized 14-membered lactone in only 15% yield.

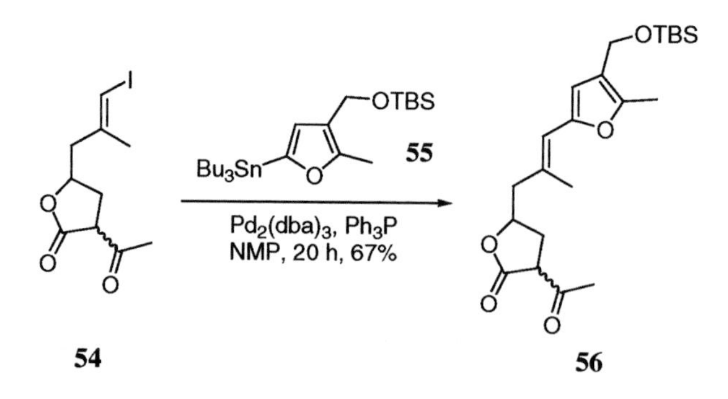


In the total synthesis of moracin M (**61**), a phytoalexin isolated from infected white mulberry, Widdowson *et al.* first prepared 2-stannylated benzofuran **58** from benzofuran **57** *via* direct metalation and treatment with $Me_3SnCl$ [57]. Stannane **58** was then coupled with aryl iodide **59** to afford adduct **60**, which was desilylated to moracin M (**61**).

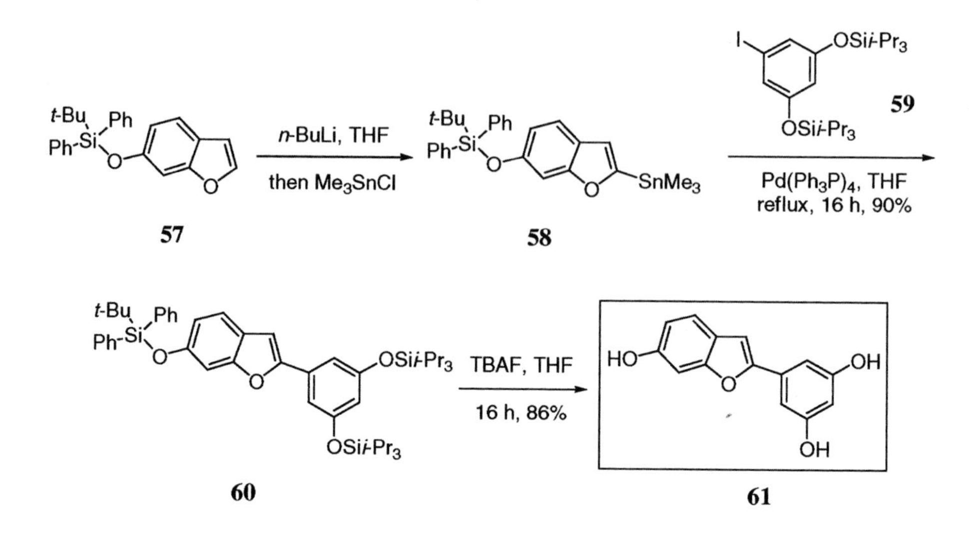


Meyers and Novachek described the Stille coupling of a chiral 2-bromooxazoline with a furylstannane to produce furyloxazoline [58]. Liebeskind and Wang conducted a benzannulation of a furylstannane using a Stille coupling with 4-chloro-2-cyclobutenone **62** to elaborate benzofuranol **63** *via* a dienyl ketene intermediate [59].

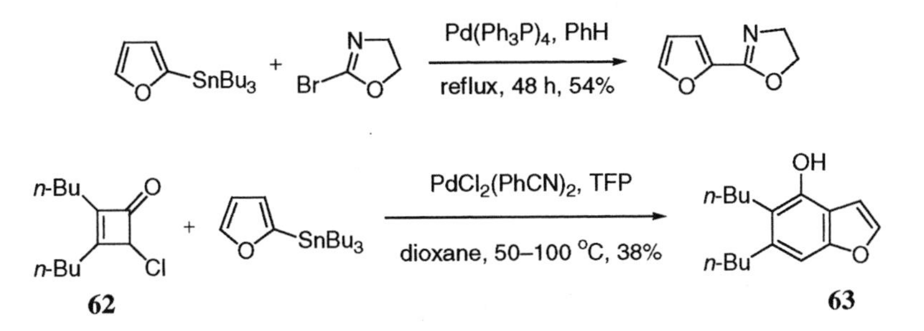


### *6.3.4.b Furan motif as an electrophile (halide or triflate)*

In the presence of hexabutylditin, Pd-catalyzed homocoupling of bromofuran **64** took place to give bifuran diester **65** [60].

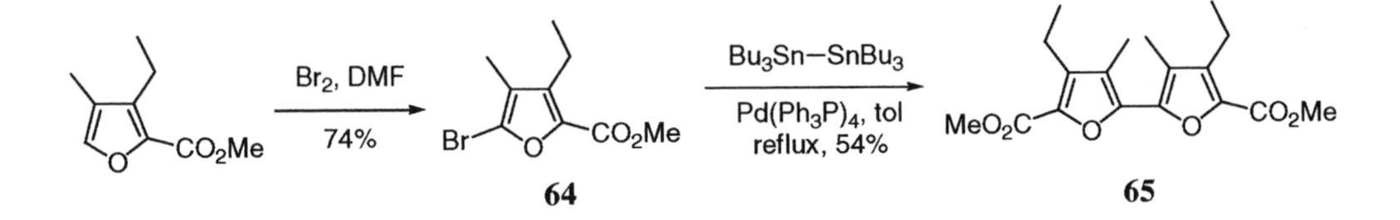


Like most aryl halides, furyl halides and furyl triflates have been coupled with a variety of organostannanes including alkenyl, aryl, and heteroaryl stannanes in the presence of catalytic palladium. Carbamoylstannane **66** was prepared by treating lithiated piperidine with carbon monoxide and tributyltin chloride sequentially. The Stille reaction of **66** and 3-bromofuran then gave rise to amide **67** [61]. In another example, lithiation of 4,4-dimethyl-2-oxazoline followed by quenching with $Me_3SnCl$ resulted in 2-(tributylstannyl)-4,4-dimethyl-2-oxazoline (**68**) in 70–80% yield [62]. Subsequent Stille reaction of **68** with 3-bromofuran afforded 2-(3'-furyl)-4,4-dimethyl-2-oxazoline (**69**).

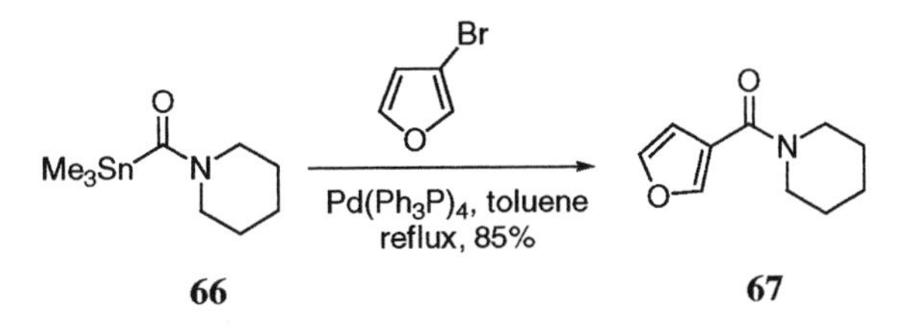

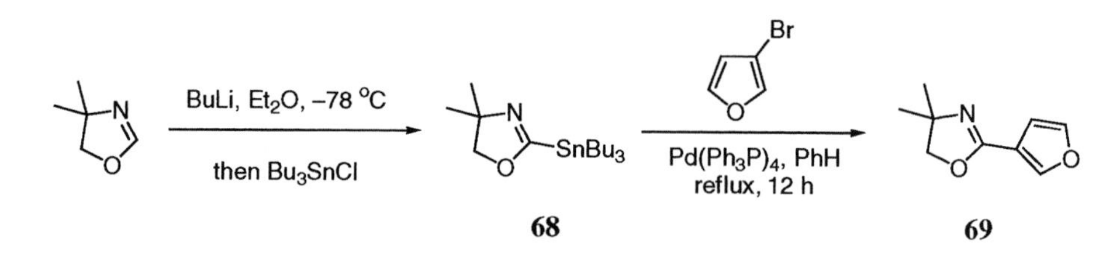


Hudkins *et al.* prepared a new reagent, 1-carboxy-2-(tributylstannyl)indole (**70**), using $CO_2$ as a temporary protecting group [63]. The Stille coupling of **70** with 2-bromobenzofuran gave the 2-substituted indole **71** while offering an advantageous temporary carboxylic acid protection, which was much easier to remove than BOC, SEM, or phenylsulfonyl groups. Snieckus and associates conducted Stille cross-coupling reactions on a solid support *via* an ester linker to make styryl, biaryl and heterobiaryl carboxylic acids [64]. Bromofuran was one of the substrates that has been successfully coupled with both vinyl- and arylstannanes.

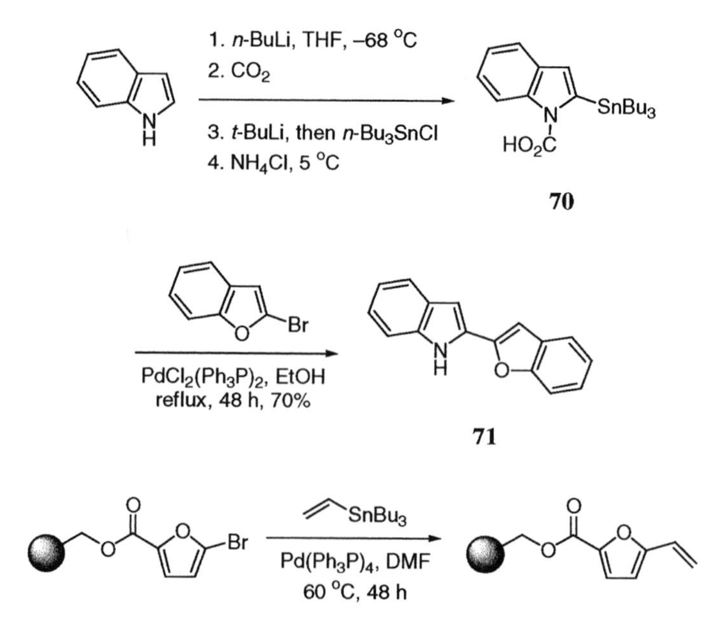


Unlike all-carbon aryl halide substrates, regioselective Stille coupling of 2,3-dihalofurans has been observed. Taking advantage of C(2) activation, regioselective Stille coupling was achieved using 2,3-dibromofuran **27** to afford **72** [35].

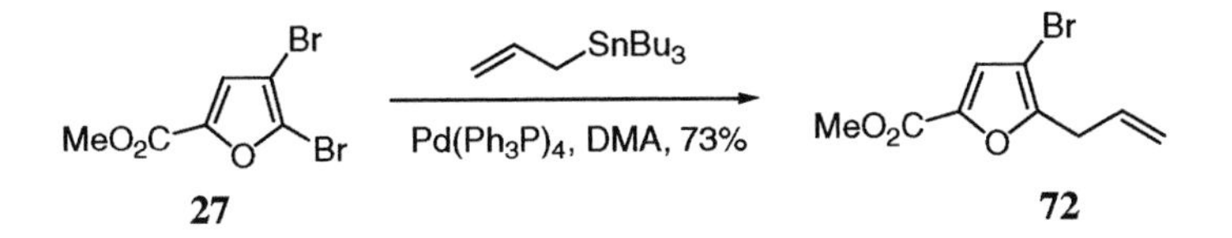


### 6.3.5 Hiyama coupling

The Hiyama coupling offers a practical alternative when selectivity and/or availability of other reagents are problematic. Hiyama *et al.* coupled alkyltrifluorosilane **74** with 2-bromofuran **73** to give the corresponding cross-coupled product **75** in moderate yield in the presence of catalytic $Pd(Ph_3P)_4$ and 3 equivalents of TBAF [65]. In this case, more than one equivalent of fluoride ion was needed to form a pentacoordinated silicate. On the other hand, alkyltrifluorosilane **74** was prepared by hydrosilylation of the corresponding terminal olefin with trichlorosilane followed by fluorination with $CuF_2$. This method provides a facile protocol for the synthesis of alkyl-substituted aromatic compounds.

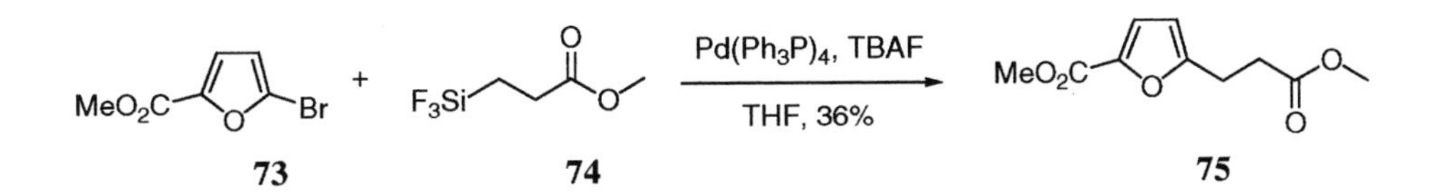


## 6.4 Sonogashira reaction

Alkynylfurans are readily prepared *via* the Sonogashira reactions of halofurans [66, 67]. Due to activation of the α-positions, regioselective Sonogashira reaction can be achieved at C(2) rather than C(3) [35, 68].

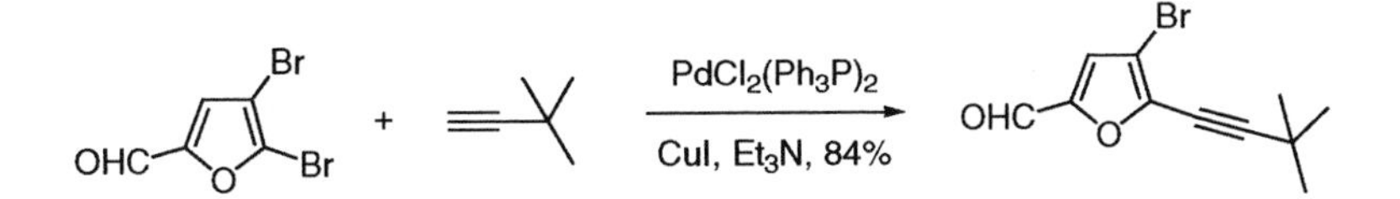


Applying a strategy akin to Yanaka's furopyrimidine synthesis [24], a Merck process group synthesized functionalized furo[2,3-*b*]pyridine **78** [69]. Pd-catalyzed coupling of iodopyridone **76** and trimethylsilylacetylene led to alkynyl pyridone **77** using $n$-$BuNH_2$ as the base, whereas $Et_3N$, $i$-$Pr_2NEt$, and $K_2CO_3$ were inferior. Subsequent cyclization of **77** provided an expeditious entry to furo[2,3-*b*]pyridine **78**. Analogously, nucleoside analogs with an unusual bicyclic base **81** were prepared from alkyne **80**, the Sonogashira adduct of iodouracil **79** [70].

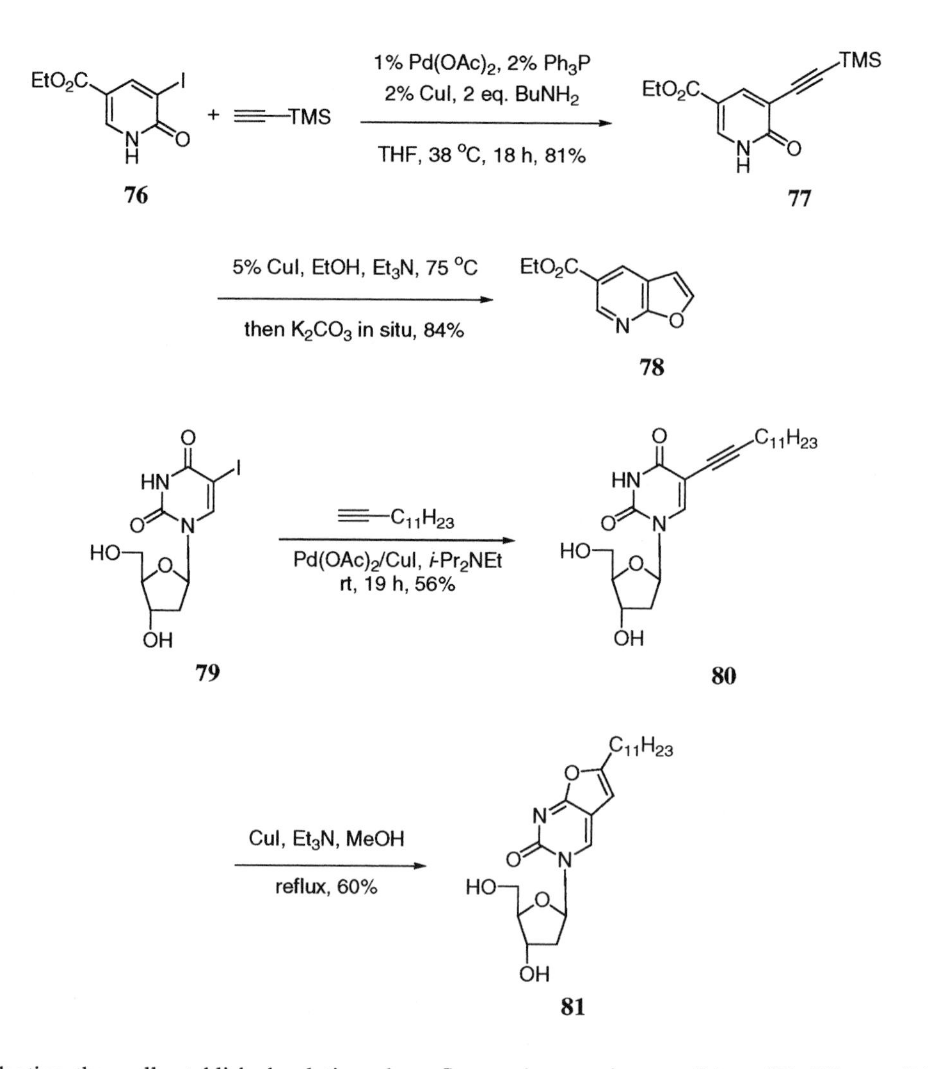


Adapting the well-established solution phase Sonogasira reaction conditions [71–73], a solid phase synthesis of 2-substituted benzofurans was accomplished *via* Pd-catalyzed heteroannulation of acetylenes [74]. The starting carboxylic acid **82** was directly linked to the hydroxy resin TentaGel™S-OH using the Mitsunobu conditions to give the resin-bound ester, which was then deacetylated by mild alkaline hydrolysis to generate phenol **83**. Treating **83** with terminal alkynes in the presence of $PdCl_2(Ph_3P)_2$, CuI and tetramethylguanidine (TMG) in DMF smoothly produced the heteroannulation products as resin-bound benzofuran **84**. Cleavage was subsequently performed with base to deliver 2-substituted benzofuran carboxylic acids **85** after neutralization.

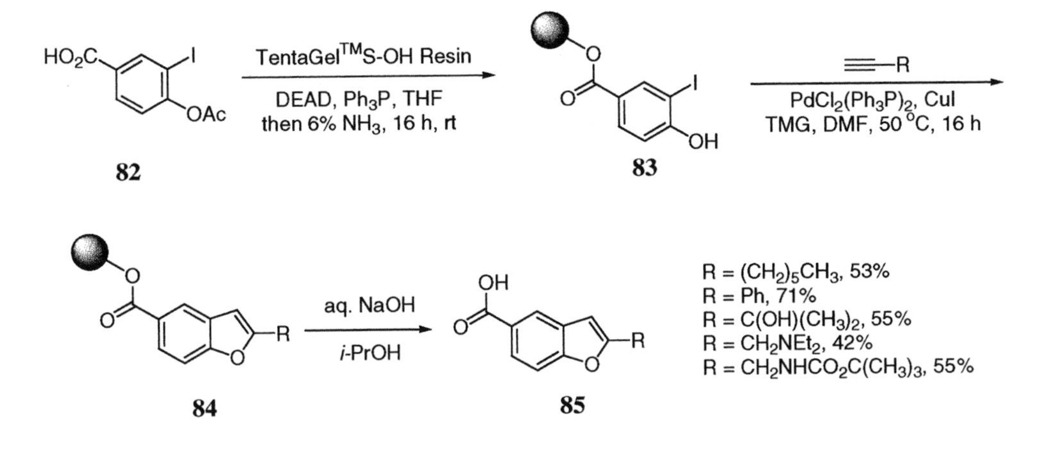


A noticeable failure of the Sonogashira reaction of 3-bromobenzofuran was recorded by Yamanaka and coworkers [75]. Only resinous materials were obtained from 3-bromobenzofuran even though the same reaction worked well (58–96% yields) for 3-iodobenzothiophene.

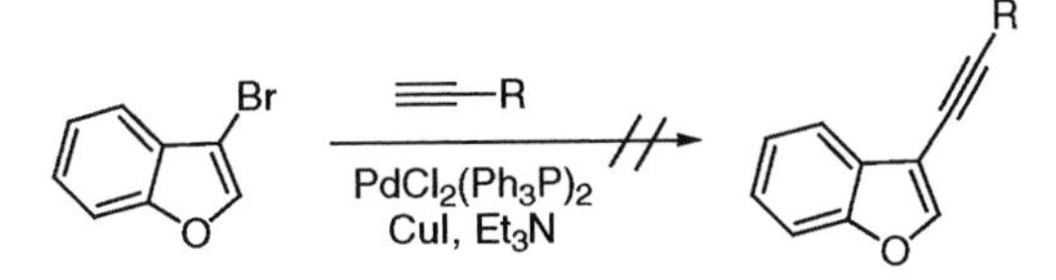


## 6.5 Heck, intramolecular Heck and heteroaryl Heck reactions

### 6.5.1 Intermolecular Heck reaction

Although the Heck reactions of heteroaryl halides are now commonplace [76], few examples are found using organohalide substrates possessing a carboxylic acid moiety [77]. However, 4,5-dibromo-2-furancarboxylic acid (**86**) underwent a Heck reaction with ethyl acrylate to afford diacrylate **87** [78].

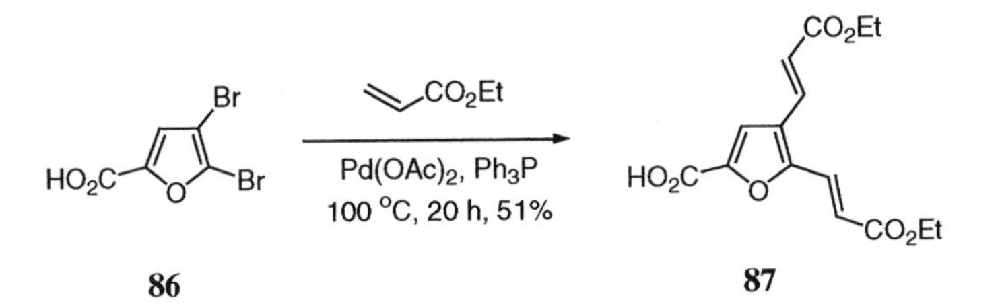

The intermolecular Heck reaction of 3-bromofuran and tosylallyamine **88** gave adduct **89** under the classical Heck conditions [79]. Subsequent Rh-catalyzed hydroformylation with ring closure occurred regioselectively to furnish the hydroxypyrrolidine, which was dehydrated using catalytic HCl to afford dihydropyrrole **90**.

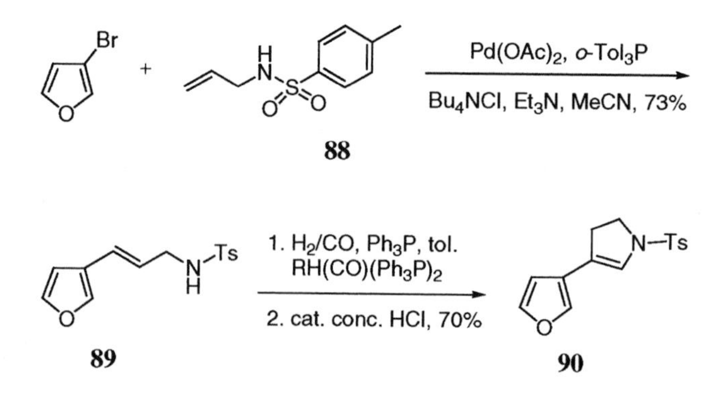


### 6.5.2 Intramolecular Heck reaction

Similar to the Pd-catalyzed pyrrole and thiophene annulations, an intramolecular Heck reaction of substrate **91** resulted in benzofuran **92** [80]. Such an approach has become a popular means of synthesizing fused furans. Muratake *et al.* exploited the intramolecular Heck cyclization to establish the tricyclic core structure *en route* to the synthesis of a furan analog of duocarmycin SA, a potent cytotoxic antibiotic [81]. Under Jeffery's phase-transfer catalysis conditions, substrate **93** was converted to tricyclic derivatives **94** and **95** as an inseparable mixture (*ca.* 4:1) of two double bond isomers.

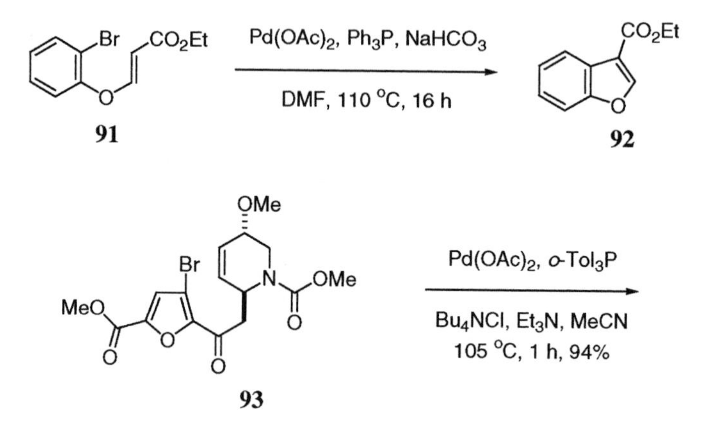

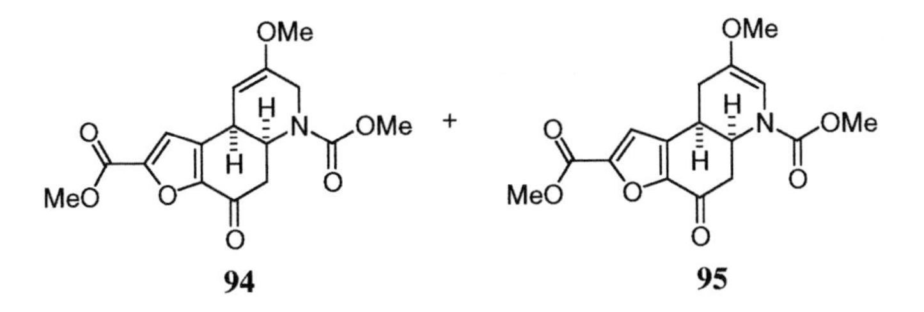


An intramolecular Heck cyclization strategy was developed for the construction of indole and benzofuran rings on solid support [82], enabling rapid generation of small-molecular libraries by simultaneous parallel or combinatorial synthesis. $S_N2$ displacement of resin-bound γ-bromocrotonyl amide **97** with *o*-iodophenol **96** afforded the cyclization precursor **98**. A subsequent intramolecular Heck reaction using Jeffery's "ligand-free" conditions furnished, after double bond tautomerization, the resin-bound benzofurans, which were then cleaved with 30% TFA in $CH_2Cl_2$ to deliver the desired benzofuran derivatives **99** in excellent yields and purity.

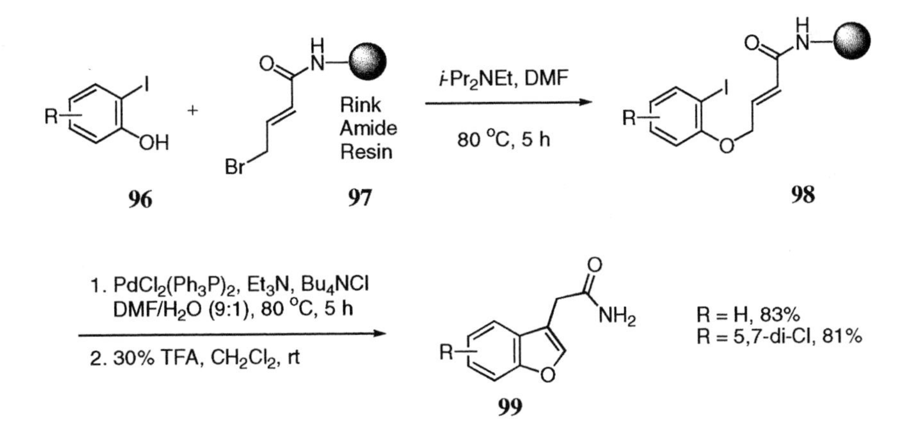


Rawal's group developed an intramolecular aryl Heck cyclization method to synthesize benzofurans, indoles, and benzopyrans [83]. The rate of cyclization was significantly accelerated in the presence of bases, presumably because the phenolate anion formed under the reaction conditions was much more reactive as a soft nucleophile than phenol. In the presence of a catalytic amount of Herrmann's dimeric palladacyclic catalyst (**101**) [84], and 3 equivalents of $Cs_2CO_3$ in DMA, vinyl iodide **100** was transformed into "*ortho*" and "*para*" benzofuran **102** and **103**. In the mechanism proposed by Rawal, oxidative addition of phenolate **104** to Pd(0) is followed by nucleophilic attack of the ambident phenolate anion on σ-palladium intermediate **105** to afford aryl-vinyl palladium species **106** after rearomatization of the presumed cyclohexadienone intermediate. Reductive elimination of palladium followed by isomerization of the exocyclic double bond furnishes **102**.

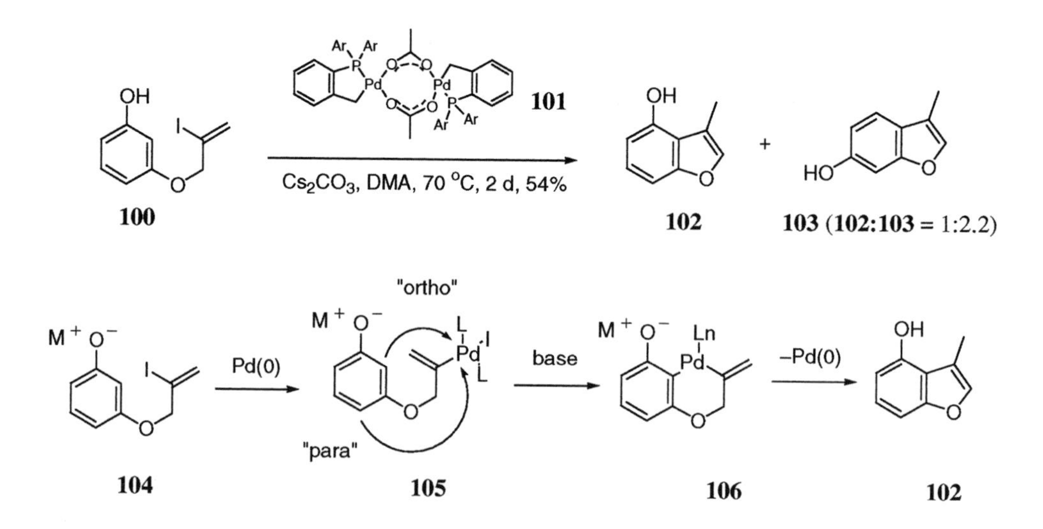


### 6.5.3 Heteroaryl Heck reaction

Ohta's group coupled aryl bromides such as 2-bromonitrobenzene with benzofuran [85]. The heteroaryl Heck reaction took place at the more electron-rich C(2) position of benzofuran. They later described the heteroaryl Heck reactions of chloropyrazines with both furan and benzofuran [86].

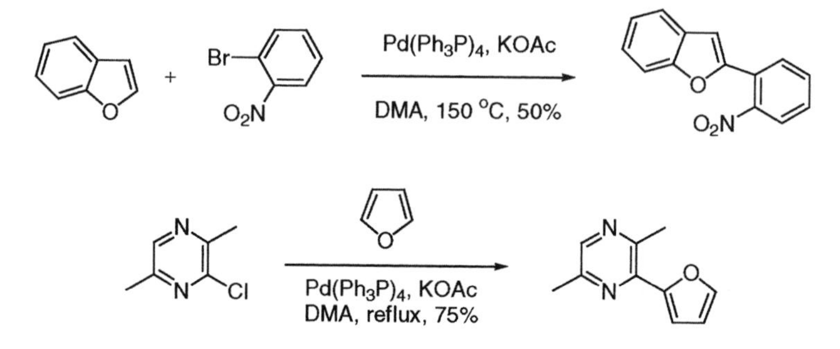


Grigg's group synthesized a unique bicyclic β-lactam **108** *via* an intramolecular Heck reaction from **107** [87, 88]. The 7-membered ring was formed *via* an unusual insertion *at C(3)* of furan, an aromatic π-system.

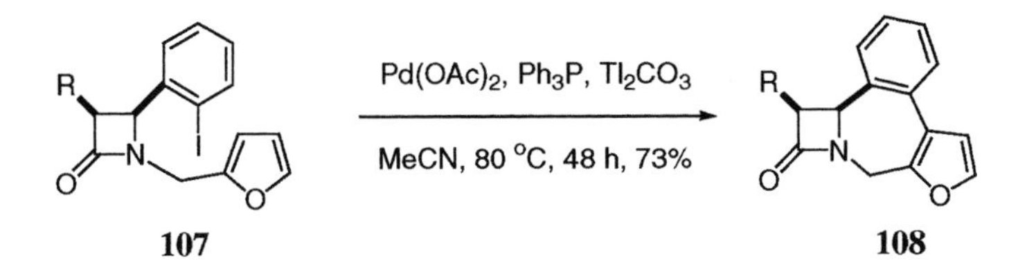


## 6.6 Heteroannulation

### 6.6.1 Propargyl carbonates

In 1985, Tsuji's group carried out a Pd-catalyzed reaction of propargyl carbonate with methyl acetoacetate as a soft carbonucleophile under *neutral conditions* to afford 4,5-dihydrofuran **109** [89–91]. The resulting unstable **109** readily isomerized to furan **110** under acidic conditions. In addition, they also reported formation of disubstituted furan **112** *via* a Pd-catalyzed heteroannulation of hydroxy propargylic carbonate **111** [92]. Presumably, an allenylpalladium complex (*cf.* **114**) was the key intermediate.

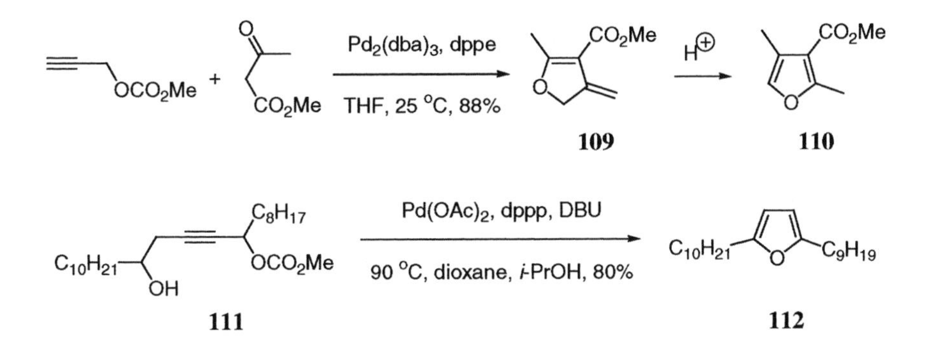


The reason why the above mentioned heteroannulation reactions of propargyl carbonates proceeded *in the absence of a base* becomes evident by the mechanism detailed below. First, oxidative addition of propargyl carbonate to Pd(0) yields allenylpalladium(II) carbonate **113**. Decarboxylation of **113** releases one molecule of $CO_2$ and a methoxide anion, along with allenylpalladium intermediate **114**. At this point, the self-generated methoxide anion serves as a base to deprotonate methyl acetoacetate to form the corresponding enolate, which attacks the *sp* carbon of **114** to give the palladium carbene complex **115**. Isomerization of **115** leads to the $\pi$-allylpalladium complex **116**, which then undergoes an *O*-alkylation with the carbonyl oxygen and gives rise to the *exo*-methylenefuran **109** following elimination of Pd(0).

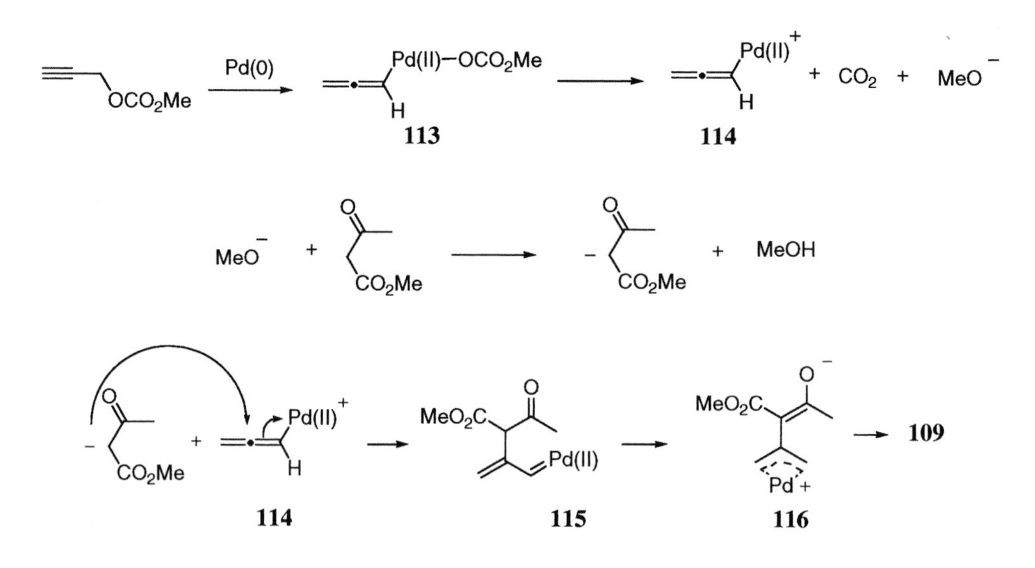


### 6.6.2 Alkynols

Analogous to the annulation of the Sonogashira adducts (see Section 6.4.), a spontaneous cyclization *via* the intramolecular alkoxylation of alkyne **117** (the coupling adduct of *o*-bromophenol and phenyl acetylene) took place under the reaction conditions to give 2-phenylbenzofuran **119** [93]. Benzofurylpalladium complex **118** was the putative intermediate during the cyclization.

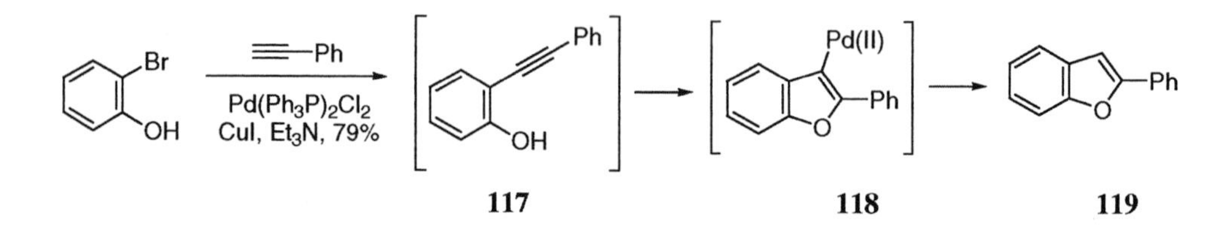


The intermediacy of benzofurylpalladium complex **118** was confirmed by trapping it with various electrophiles, including allyl halides [94], propargyl carbonates (giving rise to 3-allenylbenzofurans) [95], and carbonylating reagents [96]. Moreover, Cacchi *et al.* took advantage of the benzofurylpalladium intermediate and synthesized 2,3-disubstituted benzofurans from propargylic *o*-(alkyl)phenyl ethers [97, 98]. Scammells' synthesis of benzofuran **121** serves as an example [96]. A sequential Pd-catalyzed annulation and alkoxylcarbonylation of alkynyl phenol **120** gave **121**, an intermediate in the synthesis of XH-14, a potent antagonist of the $A_1$ adenosine receptor isolated from the plant *Salvia miltiorrhiza*.

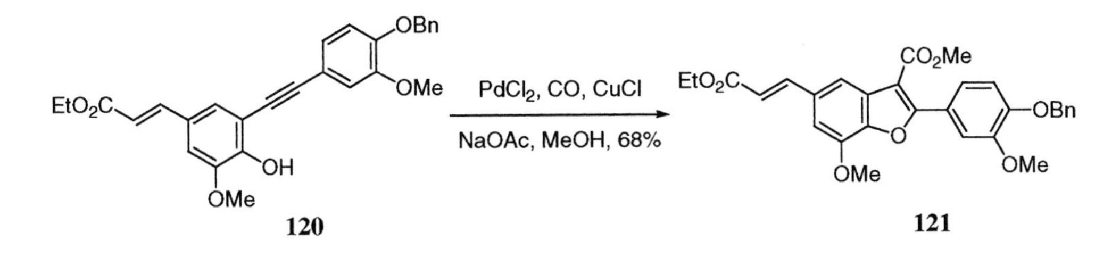


Substituting the benzene ring with a double bond, Pd-catalyzed intramolecular alkoxylation of alkyne **122** also proceeded *via* an alkenyl palladium complex to form furan **123** instead of a benzofurans [99, 100]. In addition, 3-hydroxyalkylbenzo[*b*]furans was prepared by Bishop *et al.* *via* a Pd-catalyzed heteroannulation of silyl-protected alkynols with 2-iodophenol in a fashion akin to the Larock indole synthesis, [101].

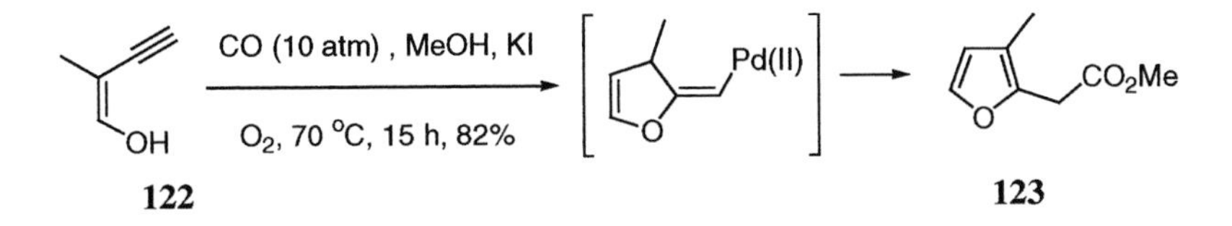


### 6.6.3 Alkynones

Pd-catalyzed isomerization of ynones to furans has been an active area of research over the last decade. Huang *et al.* described a Pd-catalyzed rearrangement of α,β-acetylenic ketones to furans in moderate yield [102]. For example, $Pd(dba)_2$ promoted the isomerization of alkyne **124** to a putative allenyl ketone intermediate **125**, which subsequently cyclized to the corresponding furan **126**.

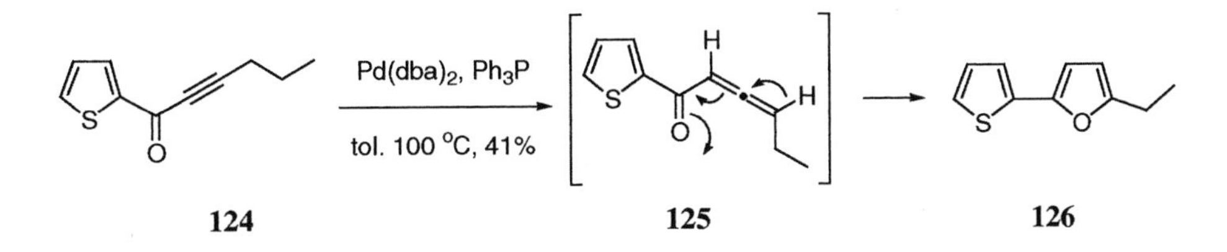


Utimoto *et al.* synthesized substituted furans using a Pd-catalyzed rearrangement of easily accessible β,γ-acetylenic ketones [103]. One plausible pathway is illustrated here using the transformation of β,γ-acetylenic ketone **127** to 2,5-disubstituted furan **128**. Enolization of **127** is followed by an intramolecular oxypalladation of the resulting enol **129** to form furylpalladium(II) species **130**, which is subsequently treated with acid to give furan **128**.

Furthermore, intercepting the furylpalladium(II) species **130** with an electrophile would result in a carbodepalladation in place of protodepalladation. Therefore, a tandem intramolecular alkoxylation of β,γ-acetylenic ketone **127** was realized to afford trisubstituted furan **131** when allyl chloride was added to the original recipe [103]. 2,2-Dimethyloxirane was used as a proton scavenger, ensuring exclusive formation of 3-allylated 2,5-disubstituted furan **131** without contamination by protonated furans.

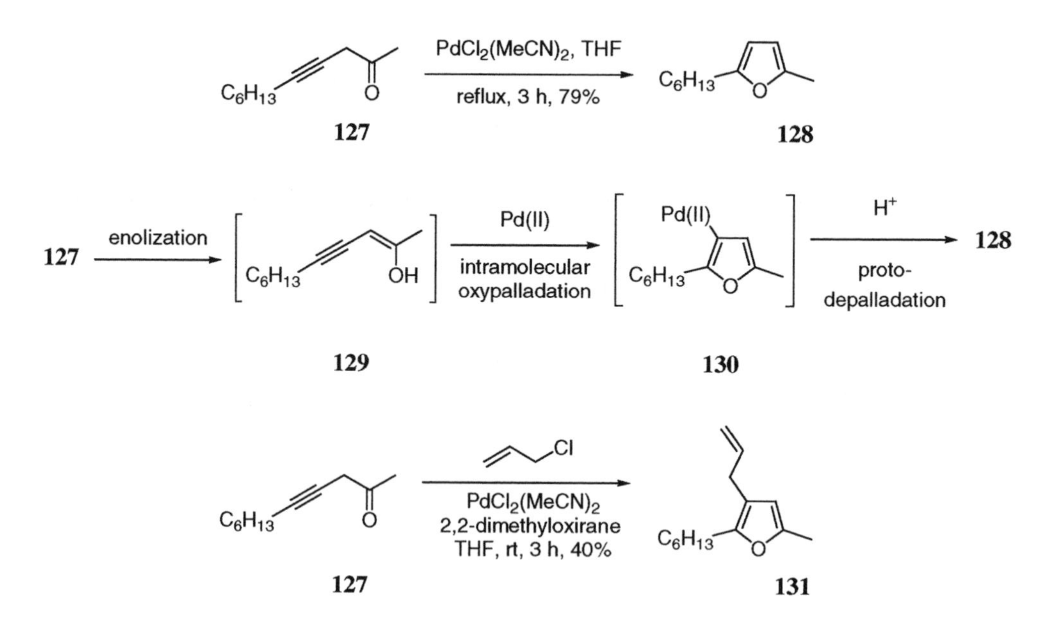


A similar substituted furan synthesis was realized *via* a Pd-catalyzed tandem carbonylation-arylation using an α,β-acetylenic ketone, carbon monoxide, and bromothiophene [104].

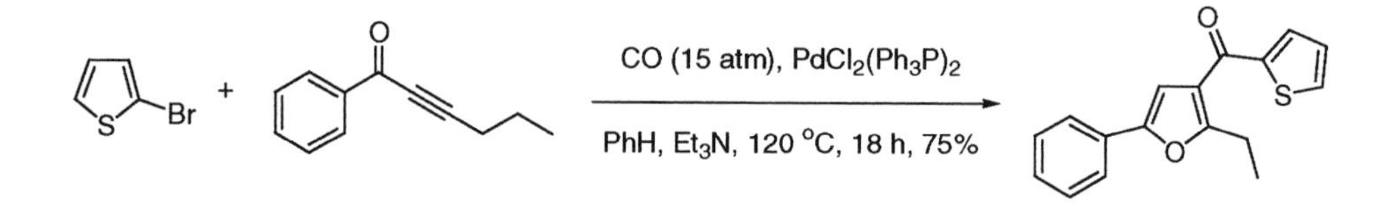


### 6.6.4 Enones

Other than the aforementioned ynones, β-iodo-β,γ-enone **132** was also converted into a 2,5-disubstituted furan, **133**, under Pd-catalyzed cyclization conditions using Herrmann's catalyst, palladacycle catalyst **101** [105].

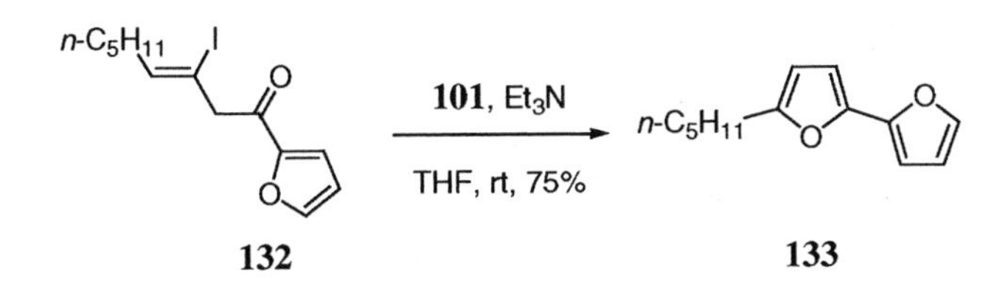


### 6.6.5 Allenones

Allenones have also been transformed into substituted furans *via* Pd-catalyzed heteroannulation. Ma and Zhang converted 1,2-dienyl ketones and organic halides to substituted furans under the agency of $Pd(Ph_3P)_4$ and $Ag_2CO_3$. Addition of 10 mol% of $Ag_2CO_3$ was crucial for the reaction of the organic halide. For example, allenone **134** and 5-bromopyrimidine were converted to pyrimidylfuran **135** in excellent yield [106].

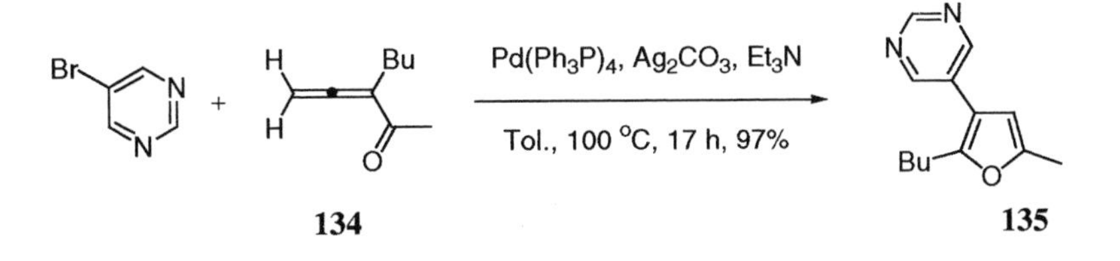


## 6.7 Carbonylation and C—N bond formation

Pd-catalyzed alkoxylcarbonylation of furan and benzofuran was achieved in the presence of $Hg(O_2CCF_3)_2$ in ethanol with low efficiency [107]. Other heterocycles including thiophene and pyrrole were also carbonylated to give the corresponding esters in low yields using the same method.

Applying Buchwald's Pd-catalyzed amination methodology, Thomas and coworkers prepared a range of bicyclic piperazine [108]. While Pd-catalyzed amination of 5-bromobenzofuran led to 5-benzofurylpiperizine **136** in 65% yield after deprotection, the corresponding reaction of 7-bromobenzofuran only gave 7-benzofurylpiperizine **137** in 20% yield. They speculated that either steric hindrance of the oxidative addition intermediate or the interaction between the oxygen lone pair and the metal center was responsible for the low yield. The debrominated benzofuran was the major by-product.

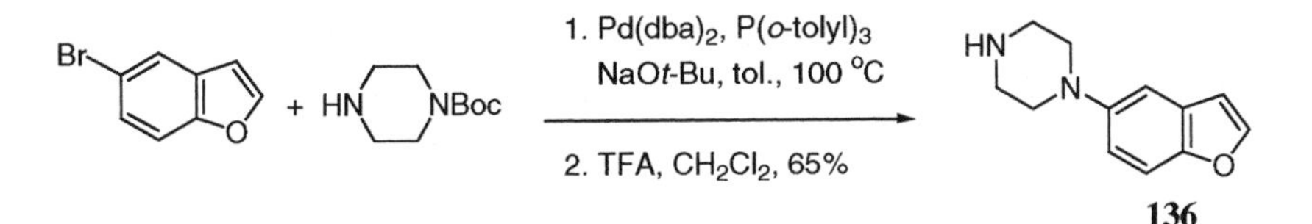

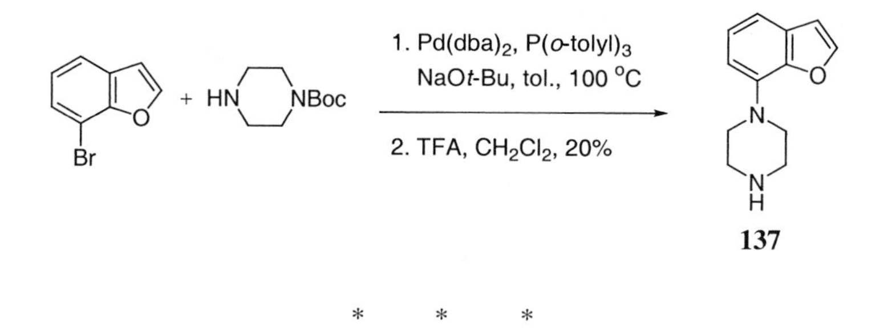


*  *  *

In conclusion, due to the activation effects stemming from the electronegativity of the oxygen atom on the α-positions of furans and benzofurans, regioselective coupling can be attained for palladium-catalyzed chemistry. Regioselective cross-couplings (Suzuki, Stille, and Sonogashira reactions) of α,β-dihalofurans can be achieved at the α-positions. Unlike 2-chloropyridines, 2-chlorofurans are not sufficiently activated to oxidatively add to Pd(0).

The most unique feature of furan synthesis using palladium chemistry is heteroannulation. Enones, ynones and ynols all have been annulated into furans and benzofurans. More importantly, trapping the reactive Pd(II) intermediates at different stages with electrophiles offers unique opportunities to synthesize substituted furans and benzofurans.

## 6.8 References

1. Zaluski, M.C.; Robba, M.; Bonhomme, M. *Bull. Soc. Chim. Fr.* **1970**, 1838–43.
2. Chadwick, D. J.; Chambers, J. Hargraves, H. E.; Meakins, G. D.; Snowden, R. C. *J. Chem. Soc., Perkin Trans. 1* **1973**, 2327–32.
3. Gorzynski, M.; Rewicki, D. *Liebigs Ann. Chem.* **1986**, 625–37.
4. Keegstra, M. A.; Klomp, A. J. A.; Brandsma, L. *Synth. Commun.* **1990**, *20*, 3371–4.
5. Sornay, R.; Meunier, J.-M.; Fourari, P. *Bull. Soc. Chim. Fr.* **1971**, 990–6.
6. Verkruijsse, H. D.; Keegstra, M. A.; Brandsma, L. *Synth. Commun.* **1989**, *19*, 1047–50.
7. Gronowitz, S.; Hörnfeldt, A. B.; Pettersson, K. *Synth. Commun.* **1973**, 213–8.
8. Mann, I. S.; Widdowson, D. A.; Clough, J. M. *Tetrahedron* **1991**, *47*, 7981–90.
9. Benincori, T.; Brenna, E.; Sannicolo, F.; Trimarco, L.; Antognazza, P.; Casarotti, E.; Demartin, F.; Pilati, T. *J. Org. Chem.* **1996**, *61*, 6244–51.
10. Tasker, A. S.; Sorensen, B. K.; Jae, H.-S.; Winn, M.; von Geldern, T. W.; *et al. J. Med. Chem.* **1997**, *40*, 322–30.
11. Barker, P.; Finke, P.; Thompson, K. *Synth. Commun.* **1989**, *19*, 257–65.
12. Kozhevnikov, I. V. *React. Kinet. Catal. Lett.* **1976**, *4*, 451–8.
13. Itahara, T.; Hashimoto, M.; Yumisashi, H. *Synthesis* **1984**, 255–6.

14. Itahara, T. *J. Org. Chem.* **1985**, *50*, 5272–5.
15. Fujiwara, Y.; Maruyama, O.; Yoshidomi, M.; Taniguchi, H. *J. Org. Chem.* **1981**, *46*, 851–5.
16. Tsuji, J.; Nagashima, H. *Tetrahedron* **1984**, *40*, 2699–702.
17. Roshchin, A. I.; Kel'chevski, S. M.; Bumagin, N. A. *J. Organomet. Chem.* **1998**, *560*, 163–7.
18. Pearlman, B. A.; MaNamara, J. M.; Hasan, I.; Hatakeyama, S.; Sekizaki, H.; Kishi, Y. *J. Am. Chem. Soc.* **1981**, *103*, 4248–51.
19. Carpita, A.; Rossi, R.; Veracini, C. A. *Tetrahedron* **1985**, *41*, 1919–29.
20. Klingstedt, T.; Frejd, T. *Organometallics* **1983**, *2*, 598–600.
21. Pelter, A.; Rowland, M.; Jenkins, I. H. *Tetrahedron Lett.* **1987**, *28*, 5213–6.
22. Amatore, C.; Jutand, A.; Negri, S.; Fauvarque, J. F. *J. Organomet. Chem.* **1990**, *390*, 389–98.
23. Brandão, M. A. F.; de Oliveira, A. B.; Snieckus, V. *Tetrahedron Lett.* **1993**, *34*, 2437–40.
24. Ennis, D. S.; Gilchrist, T. L. *Tetrahedron* **1990**, *46*, 2623–32.
25. Campbell, J. B., Jr.; Firor, J. W.; Davenport, T. W. *Synth. Commun.* **1989**, *19*, 2265–72.
26. Arcadi, A.; Burini, A.; Cacchi, S.; Delmastro, M.; Marinelli, F.; Pietrani, B. *Synlett* **1990**, 47–8.
27. Sakamoto, T.; Kondo, Y.; Watanabe, R.; Yamanaka, H. *Chem. Pharm. Bull.* **1986**, *34*, 2719–24.
28. Sakamoto, T.; Katoh, E.; Kondo, Y.; Yamanaka, H. *Heterocycles* **1988**, *27*, 1353–6.
29. Negishi, E.-i.; Takahashi, T.; King, A. O. *Org. Synth.* **1988**, *66*, 67–74.
30. Chandraratna, R. A. S.; Gillett, S. J.; Song, T. K.; Attard, J.; Vuligonda, S.; Garst, M. E.; Arefieg, T.; Gill, D. W.; Wheeler, L. *Bioorg. Med. Chem. Lett.* **1995**, *5*, 523–7.
31. Negishi, E.-i.; Xu, C.; Tan, Z.; Kotora, M. *Heterocycles* **1997**, *46*, 209–14.
32. Paterson, I.; Gardner, M.; Banks, B. J. *Tetrahedron* **1989**, *45*, 5283–92.
33. Gillet, J.-P.; Sauvêtre, R.; Normant, J.-F. *Synthesis* **1986**, 538–43.
34. Jackson, R. F. W.; Wishart, N.; Wood, A.; James, K.; Wythes, J. *J. Org. Chem.* **1992**, *57*, 3397–404.
35. Bach, T.; Krüger, L. *Synlett* **1998**, 1185–6.
36. Thompson, W. J.; Gaudino, J. *J. Org. Chem.* **1984**, *49*, 5237–43.
37. Gribble, G. W.; Conway, S. C. *Synth. Commun.* **1992**, *22*, 2129–41.
38. Bourlot, A. S.; Desarbre, E.; Mérour, J.-Y. *Synthesis* **1994**, 411–6.
39. Benoit, J.; Malapel, B.; Mérour, J.-Y. *Synth. Commun.* **1996**, *26*, 3289–95.
40. Yang, Y. *Synth. Commun.* **1989**, *19*, 1001–8.
41. Moody, C. J.; Doyle, K. J.; Elliott, M. C.; Mowlem, T. J. *J. Chem. Soc., Perkin Trans. 1* **1997**, 2413–9.
42. Song, Z. Z.; Wong, H. N. C. *J. Org. Chem.* **1994**, *59*, 33–41.

43. Yang, Y,; Martin, A. R. *Heterocycles* **1992**, *34*, 1395–8.
44. Ishikura, M.; Oda, I.; Terashima, M. *Heterocycles* **1987**, *26*, 1603–10.
45. Ishikura, M.; Oda, I.; Terashima, M. *Heterocycles* **1985**, *23*, 2375–86.
46. Wellmar, U.; Hörnfeldt, A.-B.; Gronowitz, S. *J. Heterocycl. Chem.*, **1995**, *32*, 1159–63.
47. Soós, T.; Timári, G.; Hajós, G. *Tetrahedron Lett.* **1999**, *40*, 8607–9.
48. Ishiyama, T.; Kizaki, H.; Hayashi, T.; Suzuki, A.; Miyaura, N. *J. Org. Chem.* **1998**, *63*, 4726–31.
49. Gronowitz, S.; Timari, G. *J. Heterocycl. Chem.*, **1990**, *27*, 1159–60.
50. Katsumura, S.; Fujiwara, S.; Isoe, S. *Tetrahedron Lett.* **1988**, *29*, 1173–6.
51. Arcadi, A.; Cacchi, S.; Fabrizi, G.; Marinelli, F.; Moro, L. *Synlett* **1999**, 1432–4.
52. (a) Yang, Y.; Wong, H. N. C. *J. Chem. Soc., Chem. Commun.* **1992**, 656–8; (b) Yang, Y.; Wong, H. N. C. *J. Chem. Soc., Chem. Commun.* **1992**, 656–8; (c) Yang, Y.; Wong, H. N. C. *Tetrahedron* **1994**, *50*, 9583–608.
53. Bailey, T. R. *Tetrahedron Lett.* **1986**, *27*, 4407–10.
54. Hucke, A.; Cava, M. P. *J. Org. Chem.* **1998**, *63*, 7413–7.
55. Labadie, S. S. *Synth. Commun.* **1994**, *24*, 709–19.
56. Paterson, I.; Brown, R. E.; Urch, C. J. *Tetrahedron Lett.* **1999**, *40*, 5807–10.
57. Mann, I. S.; Widdowson, D. A.; Clough, J. M. *Tetrahedron* **1991**, *47*, 7981–90.
58. Meyers, A. I.; Novachek, K. A. *Tetrahedron Lett.* **1996**, *37*, 1747–8.
59. Libeskind, L. S.; Wang, J. *J. Org. Chem.* **1993**, *58*, 3550–6.
60. Sessler, J. L.; Hoehner, M. C.; Gebauer, A.; Andrievsky, A.; Lynch, V. *J. Org. Chem.* **1997**, *62*, 9251–60.
61. Lindsay, C. M.; Widdowson, D. A. *J. Chem. Soc., Perkein Trans. 1* **1988**, 569–73.
62. Dodoni, A.; Fantin, G.; Fogagnolo, M.; Medici, A.; Pedrini, P. *Synthesis* **1987**, 693–6.
63. Hudkins, R. L.; Diebold, J. L.; Marsh, F. D. *J. Org. Chem.* **1995**, *60*, 6218–20.
64. Chamoin, S.; Houldworth, S.; Snieckus, V. *Tetrahedron Lett.* **1998**, *39*, 4175–8.
65. Matsuhashi, H.; Kuroboshi, M.; Hatanaka, Y.; Hiyama, T. *Tetrahedron Lett.* **1994**, *35*, 6507–10.
66. Garcia, J.; Lopez, M.; Romeu, J. *Synlett* **1999**, 429–31.
67. Mal'kina, A. G.; Brandsma, L.; Vasilevsky, B. A. *Synthesis* **1996**, 589–90.
68. Bach, T.; Krüger, L. *Eur. J. Org. Chem.* **1999**, 2045–57.
69. Houpis, I. N.; Choi, W. B.; Reider, P. J.; Molina, A.; Churchill, H.; Lynch, J.; Volante, R. P. *Tetrahedron Lett.* **1994**, *35*, 9355–8.
70. McGuigan, C.; Yarnold, C. J.; Jones, G.; Velézquez, S.; Barucki, H.; Brancale, A.; Andrei, G.; Snoeck, R.; DeClercq, E.; Balzarin, J. *J. Med. Chem.* **1999**, *42*, 4479–84.
71. Kundu, N. G.; Pal, M.; Mahanty, J. S.; Dasgupta, S. K. *J. Chem. Soc., Chem. Commun.* **1992**, 41–2.
72. Kundu, N. G.; Pal, M.; Mahanty, J. S.; De, M. *J. Chem. Soc., Perkin Trans. 1* **1997**, 2815–20.

73. Arcadi, A.; Cacchi, S.; Rosario, M. D.; Fabrizi, G.; Marinelli, F. *J. Org. Chem.* **1996**, *61*, 9280–8.
74. Fancelli, D.; Fagnola, M. C.; Severino, D.; Bedeschi, A. *Tetrahedron Lett.* **1997**, *38*, 2311–4.
75. Sakamoto, T.; Kondo, Y.; Watanabe, R.; Yamanaka, H. *Chem. Pharm. Bull.* **1988**, *36*, 2248–52.
76. Karabelas, K.; Hallberg, A. *J. Org. Chem.* **1986**, *51,* 5286–90.
77. Crisp, G.; O'Donoghue, A. I. *Synth. Commun.* **1989**, *19,* 1745–58.
78. Karminski-Zamola, G.; Dogan, J.; Bajić, M. *Heterocycles* **1994**, *38,* 759–67
79. Busacca, C. A.; Dong, Y. *Tetrahedron Lett.* **1996**, *37,* 3947–50.
80. Henke, B. R.; Aquino, C. J.; Birkemo, L. S.; Croom, D. K.; Dougherty, Jr., R. W.; Ervin, G. N.; *et al. J. Med. Chem.* **1997**, *40,* 2706–25.
81. Muratake, H.; Okabe, K.; Takahashi, M.; Tonegawa, M.; Natsume, M. *Chem. Pharm. Bull.* **1997**, *45*, 799–806.
82. Zhang, H.-C.; Maryanoff, B. E. *J. Org. Chem.* **1997**, *62*, 1804–9.
83. Hennings, D. D.; Iwasa, S.; Rawal, V. H. *Tetrahedron Lett.* **1997**, *38,* 6379–82.
84. Herrmann, W. A.; Brossmer, C.; Öfele, K.; Reisinger, C. P.; Priermeier, T.; Beller, M.; Fischer, H. *Angew. Chem., Int. Ed. Engl.* **1995**, *34*, 1844–8.
85. Ohta, A.; Akita, Y.; Ohkuwa, T.; Chiba, M.; Fukunaka, R.; Miyafuji, A.; Nakata, T.; Tani, N. *Heterocycles* **1990**, *31*, 1951–7.
86. Aoyagi, Y.; Inoue, A.; Koizumi, I.; Hashimoto, R.; Tokunaga, K.; Gohma, K.; Komatsu, J.; Sekine, K.; Miyafuji, A.; Konoh, J. Honma, R. Akita, Y.; Ohta, A. *Heterocycles* **1992**, *33*, 257–72.
87. Burwood, M.; Davies, B.; Diaz, I.; Grigg, R.; Molina, P.; Sridharan, V.; Hughes, M. *Tetrahedron Lett.* **1995**, *36,* 9053–6.
88. Grigg, R.; Fretwell, P.; Meerholtz, C.; Sridharan, V. *Tetrahedron Lett.* **1994**, *50,* 359–70.
89. Tsuji, J.; Watanabe, H.; Minami, I.; Shimizu, I. *J. Am. Chem. Soc.* **1985**, *107*, 2196–8.
90. Minami, I.; Yuhara, M.; Watanabe, H.; Tsuji, J. *J. Organomet. Chem.* **1987**, *334,* 225–42.
91. Greeves, N.; Torode, J. S. *Synthesis* **1993**, 1109–13.
92. Tsuji, J.; Mandai, T. in *Metal-catalyzed Cross-coupling Reactions*; Diederich, F.; Stang, P. J. Eds. Wiley-VCH: Weinhein, Germany, **1998**, 485–6.
93. Villemin, D.; Goussu, D. *Heterocycles* **1989**, *29*, 1255–61.
94. Monteiro, N.; Balme, G. *Synlett* **1998**, 746–7.
95. Monteiro, N.; Arnold, A.; Balme, G. *Synlett* **1998**, 1111–3.
96. Lütjens, H.; Scammells, P. J. *Synlett* **1999**, 1079–81.
97. Cacchi, S.; Fabrizi, G.; Moro, L. *Tetrahedron Lett.* **1998**, *39,* 5101–4.
98. Cacchi, S.; Fabrizi, G.; Moro, L. *Synlett* **1998**, 741–5.

99. Gabriele, B.; Salerno, G.; De Pascalli, F.; Scianò, G. T.; Costa, M.; Chiusoli, G. P. *Tetrahedron Lett.* **1997**, *38,* 6877–80.
100. Gabriele, B.; Salerno, G.; Lauria, E. *J. Org. Chem.* **1999**, *64,* 7687–99.
101. Bishop, B. C.; Cottrell, I. F.; Hands, D. *Synthesis* **1997**, 1315–20.
102. Sheng, H.; Lin, S.; Huang, Y. Z. *Tetrahedron Lett.* **1986**, *27,* 4893–4.
103. Fukuda, Y.; Shiragami, H.; Utimoto, K.; Nozaki, H. *J. Org. Chem.* **1991**, *56*, 5816–9.
104. Okuro, K.; Furuune, M.; Miura, M.; Nomura, M. *J. Org. Chem.* **1992**, *57*, 4754–6.
105. Luo, F.-T.; Jeevanandam, A.; Bajji, A. C. *Tetrahedron Lett.* **1999**, *27,* 4893–4.
106. Ma, S.; Zhang, J. *J. Chem. Soc., Chem. Commun.* **2000**, 117–8.
107. Jaouhari, R.; Dixneuf, P. H.; Lécolier, S. *Tetrahedron Lett.* **1986**, *27,* 6315–8.
108. Kerrigan, F.; Martin, C.; Thomas, G. H. *Tetrahedron Lett.* **1998**, *39,* 2219–22.

# CHAPTER 7

## Thiazoles and Benzothiazoles

Thiazoles play a prominent role in nature. For example, the thiazolium ring present in vitamin $B_1$ serves as an electron sink and its coenzyme form is important for the decarboxylation of α-keto-acids. Furthermore, thiazoles are useful building blocks in pharmaceutical agents as exemplified by 2-(4-chlorophenyl)thiazole-4-acetic acid, a synthetic anti-inflammatory agent.

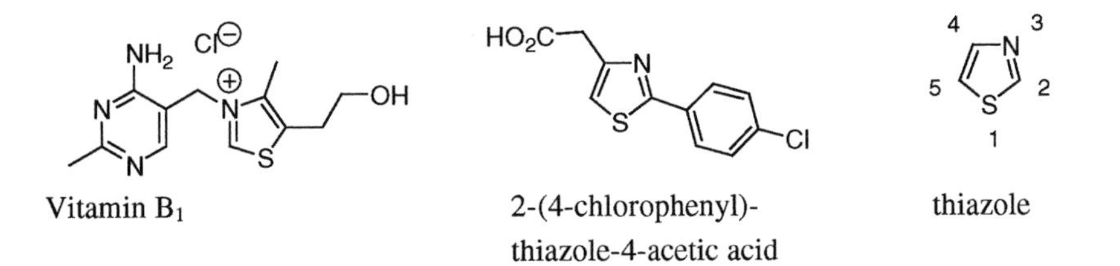


Vitamin $B_1$ | 2-(4-chlorophenyl)-thiazole-4-acetic acid | thiazole

Thiazole is a π-electron-excessive heterocycle. The electronegativity of the N-atom at the 3-position makes C(2) partially electropositive and therefore susceptible to nucleophilic attack. In contrast, electrophilic substitution of thiazoles preferentially takes place at the electron-rich C(5) position. More relevant to palladium chemistry, 2-halothiazoles and 2-halobenzothiazoles are prone to undergo oxidative addition to Pd(0) and the resulting σ-heteroaryl palladium complexes participate in various coupling reactions. Even 2-chlorothiazole and 2-chlorobenzothiazole are viable substrates for Pd-catalyzed reactions.

## 7.1 Synthesis of halothiazoles

Two of the most frequently used approaches for halothiazole synthesis are direct halogenation of thiazoles and the Sandmeyer reaction of aminothiazoles. The third method, an exchange between a stannylthiazole and a halogen, is not practical in the context of palladium chemistry simply because the stannylthiazole can be used directly in a Stille coupling.

### 7.1.1 Direct halogenation

Simple thiazole cannot be directly halogenated under standard conditions, but 2-methylthiazole can be brominated at the 5-position. If there are two substituents on the thiazole ring, the last vacant position then may be readily halogenated [1, 2].

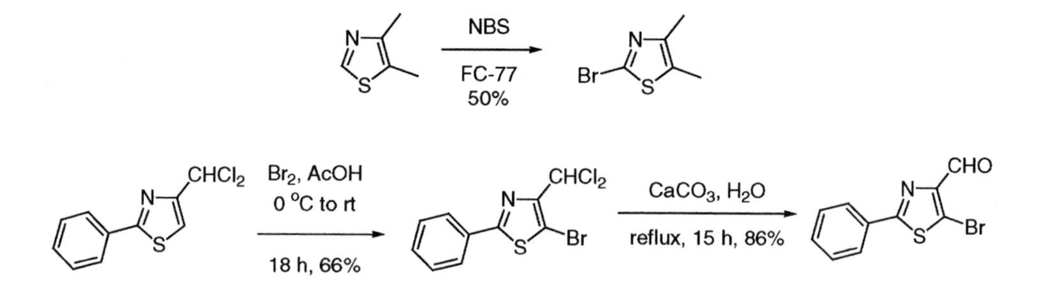


Dehydroxy-halogenation of thiazolidinedione (**1**) with phosphorus oxybromide led to 2,4-dibromothiazole (**2**) [3], whereas the same reaction conducted in DMF resulted in 2,4-dibromo-5-formylthiazole (**3**) [4].

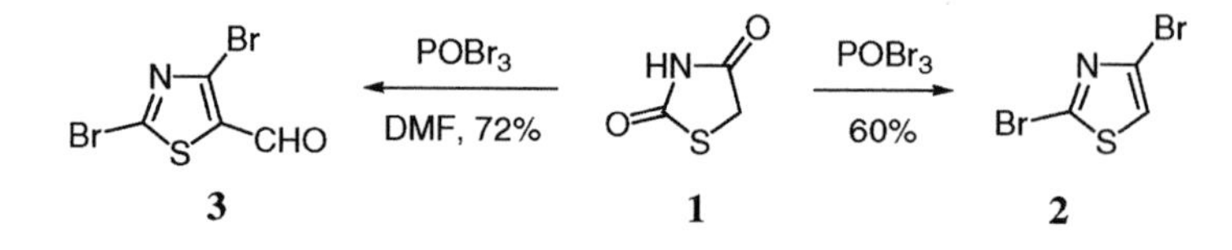


Commercially available 2,5-dibromothiazole can be alternatively generated by bromination of thiazole or 2-bromothiazole [5]. 2,4-Dibromothiazole and 2,5-dibromothiazole are among the most useful building blocks in the synthesis of thiazole-containing molecules. Regioselective bromination was achieved at C(4) when 2-amino-6-trifluoromethoxythiazole (**4**) was treated with bromine in acetic acid to afford 4-bromobenzothiazole **5** [6].

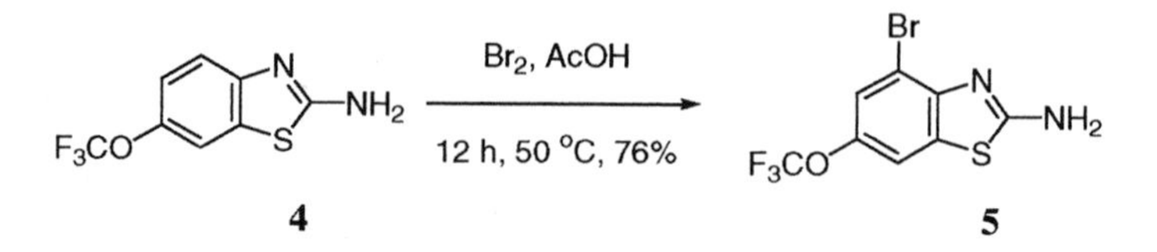


### 7.1.2 Sandmeyer reaction

The Sandmeyer reaction converts aminothiazoles, often commercially available, to halothiazoles *via* the intermediacy of diazonium salts. For instance, 2-aminothiazole was transformed into 2-iodothiazole [7, 8]. The reaction tolerates a number of functional groups including nitro and ester groups as shown below [9].

$H_2N$–(thiazole) $\xrightarrow[\text{2. KI, } H_2O\text{, 40\%}]{\text{1. } H_2SO_4\text{, } NaNO_2\text{, } H_2O}$ I–(thiazole)

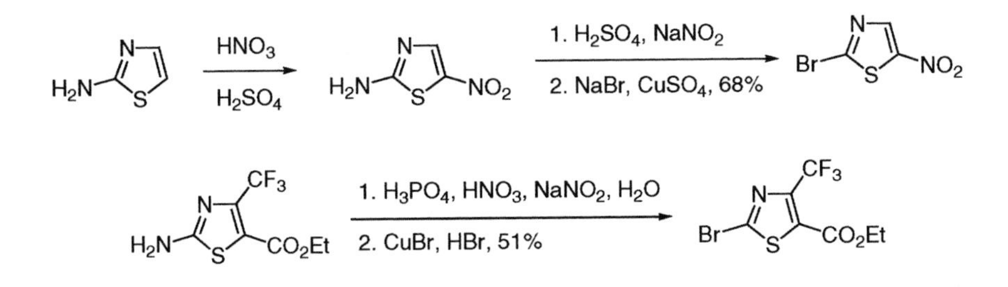


Some variants of the classic Sandmeyer conditions have been used for halothiazole synthesis. For example, sodium nitrite can be replaced with isoamyl nitrite or *tert*-butyl nitrite as illustrated by the transformation of 2-aminobenzothiazole **6** to 2-bromobenzothiazole **7** [10, 11].

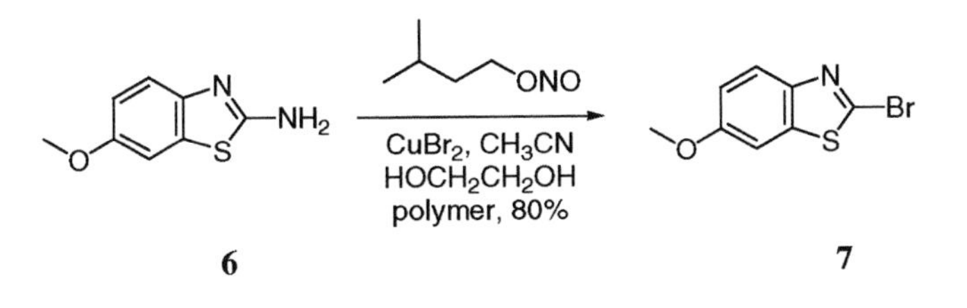


## 7.2 Coupling reactions with organometallic reagents

### 7.2.1 Negishi coupling

2-Thiazolylzinc halides have been employed as nucleophiles in Negishi couplings [12–14]. 2-Thiazolylzinc chloride was prepared by deprotonation of thiazole followed by treatment with $ZnCl_2$. Subsequent Negishi coupling of thiazolylzinc chloride with 5-iodo-2'-deoxyuridine **8** elaborated the 5-substituted pyrimidine nucleoside **9** with the assistance of one equivalent of $ZnCl_2$ [14].

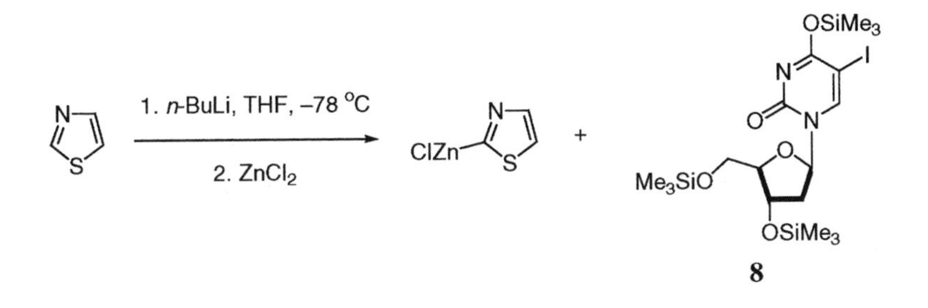

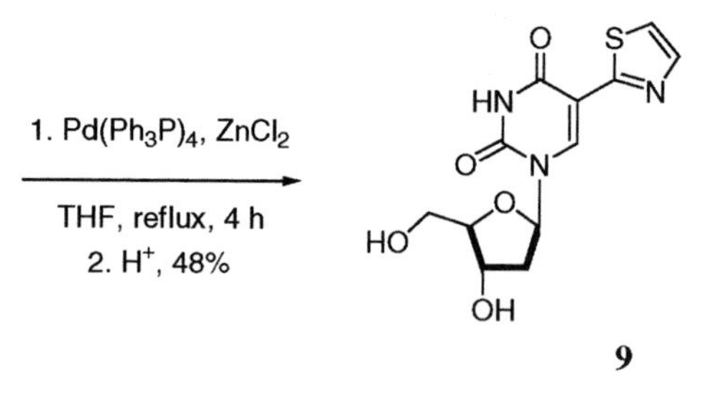


2-Thiazolylzinc bromide, formed *in situ* by quenching lithiothiazole with $ZnBr_2$, was coupled with 2-iodopyridine **10** to give thiazolylpyridine **11**. Hydrolysis of **11** then led to thiazolylpyridine acid **12**, an inhibitor of endothelin conversion enzyme-1 (ECE-1) [15].

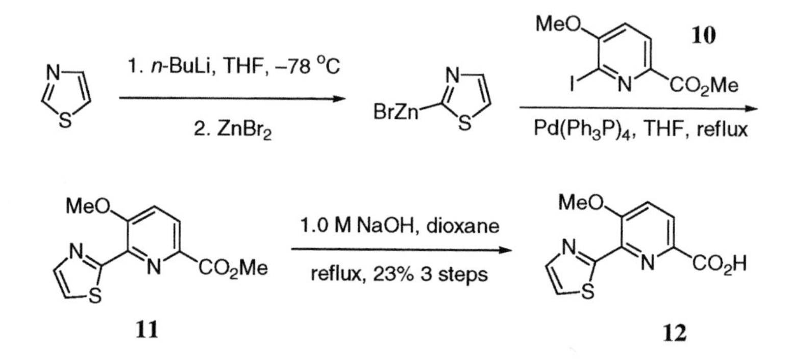


In another ECE-1-related project, the total synthesis of WS75624 B (**16**) commenced with regioselective halogen-metal exchange at C(2), and the resulting 2-lithio-4-bromothiazole was treated with thionolactone **13**. The intermediate sulfide was trapped with methyl iodide to give mixed ketal **14**. The Negishi reaction of **14** and the organozinc reagent derived from dimethoxypyridine provided adduct **15**, which was then further manipulated to accomplish the total synthesis of **16**, a non-peptide inhibitor of ECE [16a]. Recently, Bach and Heuser reported consecutive and regioselective Negishi and Suzuki couplings using 2,4-dibromothiazole [16b].

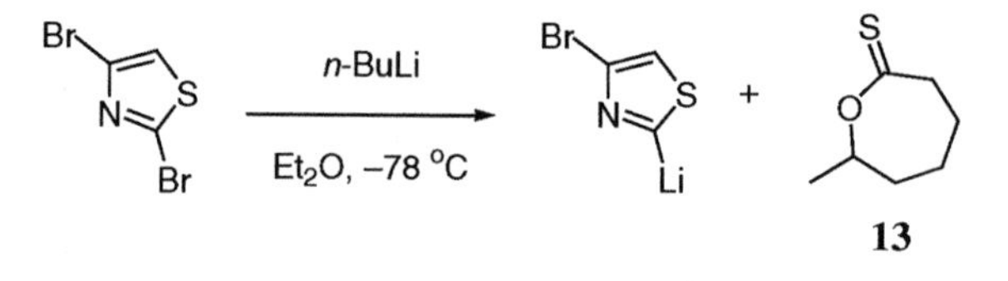

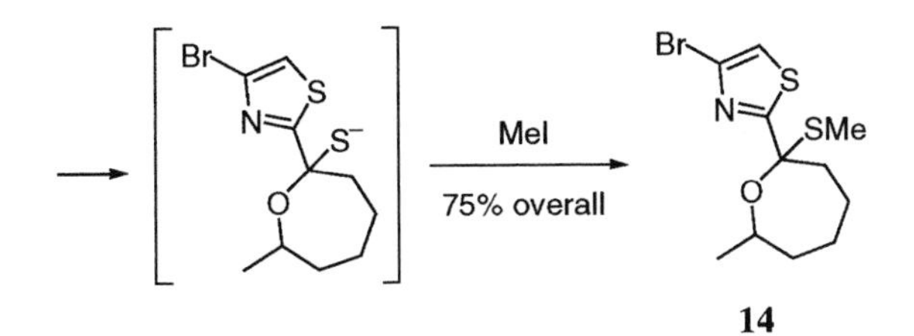


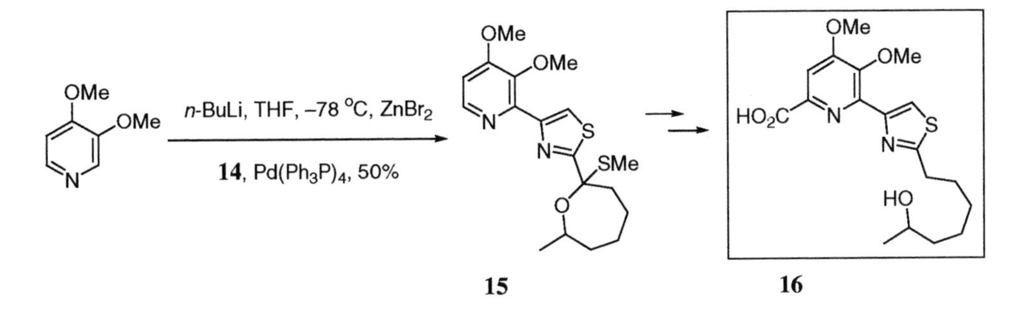


The difluorovinylzinc reagent **19** was generated by treating the corresponding vinyllithium **18**, derived from deprotonation of terminal fluoroolefin **17**, with $ZnCl_2$. Subsequent coupling of **19** with 2-iodobenzothiazole gave adduct **20** stereoselectively [17].

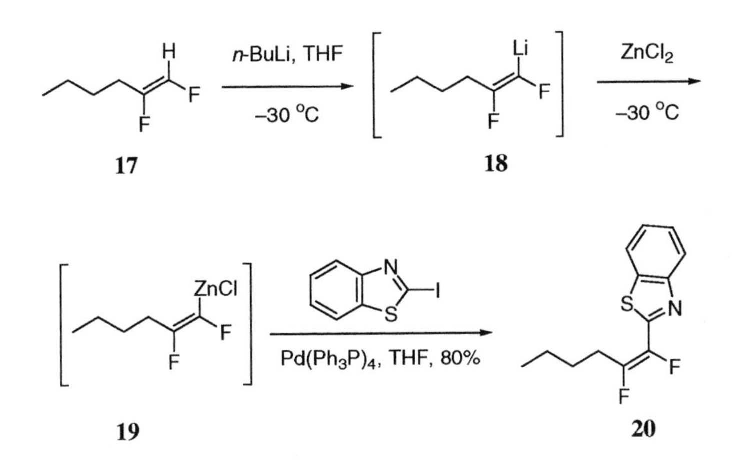


Pd(II) intermediates obtained *in situ* from the intramolecular arylation of alkynes are viable coupling partners with organozinc reagents. For instance, treatment of 4-(*o*-iodophenyl)-1-butyne (**21**) with a catalytic amount of $Pd(Ph_3P)_4$ resulted in the *cis* insertion intermediate **22** [18]. Coupling of the reactive species **22** with 2-benzothiazolylzinc chloride, derived from treating 2-benzothiazolyllithium with one equivalent of anhydrous $ZnCl_2$, gave (*Z*)-1-

indanylidene-substituted benzothiazole **23**, along with by-product **24**. Such a tandem reaction provides a quick entry to (*Z*)-1-indanylidene-substituted heteroaryls, which are otherwise not straightforward to make.

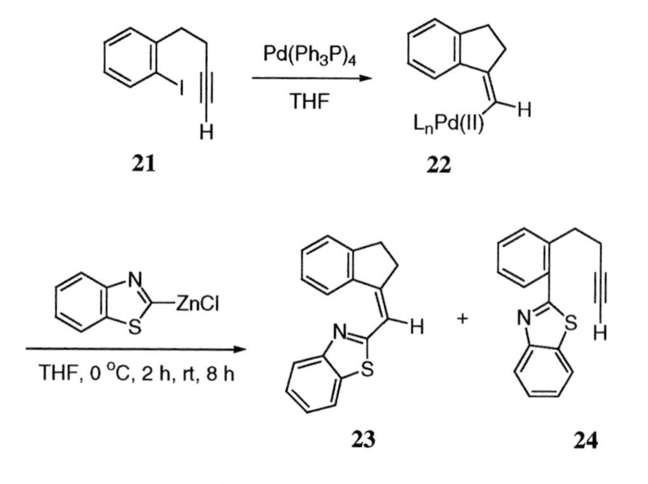


Pd-catalyzed reactions with aryl triflates have been extensively used in organic synthesis. The electron-withdrawing ability of the $CF_3SO_2$ group is essential for a rapid insertion of Pd(0) to the C—O bond of the aryl triflate. Knochel's group reported an alternative using aryl nonaflates (Nf = $SO_2(CF_2)_3CF_3$) that are readily prepared and stable to flash chromatography [19, 20]. For instance, reaction of 4-iodophenol with commercially available $FSO_2(CF_2)_3CF_3$ in the presence of $Et_3N$ in $Et_2O$ afforded aryl nonaflate **25** in excellent yield. The Negishi reaction of bifunctional **25** with 2-thiazolylzinc bromide, derived from oxidative addition of 2-bromothiazole to Zn(0), led to arylthiazole **26** in the presence of $Pd(dba)_2$ and tri-*o*-furylphosphine (TFP). The resulting arylthiazole **26** with the pendant nonaflate functional group was then subjected to another Negishi coupling reaction with 3-trifluoromethylphenylzinc bromide to afford 4-(thiazol-2-yl)-3'-(trifluoromethyl)biphenyl (**27**).

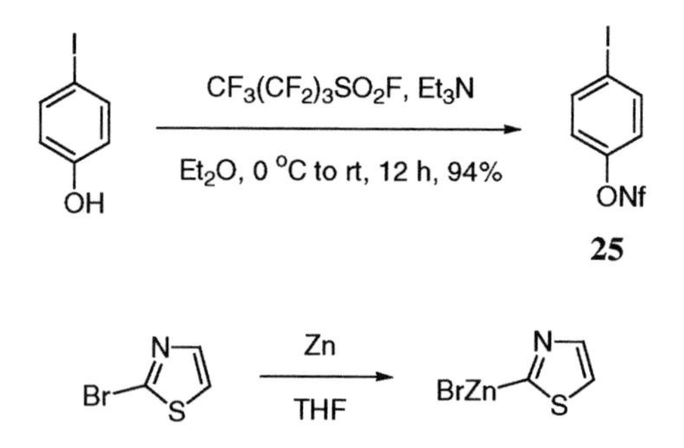

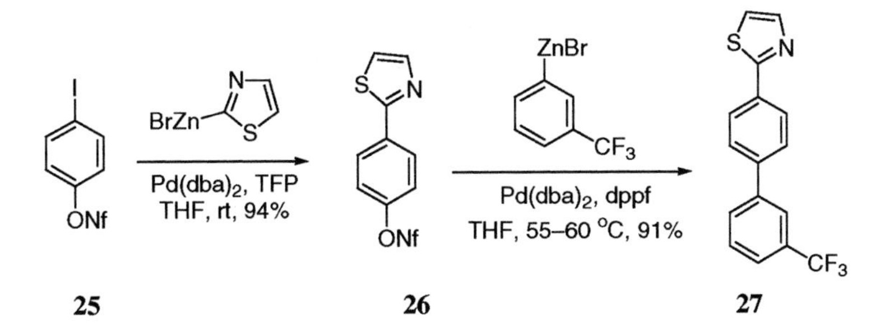


### 7.2.2 Suzuki coupling

Thiazoleboronic acids are not trivial to make. Attempts to prepare 2-thiazoleboronic acid were unsuccessful [21]. As a consequence, halothiazoles are chosen as the electrophilic coupling partners in the Suzuki reactions with other boronic acids. For instance, 2,5-di-(2-thienyl)thiazole (**28**) was installed by the union of 2,5-dibromothiazole and easily accessible 2-thiopheneboronic acid [21]. Unfortunately, the yield was poor and analogous reactions of 2,5-dibromothiazole with 3-thiopheneboronic acid and 2-selenopheneboronic acid both failed.

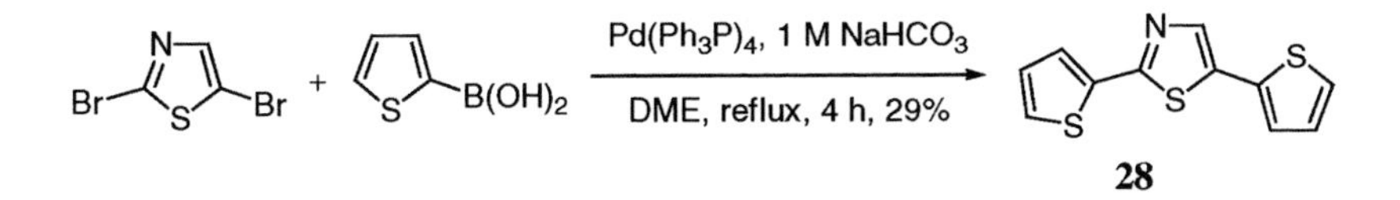


Snieckus *et al.* enlisted a combination of directed *ortho*-lithiation and Suzuki coupling to assemble some unsymmetrical heterobiaryls [22]. Carboxamidophenylboronic acid **30** was derived from sequential metalation of amide **29** and treatment with $B(OMe)_3$ followed by acidic workup. Hetero cross-coupling of **30** with 2-bromothiazole occurred smoothly to furnish phenylthiazole **31**. Similarly, a hetero cross-coupling between 2-bromothiazole and 3-formyl-4-methoxyphenylboronic acid produced a heterobiaryl as an intermediate of an orally bioavailable $NK_1$ receptor antagonist [23].

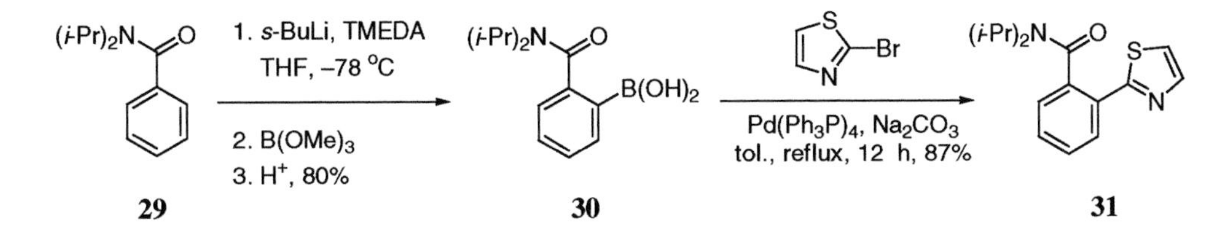

In addition to arylthiazoles, heteroarylthiazoles also have been synthesized using halothiazoles and heteroarylboronic acids. Suzuki coupling of 2-bromothiazole and 5-indolylboronic led to 5-substituted indole **32** [24]. The Suzuki coupling of 2,4-dibromothiazole with 2,4-di-*t*-butoxy-5-pyrimidineboronic acid (**33**) resulted in selective hetreoarylation at the 2-position to give pyrimidylthiazole **34** although the yield was low [25–27].

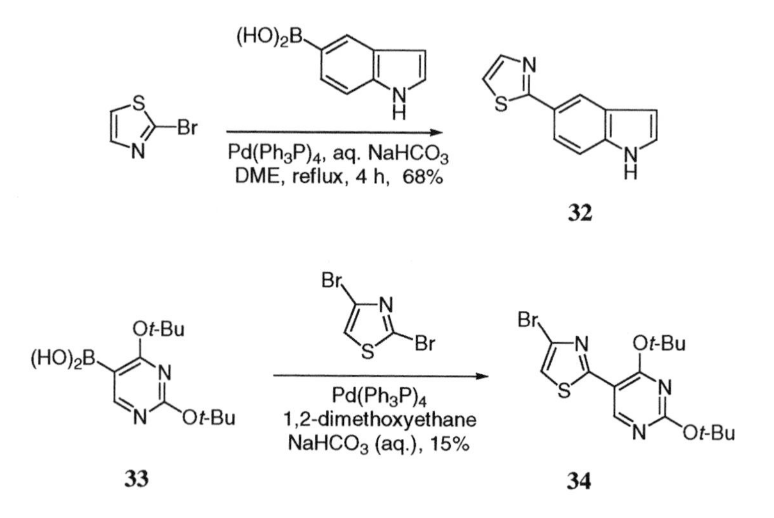


### 7.2.3 Stille coupling

Two frequently used methods for preparing stannylthiazoles involve either direct metalation or halogen-metal exchange followed by treatment with alkyltin chloride. Dondoni *et al.* described a preparation of 2-, 4- and 5-trimethylstannylthiazoles in 1986 [28]. For instance, 2-trimethylstannylthiazole was readily obtained by quenching 2-lithiothiazole, derived from direct metalation of thiazole, with trimethyltin chloride.

1. $n$-BuLi, $Et_2O$, –78 °C
2. $Me_3SnCl$, 96%
$Me_3Sn$

In order to prepare 4- and 5-trimethylstannylthiazoles selectively, C(2) was protected with an easily removable trimethylsilyl group. Regioselective halogen-metal exchange of 2,4-dibromothiazole occurred at C(2) and the resulting 2-lithio-4-bromothiazole was trapped with trimethylsilyl chloride to give 2-trimethylsilyl-4-bromothiazole (**35**). Subsequent halogen-metal exchange of **35** followed by treatment with $Me_3SnCl$ led to **36**, which was then hydrolyzed to 4-trimethylstannylthiazole. On the other hand, the preparation of 5-trimethylstannylthiazole began with selective deprotonation of 2-trimethylsilylthiazole at C(5) followed by treatment with $Me_3SnCl$, giving rise to 2-trimethylsilyl-5-trimethylstannylthiazole (**37**). Deprotection of **37** under acidic conditions then gave 5-trimethylstannylthiazole [28].

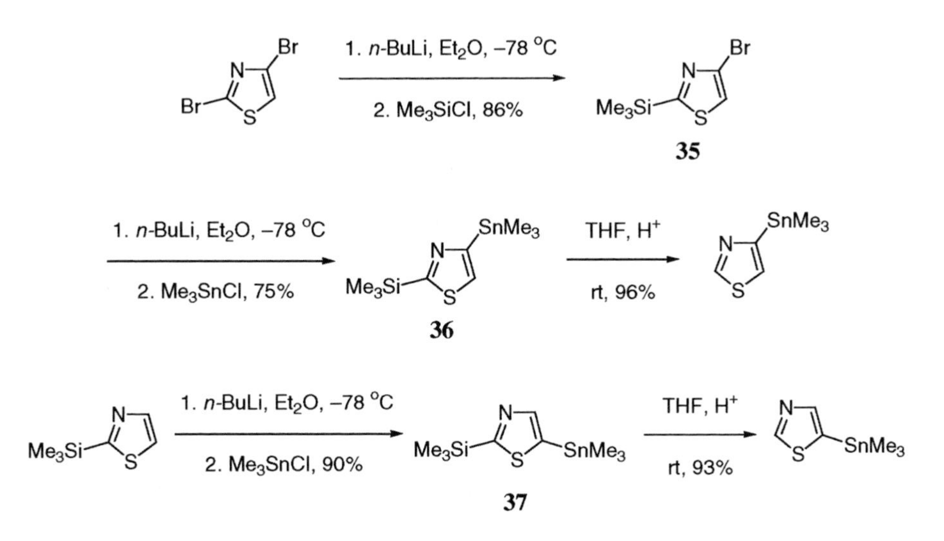


2-(Tributylstannyl)benzothiazole was prepared by deprotonation of benzothiazole with *n*-BuLi followed by the addition of tributyltin chloride [29, 30]. Subsequent treatment of the resulting 2-(tributylstannyl)benzothiazole with bromobenzene under the influence of $PdCl_2(Ph_3P)_2$ led to 2-phenylbenzothiazole (**38**).

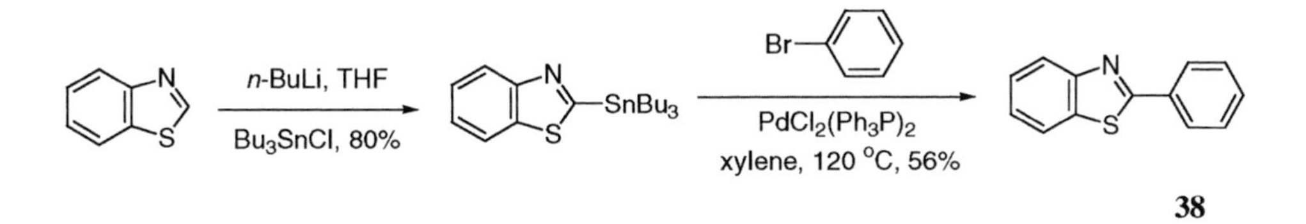


The second method to prepare stannylthiazoles involves a halogen-metal exchange followed by quenching with trimethyl- or tributyltin chloride. Gronowitz *et al.* carried out the halogen-metal exchange of 2-bromothiazole and treated the resulting 2-lithiothiazole with a solution of $Bu_3SnCl$ in $Et_2O$ to furnish 2-tributylstannylthiazole [31]. Subsequent Stille coupling of 2-tributylstannylthiazole and 5-bromopyrimidine **39** afforded 5-(2'-thiazolyl)-2,4-bis(trimethylsilyl)pyrimidine (**40**), which was hydrolyzed with acid to the corresponding uracil **41**.

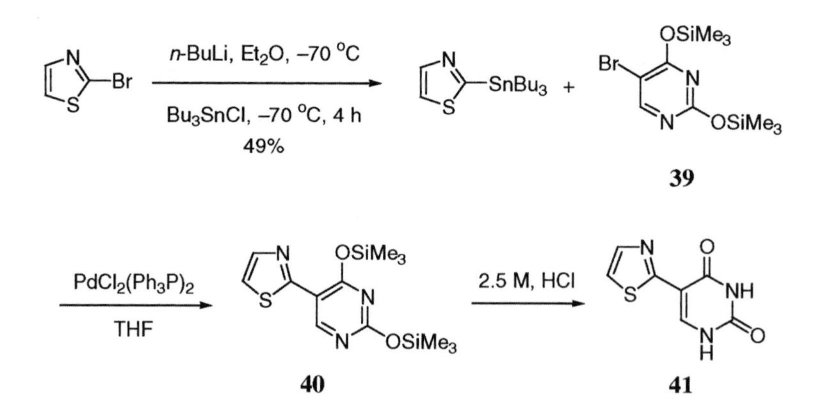


It has been demonstrated in the literature that the halogen-metal exchange of 2,4-dibromothiazole occurs predominantly at the C(2) position. 2-Stannyl-4-bromothiazole can be prepared using this strategy [32].

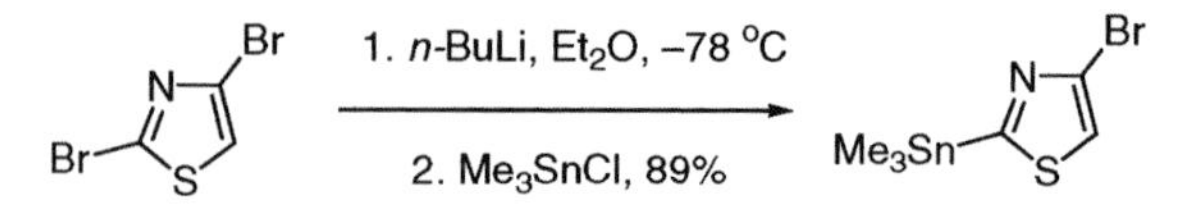


The preparation of stannylthiazoles *via* ditin chemistry has not been widely utilized. In one case, the synthesis of 4-tributylstannylthiazole **43** started with selective halogen-metal exchange at C(2) by treating 2,4-dibromothiazole with *n*-BuLi [33]. Trapping the resulting 2-lithio-4-bromothiazole with propanal and subsequent Jones oxidation secured 4-bromothiazole **42**. The Pd-catalyzed reaction of **42** with hexamethyldistannane in the presence of $PdCl_2(Ph_3P)_2$ provided 4-tributylstannylthiazole **43**.

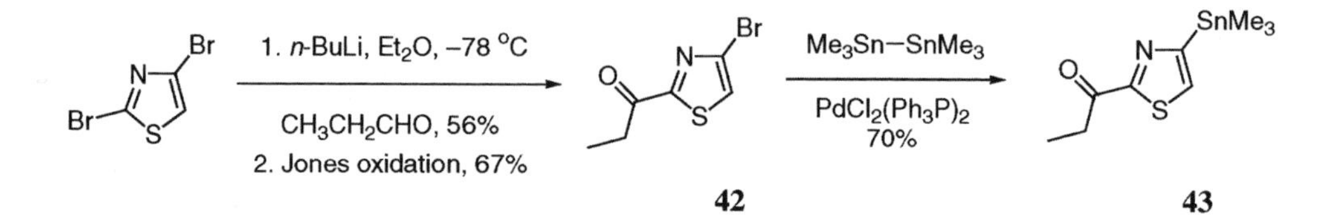


Unfortunately, application of the ditin chemistry can not be generalized. For instance, simple bromothiazole was recalcitrant towards such a method. Treatment of **44** with hexamethyldistannane in the presence of $PdCl_2(Ph_3P)_2$ gave dimerized product **45** and debrominated product **46** as the two major products [34]. Nevertheless, the both the isolation of the dimer **45** and NMR evidence indicated that the desired 2-stannylthiazole from **44** was generated during the course of the reaction. Thus, trapping the stannane *in situ* was accomplished when pyridyl triflate **47** was refluxed in the presence of $PdCl_2(Ph_3P)_2$ with slow addition of a dioxane solution of **44**, furnishing dimethyl sulfomycinamate (**48**), the methanolysis product of the antibiotic sulfomycin I [34].

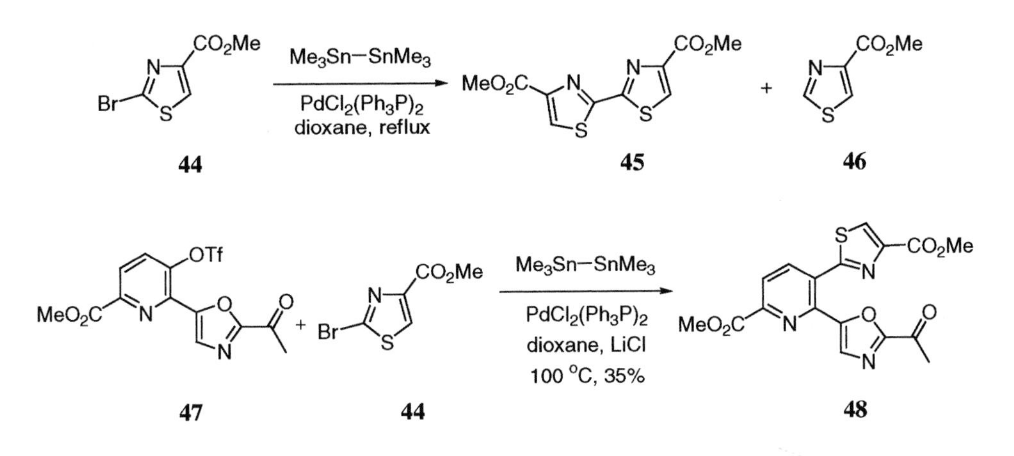


A novel homolytic substitution yielded 2-(tributylstannyl)benzothiazole [35]. Thus, 2-(alkylsulfonyl)benzothiazole **49** was allowed to react with 2 equivalents of tributyltin hydride in the presence of catalytic azobisisobutyronitrile (AIBN) in refluxing benzene, affording 2-(tributylstannyl)benzothiazole along with tributylstannylsulfinate **50**.

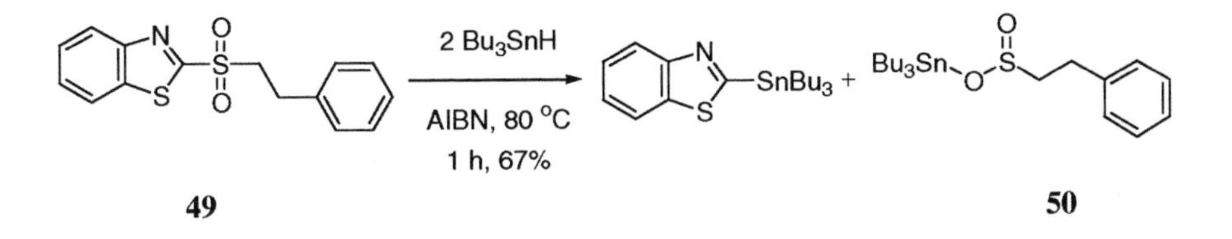


The Stille coupling of stannylthiazole with an assortment of electrophiles has been achieved. The electrophilic partners span across the whole spectrum of halides including allyl chlorides. Chloromethylcephem **51** was coupled with 2-tri-*n*-butylstannylthiazole to give **52**, which was then manipulated into a C(3) thiazole analog of cephalosporin [36].

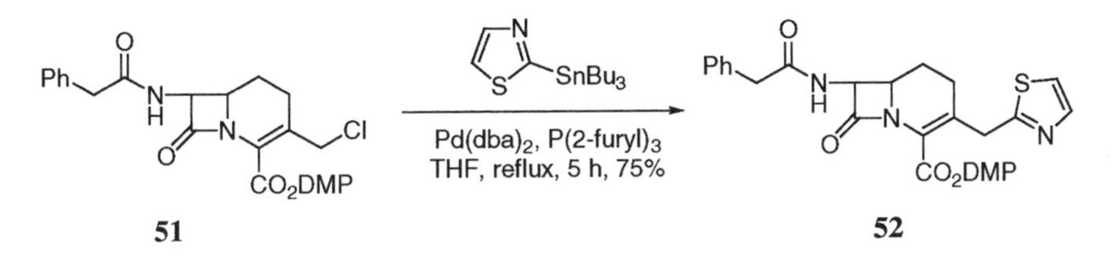

With respect to the coupling reactions of stannylthiazoles with aryl halides, the union of 4-chlorobromobenzene and 2-tributylstannylthiazole constructed arylthiazole **53** [37]. The Stille reaction of 3-bromobenzylphosphonate (**54**) and 2-tributylstannylthiazole led to heterobiaryl phosphonate **55**, which may be utilized as a substrate in a Wadsworth–Horner–Emmons reaction or a bioisosteric analog of a carboxylic acid [38]. The phosphonate did not interfere with the reaction. In addition, the coupling of 5-bromo-2,2-dimethoxy-1,3-indandione (**56**) and 2-tributylstannylbenzothiazole resulted in adduct **57**, which was then hydrolyzed to 5-(2'-benzothiazolyl)ninhydrin [39].

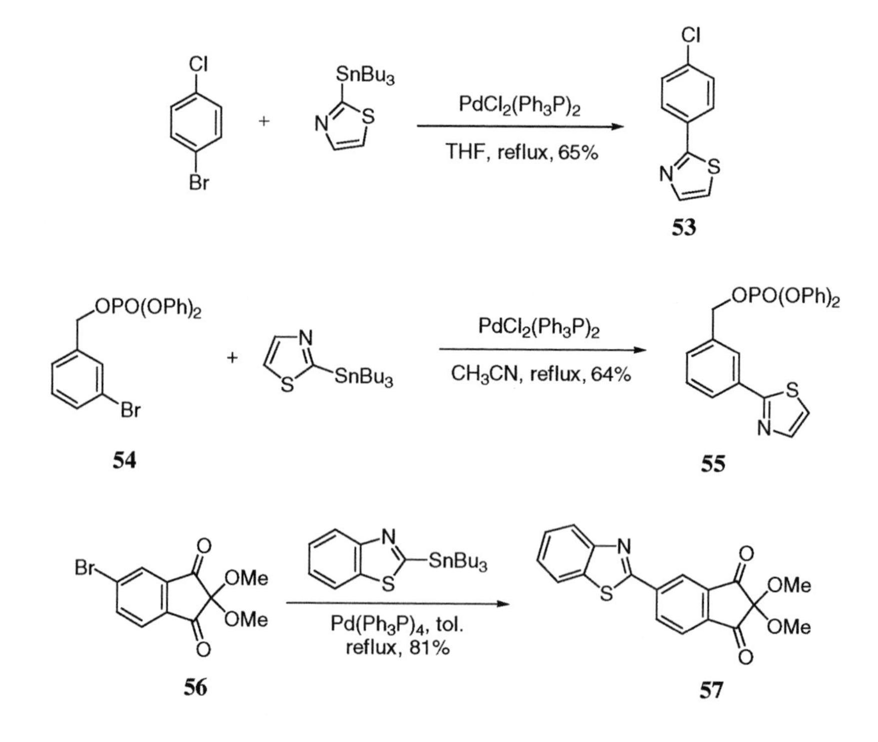


In addition, Unheim *et al.* coupled 2-bromothiazole with 2-methylthio-5-stannylpyrimidine (**58**) to assemble the pyrimidinethiazole **59** [43]. Stannane **58** was prepared by Pd-catalyzed coupling between 2-methylthio-5-bromopyrimidine and hexamethyldistannane in the presence of fluoride ion at ambient temperature. In another case, 2-tributylstannylthiazole was coupled with 2'-deoxyuridine iodide **60** (DMT = dimethoxytrityl) to furnish 5-thiazolyl-2'-deoxyuridine **61** [44].

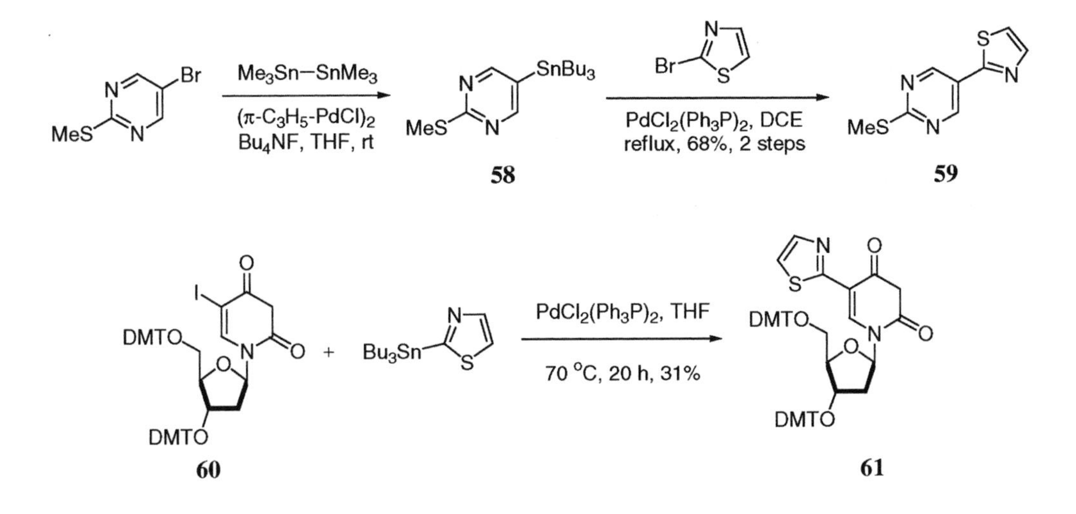


Stannylthiazoles have also been coupled with an array of heterocyclic halides. Reaction of 2.4 equivalents of 2-trimethylstannylthiazole with 2,5-dibromothiophene gave 2,5-di-(2'-thiazolyl)-thiophene (**62**) [40]. Similarly, Dondoni's group coupled 2,5-dibromothiazole with 2 equivalents of 2-trimethylstannylthiazole to assemble trithiazole **63** [41].

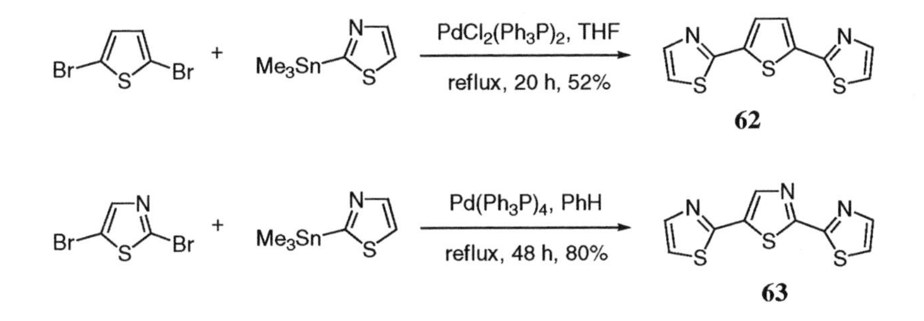


6-Substituted thiazolylindole **66** was formed *via* the Stille coupling of 6-iodoindole **64** and 5-tributylstannylthiazole **65** [42]. Stannane **65** was not stable at temperatures higher than 60 °C and decomposition occurred presumably by amidic hydrogen transfer on the 5-position. Simply blocking NH with a methyl group greatly improved the yield.

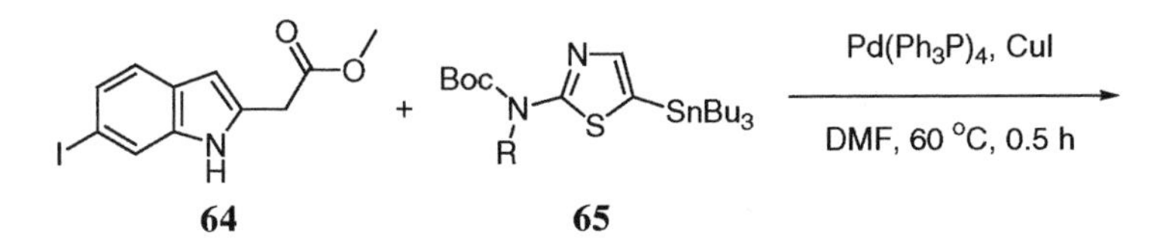

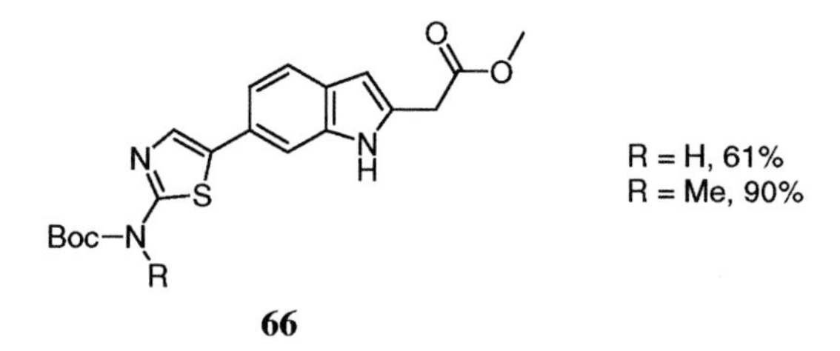


Synthesis of a thiazolylpyridine *via* the Stille coupling has been reported [45, 46]. Also described was a Stille coupling of a bromoquinolizinium salt **67** and 2-tributylstannylthiazole [47]. The reaction was conducted at room temperature to give a substituted quinolizinium salt **68**.

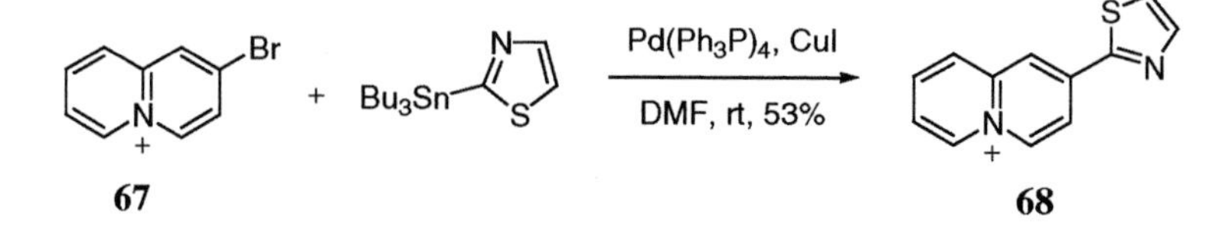


The Liebeskind group cross-coupled 4-chloro-2-cyclobutenone **69** with 2-tribuylstannyl-benzothiazole to synthesize α-pyridone-based azaheteroaromatics [48]. The adduct **70** underwent a thermal rearrangement to afford a transient vinylketene **71**, which then intramolecularly cyclized onto the C—N double bond of benzothiazole, giving rise to thiazolo[3,2-*a*]pyridin-5-one **72**. In another case, 2-acetyl-4-trimethylstannylthizaole (**73**) was coupled with an acid chloride **74** to form the desired ketone **75** [49].

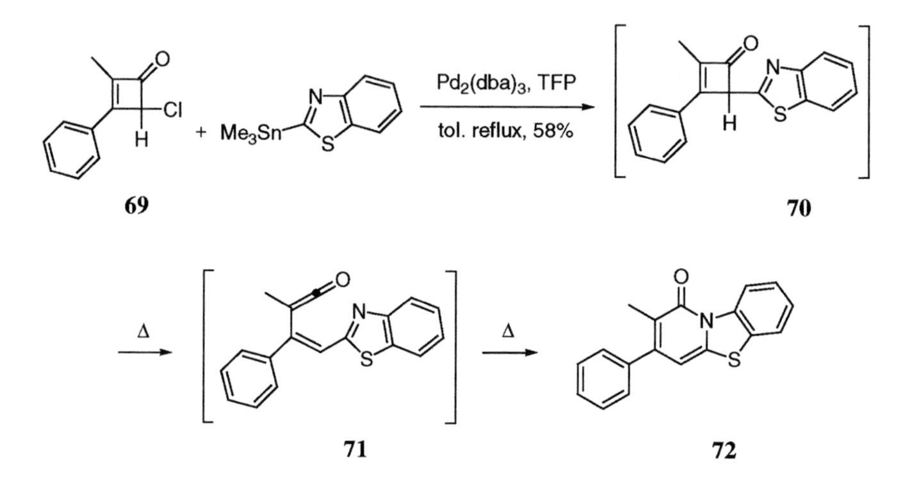

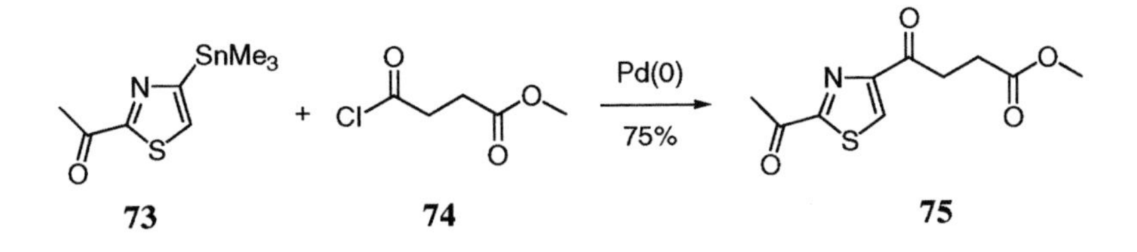


Nicolaou *et al.* took advantage of the Stille coupling to install the thiazole motif of epothilones and to conduct analog synthesis using isosteres of thiazole [50]. Thus, the union of vinyl iodide **76** and stannane **77** led efficiently to adduct **78** as the precursor to epothilone E. All the functional groups including hydroxyl groups, ketone, ester and olefins were preserved. This is a good example of functional group tolerance in the Stille coupling.

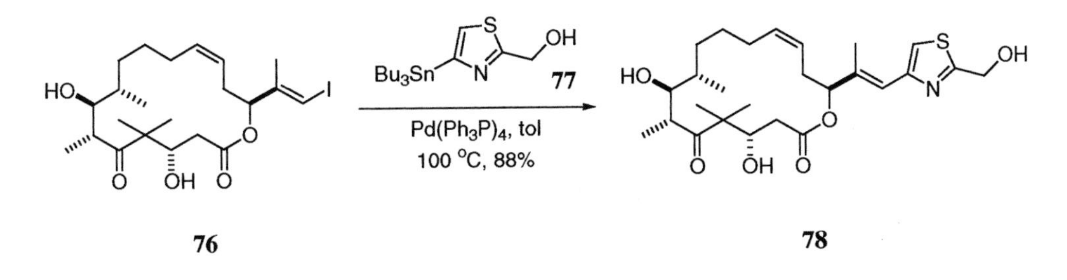


Kelly's total synthesis of micrococcinic acid (**88**) is probably the best example to showcase the utility of the Stille reaction in the synthesis of thiazole-containing molecules [33]. The first Stille coupling was carried out between bromothiazole **79** and trimethylstannylpyridine **80** to form thiazolylpyridine **81**. Similarly, the second coupling between trimethylstannylthiazole **83** and bromothiazole **79** afforded dithiazole **84**. The ethoxy group in **81** and the trimethysilyl unit in **84** were modified to give bromides **82** and **85**. Subsequently, the Stille–Kelly reaction featuring a 1:1 mixture of **82** and **85** with $(Me_3Sn)_2$ and $PdCl_2(Ph_3P)_2$ furnished the desired cross-coupled **86**. Conventional functional group transformations led to triflate **87**, which was subjected to the fourth Stille coupling with stannane **43** to secure the pentacycle which was converted to micrococcinic acid (**88**) *via* acidic cleavage of the two amides. Kelly's total synthesis of micrococcinic acid (**88**) outlined here is a strong testmony to the utility of the Stille coupling reaction even for very complex molecules.

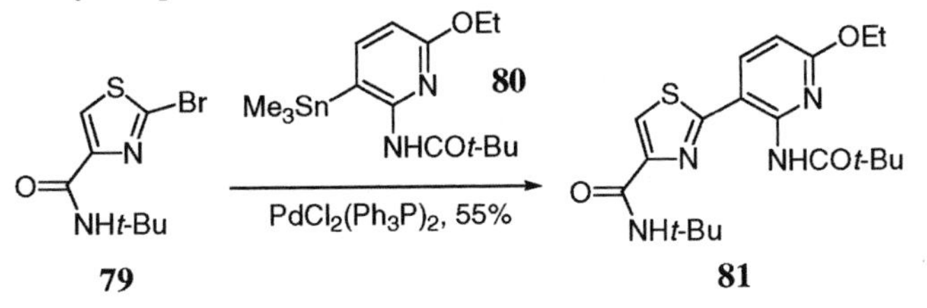

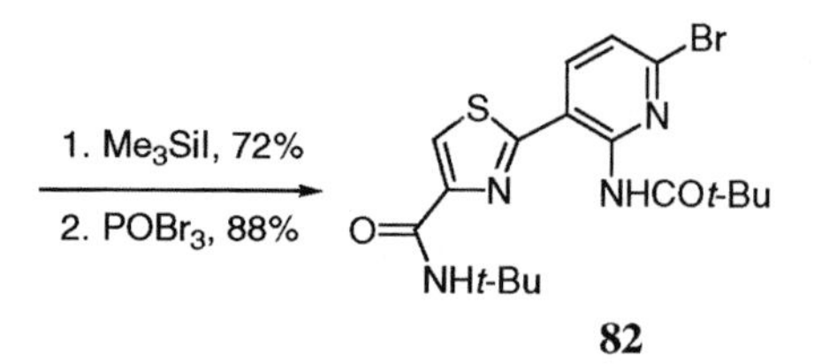
1. Me3SiI, 72%
2. POBr3, 88%
Br
S
N
N
O
NHt-Bu
NHCOt-Bu
82

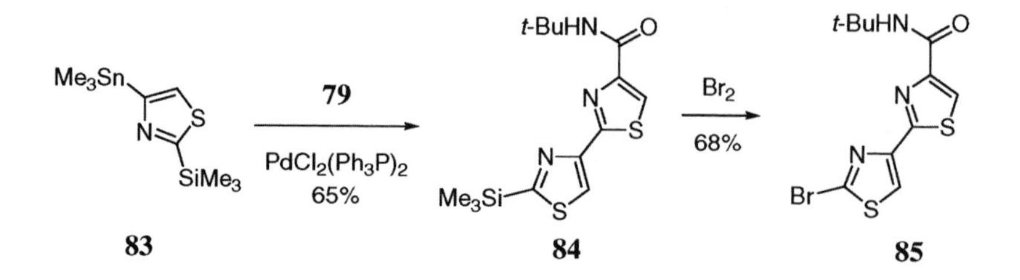
Me3Sn
S
N
SiMe3
83
79
PdCl2(Ph3P)2
65%
t-BuHN
O
N
S
N
Me3Si
S
84
Br2
68%
t-BuHN
O
N
S
N
Br
S
85

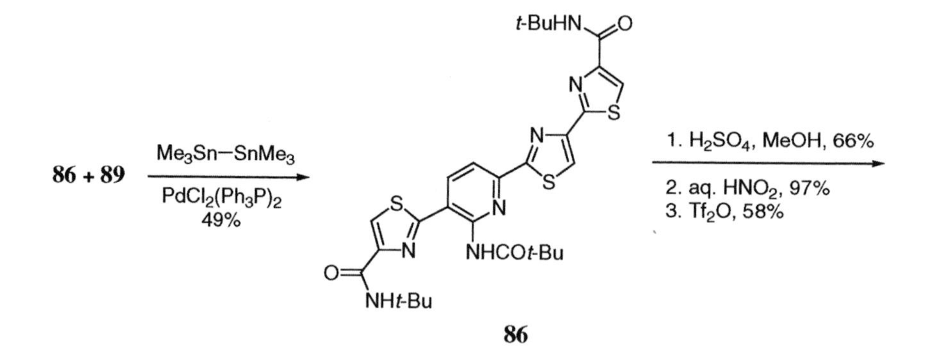
86 + 89
Me3Sn–SnMe3
PdCl2(Ph3P)2
49%
t-BuHN
O
N
S
N
S
S
N
N
O
NHt-Bu
NHCOt-Bu
86
1. H2SO4, MeOH, 66%
2. aq. HNO2, 97%
3. Tf2O, 58%

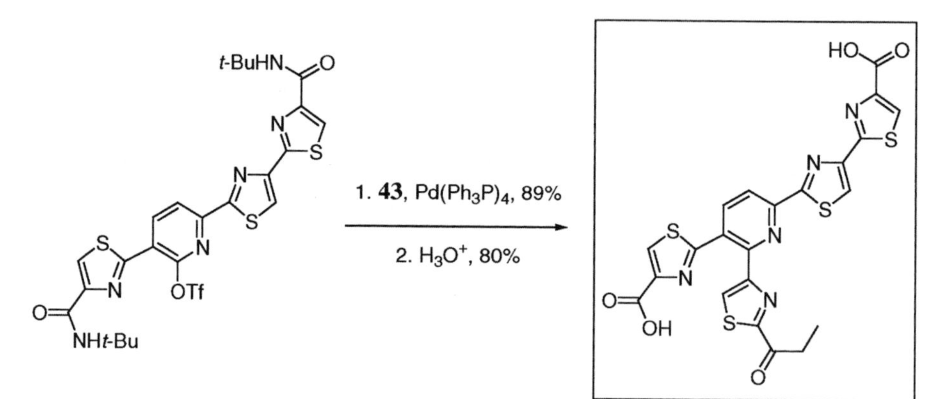
t-BuHN
O
N
S
N
S
S
N
N
O
NHt-Bu
OTf
87
1. 43, Pd(Ph3P)4, 89%
2. H3O+, 80%
HO
O
N
S
N
S
S
N
N
O
OH
N
S
O
88

## 7.3 Sonogashira reaction

In 1987, Yamanaka's group described a Pd-catalyzed reaction of halothiazoles with terminal acetylenes [51]. While the yield for the Sonogashira reaction of 2-bromo-4-phenylthiazole (**89**) with phenylacetylene to afford **90** was moderate (36% after desilylation), the coupling of 4-bromothiazole and 5-bromo-4-methylthiazole with phenylacetylene gave the desired internal acetylenes **91** and **92** in 71% and 65% yield, respectively.

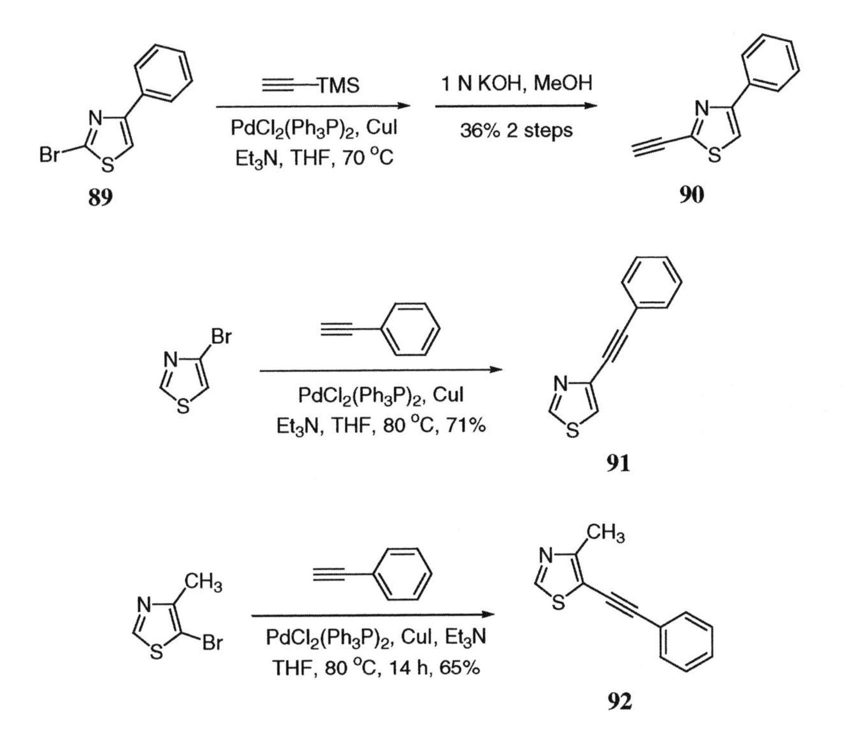


2-Iodobenzothiazole was coupled with trimethylsilylacetylene to give adduct **93** which was readily desilylated to furnish 2-ethynyl-1,3-benzothiazole (**94**) [52].

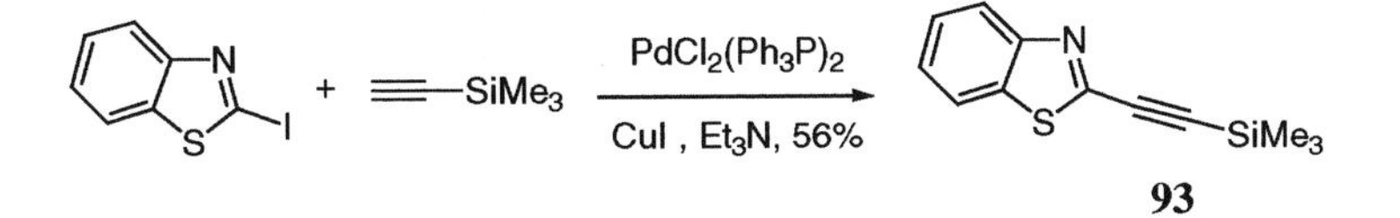

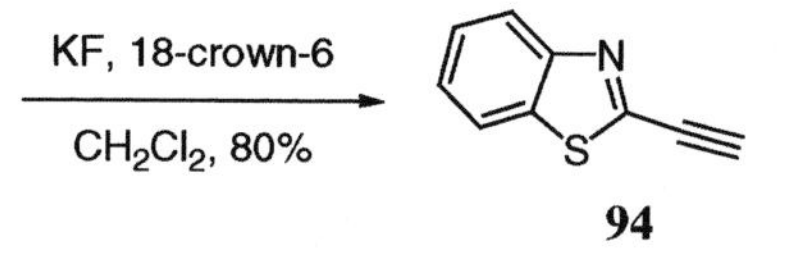


Dimethyl propargyl alcohol **95** serves as a mask for the corresponding terminal acetylene. Therefore, basic cleavage of **95** unveiled the terminal acetylene, which was coupled *in situ* with 2-bromobenzothiazole in the presence of a phase-transfer catalyst to afford the unsymmetrical diarylbutadiyne **96** [53].

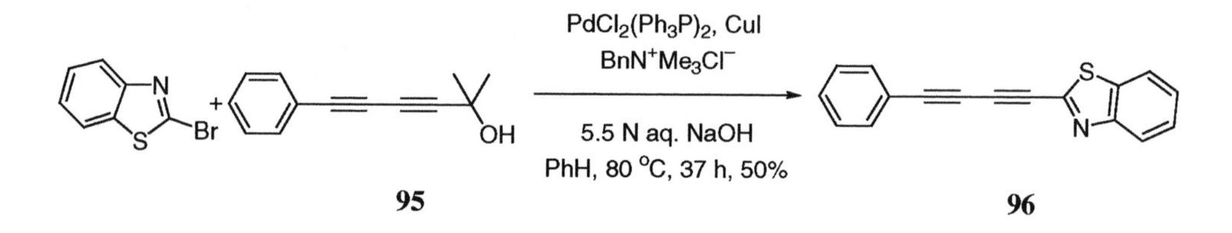


The Sonogashira reaction of 2-substituted-5-acetyl-4-thiazolyl triflate **97** and phenylacetylene led to 3-alkynylthiazole **98**, which subsequently underwent a *6-endo-dig* annulation in the presence of ammonia to produce pyrido[3,4-*c*]thiazole **99** [54].

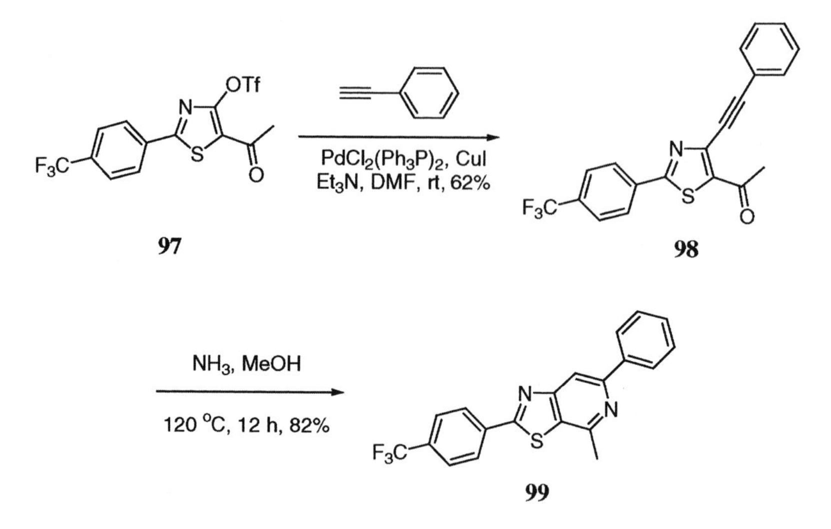

## 7.4 Heck and heteroaryl Heck reactions

Heck *et al.* reported that Pd-catalyzed reaction of 2-bromothiazole with methyl acrylate using tri-(*o*-tolyl)phosphine as the ligand failed to give significant amounts of the adduct [55]. During the investigation on 2-halo-1,3-azoles conducted by Yamanaka's group, they discovered that both a Sonogashira reaction with acetylenes and a Heck reaction with terminal olefins gave a large amount of resinous substance, probably due to ring-cleavage caused by palladation at the 2-position of the substrates [56]. Similarly, the Heck reaction of 4-bromothiazole with terminal olefins gave the adducts in only 8–19% yields. The results for the Heck reactions of 5-bromothiazoles were idiosyncratic as well. Very different results were observed for different substrates. As illustrated below, 5-bromothiazole **100** was subjected to the Heck reaction conditions with ethyl acrylate to afford **101** in 61% yield, whereas the same reaction using styrene led to almost exclusively the reduction product **102** along with only 1% of the Heck adduct.

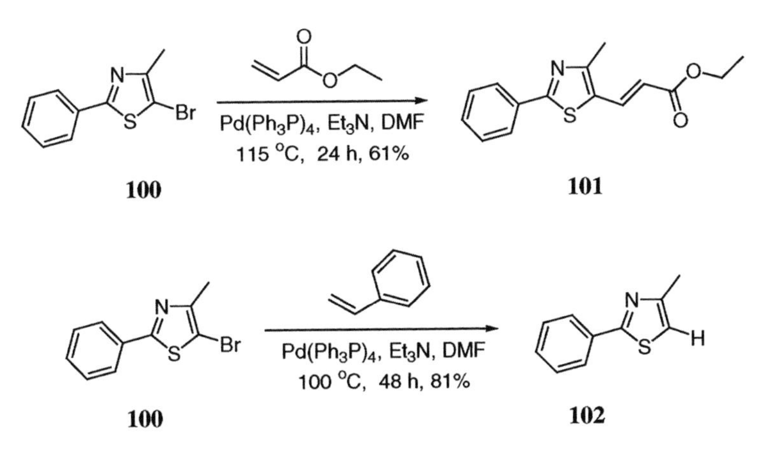


In contrast, thiazoles and benzothiazoles are suitable recipients for the "heteroaryl Heck" reaction. Treatment of 2-chloro-3,6-diisobutylpyrazine (**103**) with thiazole led to regioselective addition at C(5), giving rise to adduct **104** [57]. A similar reaction between 2-chloro-3,6-diethylpyrazine (**105**) and benzo[*b*]thiazole took place at C(2) exclusively to afford pyrazinylbenzothiazole **106** [57].

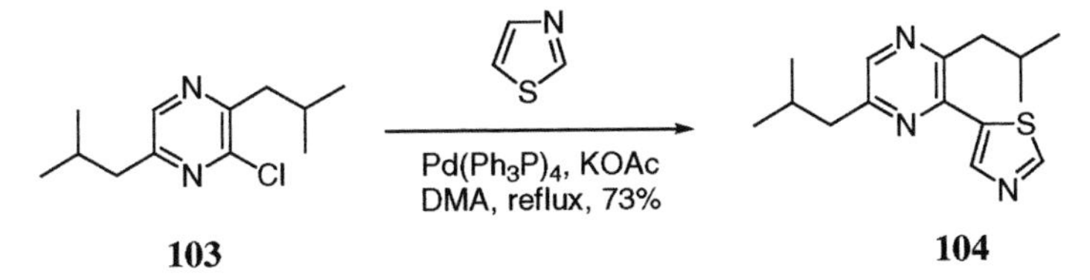

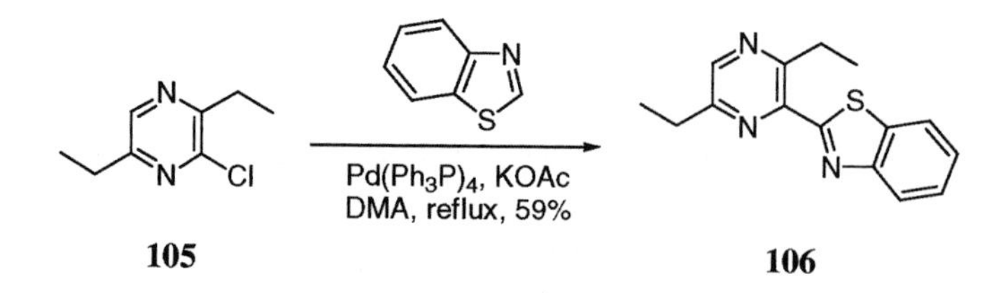


The same result held true for the heteroaryl Heck reaction of a 2-pyridyl triflate and thiazole in which the addition also occurred at the electron-rich C(5) position of thiazole [58]. An analogous result was observed for the reaction of iodobenzene and thiazole to furnish 5-phenylthiazole (**107**) [59]. The mechanism for the heteroaryl Heck reaction may be exemplified by the formation of **107**. Oxidative addition of iodobenzene to Pd(0) gives intermediate **108**, which subsequently inserts onto thiazole regioselectively at the electron-rich C(5) position to form **109**. The regioselectivity of the Heck reaction is analogous to that seen with electrophilic substitution on thiazole, which is known to be C(5) > C(4) > C(2) [60]. The base present deprotonates the insertion adduct **109**, giving rise to aryl(thiazolyl)palladium(II) intermediate **110**, which then undergoes a reductive elimination to afford 5-phenylthiazole (**107**) and regenerates Pd(0) for next catalytic cycle.

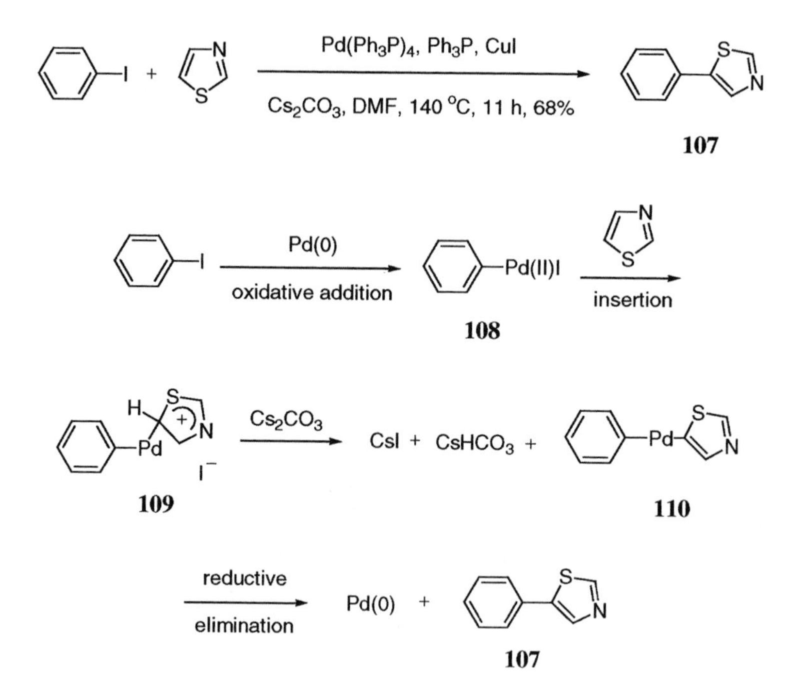

## 7.5 Carbonylation

The Pd-catalyzed alkoxycarbonylation of 5-bromothiazole using $PdCl_2(Ph_3P)_2$ as the catalyst in the presence of CO, EtOH and $Et_3N$ led to smooth formation of 5-ethoxycarbonylthiazole (**111**) [61]. $Et_3N$ served as an HBr scavenger, facilitating the transformation. This is one of the first examples of the Pd-catalyzed alkoxycarbonylation of a haloheterocycle, providing a variety of heterocyclic esters.

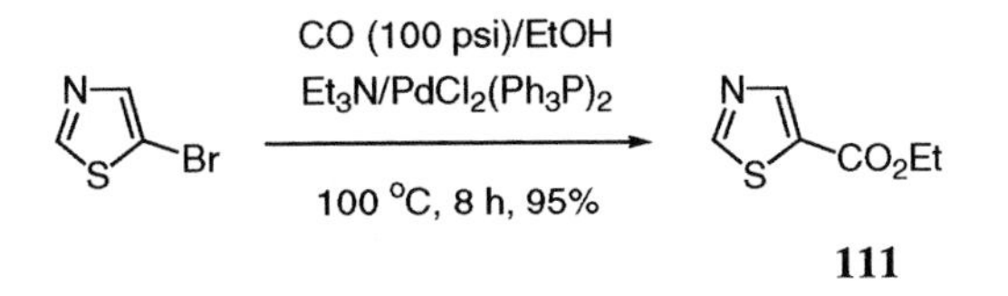


## 7.6 C—N bond formation

An extension of Buchwald's Pd-catalyzed amination led to amination of 2-chlorobenzothiazole with piperidine **112** to form **113** [62]. The marked reactivity enhancement of the chloride is attributed to the polarization at C(2). The methodology is also suitable for 2-chlorobenzoxazole and 2-chlorobenzoimidazole.

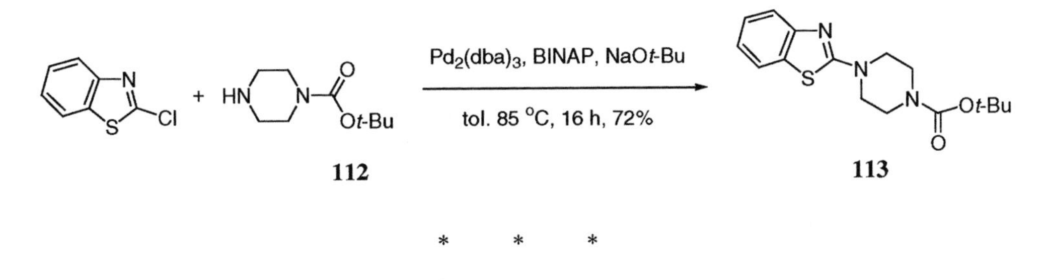


* * *

To summarize, among the Pd-catalyzed cross-coupling reactions involving thiazoles, the Stille coupling has once again proven to be the most robust among all organometallics, displaying better tolerance of a wide variety of functional groups. In an academic research laboratory or a drug discovery setting, it is the method of choice for preparing substituted thiazoles. However, an alternative route, such as Negishi coupling should be pursued first in a large-scale preparation simply because of stannane toxicity.

For the Heck reactions involving halothiazoles, 2-bromo- and 4-bromothiazoles tend to give resinous products, whereas some 5-bromothiazoles may form the desired Heck adduct with appropriate olefins. With respect to the heteroaryl Heck reaction using a thiazole as a coupling partner, the addition occurs regioselectively at the electron-rich C(5) position, whereas it occurs at C(2) for benzothiazole.

## 7.7 References

1. Al Hariri, M.; Galley, O.; Pautel, F.; Fillion, H. *Eur. J. Org. Chem.* **1998**, *4*, 593–4.
2. Ceulemans, E.; Dyall, L. K.; Dehaen, W. *Tetrahedron* **1999**, *55,* 1977–88.
3. Reynaud, P.; Robba, M.; Moreau, R. C. *Bull. Soc. Chim. Fr.* **1962**, 1735–8.
4. Kerdesky, F. A. J.; Seif, L. S. *Synth. Commun.* **1995**, *25*, 2639–15.
5. Klein, P. *Helv. Chim. Acta* **1954**, *37*, 2057–67.
6. Mignani, S.; Audiu, F.; Le Belvec, J.; Nemecek, C.; Barreau, M.; Jimonet, P.; Gueremy, C. *Synth. Commun.* **1992**, *22*, 2769–80.
7. Neenan, T.; Whitesides, G. M. *J. Org. Chem.* **1988**, *53*, 2489–96.
8. Gellis, A.; Vanelle, P.; Maldonado, J.; Crozet, M. P. *Tetrahedron Lett.* **1997**, *38*, 2085–6.
9. Lee, L. F.; Schleppnik, F. M.; Howe, R. K. *J. Heterocycl. Chem.* **1985**, *22*, 1621–30.
10. Suzuki, N.; Nomoto, T.; Toya, Y.; Yoda, B.; Saeki, A. *Chem. Express* **1992**, *7*, 717–20.
11. Suzuki, N.; Nomoto, T.; Toya, Y.; Kanamori, N.; Yoda, B.; Saeki, A. *Biosci. Biotechnol. Biochem.* **1993**, *57*, 1561–2.
12. Vincent, P.; Beaucourt, J.; Pichat, L. *Tetrahedron Lett.* **1984**, *25*, 201–2.
13. Bell, A.; Roberts, D.; Ruddoch, K. *Tetrahedron Lett.* **1988**, *29*, 5013–6.
14. Wigerinck, P.; Pannecouque, C.; Snoeck, R.; Claes, P.; De Clercq, E.; Herdewijn, P. *J. Med. Chem.* **1991**, *34*, 2383–9.
15. Massa, M. A.; Patt, W. C.; Ahn, K.; Sisneros, A. M.; Herman, S. B.; Doherty, A. *Bioorg. Med. Chem. Lett.* **1998**, *8*, 2117–22.
16. (a) Huang, S.-T.; Gordon, D. M. *Tetrahedron Lett.* **1998**, *39*, 9335–8. (b) Bach, T.; Heuser, S. *Tetrahedron Lett.* **2000**, *41*, 1707–10.
17. Gillet, J.-P.; Sauvétre, R.; Normant, J.-F. *Tetrahedron Lett.* **1985**, *26*, 3999–4002.
18. Luo, F.-T.; Wang, R.-T. *Heterocycles* **1990**, *31*, 1543–8.
19. Rottländer, M.; Knochel, P. *J. Org. Chem.* **1998**, *63*, 203–8.
20. Prasad, A. S. B.; Stevenson, T. M.; Citineni, J. R.; Nyzam, V.; Knochel, P. *Tetrahedron* **1997**, *53*, 7237–54.
21. Gronowitz, S.; Peters, D. *Heterocycles* **1990**, *30*, 645–8.
22. Sharp, M. J.; Snieckus, V. *Tetrahedron Lett.* **1985**, *26*, 5997–6000.
23. Ward, P.; Armour, D. R.; Bays, D. E.; Evans, B.; Giblin, G. M. P. *et al. J. Med. Chem.* **1995**, *38*, 4985–92.
24. Yang, Y.; Martin, A. R. *Heterocycles* **1992**, *34*, 1395–8.

25. Peters, D.; Hörnfeldt, A.-B.; Gronowitz, S. *J. Heterocycl. Chem.* **1990**, *27*, 2165–73.
26. Peters, D.; Hörnfeldt, A.-B.; Gronowitz, S. *J. Heterocycl. Chem.* **1991**, *28*, 529–31.
27. Wellmar, U.; Hörnfeldt, A.-B.; Gronowitz, S. *J. Heterocycl. Chem.* **1995**, *32*, 1159–63.
28. Dondoni, A.; Mastellari, A. R.; Medici, A.; Negrini, E. *Synthesis* **1986**, 757–60.
29. Kosugi, M.; Koshiba, M.; Atoh, A.; Sano, H.; Migita, T. *Bull. Chem. Soc. Jpn.* **1986**, *59*, 677–9.
30. Molloy, K. C.; Waterfield, P. C.; Mahon, M. F. *J. Organomet. Chem.* **1989**, *365*, 61–73.
31. Peters, D.; Hörnfeldt, A.-B.; Gronowitz, S. *J. Heterocycl. Chem.* **1990**, *27*, 2165–73.
32. Kelly, T. R.; Lang, F. *Tetrahedron Lett.* **1995**, *36*, 9293–6.
33. Kelly, T. R.; Jagoe, C. T.; Gu, Z. *Tetrahedron Lett.* **1991**, *32*, 4263–6.
34. Kelly, T. R.; Lang, F. *J. Org. Chem.* **1996**, *61*, 4623–4633.
35. Watanabe, Y.; Ueno, Y.; Araki, T.; Endo, T.; Okawara, M. *Tetrahedron Lett.* **1986**, *27*, 215–8.
36. Park, H.; Lee, J. Y.; Lee, Y. S.; Park, J. O.; Koh, S. B.; Ham, W-H. *J. Antibiotics* **1994**, *47*, 606–8.
37. Bailey, T. R. *Tetrahedron Lett.* **1987**, *27*, 4407–10.
38. Kennedy, G.; Perboni, A. D. *Tetrahedron Lett.* **1996**, *37*, 7611–4.
39. Hark, R. R.; Hauze, D. B.; Petrovskaia, O.; Joullié, M. M.; Jaouhari, R.; McComisky, P. *Tetrahedron* **1994**, *35*, 7719–22.
40. Gronowitz, S.; Peters, D. *Heterocycles* **1990**, *30*, 645–58.
41. Dondoni, A.; Fogagnolo, M.; Medici, A.; Negrini, E. *Synthesis* **1987**, 185–6.
42. Benhida, R.; Lecubin, F.; Fourrey, J.-L.; Castellasnos, L. R.; Quintro, L. *Tetrahedron Lett.* **1999**, *40*, 5701–3.
43. Sandosham, J.; Undheim, K. *Acta Chem. Scand.* **1989**, *43*, 684–9.
44. Gutierrez, A. J.; Terhorst, T. J.; Matteucci, M. D.; Froehler, B. C. *J. Am. Chem. Soc.* **1994**, *116*, 5540–4.
45. Malm, J.; Hörnfeldt, A-B.; Gronowitz, S. *Tetrahedron Lett.* **1992**, *33*, 2199–202.
46. Gronowitz, S.; Björk, P.; Malm, J.; Hörnfeldt, A-B. . *J. Organomet. Chem.* **1993**, *460*, 127–9.
47. Barchín, B. M.; Valenciano, J.; Cuadro, A. M.; Alvarez-Builla, J.; Vaquero, J. *Org. Lett.* **1999**, *1,* 545–7.
48. Birchler, A. G.; Liu, F.; Liebeskind, L. S. *J. Org. Chem.* **1994**, *59*, 7737–45.
49. Dondoni, A.; Fantin, G.; Fogagnolo, M.; Mastellari, A.; Medici, A. *Gazz. Chim. Ital.* **1988**, *118*, 211–32.
50. Nicolaou, K. C.; He, Y.; Roschangar, F.; King, N. P.; Vourloumis, D.; Li, T. *Angew. Chem., Int. Ed.* **1998**, *37,* 84–7.
51. Sakamoto, T.; Nagata, H.; Kondo, Y.; Shiraiwa, M.; Yamanaka, H. *Chem. Pharm. Bull.* **1987**, *35*, 823–8.
52. Schlegel, J.; Maas, G. *Synthesis* **1999**, 100–6.

53. Nye, S. A.; Potts, K. T. *Synthesis* **1988**, 375–7.
54. Arcadi, A.; Attanasi, O. A.; Guidi, B.; Rossi, E.; Santeusanio, S. *Chem. Lett.* **1999**, 59–60.
55. Frank, W. C.; Kim, Y.; Heck, R. F. *J. Org. Chem.* **1978**, *43*, 2947–9.
56. Sakamoto, T.; Nagata, H.; Kondo, Y.; Shiraiwa, M.; Yamanaka, H. *Chem. Pharm. Bull.* **1987**, *35*, 823–8.
57. Aoyagi, Y.; Inoue, A.; Koizumi, I.; Hashimoto, R.; Tokunaga, K.; Gohma, K.; Komatsu, J.; Sekine, K.; Miyafuji, A.; Konoh, J. Honma, R. Akita, Y.; Ohta, A. *Heterocycles* **1992**, *33*, 257–72.
58. Proudfoot, J. R. *et al. J. Med. Chem.* **1995**, *38*, 4930–8.
59. Pivsa-Art, S.; Satoh, T.; Kawamura, Y.; Miura, M.; Nomura, M. *Bull. Chem. Soc. Jpn.* **1998**, *71*, 467–73.
60. Potts, K. T. *Comprehensive Heterocyclic Chemistry* Pergamon Press, Oxford, **1984**, *Vols* 5 and 6.
61. Head, R. A.; Ibbotson, A. *Tetrahedron Lett.* **1984**, *25*, 5939–42.
62. Hong, Y.; Tanoury, G. J.; Wilkinson, H. S.; Bakale, R. P.; Wald, S. A.; Senanayake, C. H. *Tetrahedron Lett.* **1997**, *38*, 5607–10.

# CHAPTER 8

## Oxazoles and Benzoxazoles

During the last decade, several oxazole-containing natural products have been isolated and found to be biologically active. Much synthetic effort has been expended in their total synthesis. Two prominent examples are hennoxazole A and diazonamide A, both of which embody a bis-oxazole entity. Hennoxazole A was isolated from the sponge *Polyfibrospongia* sp. It shows activity against Herpes simplex virus type 1 ($IC_{50}$ = 0.6 μg/mL) and is a peripheral analgesic. Diazonamide A, in turn, is a secondary metabolite of the colonial ascidian, *Diazona chinensis*, a marine species collected from the ceilings of small caves in the Philippines. It is potent *in vitro* against HCT-116 human colon carcinoma and B-16 murine melanoma cancer cell lines ($IC_{50}$ < 15 ng/mL). Conversely, zoxazoleamine, a sedative and muscle-relaxant, is an example of synthetic pharmaceutical agents.

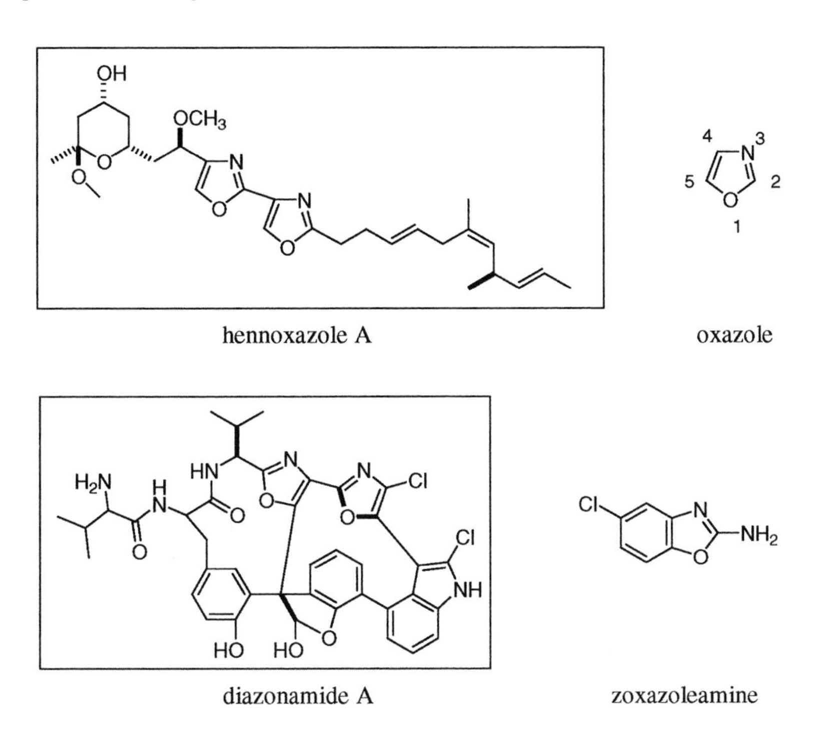


hennoxazole A

oxazole

diazonamide A

zoxazoleamine

Like thiazole, oxazole is a π-electron-excessive heterocycle. The electronegativity of the N-atom attracts electrons so that C(2) is partially electropositive and therefore susceptible to nucleophilic attack. However, electrophilic substitution of oxazoles takes place at the electron-rich position C(5) preferentially. More relevant to palladium chemistry, 2-halooxazoles or 2-halobenzoxazoles are prone to oxidative addition to Pd(0). Even 2-chlorooxazole and 2-chlorobenzoxazole are viable substrates for Pd-catalyzed reactions.

## 8.1 Synthesis of halooxazoles and halobenzoxazoles

### 8.1.1 Direct halogenation

Condensation of 2-bromoethylamine hydrobromide with benzoyl chloride in benzene in the presence of 5 equivalents of $Et_3N$ gave 2-phenyl-4,5-dihydrooxazole (**1**) in 67% yield [1]. Treatment of **1** with 3 equivalents of NBS in boiling $CCl_4$ in the presence of AIBN led to 5-bromo-2-phenyloxazole (**2**). Presumably, sequential bromination and dehydrobromination of **1** led to 2-phenyloxazole, which underwent further bromination to afford **2**.

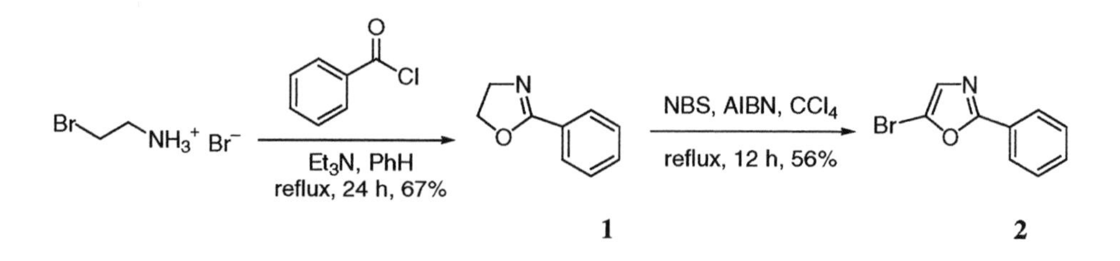


### 8.1.2 Metalation and halogen quench

Barrett and Kohrt deprotonated oxazole **3** using *n*-BuLi and then quenched the resulting oxazol-2-yllithium with iodine to prepare the desired 2-iodooxazole **4** [2].

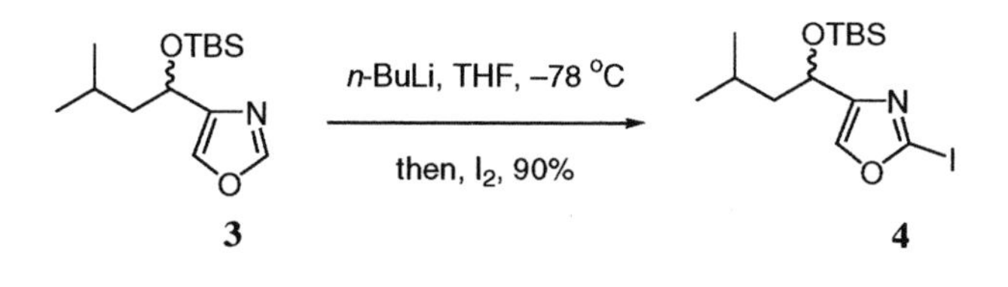

The regiochemistry for trapping lithiooxazole depends upon the oxazole substituents as well as the nature of the electrophile. Hodges, Patt and Connolly observed that the major product of reaction between lithiated oxazole (**5** + **6**) and benzaldehyde was the C(4)-substituted oxazole **7**, resulting from reaction of the dominant acyclic valence bond tautomer **5** *via* the initial aldol adduct **6** followed by proton transfer and recyclization [3].

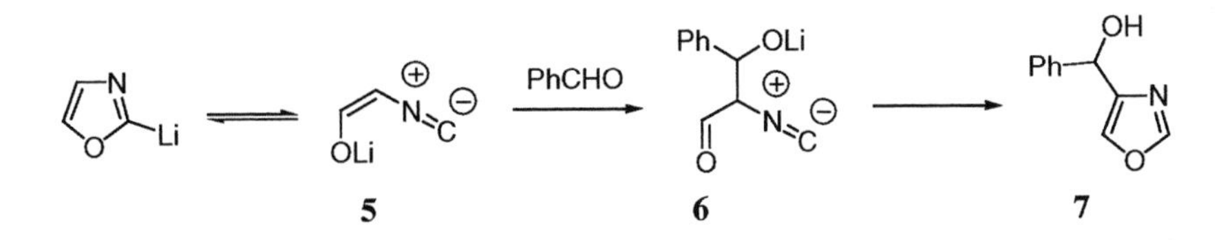


Vedejs *et al.* developed a method for the iodination of oxazoles at C(4) *via* 2-lithiooxazoles by exploiting the aforementioned equilibrium between cyclic (**5**) and acyclic (**6**) valence bond tautomers of 2-lithiooxazole [4]. When 5-(*p*-tolyl)oxazole (**8**) was treated with lithium hexamethyldisilazide (LiHMDS) in THF followed by treatment with 1,2-diiodoethane as the electrophile, 2-iodooxazole **9** was obtained exclusively. On the other hand, when 50 volume% of DMPU was added *prior to* the addition of the base, 4-iodooxazole **10** was isolated as the predominant product (73%) with ca. 2% of **9** and ca. 5% of the 2,4-diiodooxazole derivative.

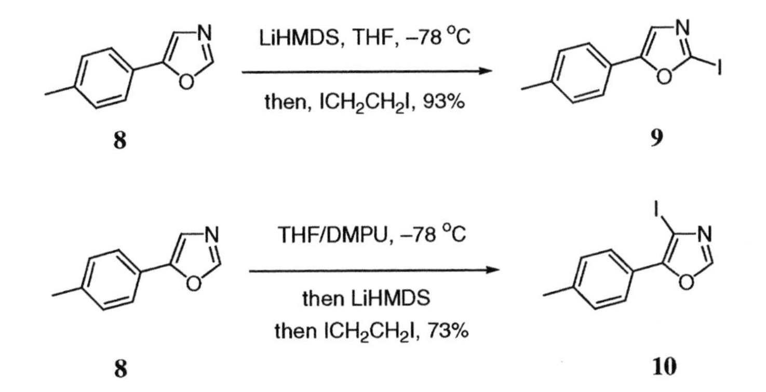


### 8.1.3 Sandmeyer reaction

Aminooxazole **11**, readily obtained by reaction of *N*-Boc-L-Val-OH with aminomalononitrile *p*-toluenesulfonate and EDC in pyridine [5], was converted directly to bromooxazole **12** by *in situ* bromination *via* a nitrosamine intermediate.

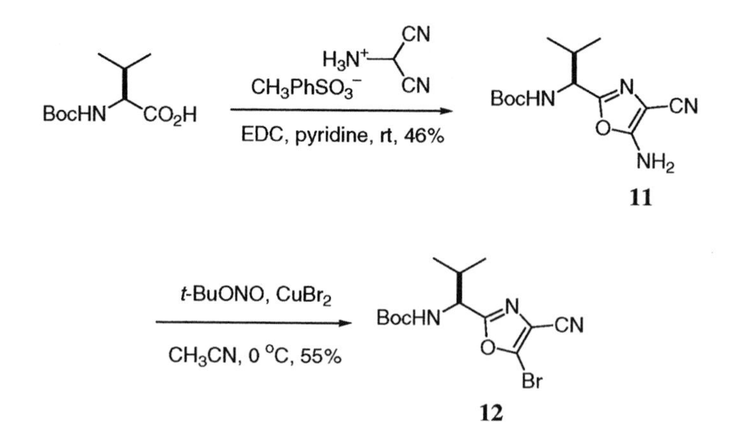


While halogenation and Sandmeyer reactions are suitable for preparation of oxazolyl halides, benzoxazolyl halides with halogen on the benzene ring moiety may be synthesized *via* other approaches. For instance, 5-halobenzoxazoles were prepared by treating 4-halo-2-aminophenols with trimethyl orthoformate and concentrated aqueous HCl [6].

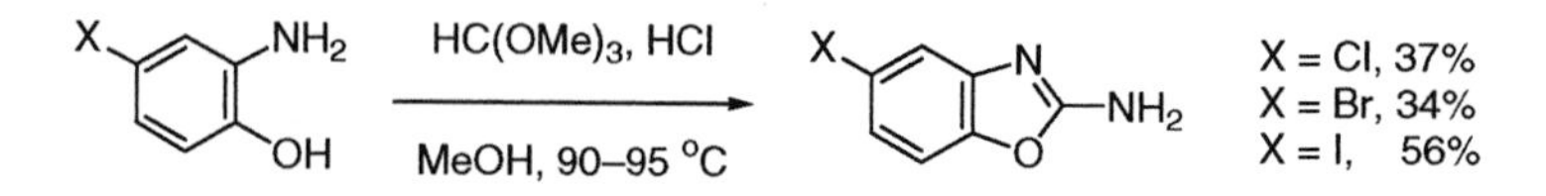


## 8.2 Coupling reactions with organometallic reagents

There are limited precedents for elaborating the oxazole nucleus using Pd-catalyzed cross-coupling reactions, partially because of the ring-opening tendency of oxazol-2-yllithium. Nonetheless, the rapid progress in palladium chemistry compounded with the biological importance of oxazole-containing entities will surely spur more interest in this area.

Condensation of aryl halides with various active methylene compounds is readily promoted by catalytic action of palladium to give the corresponding arene derivatives containing a functionalized ethyl group [7]. Yamanaka *et al.* extended this chemistry to haloazoles including oxazoles, thiazoles and imidazoles [8]. Thus, in the presence of $Pd(Ph_3P)_4$, 2-chlorooxazole was refluxed with phenylsulfonylacetonitrile and NaH to form 4,5-diphenyl-α-phenylsulfonyl-2-oxazoloacetonitrile, which existed predominantly as its enamine tautomer. In a similar fashion, 4-bromooxazole and 5-bromooxazole also were condensed with phenylsulfonylacetonitrile under the same conditions.

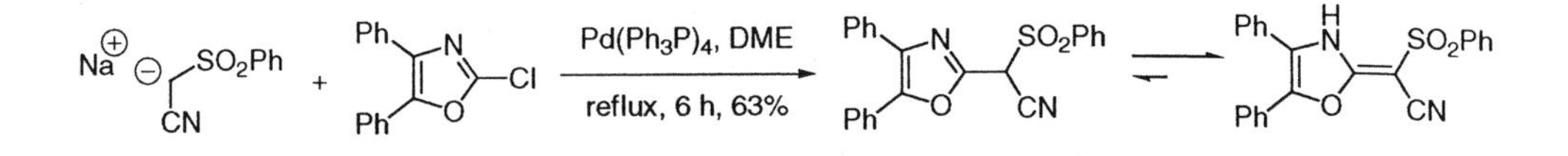


### 8.2.1 Negishi coupling

Fluorovinylzinc, generated by treating the corresponding vinyllithium with $ZnCl_2$, was coupled with 2-iodobenzoxazole to give the adduct stereoselectively [9, 10].

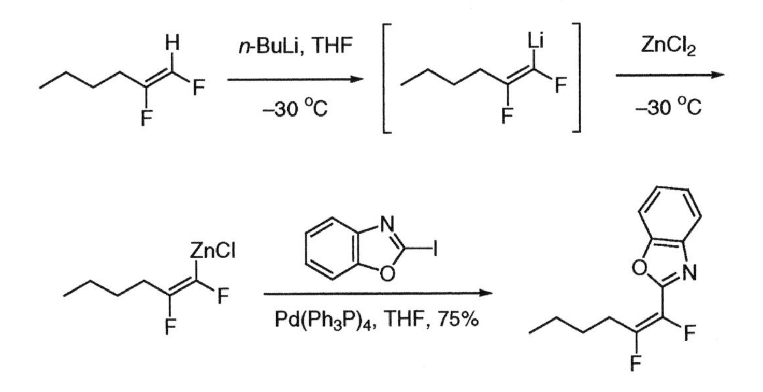


Oxazol-2-ylzinc chloride reagents can be prepared from transmetalation of oxazol-2-yllithium with $ZnCl_2$ [11]. A strong covalent bond with carbon and relatively low oxophilicity suggested that the equilibrium for zinc derivatives should sufficiently favor the metalated oxazol-2-ylzinc species to yield a productive reaction. Indeed, Anderson *et al.* successfully carried out Negishi couplings of oxazol-2-ylzinc chloride reagents with aryl halides and acid chlorides as well as organotriflates [12–14]. Therefore, deprotonation of benzoxazole using *n*-BuLi followed by transmetalation with $ZnCl_2$ gave 2-chlorozincbenzoxazole, which was cross-coupled with 1-naphthalene triflate to provide the expected naphthalenylbenzoxazole. The Pd(0) here was derived from the treatment of $PdCl_2(Ph_3P)_2$ with *n*-BuLi. In like fashion, the cross-coupling of iodoindoline **13** with 2-oxazol-ylzinc chloride led to an oxazole-containing partial ergot alkaloid **14**, a potent 5-$HT_{1A}$ agonist.

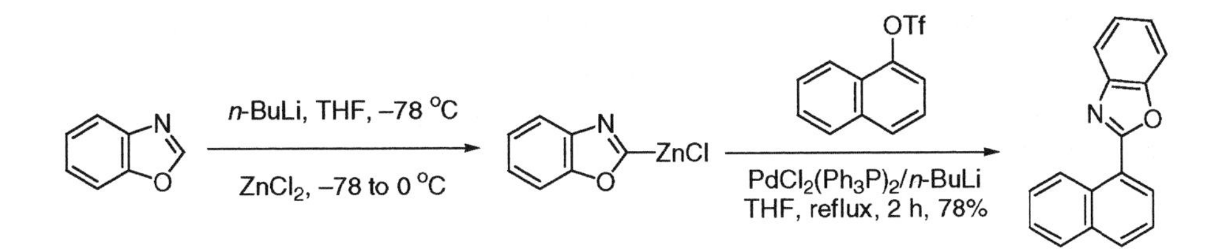

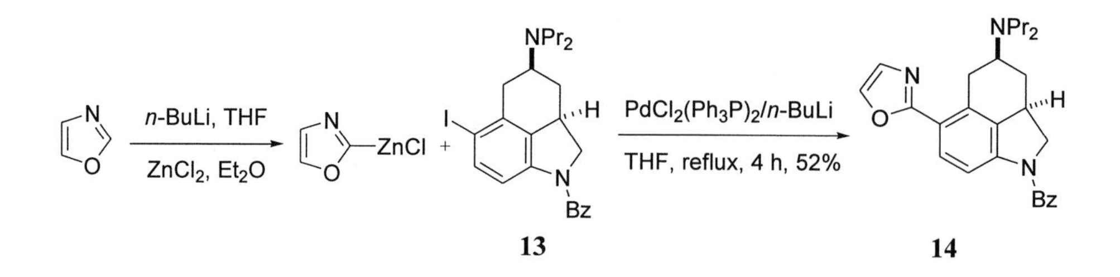


4-Iodooxazole **15**, prepared from the corresponding 2-lithiooxazole, underwent a Negishi reaction with 2-oxazolylzinc chloride to give bis-oxazole **16** in the presence of $Pd(dba)_2$ and trifuranylphosphine (TFP) [4].

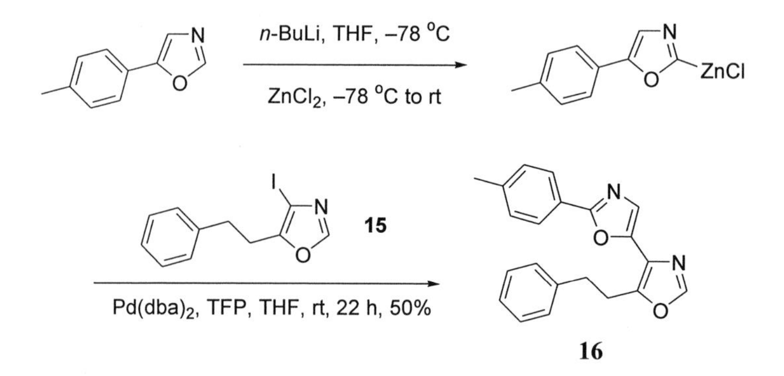


### 8.2.2 Suzuki coupling

Attempts to make oxazolylboronic acids failed probably due to the equilibrium between cyclic and acyclic valence bond tautomers of the lithiooxazoles. A somewhat relevant Suzuki coupling involved the Pd-catalyzed cross-coupling of 6-bromo-2-phenyloxazolo[4,5-*b*]pyridine with phenylboronic acid to provide 6-phenyl-2-phenyloxazolo[4,5-*b*]pyridine [15].

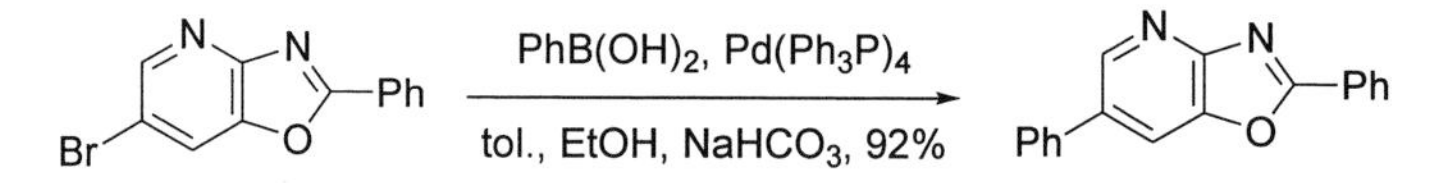


### 8.2.3 Stille coupling

2-Benzoxazolyllithium undergoes ring-opening tautomerization. However, intercepting the cyclic lithium species with $R_3SnCl$ furnishes the desired 2-trialkylstannylbenzoxazole. Interestingly, trapping the oxazolyllithium system with $Me_3SiCl$ resulted in predominantly the acyclic product [16]. The Stille coupling of 2-(tributylstannyl)benzoxazole with bromobenzene under the influence of $PdCl_2(Ph_3P)_2$ led to 2-phenylbenzooxazole [17].

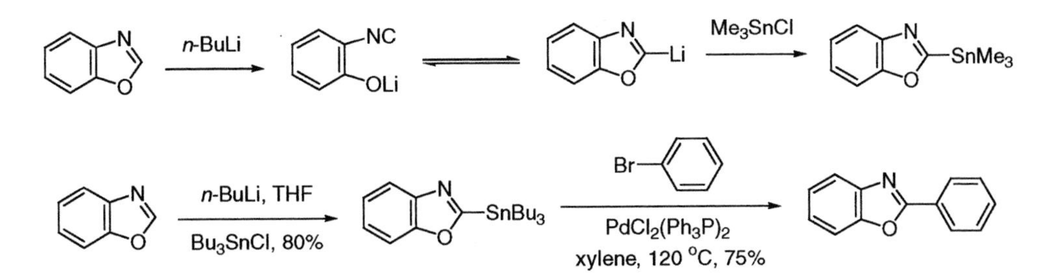


By lithiation of 4-methyloxazole with *n*-BuLi and subsequent quenching with trimethyltin chloride, Dondoni *et al.* prepared 2-trimethylstannyl-4-methyloxazole [18], which was then coupled with 3-bromofuran to provide the expected furylthiazole.

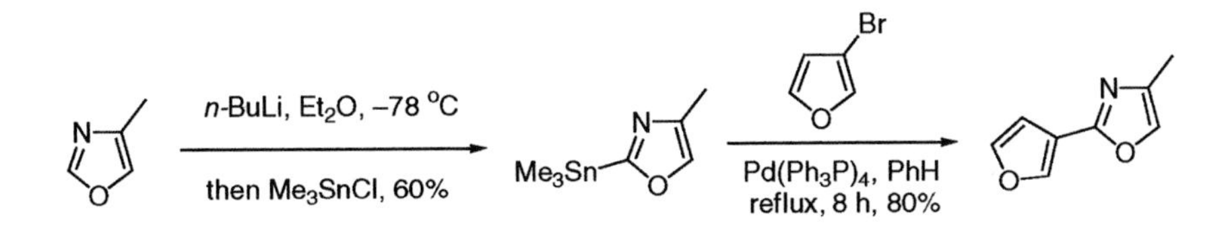


In the model studies toward the total synthesis of dimethyl sulfomycinamate, Kelly *et al.* successfully carried out the Stille couplings of oxazolyl triflate **18** with an array of organostannanes [19, 20]. Thus, 2-aryl-4-oxalone **17** was transformed into the corresponding triflate **18**, which was then coupled with 2-trimethylstannylpyridine under the agency of $Pd(Ph_3P)_4$ and LiCl to provide adduct **19**. The couplings of triflate **18** with phenyl-, vinyl- and phenylethynyl trimethyltin all proceeded in excellent yields. Unfortunately, application to the more delicate system in the natural product failed and the oxazole moiety was installed from acyclic precursors.

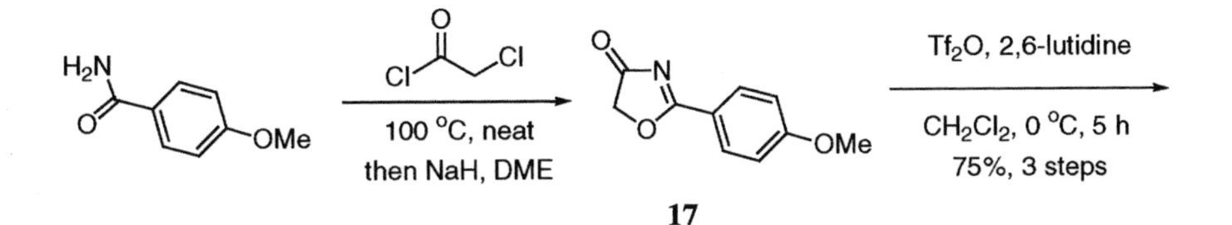

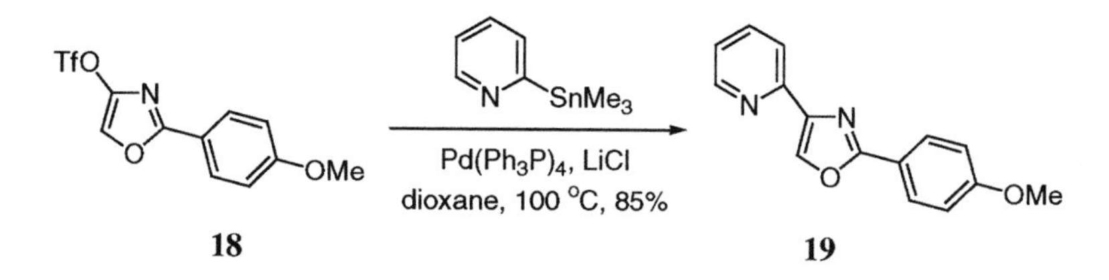


Analogously, Barrett and Kohrt transformed 2-phenyl-4-oxalone into triflate **20**, which was then converted to the corresponding stannane **21** using Pd-catalyzed coupling with hexamethyldistannane. Subsequent coupling with 2-iodooxazole **4** elaborated bis-oxazole **22** [2].

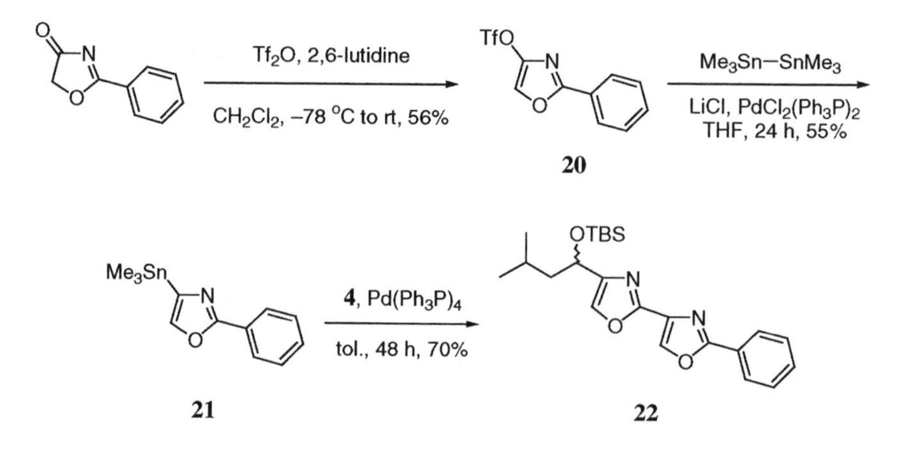


Stille couplings also have been exploited in the synthesis of the aromatic macrocyclic core of diazonamide A (**2**) [5, 20]. Pattenden's group utilized the Pd-catalyzed coupling between the 3-stannyl substituted indole **23** and the 3-bromooxazole **24** to provide a particularly expeditious route to the ring system **25** [20]. In addition, Harran's group secured the connection between bromooxazole **12** and vinylstannane **26** also using a Stille coupling [5].

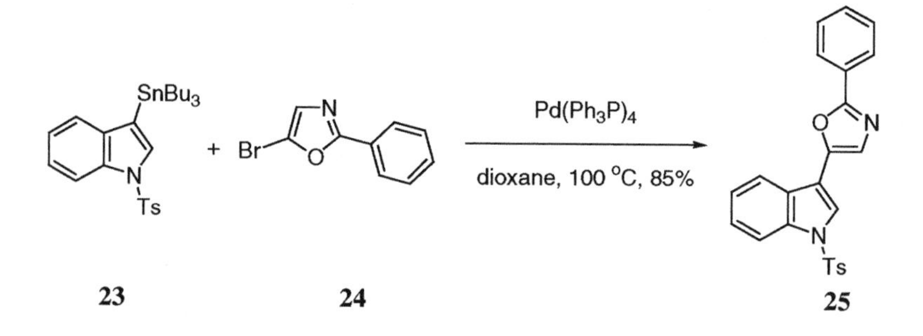

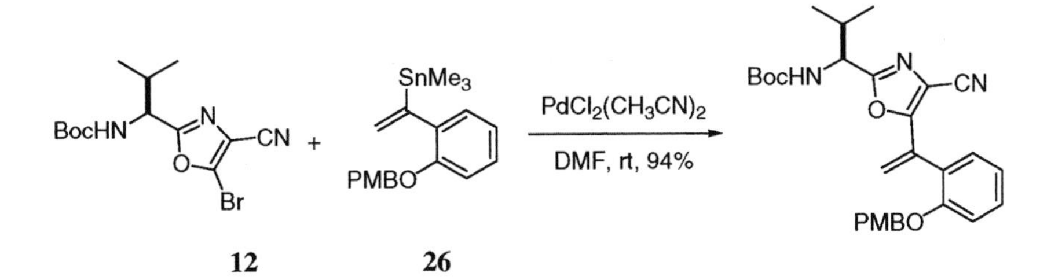


In synthetic studies toward the synthesis of phorboxazole described by Schaus and Panek, although their initial attempts of carboalumination of terminal alkynes followed by Pd-catalyzed cross-coupling of the resulting vinylmetallic intermediates with **20** were successful, the strategy failed to install the C(27)–C(29) olefin of phorboxazole. However, employing a Stille coupling of oxazole-4-triflate **20** and a vinyl stannane, they constructed the C(26)–C(31) subunit of phorboxazole [21]. The trimethylstannane was found to be advantageous over the corresponding tributylstannane presumably due to the ease of transmetalation.

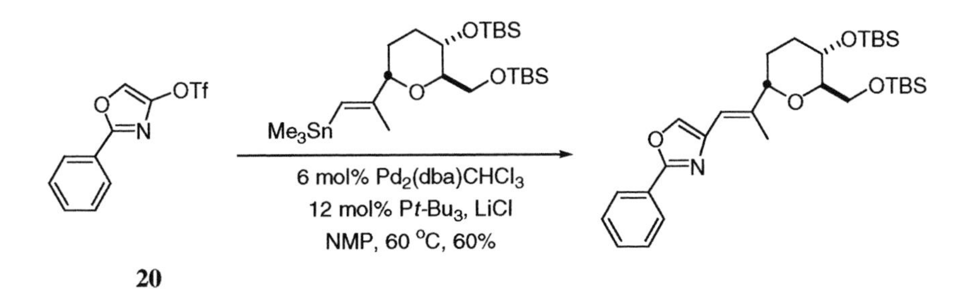


## 8.3 Sonogashira reaction

In 1987, Yamanaka's group described the Pd-catalyzed reactions of halothiazoles with terminal acetylenes [22a]. Submission of 4-bromo- and 5-bromo-4-methyloxazoles to the Sonogashira reaction conditions with phenylacetylene led to the expected internal acetylenes.

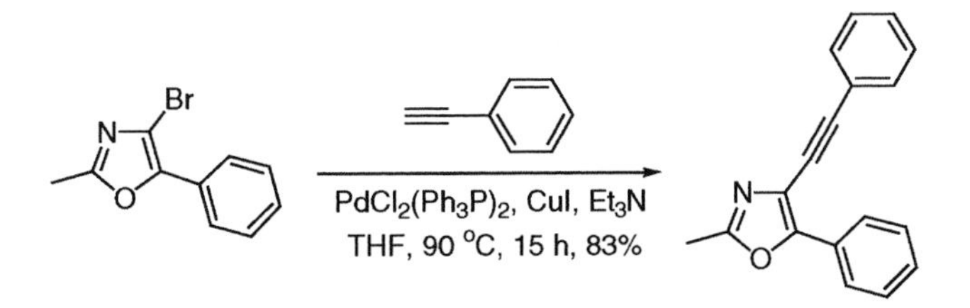

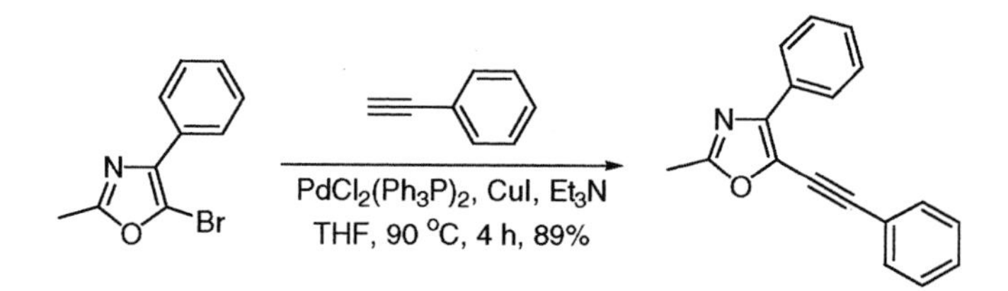


## 8.4 Heck and heteroaryl Heck reactions

Yamanaka's group reported the Heck reactions of 4-bromo- and 5-bromo-4-methyloxzoles with ethyl acrylate and acrylonitrile, respectively [22a].

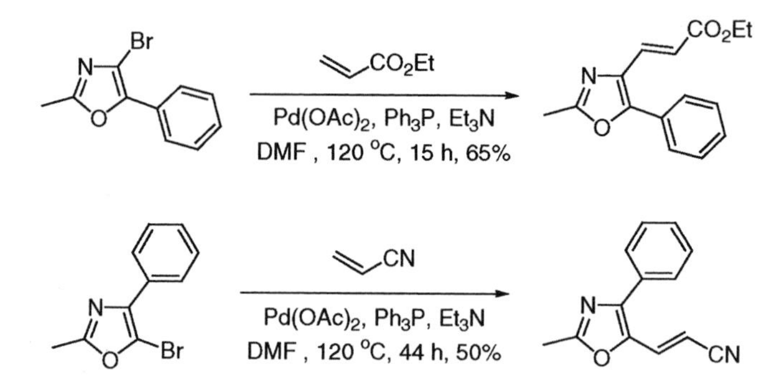


Oxazoles and benzoxazoles are viable participants in the heteroaryl Heck reactions. In their monumental work published in 1992, Ohta and colleagues demonstrated that oxazoles and benzoxazoles, along with other π-sufficient aromatic heterocycles such as furans, benzofurans, thiophenes, benzothiophenes, pyrroles, thiazole and imidazoles, are acceptable recipient partners for the heteroaryl Heck reactions of chloropyrazines [22b]. Therefore, treatment of 2-chloro-3,6-diethylpyrazine (**27**) with oxazole led to regioselective addition at *C(5),* giving rise to adduct **28**. By contrast, a similar reaction between 2-chloro-3,6-diisobutylpyrazine (**29**) and benz[*b*]oxazole took place at *C(2)* exclusively to afford pyrazinylbezoxazole **30**.

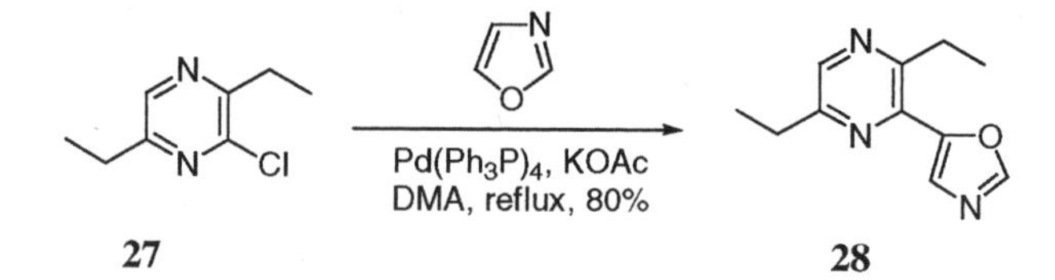

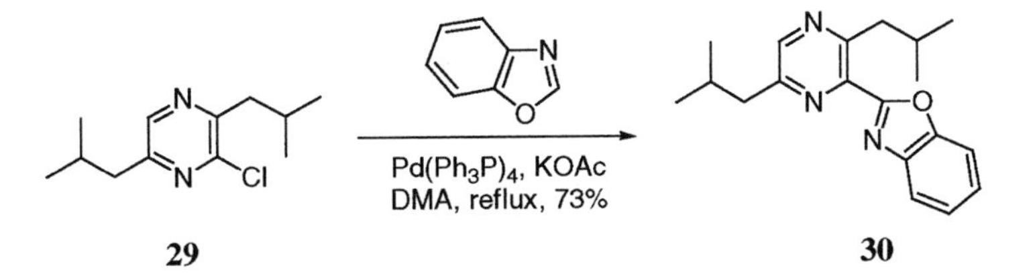


Similar results were obtained for the heteroaryl Heck reaction of iodobenzene or bromobenzene with oxazole and benzoxazole [23].

## 8.5 Carbonylation

2-Arylbenzoxazoles were prepared by Pd-catalyzed three-component condensation of aryl halides with *o*-aminophenols and carbon monoxide followed by dehydrative cyclization [24]. A variant of such methodology using *o*-fluorophenylamines in place of *o*-aminophenols was used to synthesize arylbenzoxazoles [25].

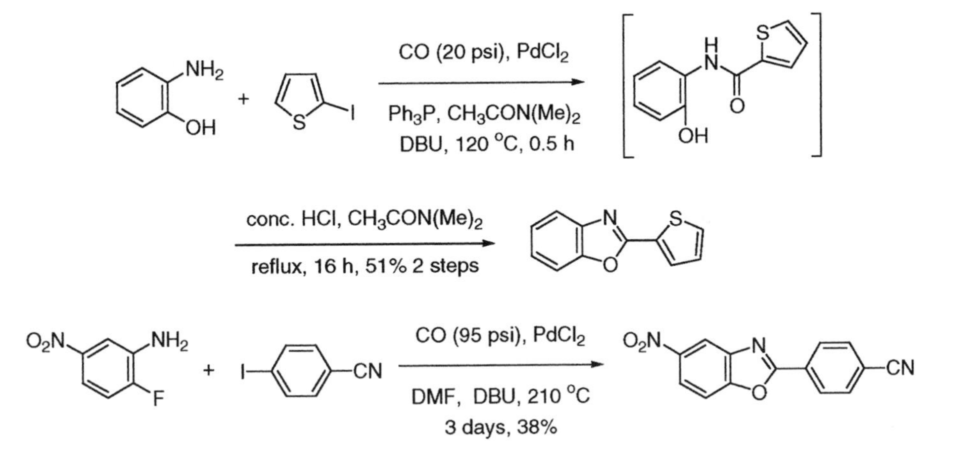


A unique Pd(II)-promoted *ortho*-esterification of 2,5-diphenyloxazole has been described [26]. When 2,5-diphenyloxazole was heated with 2.5 equivalents of $Pd(OAc)_2$ in acetic acid and $CCl_4$, regioselective palladation took place, giving rise to arylpalladium(II) σ-complex **31** in almost quantitative yield. The regioselectivity observed reflects the strong coordination ability of the oxazole nitrogen atom to the palladium atom. Complex **31** was then dissolved in MeOH–THF (1:1) and the solution was stirred at 0 °C to produce 2-(5'-phenylthiazol2-yl)-benzoate **32**. A drawback of this method is the use 2.5 equivalents of $Pd(OAc)_2$, making it less useful for preparative purposes.

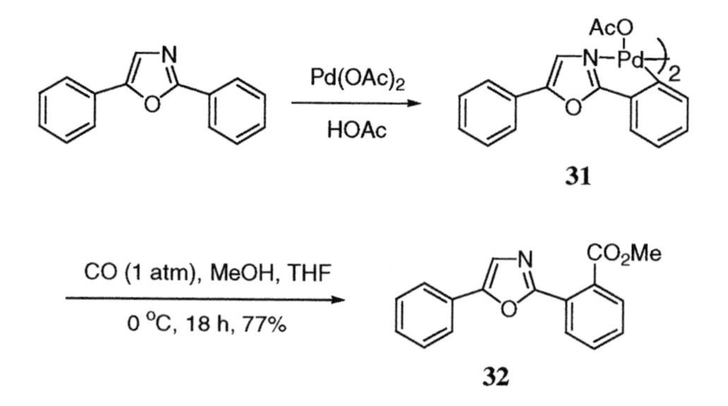


## 8.6 C—N bond formation

An extension of Buchwald's catalytic amination using palladium led to amination of 2-chlorobenzoxazole [27]. The marked reactivity enhancement of the chloride is attributed to the polarization at C(2).

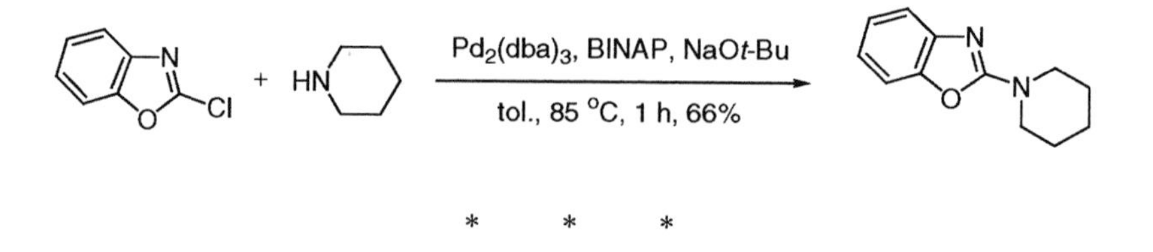


*    *    *

To summarize: the ring-opening process of oxazol-2-yllithium to the acyclic isonitrile tautomer is well documented. Oxazol-2-ylzinc chloride reagents can be prepared from the transmetalation of oxazol-2-yllithium with $ZnCl_2$, whereas oxazol-2-ylboronic acid is unprecedented. Stannyloxazoles are readily prepared and the Stille reaction is still the method of choice for synthesizing oxazole-containing molecules if the toxicity is not of great concern. 4-Bromo- and 5-bromo-4-methyloxazoles are viable substrates for both the Sonogashira and Heck reactions. The heteroaryl Heck reaction using oxazole as a recipient partner takes place predominantly at the electron-rich position C(5), whereas it occurs at C(2) for benzoxazole.

## 8.7 References


1. Kashima, C.; Arao, H. *Synthesis* **1989**, 873–4.
2. Barrett, A. G. M.; Kohrt, J. T. *Synlett* **1995**, 415–6.
3. Hodges, J. C.; Patt, W. C.; Connolly, C. J. *J. Org. Chem.* **1991**, *56*, 449–52.
4. Vedejs, E.; Luchetta, L. M. *J. Org. Chem.* **1999**, *64*, 1011–4.
5. Jeong, S.; Chen, X.; Harran, P. G. *J. Org. Chem.* **1998**, *63*, 8640–1.
6. Kunz, K. R.; Taylor, E. W.; Hutton, H. M.; Blackburn, B. J. *Org. Prep. Proc. Int.* **1990**, *22*, 613–8.
7. Sakamoto, T.; Katoh, E.; Kondo, Y.; Yamanaka, H. *Chem. Pharm. Bull.* **1990**, *38*, 1513–7.
8. Sakamoto, T.; Kondo, Y.; Suginome, T.; Ohba, S.; Yamanaka, H. *Synthesis* **1992**, 552–4.
9. Gillet, J.-P.; Sauvêtre, R.; Normant, J.-F. *Tetrahedron Lett.* **1985**, *26*, 3999–4002.
10. Gillet, J.-P.; Sauvêtre, R.; Normant, J.-F. *Synthesis* **1986**, 538–43.
11. Miller, R. D.; Lee, V. Y.; Moylan, C. R. *Chem. Mater.* **1994**, *6*, 1023–32.
12. Harn, N. K.; Gramer, C. J.; Anderson, B. A. *Tetrahedron Lett.* **1995**, *36,* 9453–6.
13. Anderson, B. A.; Harn, N. K. *Synthesis* **1996**, 583–5.
14. Anderson, B. A.; Becke, L. M.; Booher, R. N.; Flaogh, M. F.; Harn, N. K.; Kress, T. J.; Varie, D. L.; Wepsiec, J. P. *J. Org. Chem.* **1997**, *62*, 8634–9.
15. Viaud, M.-C.; Jamoneau, P.; Savelon, L.; Guillaumet, G. *Heterocycles* **1995**, *41*, 2799–809.
16. Jutzi, P.; Gilge, U. *J. Organomet. Chem.* **1983**, *246*, 159–62.
17. Kosugi, M.; Koshiba, M.; Atoh, A.; Sano, H.; Migita, T. *Bull. Chem. Soc. Jpn.* **1986**, *59*, 677–9.
18. (a) Kelly, T. R.; Lang, F. *Tetrahedron Lett.* **1995**, *36*, 5319–22. (b) Cacchi, S.; Ciattini, P. G.; Morera, E.; Ortar, G. *Tetrahedron Lett.* **1986**, *27*, 5541–4.
19. Kelly, T. R.; Lang, F. *J. Org. Chem.* **1996**, *61*, 4623–33.
20. Boto, A.; Ling, M.; Meek, G.; Pattenden, G. *Tetrahedron Lett.* **1998**, *39*, 8167–70.
21. Schaus, J. V.; Panek, J. S. *Org. Lett.* **2000**, *2*, 469–71.
22. (a) Sakamoto, T.; Nagata, H.; Kondo, Y.; Shiraiwa, M.; Yamanaka, H. *Chem. Pharm. Bull.* **1987**, *35*, 823–8. (b) Aoyagi, Y.; Inoue, A.; Koizumi, I.; Hashimoto, R.; Tokunaga, K.; Gohma, K.; Komatsu, J.; Sekine, K.; Miyafuji, A.; Konoh, J. Honma, R. Akita, Y.; Ohta, A. *Heterocycles* **1992**, *33*, 257–72.
23. Pivsa-Art, S.; Satoh, T.; Kawamura, Y.; Miura, M.; Nomura, M. *Bull. Chem. Soc. Jpn.* **1998**, *71*, 467–73.
24. Perry, R. J.; Wilson, B. D.; Miller, R. J. *J. Org. Chem.* **1992**, *57*, 2883–7.
25. Perry, R. J.; Wilson, B. D. *J. Org. Chem.* **1992**, *57*, 6351–4.
26. Sakakibara, T.; Kume, T.; Ohyabu, T.; Hase, T. *Chem. Pharm. Bull.* **1989**, *37*, 1694–7.

27. Hong, Y.; Tanoury, G. J.; Wilkinson, H. S.; Bakale, R. P.; Wald, S. A.; Senanayake, C. H. *Tetrahedron Lett.* **1997**, *38*, 5607–10.

# CHAPTER 9

# Imidazoles

The imidazole ring is present in a number of biologically important molecules as exemplified by the amino acid histidine. It can serve as a general base (p*K*a = 7.1) or a ligand for various metals (e.g., Zn, *etc.*) in biological systems. Furthermore, the chemistry of imidazole is prevalent in protein and DNA biomolecules in the form of histidine or adenine/guanine, respectively.

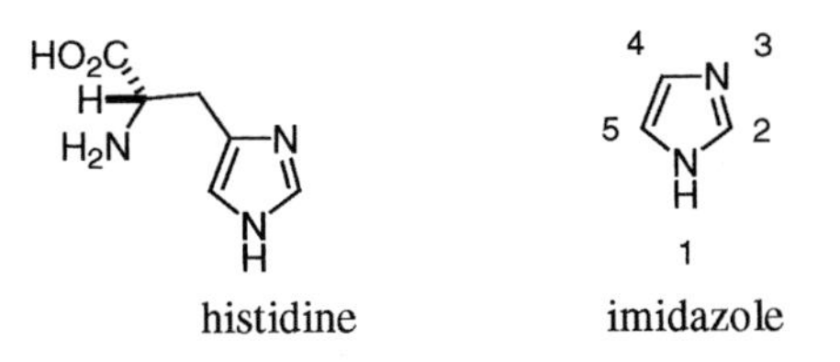


Imidazole is a π-electron-excessive heterocycle. Electrophilic substitution normally occurs at C(4) or C(5), whereas nucleophilic substitution takes place at C(2). The order of reactivity for electrophilic substitution for azoles is:

imidazole > thiazole > oxazole

## 9.1 Synthesis of haloimidazoles

The H(2) position on 1-alkyl or 1-arylimidazole is the most acidic hydrogen, and a carbanion is readily generated at C(2) with *n*-butyllithium. The resulting carbanion can be treated by a halogen electrophile to furnish 2-haloimidazoles. Kirk selectively deprotonated 1-tritylimidazole with *n*-butyllithium and quenched the resulting 2-lithioimidazole species with NIS, NBS, or *tert*-butyl hypochlorite, respectively, to make the corresponding 2-iodo-, 2-bromo-, or 2-chloroimidazole [1]. An extension of Kirk's method to 1,4-disubstituted imidazole **1** provided an entry to the corresponding 2-iodoimidazole **2** [2].

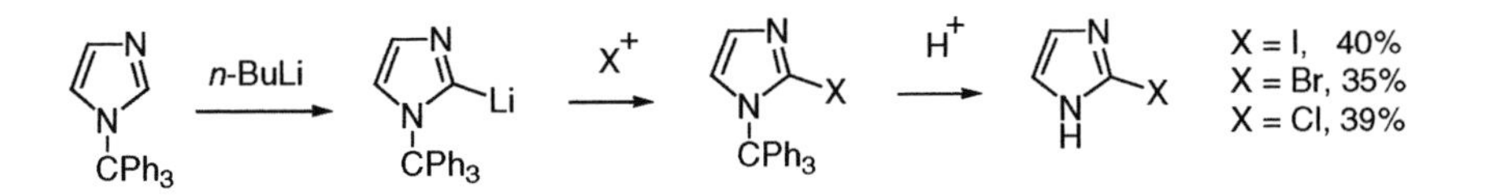

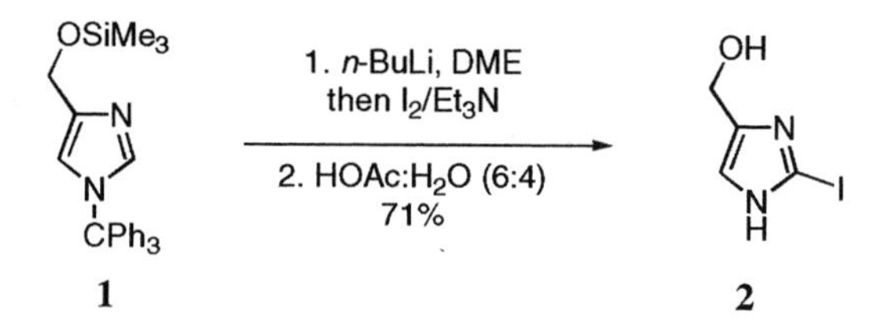


Iodination of 2-substituted imidazoles using NIS in DMF led to 4,5-diiodoimidazoles [3, 4].

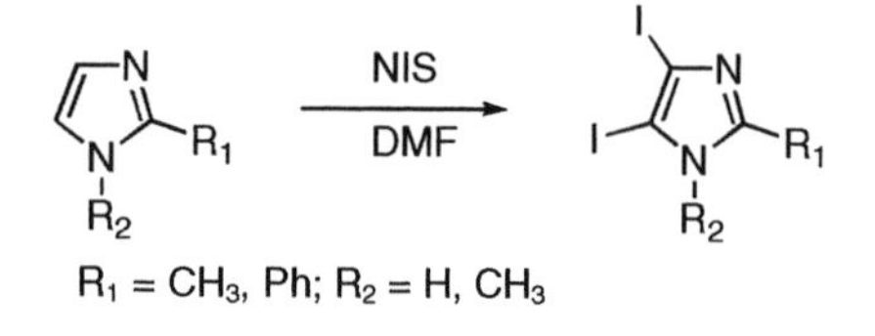


Treating imidazole with iodine in an NaOH solution afforded the complete ring iodination product, 2,4,5-triiodoimidazole. 4-Iodoimidazole was then produced in 75% yield upon selective reductive deiodination, [5]. The same procedure converted 2-methylimidazole and 1-methylimidazole to 4(5)-iodo-2-methylimidazole [6] and 4(5)-bromo-1-methylimidazole [7], respectively.

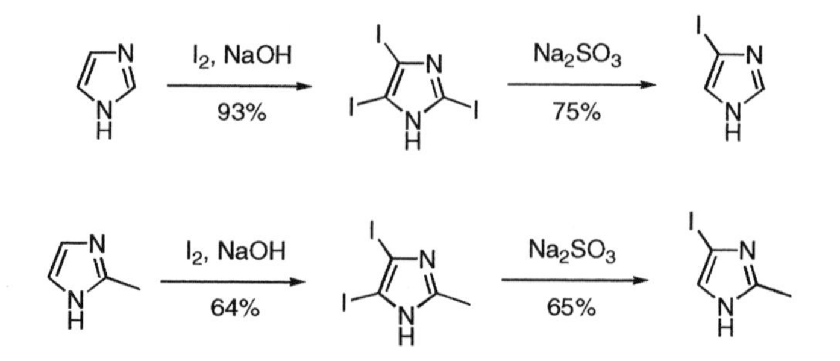


By carefully monitoring the pH of the reaction, regioselective diiodination was achieved. Treating imidazole with iodine at pH 12 furnished 2,5-diiodoimidazole. Selective reductive deiodination secured 2-iodoimidazole, which upon bromination afforded 4,5-dibromo-2-iodoimidazole [6].

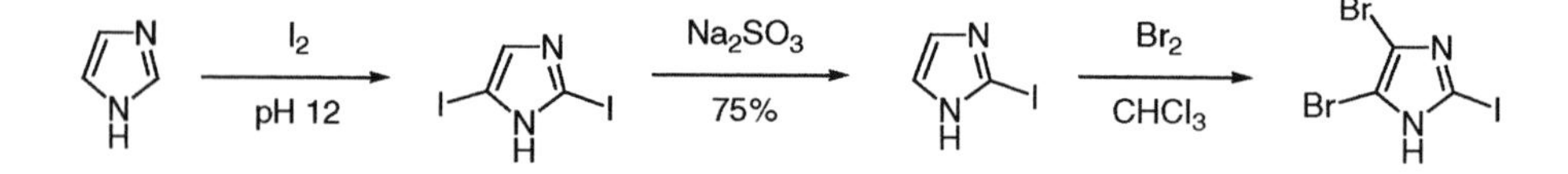

Having appropriate substituents on the imidazole ring, Haseltine *et al.* were able to carry out a regioselective bromination of 1-benzyl-5-methyimidazole in excellent yield [8].

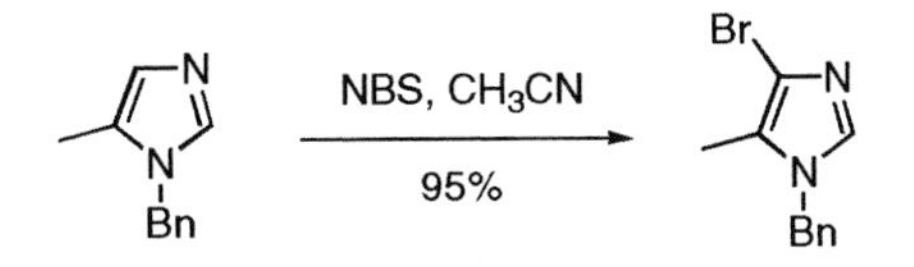


A variant of the Sandmeyer reaction transformed 5-aminoimidazole **3** into the corresponding 5-iodoimidazole **4**. Here the diazotization was accomplished using isoamyl nitrite [9, 10].

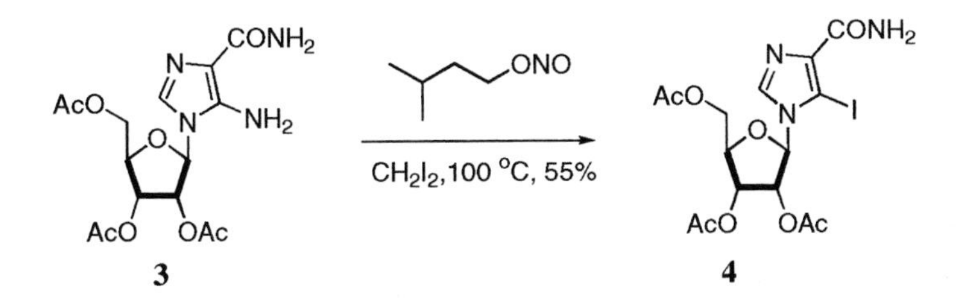


While dealing with imidazoles, an important characteristic is their annular tautomerism. A tautomeric equilibrium for many imidazoles is rapidly achieved at room temperature. In some tautomeric pairs, though, one tautomer often predominates over the other. For instance, 4(5)-bromoimidazole favors the 4-bromo-tautomer in a 30:1 ratio, whereas 4(5)-nitroimidazole exists predominantly as the 4-nitro tautomer (700:1) [11]. 4(5)-Methoxyimidazole has a ratio of 2.5:1 for the 4- and 5-methoxy tautomers.

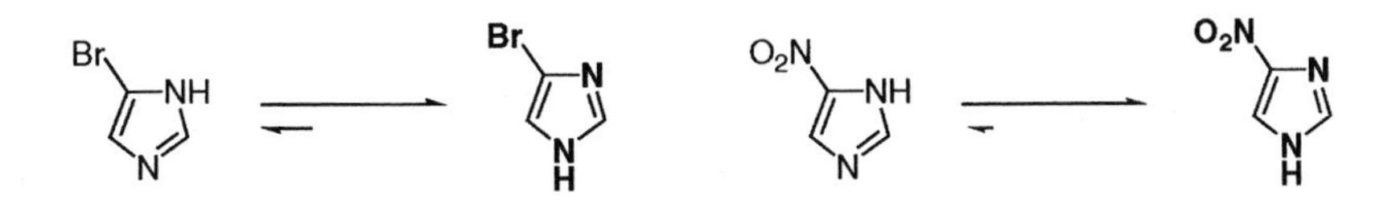


## 9.2 Homocoupling reaction

Whereas the synthesis of 2,2'-biimidazoles has been very well developed, that of 4,4'-biimidazoles has not. Pyne *et al.* revealed that the Pd(0)-catalyzed homocoupling of 4-iodo-1-(triphenylmethyl)imidazole (**5a**) or its 2-methyl analog **5b** delivered the 4,4'-biimidazoles **6a** and **6b**, respectively [6].

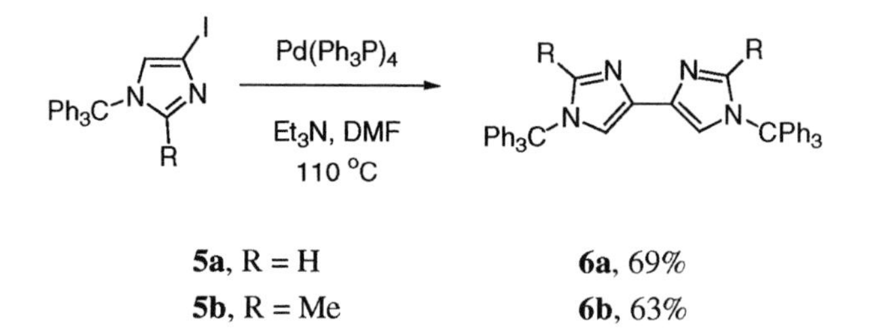


## 9.3 Coupling reactions with organometallic reagents

### 9.3.1 Negishi coupling

There are several ways to prepare imidazol-2-ylzinc reagents. First, direct metalation followed by $ZnCl_2$ quench is suitable for preparation of imidazol-2-ylzinc chlorides. Treating 1-dimethylaminosulfonylimidazole with *n*-BuLi and quenching the resulting 2-lithioimidazole species with $ZnCl_2$ generated the organozinc reagent *in situ*. The organozinc reagent was then cross-coupled with 2-bromopyridine to give the unsymmetrical pyridylimidazole [12].

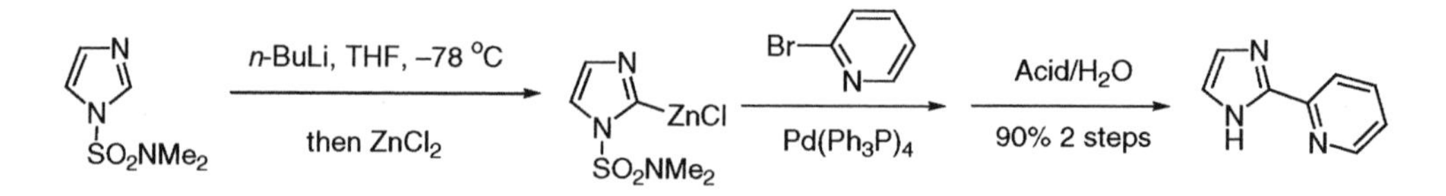


Second, halogen-metal exchange followed by quenching with $ZnCl_2$ works for all three possible haloimidazoles. Bromide **8** was derived from direct bromination at C(4) from 2-(4-fluorophenyl)-5*H*-pyrrolo[1,2-a]imidazole (**7**) using bromine. Halogen-metal exchange of **8** with *n*-BuLi followed by quenching with $ZnCl_2$ furnished organozinc reagent **9** *in situ*, which then was coupled with vinyl iodide **10** in the presence of $Pd(Ph_3P)_4$ to form tetrahydropyridine derivative **11** [13]. Adduct **11** was an intermediate to a potent inhibitor of cyclooxygenase and 5-lipoxygenase enzymes. In addition, imidazol-4-ylzinc reagent **12** was easily generated by treating 1-trityl-4-iodoimidazole with ethyl Grignard reagent followed by the addition of $ZnCl_2$ [14]. The corresponding imidazol-4-ylzinc bromide was also made in the same fashion using $ZnBr_2$ [15]. The Negishi reaction of **12** with triflate **13** in the presence of LiCl led to quinolinylimidazole **14**.

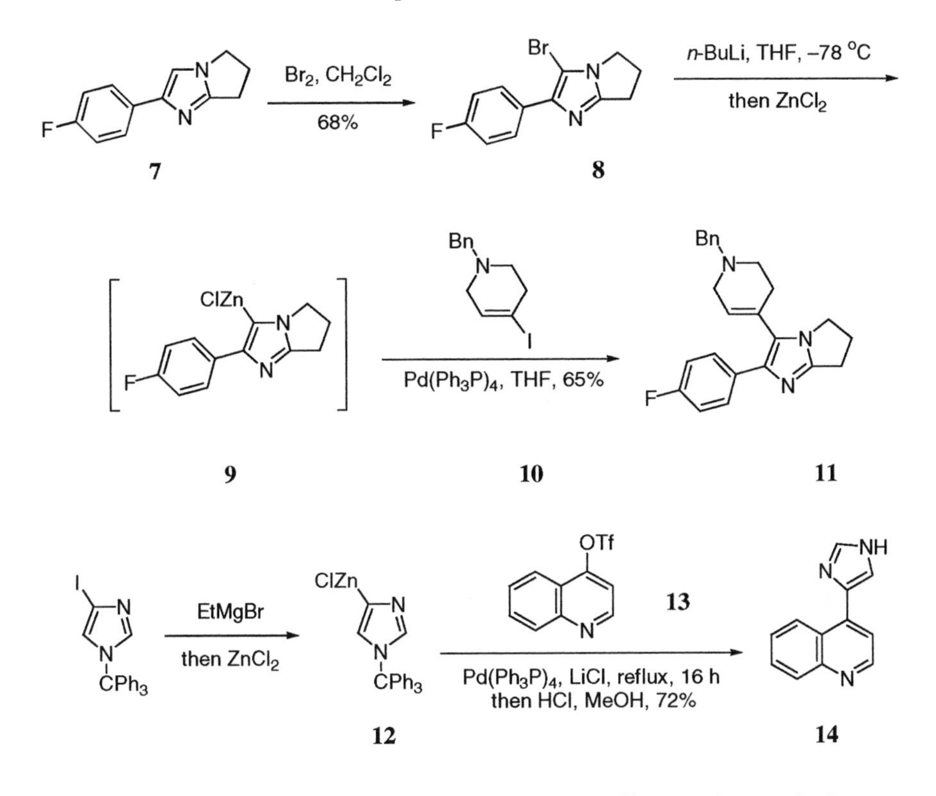


Third, direct oxidative addition of organohalides to Zn(0) offers another synthetic strategy of preparing imidazol-2-ylzinc reagents. This method is advantageous over the usual transmetalations using organolithium or Grignard reagents because of better tolerance of functional groups. Treating 1-methyl-2-iodoimidazole with readily available zinc dust, the Knochel group produced the expected imidazol-2-zinc iodide [16]. The subsequent Negishi reaction was carried out with vinyl-, aryl- and heteroaryl halides.

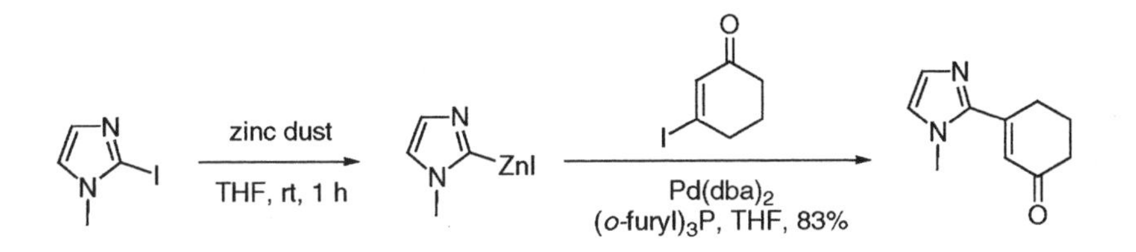


Evans *et al.* prepared organozinc reagent **16** from quantitative metalation of iodide **15** by using an activated Zn/Cu couple in the presence of an ester functionality. Treating 2-iodoimidazole **17** with three equivalents of **16** in the presence of $PdCl_2(Ph_3P)_2$ afforded adduct **18**, which was then transformed into diphthine (**19**) in five additional steps [17].

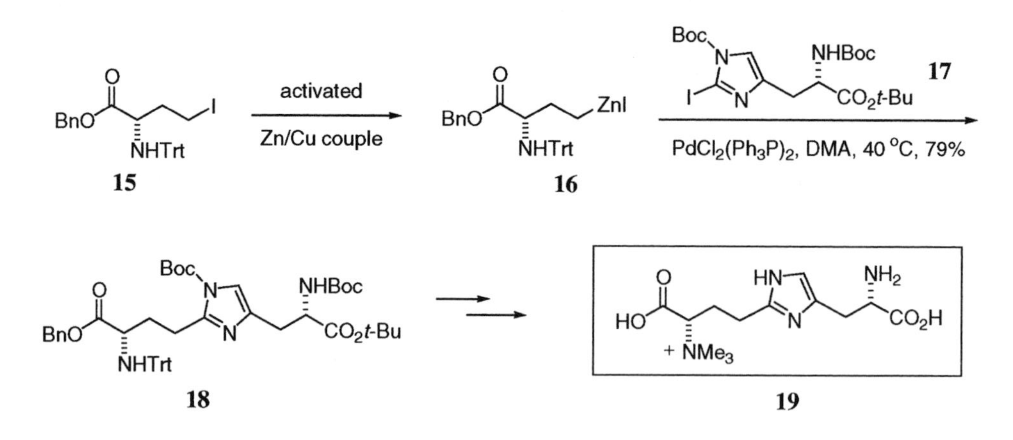


### 9.3.2 Suzuki coupling

Imidazolylboronic acids are not well known. 5-Imidazolylboronic acid was prepared *in situ* by Ohta's group using selective lithiation at C(5) from 1,2-protected imidazole. However, the subsequent Suzuki coupling with 3-bromoindole gave the expected indolylimidazole in only 7% yield [18]. This observation indicated that more investigation into imidazolylboronic acids is necessary to better understand their synthesis and behavior.

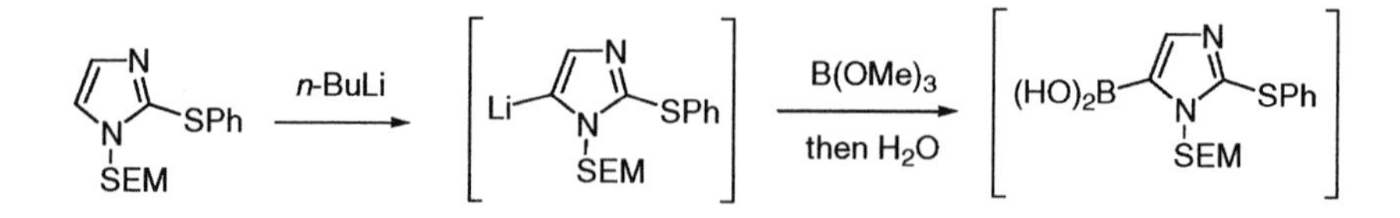


Nonetheless, to make imidazole-containing molecules, haloimidazoles may serve as electrophilic coupling partners for the Suzuki reaction. As described in section 9.1, regioselective bromination at the C(4) position could be achieved for a 1,5-dialkylimidazole using NBS in $CH_3CN$. The Suzuki coupling of 1-benzyl-4-bromo-5-methylimidazole with phenylboronic acid assembled the unsymmetrical heterobiaryl in 93% yield, whereas the corresponding Stille reaction with phenyltrimethyltin proceeded in only moderate yield (51%) [8].

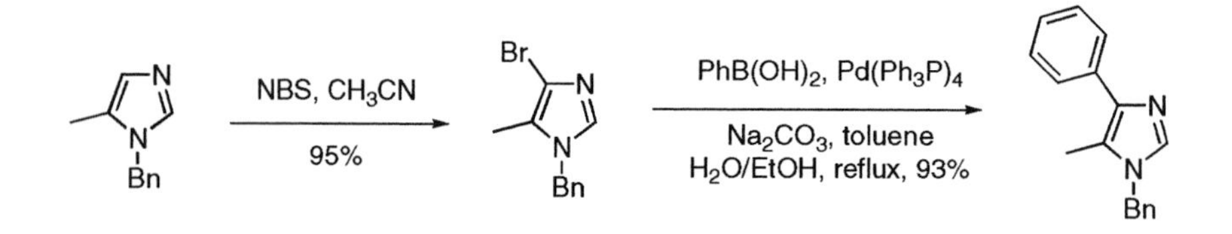

Nortopsentin C (**20**), along with nortopsentins A and B, was isolated from the marine sponge *Spongosorites ruetzleri*. They all possess a characteristic 2,4-bisindolylimidazole skeleton and exhibit cytotoxic and antifungal activities. Successive and regioselective diarylation *via* Suzuki coupling reactions using halogenated imidazole to make nortopsentins have been disclosed in the total synthesis of **20** by Ohta's group [19, 20]. The *N*-protected 2,4,5-triiodoimidazole (**21**) was coupled with one equivalent of the 3-indolylboronic acid **22** to install indolylimidazole **23**. The Suzuki reaction occurred regioselectively at C(2) of the imidazole ring. Subsequently, a regioselective halogen-metal exchange reaction took place predominantly at C(5) to access **24**. A second Suzuki coupling reaction at C(4) of **24** with 6-bromo-3-indolylboronic acid **25** resulted in the assembly of the entire skeleton. Two consecutive deprotection reactions removed all three silyl groups to deliver the natural product (**20**). This synthesis elegantly exemplifies the chemoselectivity of the Suzuki reaction between an arylbromide and an aryliodide. The desired regioselectivity was achieved *via* manipulation of the substrate functionalities.

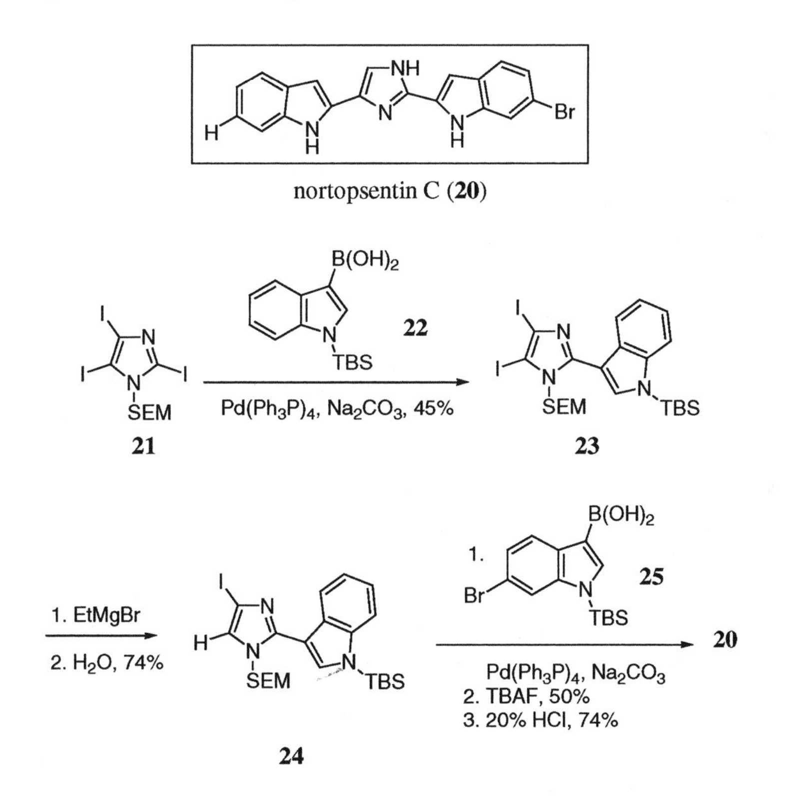

### 9.3.3 Stille coupling

In contrast to the scarcity of precedents in the imidazolylboronic acids, abundant reports exist on imidazolylstannanes. In the Stille reaction, the imidazole nucleus may serve as either the nucleophile as an imidazole halide or the electrophile as an imidazolylstannane. One advantage of the Stille reaction is that it often tolerates delicate functionalities in both partners.

One method for preparing imidazolylstannanes is direct metalation followed by treatment with $R_3SnCl$ [21]. 1-Methyl-2-tributylstannylimidazole, derived in such manner, was coupled with 3-bromobenzylphosphonate (**26**) to furnish heterobiaryl phosphonate **27** [22]. Under the same reaction conditions, 4-bromobenzylphosphonate led to the adduct in 69% yield, whereas only 24% yield was obtained for 2-bromobenzylphosphonate. The low yield encountered for the *ortho* derivative may be attributed to the steric factors to which the Stille reaction has been reported to be sensitive [23]. Heterobiaryl phosphonates such as **27** are not only substrates for the Wadsworth–Horner–Emmons reaction, but also bioisosteric analogs of the carboxylic acid group.

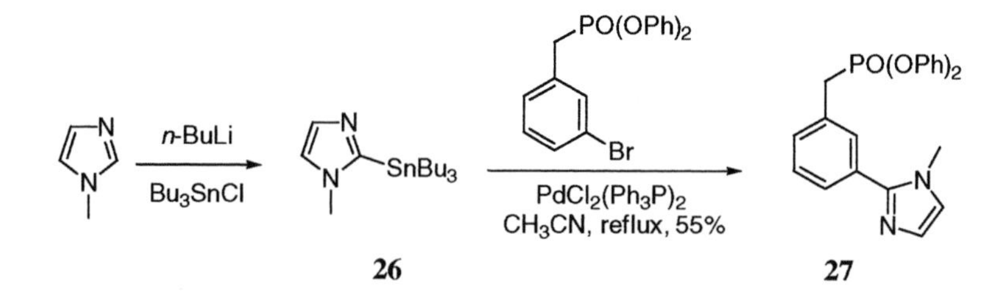


Analogously, 5-tributylstannylimidazole **29** was easily obtained from the regioselective deprotonation of 1,2-disubstituted imidazole **28** at C(5) followed by treatment with tributyltin chloride [24]. In the presence of 2.6 equivalents of LiCl, the Stille reaction of **29** with aryl triflate **30** afforded the desired 1,2,5-trisubstituted imidazole **31** with 2,6-di-*tert*-butyl-4-methylphenol (BHT) as a radical scavenger. Reversal of the nucleophile and electrophile of the Stille reaction also provided satisfactory results. For example, the coupling reaction of 5-bromoimidazole **33**, derived from imidazole **32** *via* a regioselective bromination at C(5), and vinylstannane **34** produced adduct **35** [24].

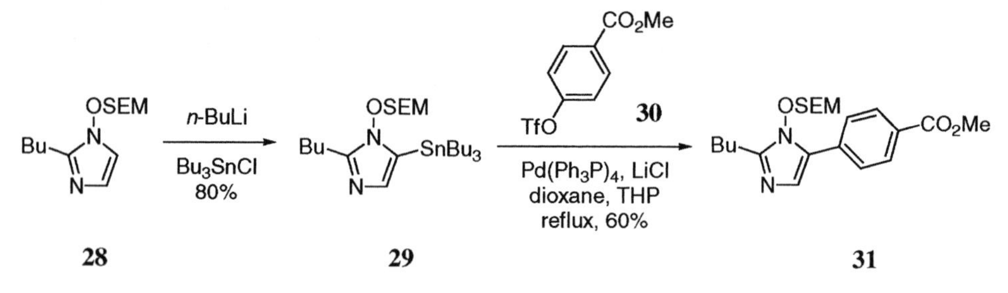

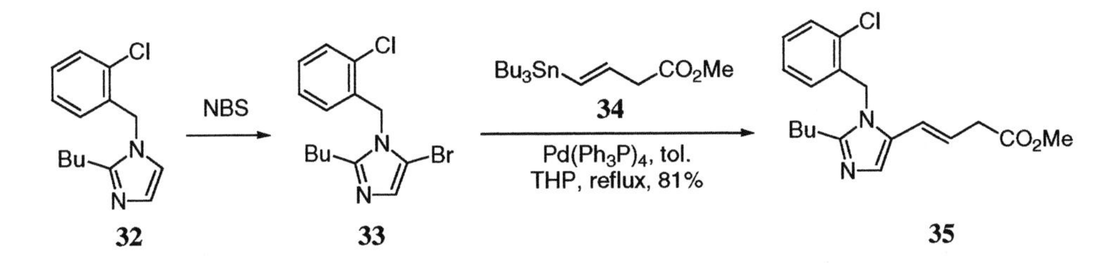


Steglich's group coupled *N*-methyl bromoindolylmaleimide **36** with 5-tributylstannyl-1-methyl-1*H*-imidazole to afford **37** and then converted **37** to didemnimide C (**38**), an alkaloid isolated from the Caribbean ascidian *Didemnum conchyliatum* [25]. 5-Tributylstannyl-1-methyl-1*H*-imidazole, in turn, was prepared according to Undheim's direct metalation method from 1-methyl-1*H*-imidazole [26].

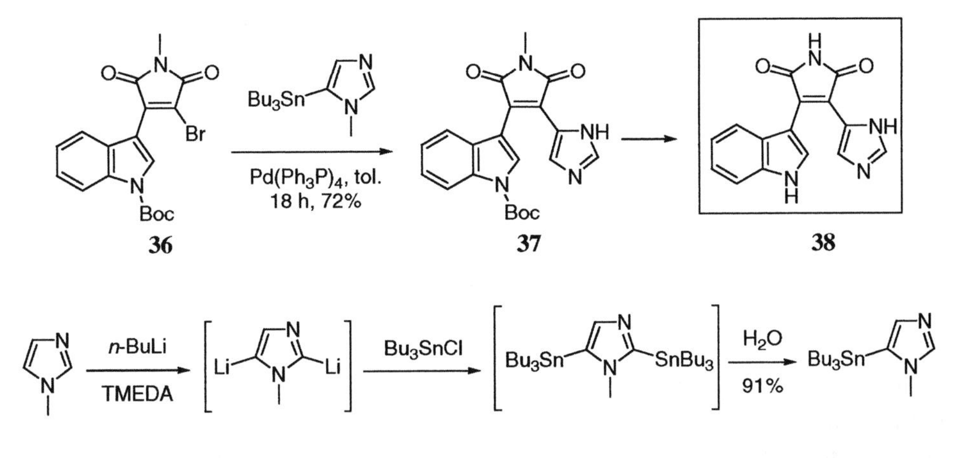


Achad conducted a Stille coupling using 4-iodoimidazole **39** as the electrophilic coupling partner in his synthesis of topsentin. The connection between an indole and an imidazole was realized by the Stille reaction of **39** with indolylstannane **40** to furnish adduct **41**. Global deprotection then transformed **41** into topsentin (**42**), a marine bis-indole alkaloid [27]. In another case, Hibino *et al.* carried out the Stille reaction of stannylimidazole **43** with 3-iodoindole **44** to assemble indolylimidazole **45**, an important intermediate for the total synthesis of grossularines-1 and -2 [28]. Stannane **43**, in turn, was synthesized from the corresponding 5-bromoimidazole *via* halogen-metal exchange followed by treatment with $Me_3SnCl$.

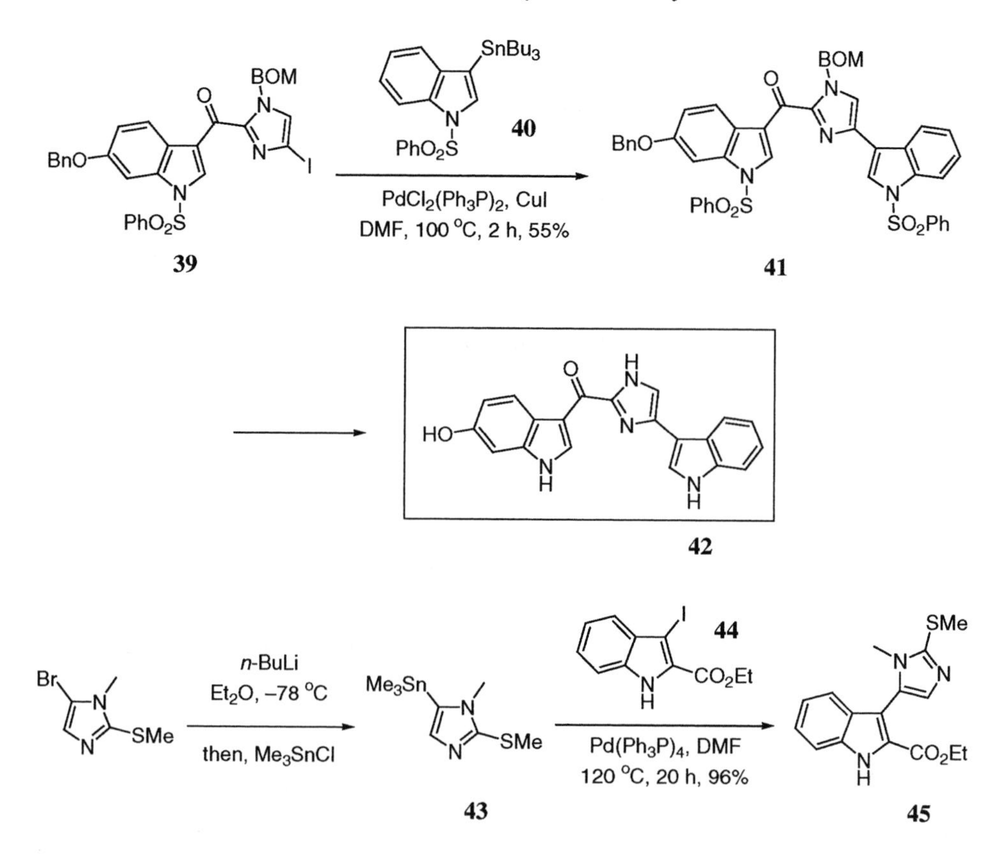


Interestingly, with only slight difference in the imidazole *N*-protection, an anomalous Stille reaction occurred for the stannane **43** [29]. When the protecting group of the 5-stannylimidazole was changed from methyl into 1-ethoxymethyl, the reaction between 3-iodoindole **44** and 1-ethoxymethyl-5-bromo-2-methylthioimidazole (**46**) gave not only the normal *ipso* product **47** but also the *cine* product **48**.

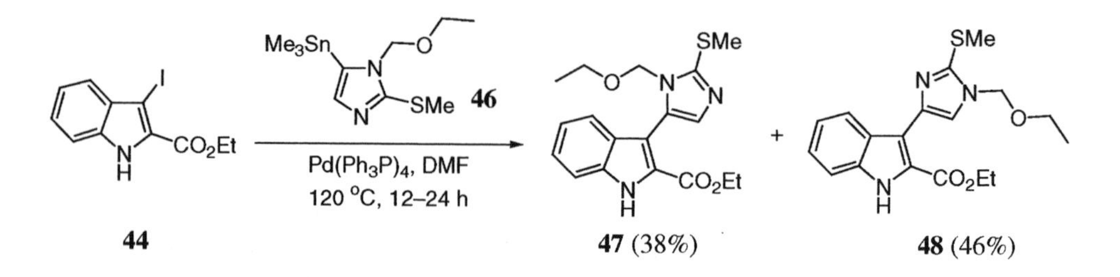


Additional literature precedents for the Stille reactions in which imidazolylstannanes are used are listed below [30–32].

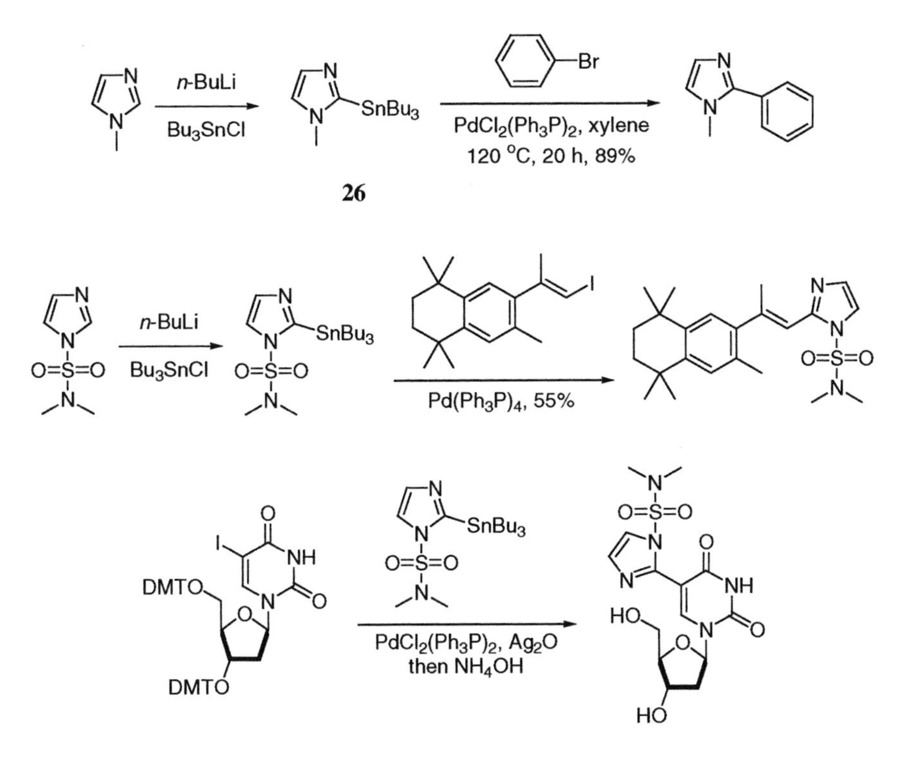


Most literature precedents assemble the heterobiaryls at the C(4) position. In one instance, however, the substitution took place at the C(5) position. As shown below, treatment of 5-iodoimidazole derivative **4** with (*E*)-1-tributylstannylprop-1-en-3-ol (**49**) in the presence of $PdCl_2(PhCN)_2$ led to adduct **50** in 68% yield [33].

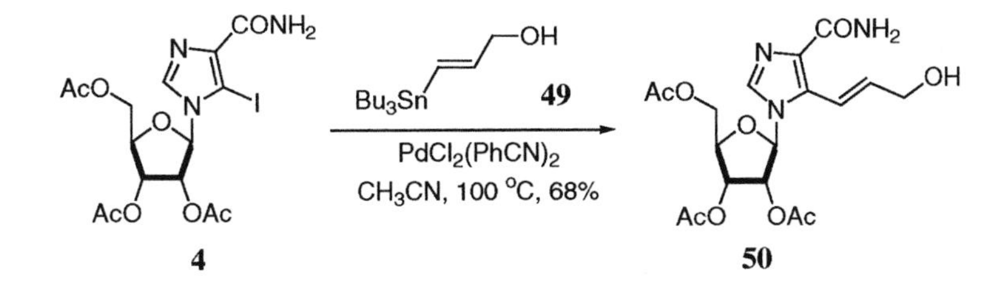


*N*-Protected 4-iodoimidazoles have been employed to synthesize an assortment of imidazole–heteroaryls [34, 35]. By taking advantage of the Stille reaction, **55**, a naturally occurring imidazole alkaloid was synthesized [36–39]. Pyne *et al.* protected 4-iodoimidazole with chloromethylethyl ether, forming two tautomers **51** and **52** in a 1:9 ratio. After isolation, the Stille coupling of **52** with vinylstannane **53** [(*E*) : (*Z*) = 88 : 22] afforded isomerically pure alkene **54**, which was then manipulated in five additional steps into (1*R*, 2*S*, 3*R*)-2-acetyl-4(5)-(1,2,3,4-tetrahydroxybutyl)imidazole (THI, **55**), a constituent of Caramel Colour III.

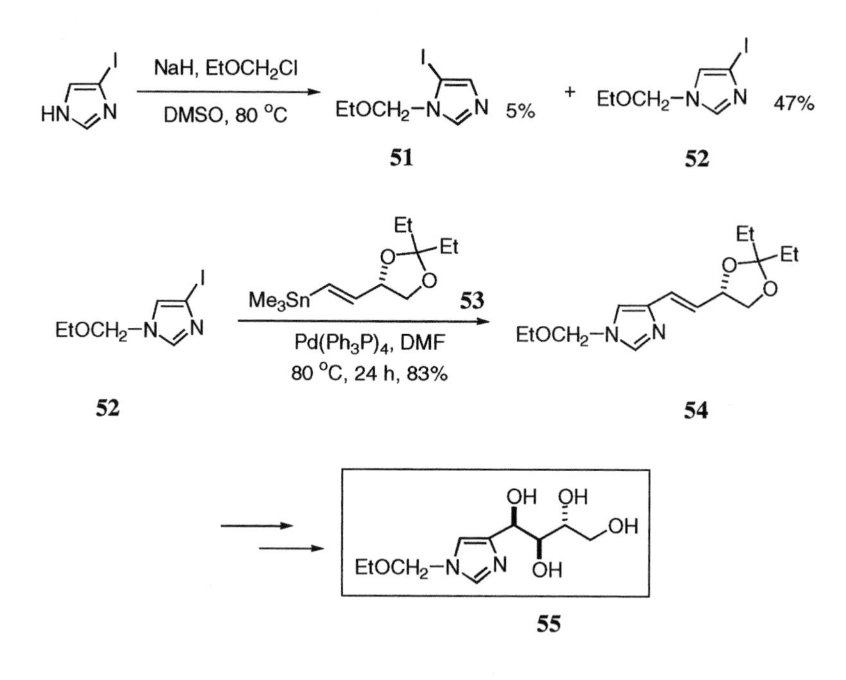


## 9.4 Sonogashira reaction

Regardless of the position of the iodide [C(2), C(4), or C(5)] on the imidazole ring, the Sonogashira reaction of an iodo-*N*-methylimidazole and simple acetylenes gave unsatisfactory results during initial investigations [40]. However, when diiodoimidazole **56**, derived from iodination of 1,2-dimethylimidazole, was submitted to the classic Sonogashira reaction conditions with trimethylsilylacetylene, diyne **57** was isolated in 68% yield. The same reaction using propargyl alcohol did not work as well (17%, 34% yields) for substrate **56** [4].

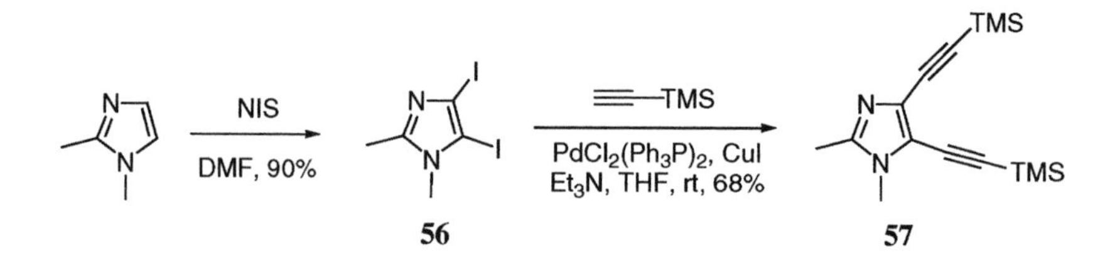

In contrast, an excellent yield was obtained for the reaction between 5-iodoimidazole and propargyl alcohol *in the absence of CuI* to install the 5-alkynyl derivative [10, 41]. The adduct was then deprotected to 5-alkynyl-1-β-*D*-ribofuranosylimidazole-4-carboxamide **58**, an antileukemic agent. It is noteworthy that if the Sonogashira reaction was conducted in the presence of CuI, the yield dropped to 19%.

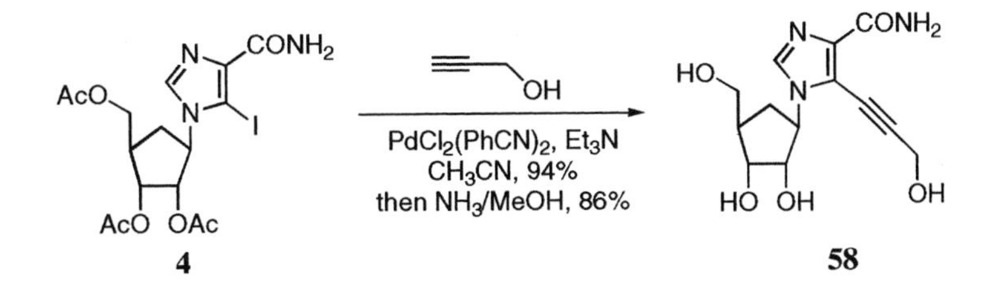


Regioselective functionalization of histidine at the C(2) position of the imidazole nucleus has been a synthetic challenge. The Evans group observed an anomalous phenomenon when diiodide **59** was subjected to standard Sonogashira reaction conditions [17]. When the Sonogashira reaction of diiodoimidazole **59** was carried out under normal conditions using phosphine ligands, adduct **60** was isolated in 52% yield. This pivotal reaction thus solved the dual problems of the dehalogenation at the C(4) atom and the bond construction at the C(2) atom in a single operation. More interestingly, when the same reaction was effected in the presence of $Pd(dba)_2$, phenylacetylene served as *a reducing agent* to afford the C(2) monoiodoimidazole by a selective H—I exchange at the C(4) atom.

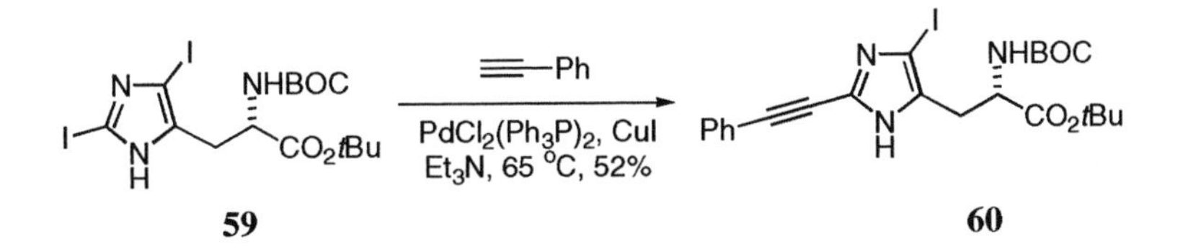


## 9.5 Heck and heteroaryl Heck reactions

Generally, the intermolecular Heck reaction between 2-iodo-, 4-iodo- and 5-iodo-1-methylimidazoles and olefins suffers from low yields (< 25%). Therefore, these transformations are of limited synthetic utility [29]. In one case, variable yields for adduct **62** (15–58%) were observed for the Heck reaction of 5-bromo-1-methyl-2-phenylthio-1*H*-imidazole (**61**) and a large excess of methyl acrylate [42].

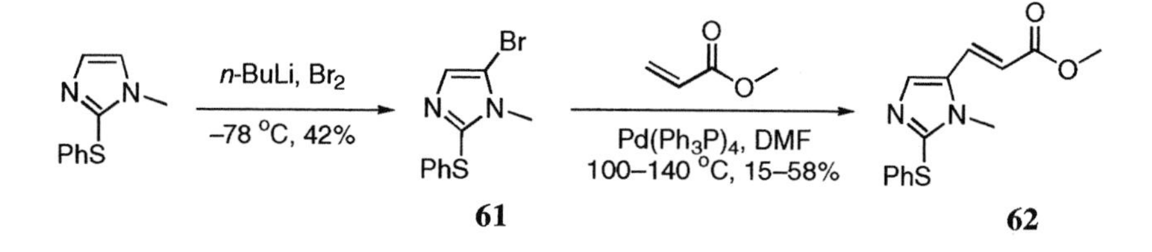


Ohta's group thoroughly studied the heteroaryl Heck reactions of chloropyrazines and π-electron-rich heteroaryls [42–44]. The substitution occurred at the electron-rich C(5) position of the imidazole ring for the heteroaryl Heck reaction of 2-chloro-3,6-dimethylpyrazine and *N*-methylimidazole.

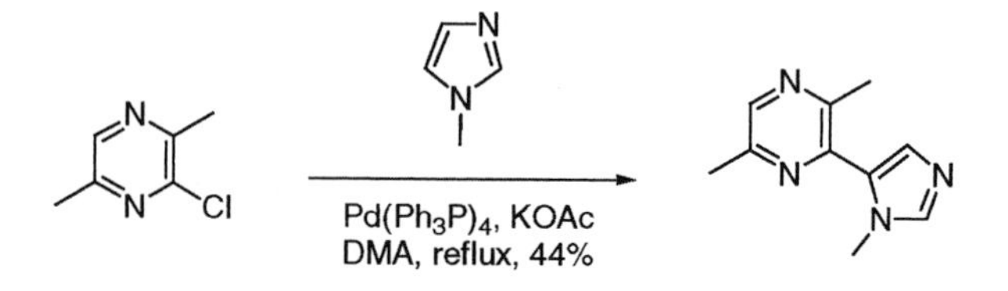


Miura's group carried out a heteroaryl Heck reaction of bromobenzene and 1-methylimidazole and isolated both mono-arylation (53%) and bis-arylation products [45]. In accord with Ohta's observation, the first arylation took place at the electron-rich C(5) and the second arylation occurred at the more electron-poor C(2).

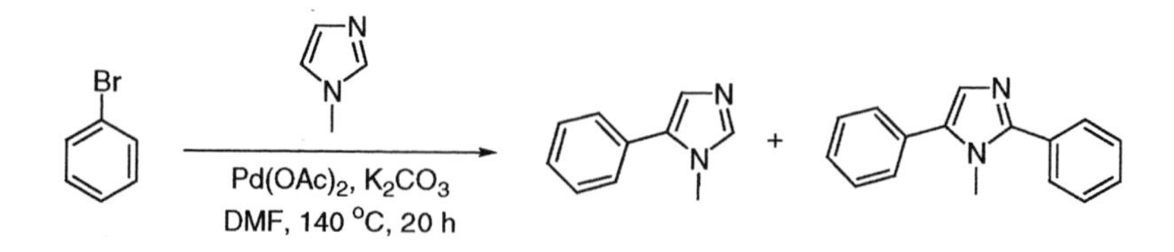


An intramolecular heteroaryl Heck was the pivotal step in the synthesis of 5-butyl-1-methyl-1*H*-imidazo[4,5-*c*]quinolin-4(5*H*)-one (**63**), a potent antiasthmatic agent [46]. The optimum yield was obtained under Jeffery's "ligand-free" conditions, echoing Ohta's observation for the intermolecular version. Once again, the $C_{aryl}$—$C_{aryl}$ bond was constructed at the C(5) position of the imidazole ring. Another intramolecular heteroaryl Heck cyclization of pyrrole and imidazole derivatives was also reported to assemble annulated isoindoles [47].

## 9.6 Tsuji–Trost reaction

The Tsuji–Trost reaction is the Pd(0)-catalyzed allylation of a nucleophile [48–51]. The NH group in imidazole can take part as a nucleophile in the Tsuji–Trost reaction, whose applications are found in both nucleoside and carbohydrate chemistry. Starting from cyclopentadiene and paraformaldehyde, cyclopentenyl allylic acetate **64** was prepared in diastereomerically-enriched form *via* a Prins reaction [52]. Treating **64** with imidazole under Pd(0) catalysis provided the *N*-alkylated imidazole **65**.

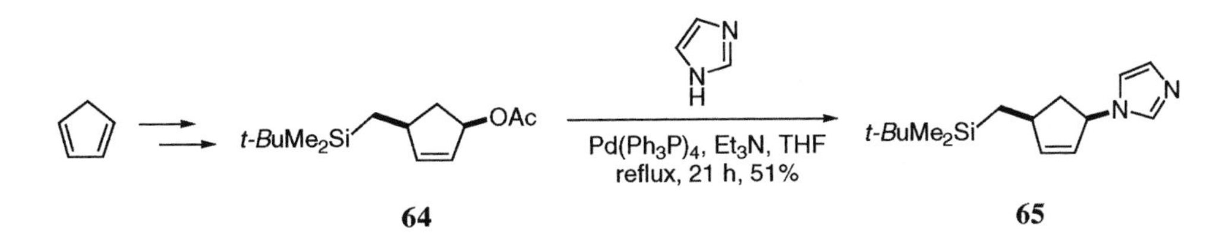


Extending the aforementioned methodology from imidazole to adenine, the Tsuji–Trost reaction between the sodium salt of adenine and allylic acetate **66** gave **67** as a 82:18 mixture of *cis:trans* isomers. Carbocyclic nucleoside **67** was advantageous over normal nucleosides as a drug candidate because it was not susceptible to degradation *in vivo* by nucleosidases and phosphorylases [52].

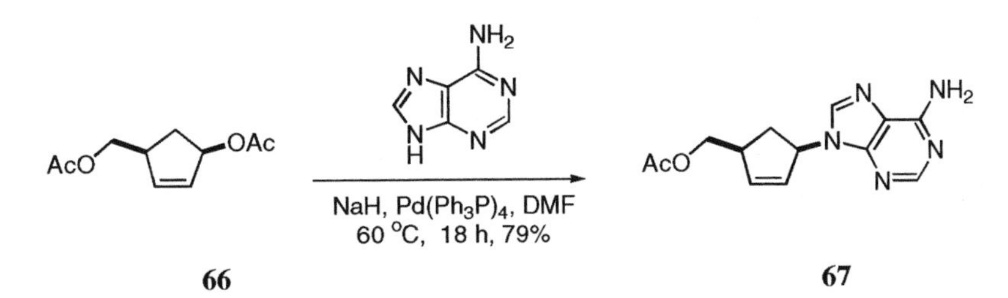

In the carbohydrate chemistry arena, the Tsuji–Trost reaction has been applied to construct *N*-glycosidic bonds [53]. In the presence of $Pd_2(dba)_3$, the reaction of 2,3-unsaturated hexopyranoside **68** and imidazole afforded *N*-glycopyranoside **69** regiospecifically at the anomeric center with retention of configuration. In terms of the stereochemistry, the oxidative addition of allylic substrate **68** to Pd(0) formed the $\pi$-allyl complex with inversion of configuration, then nucleophilic attack by imidazole proceeded with another inversion of the configuration. Therefore, the overall stereochemical outcome is retention of configuration.

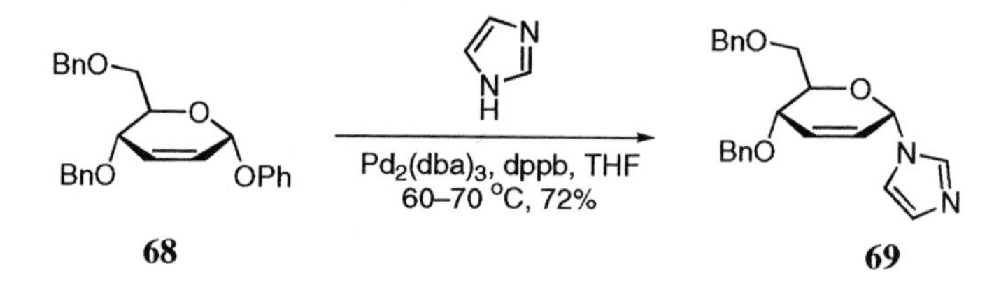


Moreover, Pd(0)-catalyzed allylations of imidazole with cyclopentadiene monoepoxide led to imidazole-substituted cyclopentenol in moderate yield [54].

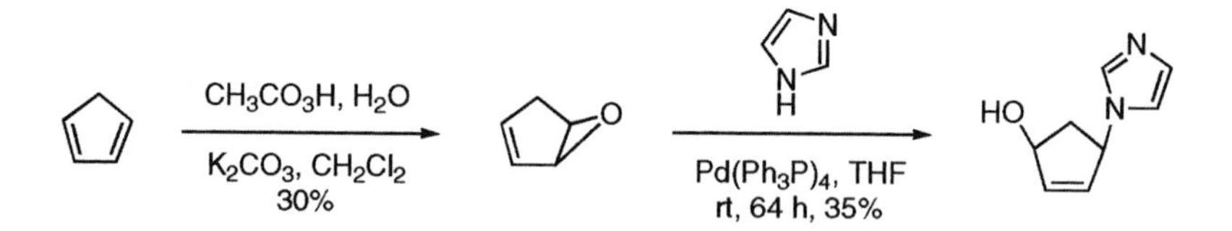


Pd(0)-catalyzed allylations of 4(5)-nitroimidazole, 2-methyl-4(5)-nitroimidazole, 4(5)-bromoimidazole and 4(5)-methoxyimidazole resulted in complicated mixtures, which did not necessarily reflect the tautomeric ratios of the starting material [7]. For example, poor regioselectivity for the products (**70** and **71**) was observed in the Tsuji–Trost reaction of 4(5)-bromoimidazole with cinnamyl carbonate. However, the same reaction with 4(5)-nitroimidazole and 2-methyl-4(5)-nitroimidazole led predominantly to the 1-allylation products. In addition, removal of the *N*-imidazole allyl groups can be selectively effected under mild conditions by Pd-catalyzed $\pi$-allyl chemistry [55].

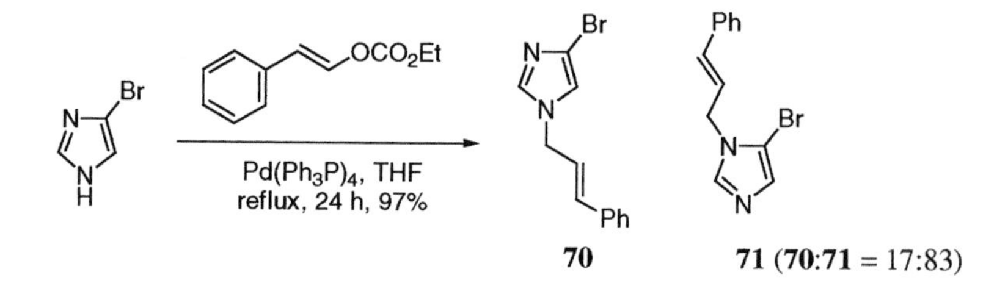

## 9.7 Phosphonylation

Phosphorus esters are important isosteres for carboxylic acid derivatives possessing similar structural and electronic properties. Although formation of a $C_{sp3}$—P bond is relatively straightforward, making a $C_{sp2}$—P bond is more cumbersome. Recently, Pd(0)-catalyzed coupling reactions have opened up a new venue for $C_{sp2}$—P bond creation.
In order to synthesize biologically relevant phosphonylimidazole **73**, bromoimidazole **72** was derived from radical-initiated bromination of methyl 1-*p*-methoxybenzyl-2-thiomethyl-5-imidazolylcarboxylate (**71**) [56]. The thiomethyl group served to block the C(2) position, which would otherwise undergo preferential halogenation under these conditions. As expected, a variety of Arbusov-Michaelis reaction conditions failed even under forcing conditions. On the other hand, Pd-catalyzed phosphorylation of **72** with diethyl phosphite led to methyl-4-diethylphosphonyl-1-*p*-methoxybenzyl-2-thiomethyl-5-imidazolylcarboxylate (**73**). After further manipulations, the desired phosphonic acid-linked aminoimidazoles, which resembled intermediates formed during purine biosynthesis, were accessed.

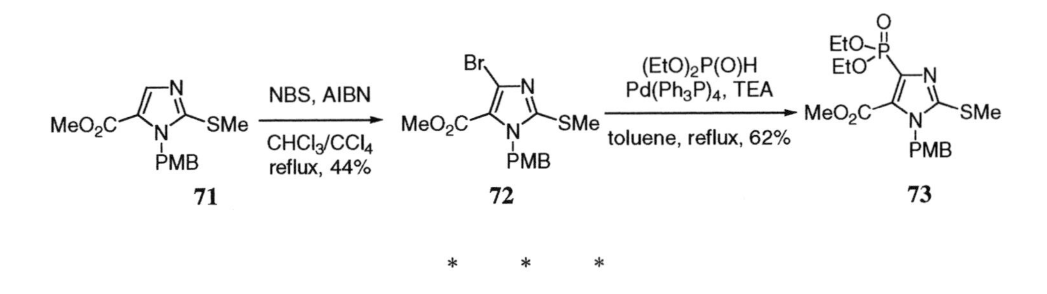


* * *

In conclusion, the imidazolylzinc and imidazolyltin reagents are undoubtedly easier to prepare than the corresponding imidazolylboronic acids. As a consequence, if the imidazole fragment is to serve as a nucleophile in a Pd-catalyzed cross-coupling reaction, the Negishi and Stille reactions are better choices than the Suzuki coupling. However, if the imidazole fragment is to serve as an electrophile, the choice is not that crucial. In other words, most imidazole-containing molecules may be assembled by judiciously choosing appropriate coupling partners.
In triiodoimidazole **21**, a selective Suzuki coupling was achieved at C(2), whereas a selective halogen-metal exchange in a diiodoimidazole **23** was realized at C(5). In addition, the Sonogashira reactions of haloimidazoles appear to be very substrate-dependent.

## 9.8 References


1. Kirk, K. L. *J. Org. Chem.* **1978**, *43*, 4381–3.
2. Bond, R. F.; Bredenkamp, M. W.; Holzafel, C. W. *Synth. Commun.* **1989**, *19*, 2551–66.
3. Bell, A. S.; Campbell, S. F.; Morris, D. S.; Roberts, D. A.; Stefaniak, M. H. *J. Med. Chem.* **1989**, *32*, 1552–8.
4. Kim, G.; Kang, S.; Ryu, Y.; Keum, G.; Seo, M. J. *Synth. Commun.* **1999**, *29*, 507–12.
5. Stensiö, K. E.; Wahlberg, K.; Wahren, R. *Acta Chem. Scand., Sect. B.* **1973**, *27*, 2179–83.
6. Cliff, M. D.; Pyne, S. G. *Synthesis* **1994**, 681–2.
7. Katritzky, A. R.; Slawinski, J. J.; Brunner, F.; Gorun, S. *J. Chem. Soc., Perkin Trans. 1*, **1989**, 1139–45.
8. Wang, D.; Haseltine, J. *J. Heterocycl. Chem.* **1994**, *31*, 1637–9.
9. Matsuda, A.; Minakawa, N.; Sasaki, T.; Ueda, T. *Chem. Pharm. Bull.* **1988**, *36*, 2730–3.
10. Minakawa, N.; Takeda, T.; Sasaki, T.; Matsuda, A.; Ueda, T. *J. Med. Chem.* **1991**, *34*, 778–86.
11. Arnau, N.; Arredondo, Y.; Moreno–Mañas, M.; Pleixats, R.; Villarroya, M. *J. Heterocycl. Chem.* **1995**, *32*, 1325–34.
12. Bell, A. S.; Roberts, D. A.; Ruddock, K. S. *Tetrahedron Lett.* **1988**, *29*, 5013–6.
13. Heys, J. R.; Villani, A. J.; Mastrocola, A. R. *J. Labeled Compds. Radiopharm.* **1996**, *38*, 761–9.
14. Jetter, M. C.; Reitz, A. B. *Synthesis* **1998**, 829–31.
15. Turner, R. M.; Ley, S. V.; Lindell, S. D. *Synlett* **1993**, 748–50.
16. Prasad, A. S.; B.; Stevenson, T. M.; Citineni, J. R.; Zyzam, V.; Knochel, P. *Tetrahedron* **1997**, *53*, 7237–54.
17. Evans, D. A.; Bach, T. *Angew. Chem., Int. Ed. Engl.* **1993**, *32*, 1326–7.
18. Kawasaki, I.; Yamashita, M.; Ohta, S. *J. Chem. Soc., Chem. Commun.* **1994**, *18*, 2085–6.
19. Kawasaki, I.; Yamashita, M.; Ohta, S. *Chem. Pharm. Bull.* **1996**, *44*, 1831–9. and references cited therein.
20. Kawasaki, I.; Katsuma, H.; Nakayama, Y.; Yamashita, M.; Ohta, S. *Heterocycles* **1998**, *48*, 1887–901.
21. Molloy, K. C.; Waterfield, P. C.; Mahon, M. F. *J. Organomet. Chem.* **1989**, *365*, 61–73.
22. Kennedy, G.; Perboni, A. D. *Tetrahedron Lett.* **1996**, *37*, 7611–4.
23. Saá, J. M.; Martel, J. M.; García-Rosa, G. *J. Org. Chem.* **1992**, *57*, 678–85.
24. Keenan, R. M.; Weinstock, J.; Finkelstein, J. A.; Franz, R. G.; Gaitanopoulos, D. E.; Girard, G. R.; Hill, D. T.; Morgan, T. M.; Samanen, J. M.; *et al. J. Med. Chem.* **1992**, *35*, 3858–72.
25. Terpin, A.; Winklhofer, C.; Schumann, S.; Steglich, W. *Tetrahedron* **1998**, *54*, 1745–52.
26. Gaare, K.; Repstad, T.; Benneche, T.; Undheim, K. *Acta Chem. Scand.* **1993**, *47*, 57–62.

27. S. Achad, *Tetrahedron Lett.* **1996**, *37*: 5503–6.
28. Choshi, T.; Yamada, Nobuhiro, J.; Mihara, Y.; S.; Sugino, E.; Hibino, S. *Heterocycles* **1998**, *48*, 11–4.
29. Choshi, T.; Yamada, S.; Sugino, E.; Kuwada, T.; Hibino, S. *J. Org. Chem.* **1995**, *60*, 5899–904.
30. Kosugi, M.; Koshiba, M.; Atoh, A.; Sano, H.; Migita, T. *Bull. Chem. Soc. Jpn.* **1986**, *59*, 677–9.
31. Beard, R. L.; Colon, D. F.; Klein, E. S.; Vorse, K. A.; Chandraratna, R. A. S. *Biorg. Med. Chem. Lett.* **1995** *5*, 2729–34.
32. Gutierrez, A. J.; Terhorst, T. J.; Matteucci, M. D.; Froehler, B. C. *J. Am. Chem. Soc.* **1994**, *116*, 5540–4.
33. Minakawa, N.; Sasaki, T.; Matsuda, A. *Biorg. Med. Chem. Lett.* **1993**, *3*, 183–6.
34. Shi, G.; Cao, Z.; Zhang, X. *J. Org. Chem.* **1995**, *60*, 6608–11.
35. Hutchinson, J. H.; Cook, J. J.; Brashear, K. M.; Breslin, M. J.; Glass, J. D.; Gould, R. J.; Halczenko, W.; Holahan, M. A.; Lynch, R. J.; *et al. J. Med. Chem.* **1996**, *39*, 4583–91.
36. Cliff, M. D.; Pyne, S. G. *J. Org. Chem.* **1995**, *60*, 2378–83.
37. Cliff, M. D.; Pyne, S. G. *Tetrahedron Lett.* **1995**, *36*, 5969–72.
38. Cliff, M. D.; Pyne, S. G. *Tetrahedron* **1996**, *52*, 13703–12.
39. Cliff, M. D.; Pyne, S. G. *J. Org. Chem.* **1997**, *62*, 1023–32.
40. Sakamoto, T.; Nagata, H.; Kondo, Y.; Shiraiwa, M.; Yamanaka, H. *Chem. Pharm. Bull.* **1987**, *35*, 823–8.
41. Matsuda, A.; Minakawa, N.; Ueda, T. *Nucleic Acids Symp. Ser.* **1988**, *20*, 13–4.
42. Yamashita, M.; Oda, M.; Hayashi, K.; Kawasaki, I.; Ohta, S. *Heterocycles* **1998**, *48*, 2543–50.
43. Ohta, A.; Akita, Y.; Ohkuwa, T.; Chiba, M.; Fukunaka, R.; Miyafuji, A.; Nakata, T.; Tani, N. *Heterocycles* **1990**, *31*, 1951–7.
44. Aoyagi, Y.; Inoue, A.; Koizumi, I.; Hashimoto, R.; Tokunaga, K.; Gohma, K.; Komatsu, J.; Sekine, K.; Miyafuji, A.; Konoh, J. Honma, R. Akita, Y.; Ohta, A. *Heterocycles* **1992**, *33*, 257–72.
45. Pivsa-Art, S.; Satph, T.; Yawamura, Y.; Miura, M.; Nomura, M. *Bull. Chem. Soc. Jpn.* **1998**, *71*, 467–73.
46. Kuroda, T.; Suzuki, F. *Tetrahedron Lett.* **1991**, *32*, 6915–8.
47. Huang, J.; Du, M. *Youji Huaxue* **1994**, *14,* 604–8.
48. Tsuji, J.; Takahashi, H.; Morikawa, M. *Tetrahedron Lett.* **1965**, 4387–90.
49. Tsuji, J. *Acc. Chem. Res.* **1969**, *2*, 144–52;
50. Godleski, S. A. *in "Comprehensive Organic Synthesis*" (Trost, B. M. and Fleming, I. eds.), vol. 4. Chapter 3.3. Pergamon, Oxford, **1991**.
51. Moreno-Mañas, M.; Pleixats, R. in "*Advances in Heterocyclic Chemistry*" (Katritzky, A. R., ed.), **1996**, *66*, 73–129, Academic Press.

52. Saville-Stones, E. A.; Lindell, S. D.; Jennings, N. S.; Head, J. C.; Ford, M. J. *J. Chem. Soc., Perkin Trans. 1* **1991**, 2603–4.
53. Bolitt, V.; Chaguir, B.; Sinou, D. *Tetrahedron Lett.* **1992**, *33*, 2481–4.
54. Arnau, N.; Cortés, J.; Moreno-Mañas, M.; Pleixats, R.; Villarroya, M. *J. Heterocycl. Chem.* **1997**, *34*, 233–9.
55. Kimbonguila, A. M.; Boucida, S.; Guibe, F.; Loffet, A. *Tetrahedron* **1997**, *53*, 12525–38.
56. Lin, J.; Thompson, C. M. *J. Heterocycl. Chem.* **1994**, *31*, 1701–5.

# CHAPTER 10

# Pyrazines and Quinoxalines

## 10.1 Pyrazines

Minuscule quantities of naturally-occurring pyrazines have been found in some foodstuffs and are largely responsible for their flavor and aroma. For example, 3-isopropyl-2-methoxypyrazine is isolated from green peas and wine and a seasoned wine connoisseur can identify a ppt quantity. In addition, 2-methyl-6-vinylpyrazine exists in coffee.

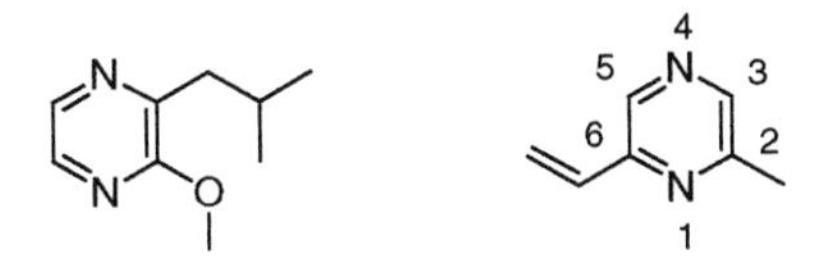


Pyrazine is an electron-deficient, 6π-electron heteroaromatic compound. The inductive effects of the nitrogen atoms induce a partially positive charge on the carbon atoms. As a consequence, oxidative addition of chloropyrazine takes place more readily than chlorobenzene although the C—Cl bond lengths for both chloropyrazine and chlorobenzene are virtually the same—1.733 Å [1] and 1.745 Å [2], respectively. Simple chloropyrazines and their *N*-oxides undergo a wide range of palladium-catalyzed carbon–carbon bond forming reactions under standard palladium-catalyzed reaction conditions. At this point, it is worth noting that the oxidative addition of chlorobenzene to Pd(0) does occur using sterically hindered, electron-rich phosphine ligands as reported by Reetz, Fu and Buchwald [3, 4]. This enhanced reactivity may be ascribed to the observation that oxidative addition of an aryl chloride is greatly promoted by electron-rich palladium complexes.

### 10.1.1 Coupling reactions with organometallic reagents

Pd-catalyzed coupling reactions between chloropyrazines and organometallic reagents *without β-sp³-hydrides* are straightforward and generally proceed in good yields. In 1981, Ohta *et al.* introduced a cyano group by refluxing chloropyrazine **1** with KCN in DMF in the presence of catalytic amount of $Pd(Ph_3P)_4$ to give cyanopyrazine **2** [5]. Analogously, cyanation of 2-amino-5-bromopyrazine with KCN in 18-crown-6 was facilitated by $Pd(Ph_3P)_4$ and CuI [6].

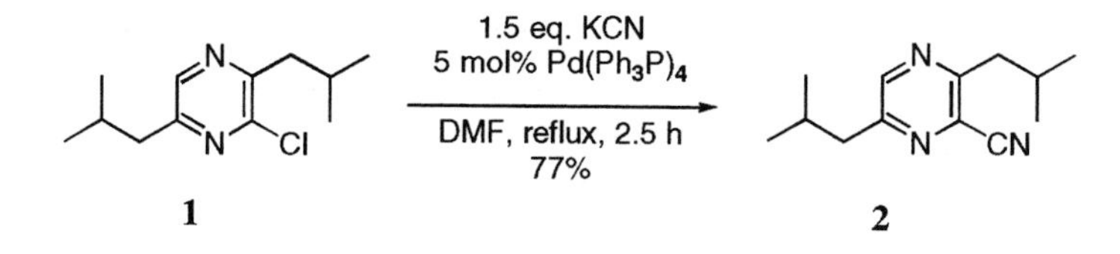


With palladium catalysis, both 2,6-dichloropyrazine **3** and chloropyrazine *N*-oxide **5** were methylated using trimethylaluminum to give adducts **4** and **6**, respectively [7, 8].

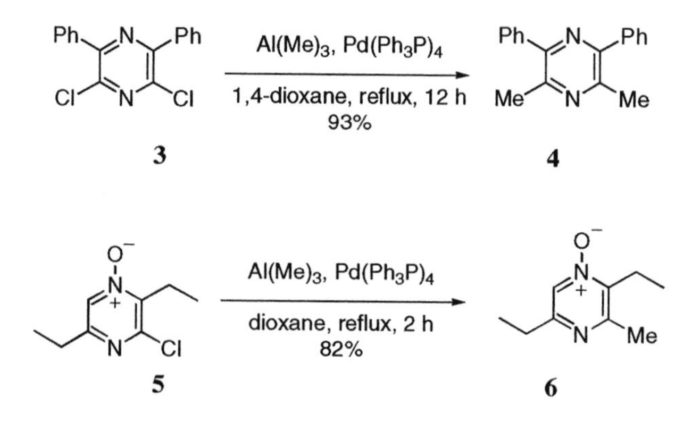


The Stille reaction of 2-chloro-3,6-diisopropylpyrazine (**7**) and 2-chloro-3,6-diisopropylpyrazine 4-oxide (**9**) with tetra(*p*-methoxyphenyl)stannane (readily prepared *in situ* from the corresponding Grignard reagent and $SnCl_4$) led to the corresponding arylation products **8** and **10**, respectively [9]. Additional Stille coupling reactions of chloropyrazines and their *N*-oxides have been carried out with tetraphenyltin [10] and aryl-, heteroaryl-, allyl- and alkylstannanes [11].

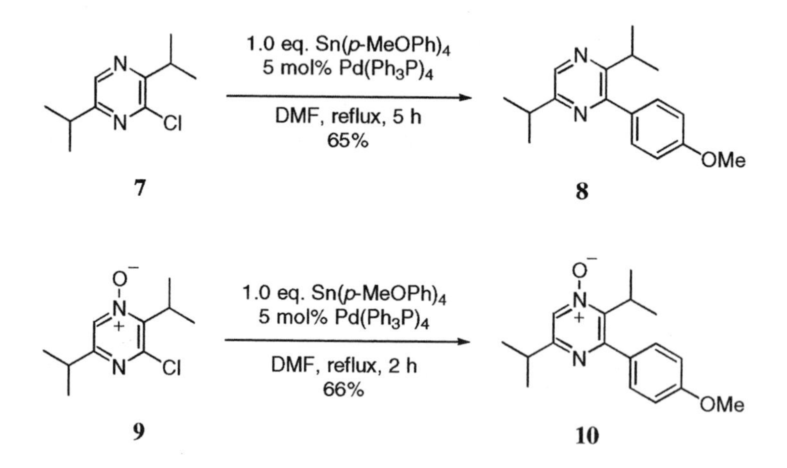

Intriguingly, the Stille coupling of quaternary pyridylstannane **12** with 2-chloropyrazine (**13**) proceeded to afford adduct **14** [12]. *N*-Methylated 3-(tributylstannyl)pyridine **12** was easily prepared by refluxing 3-(tributylstannyl)pyridine (**11**) with methyl tosylate in EtOAc. By contrast, only 29% yield of the coupling adduct was isolated from the Stille reaction of 3-(tributylstannyl)pyridine *N*-oxide and **13**.

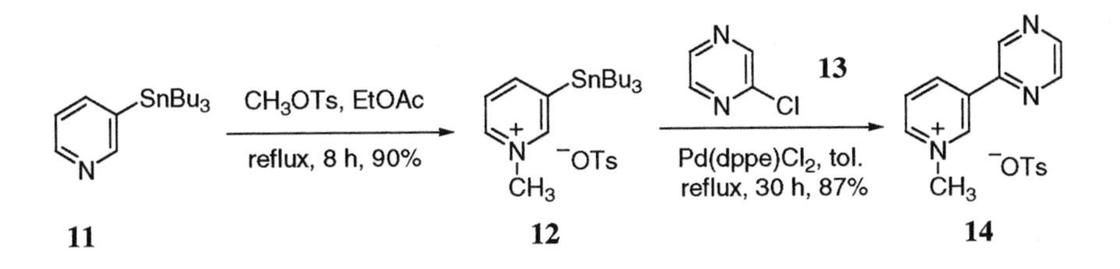


The Stille coupling of tetraethyltin and chloropyrazine **15** led to ethylpyrazine **16**, which was an important intermediate for preparing quinuclidinylpyrazine derivatives as muscarinic agonists [13]. In this particular case, the reductive elimination took place faster than β-hydride elimination.

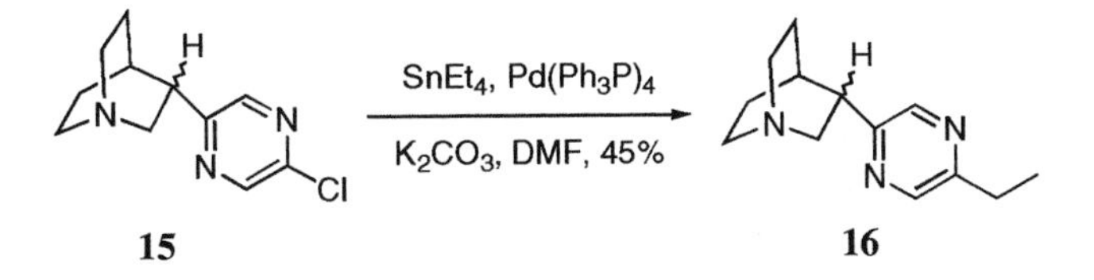


Although pyrazinylboron reagents are unknown, the Suzuki couplings have been conducted using halopyrazines and other boronic acids. As one of the early examples, 6-bromo-3-aminopyrazinoate **17** was coupled with 2-furylboronic acid to furnish furylpyrazine **18** [14]. Among three palladium catalysts examined, Pd(dppf)$(OAc)_2$ was found to be the best choice. In the same manner, 3-furyl and 4-pyridyl substituents were also introduced onto **16** in good yields. Moreover, bromopyrazine **19** and 2-thiopheneboronic acid were coupled to deliver thienylpyrazine **20** [15]. Also reported was a simpler version of a similar Suzuki coupling between 2-chloropyrazine and phenylboronic acid [16, 17].

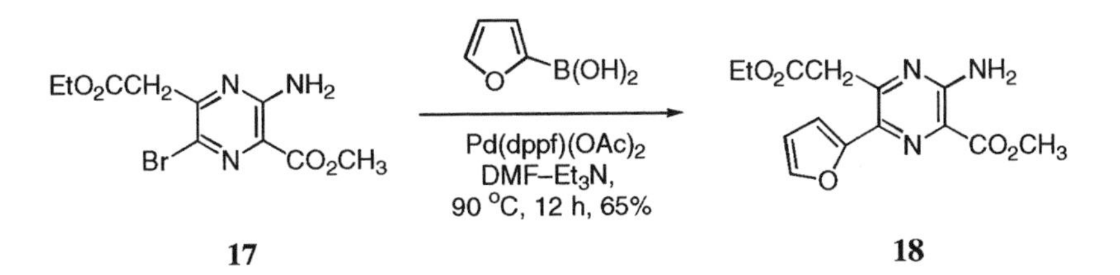

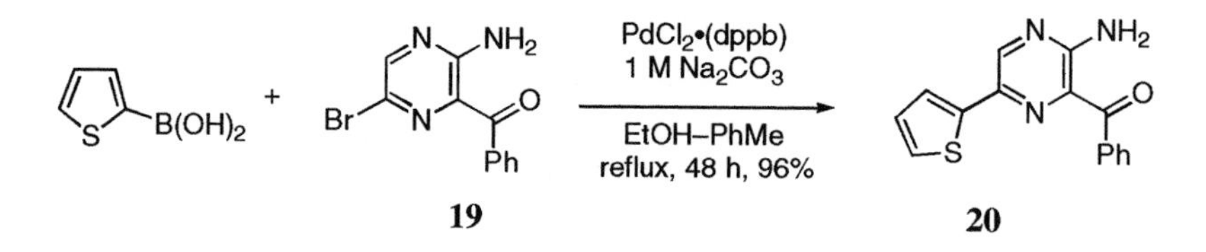


5-Indolylboronic acid (**21**), easily obtained from commercially available 5-bromoindole, was coupled with 2-chloropyrazine (**13**) to efficiently furnish 5-(2-pyrazinyl)indole (**22**) [18].

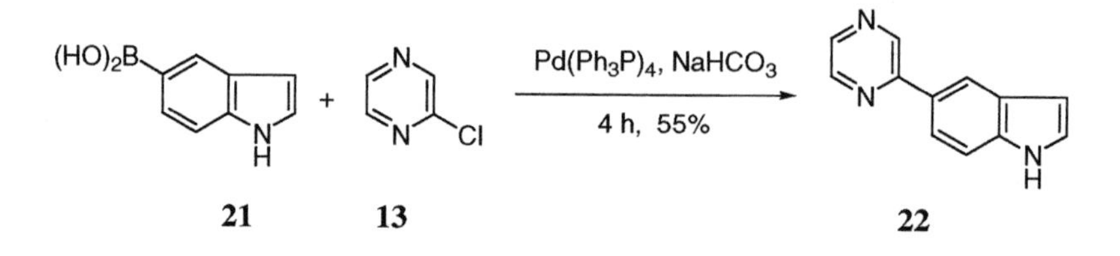


In contrast to aforementioned arylation with heteroaryls, palladium-catalyzed alkylation reactions of chloropyrazines with alkyl organometallic reagents bearing β-$sp^3$-hydrides have sometimes proven to be quite troublesome. In addition to coupling, concomitant dehalogenation may also take place [19, 20]. When 2-chloro-3,6-dimethylpyrazine (**23**) was heated with triethylaluminum in the presence of $K_2CO_3$ and $Pd(Ph_3P)_4$, the desired ethylation product **24** was obtained in only 23% yield along with 49% of the dechlorinated product, 2,5-dimethylpyrazine (**25**). Similar results were also observed when the alkylmetal was diethylzinc. The formation of dechlorinated product for this substrate was therefore unavoidable but could alleviated by the use of organoboron reagents. The Suzuki coupling of **23** with triethylborane gave adduct **24** in 74% yield along with only 5% of **25**.

In a similar fashion, ethylation reactions of the corresponding 1-oxide and 4-oxide of **23** have been described using triethylaluminum, diethylzinc and triethylborane. Pentylation and octylation of 2-chloropyrazines and its 4-oxides were also feasible using pentylstannane and octylstannane, but pentylation and octylation of 2-chloropyrazine 1-oxide failed [9].

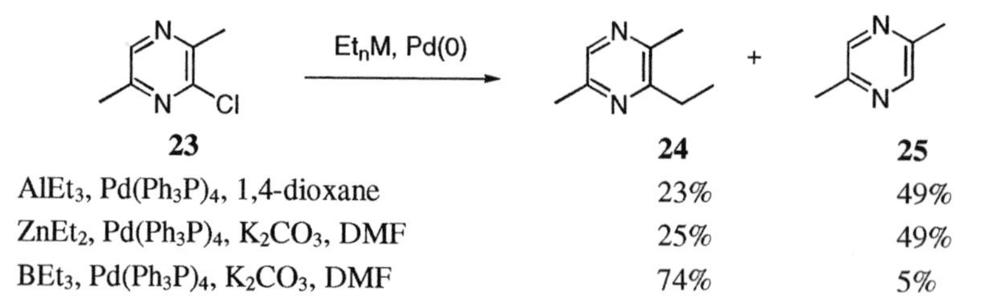


| | **24** | **25** |
|---|---|---|
| $AlEt_3$, $Pd(Ph_3P)_4$, 1,4-dioxane | 23% | 49% |
| $ZnEt_2$, $Pd(Ph_3P)_4$, $K_2CO_3$, DMF | 25% | 49% |
| $BEt_3$, $Pd(Ph_3P)_4$, $K_2CO_3$, DMF | 74% | 5% |

What is the mechanism for formation of dechlorinated product **25**? Presumably, oxidative addition of **23** to Pd(0) generates **26**, which undergoes a transmetalation process to give the ethylpalladium(II) species **27**. Intermediate **27** has two competing destinations: (a) reductive elimination to deliver the desired adduct **24**, and (b) β-hydride elimination to furnish palladium(II) hydride **28**, which subsequently undergoes a reductive elimination, giving rise to the dechlorinated product **25**.

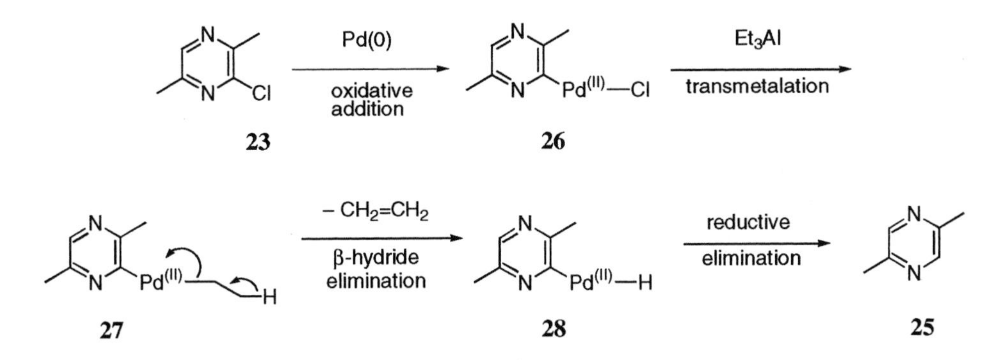


Dechlorination of chloropyrazines and their *N*-oxides was accomplished using either $Pd(Ph_3P)_4$ and $HCO_2Na$ in DMF [21, 22] or hydrogenation with $Pd(Ph_3P)_4$ and KOAc in DMF [23]. The *N*-oxide functionality was not reduced under these conditions.

### 10.1.2 Sonogashira reaction

Akita and Ohta disclosed one of the earliest Sonogashira reactions of chloropyrazines and their *N*-oxides [24, 25]. The union of 2-chloro-3,6-dimethylpyrazine (**23**) and phenylacetylene led to 2,5-dimethyl-3-phenylethynylpyrazine (**29**). Subsequent Lindlar reduction of adduct **29** then delivered (*Z*)-2,5-dimethyl-3-styrylpyrazine (**30**), a natural product isolated from mandibular gland secretion of the Argentine ants, *Iridomyrmex humilis*.

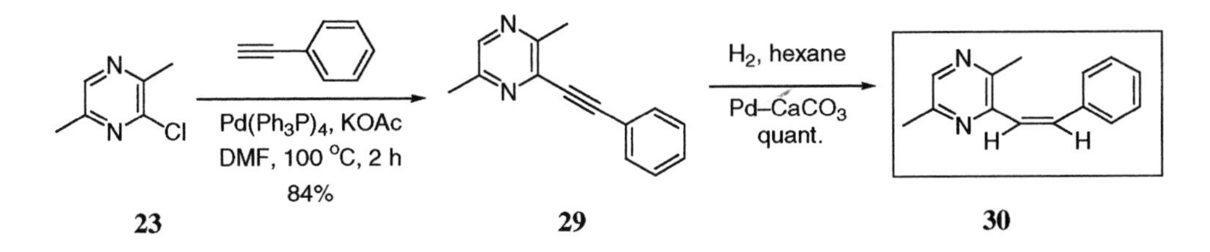


Initially, **31** was obtained as the Sonogashira adduct of 2-chloro-3,6-diisobutylpyrazine and trimethylsilylacetylene. Interestingly, **31** underwent an additional Sonogashira coupling with 2-chloropyrazine (**13**) to afford unsymmetrical 1,2-bispyrazinylacetylene **32** in excellent yield [26]. Here, desilylation occurred *in situ*, and the resulting terminal alkyne was then coupled with **13**.

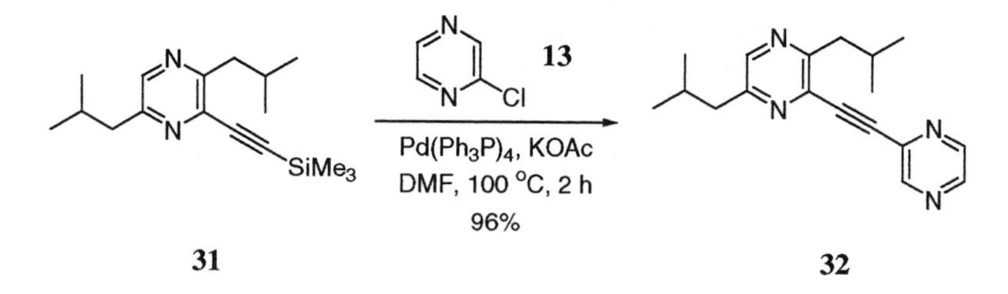


Under standard Sonogashira reaction conditions [$Pd(Ph_3P)_2Cl_2$, CuI, $Et_2NH$], many alkynylpyrazines have been synthesized [27–31]. Taylor *et al.* coupled 2-amino-3-cyano-5-bromopyrazine (**33**) with *tert*-butyl 4-ethynylbenzoate (**34**) to give **35** [32]. Alkynylpyrazine **35** was an intermediate in the synthesis of methotrexate analogs as chemotherapeutic agents.

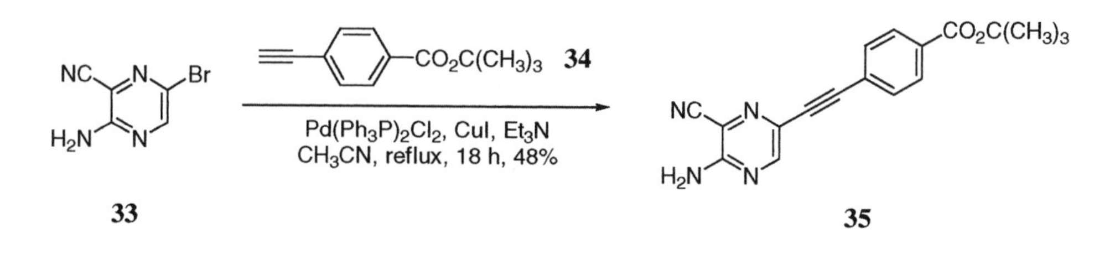


Having pyrazinylacetylenes in hand, one could convert the alkynyne functionality into the corresponding ketone *via* hydration [33]. Thus, the coupling of iodide **36** and acetylene **37** produced pyrazinylalkyne **38**. Subsequent exposure of **38** to aqueous sodium sulfide and aqueous hydrochloric acid in methanol led to ketone **39**. Such a maneuver provides additional opportunities for further manipulation of the alkynes derived from the Sonogashira coupling reactions.

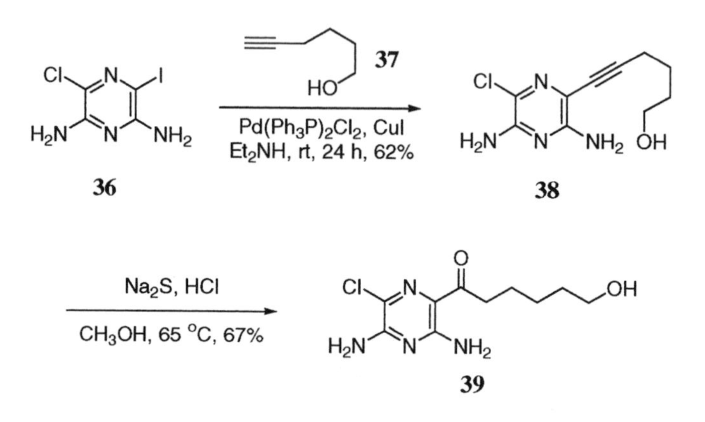

The standard Sonogashira reaction conditions were not successful for the coupling reaction of 3-chloropyrazine 1-oxide (**40**) and 1-hexyne. In contrast, treatment of **40** and 1-hexyne with $Pd(Ph_3P)_4$ and KOAc produced 3-(1-hexynyl)pyrazine 1-oxide (**41**), together with the co-dimeric product, (*E*)-enyne **42** [34]. Presumably, the co-dimerization product **42** resulted from the *cis* addition of 1-hexyne to adduct **41**.

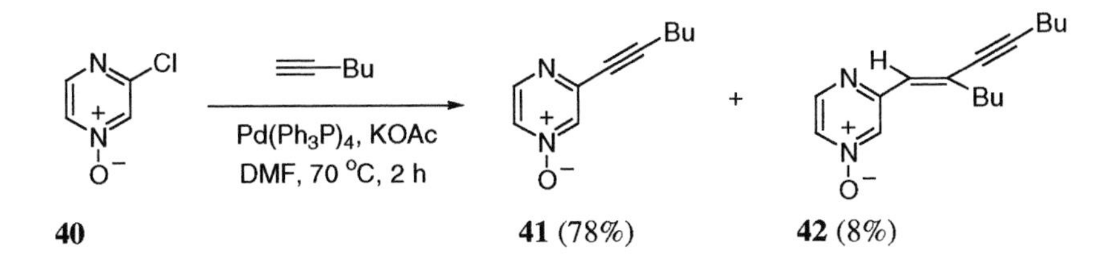


The Sonogashira reaction of 2-chloropyrazine 1-oxide gave only recovered starting material. Pentylation and octylation of 2-chloropyrazine 1-oxide also failed [9]. Possible explanations for these results were either catalyst agglomeration or metal formation from pyrazinylpalladium complex.

Quéguiner's group lithiated a *sym*-disubstituted pyrazine, 2,6-dimethoxypyrazine (**43**), with lithium 2,2,6,6-tetramethylpiperidine (LTMP). The resulting lithiated intermediate was quenched with $I_2$ to give 3-iodo-2,6-dimethoxypyrazine (**44**) and 3,5-diiodo-2,6-dimethoxypyrazine (**45**) [35]. Iodide **44** was then coupled with phenylacetylene to provide adduct **46**.

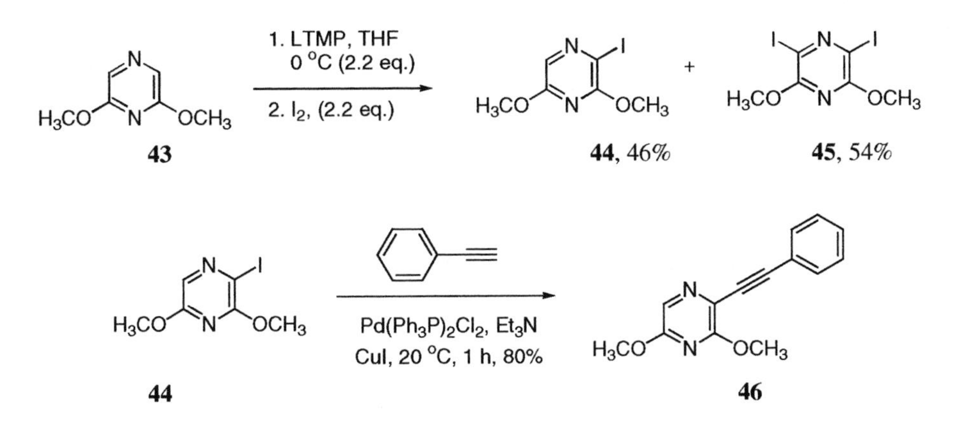


Quéguiner's metalation/cross-coupling strategy was applied to the total synthesis of arglecin (**49**), an antiarythmic pyrazine natural product extracted from cultures of *Streptomyces* and *Lavandurae*. Starting from commercially available 2,6-dichloropyrazine (**47**), iodopyrazine **48** was prepared in four steps comprising of Cl—I exchange, $S_NAr$ with sodium methoxide, metalation followed by quenching with isobutanal and dehydration. The Sonogashira reaction of **47** with propargyl alcohol provided **49**, which was subsequently transformed into arglecin (**50**) in five steps [36].

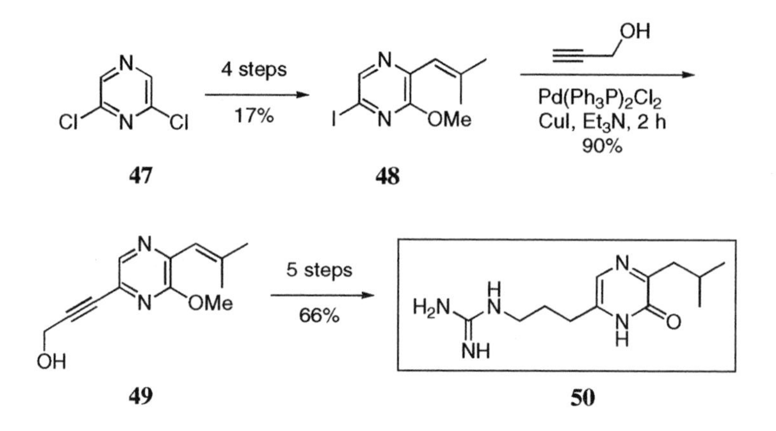


### 10.1.3 Heck reaction

Akita and Ohta revealed one of the early Heck reactions of halopyrazines [23]. They reacted 2-chloro-3,6-dimethylpyrazine (**23**) with styrene in the presence of $Pd(Ph_3P)_4$ and KOAc using *N,N*-dimethylacetamide (DMA) as solvent to make (*E*)-2,5-dimethyl-3-styrylpyrazine (**51**). This methodology was later extended to 2-chloropyrazine *N*-oxides although the yields were modest (28–38%) [37].

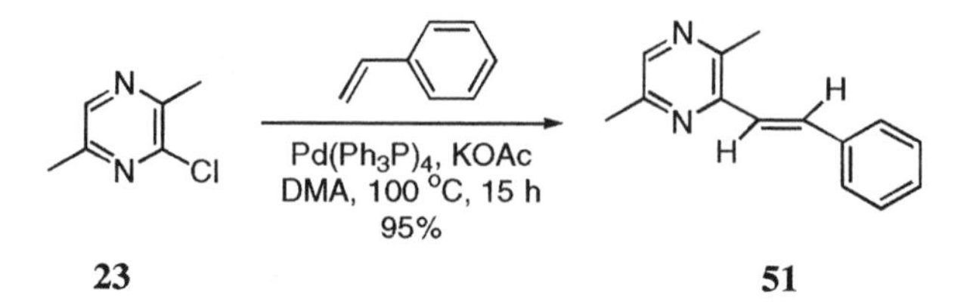


Bearing a structural resemblance to the *N*-linked nucleosides, *C*-nucleosides possess different physicochemical and biochemical properties. By a stereospecific Heck cross-coupling reaction between ribofuranoid glycal **52** and an iodopyrazine aglycon **53**, the Townsend group synthesized a novel 2'-deoxy-β-D-ribofuranosylpyrazine *C*-nucleoside **55** [38, 39]. While both hydroxyl groups in **52** were protected as their TBDMS ethers, the Heck reaction between **52** and **53** was followed by desilylation to produce **54** with exclusively the β-configuration. Protection of the two amino groups in **53** was not needed possibly due to the weak basicity of the amino nitrogen on the electron-deficient pyrazine ring. Finally, **54** was stereospecifically reduced using sodium triacetoxyborohydride to deliver 2'-deoxy-β-D-ribofuranosylpyrazine *C*-nucleoside **55**.

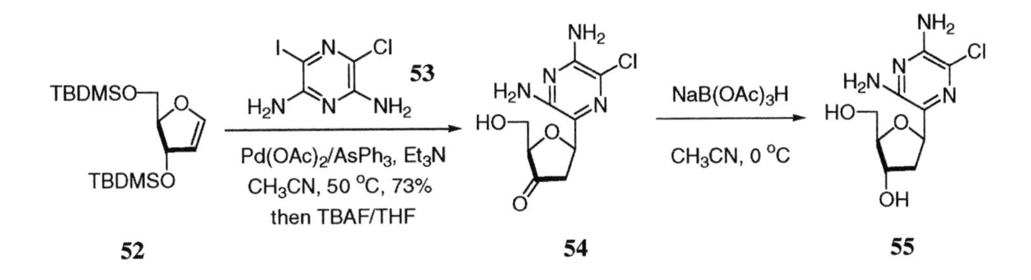


Ohta *et al.* thoroughly studied the heteroaryl Heck reactions of halopyrazines with both π-electron-rich and π-electron-deficient heteroaryls. These Pd-catalyzed direct couplings are advantageous over coupling with organometallics because the conversion of the heteroaryls to the corresponding organometallic reagents is obviated. A heterobiaryl, 2-(pyrazin-2-yl)indole **56**, was obtained from the heteroaryl Heck reaction of 2-chloro-3,6-diisobutylpyrazine (**1**) and indole [40]. Ohta *et al.* also coupled other 2-chloro-3,6-dialkylpyrazines with indole in moderate to good yields under the same conditions. In these cases, the couplings took place regioselectively at the C(2) position of indole. However, when the coupling reaction of **1** was conducted with 1-tosylindole, 2% of the coupling occurred at the C(3) position of 1-tosylindole [41]. Moreover, when the halopyrazines were 2-chloro-3,6-diethylpyrazine and 2-chloro-3,6-dimethylpyrazine, the ratio of the C(3)/C(2) coupling increased to 7% and 12%, respectively.

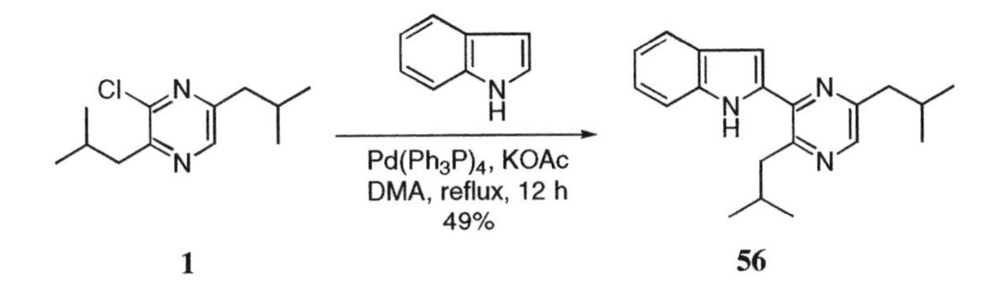


Furthermore, Ohta's group successfully conducted heteroaryl Heck reactions of chloropyrazines with many π-electron-rich heteroaryls including furan, thiophene, benzo[*b*]furan and benzo[*b*]thiophene [42, 43]. In reactions of chloropyrazines with furan, thiophene and pyrrole, disubstituted heterocycles were also isolated albeit in low yields.

Along with some disubstituted furan **59**, mono-arylation product **58** was isolated when 2-chloro-3,6-diethylpyrazine (**57**) and furan were refluxed in the presence of $Pd(Ph_3P)_4$ and KOAc. In the case of 2-chloro-3,6-dimethylpyrazine (**23**) and thiophene, monothienylpyrazine **60** was the sole product. When 2-chloro-3,6-diisobutylpyrazine was used as substrate, 9% of the disubstituted thiophene was detected. Analogous to the couplings with furan and thiophene, the heteroaryl

Heck reactions of chloropyrazine **57** with benzo[*b*]furan and benzo[*b*]thiophene produced **61** and **62**, respectively.

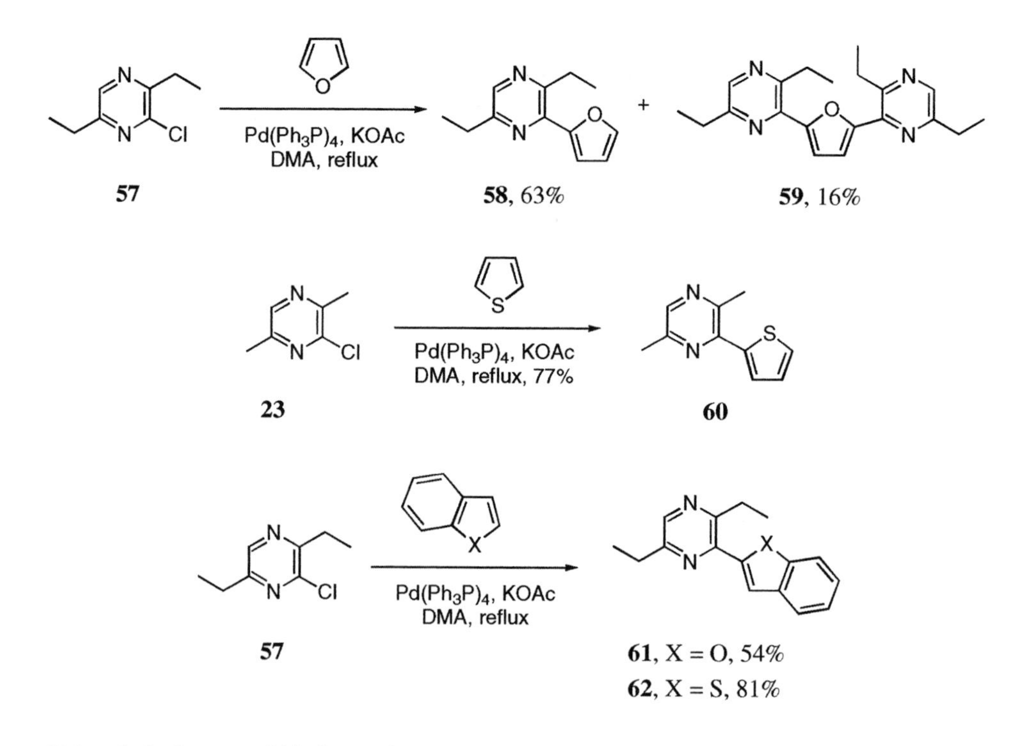


Although the heteroaryl Heck reactions of chloropyrazines with pyrrole itself were low-yielding for both mono- and bis-arylation products, better yields were obtained for *N*-phenylsulfonylpyrrole. Bulkier alkyl substituents on the pyrazine ring promoted the formation of C(3)-substituted pyrroles. The C(3)-substituted pyrrole **64** was the major product (62%) for the coupling of **1** and *N*-phenylsulfonylpyrrole, while C(2)-substituted pyrrole **63** was a minor product (15%).

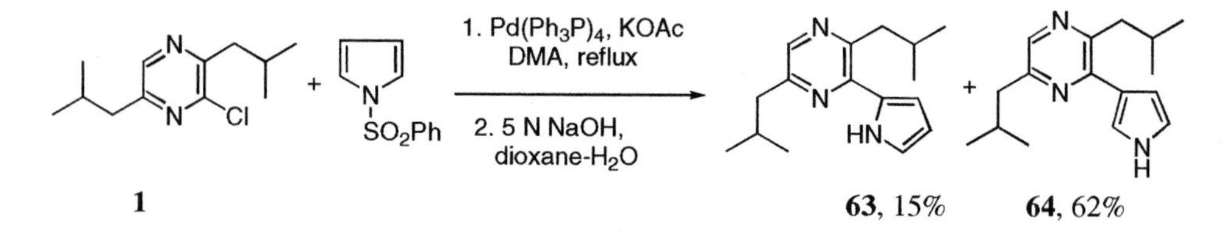


Similar results were observed for the heteroaryl Heck reactions of chloropyrazines with $\pi$-electron-rich heteroaryls including oxazole, thiazole, benz[*b*]oxazole, benz[*b*]thiazole and *N*-

methylimidazole.

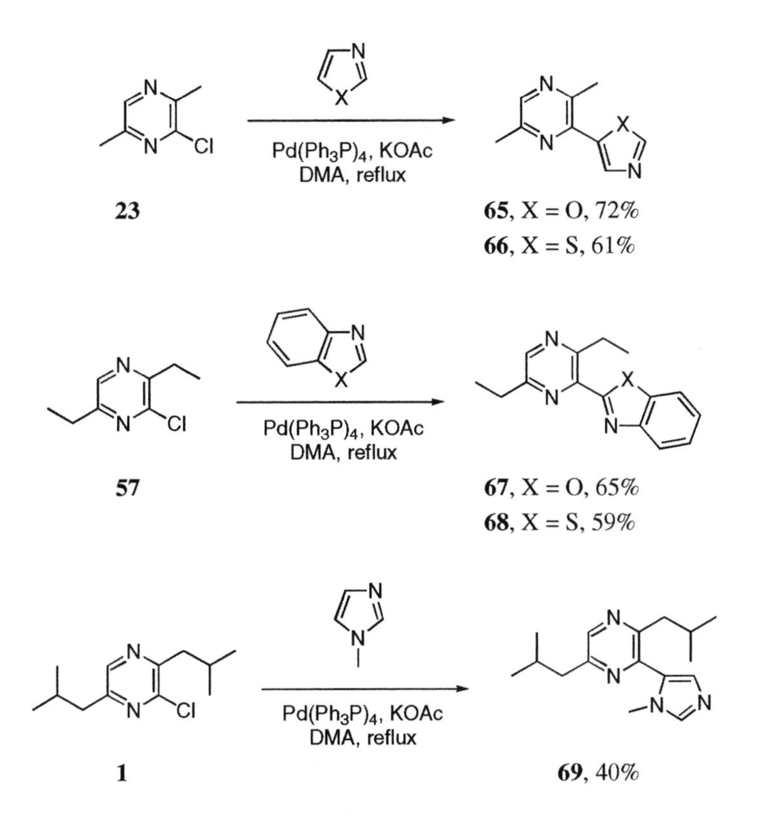


### 10.1.4 Carbonylation reactions

To synthesize pyrazinecarboxylic esters and pyrazinecarboxamides, chloropyrazines were subjected to Pd-catalyzed carbonylation reactions in either alcohols or dialkylamines [44, 45]. 2-Chloro-3,6-dimethylpyrazine (**23**) was smoothly carbonylated in methanol containing a catalytic amount of $Pd(dba)_2$ and $Ph_3P$ to give 2-methoxycarbonyl-3,6-dimethylpyrazine (**70**). Somewhat lower yields were observed in the preparation of pyrazinecarboxylic diesters under the same conditions.

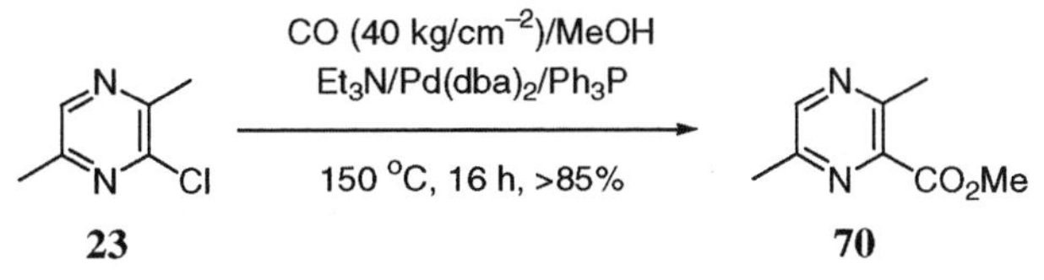


In comparison to the ease of alkoxycarbonylation of halopyrazines, the aminocarbonylation was dramatically influenced by both the phosphorus ligands and carbon monoxide pressure. Pyrazinecarboxamide was prepared *via* aminocarbonylation using diethylamine as solvent. For example, 2-chloropyrazine (**13**) was converted into 2-(diethylaminocarbonyl)pyrazine (**71**) in 85% yield along with 8% of the nucleophilic substitution product **72**. Intriguingly, when butylamine was used, the corresponding $S_NAr$ product similar to **72** was the major product (72% yield), whereas the amide was isolated as a minor product (28%).

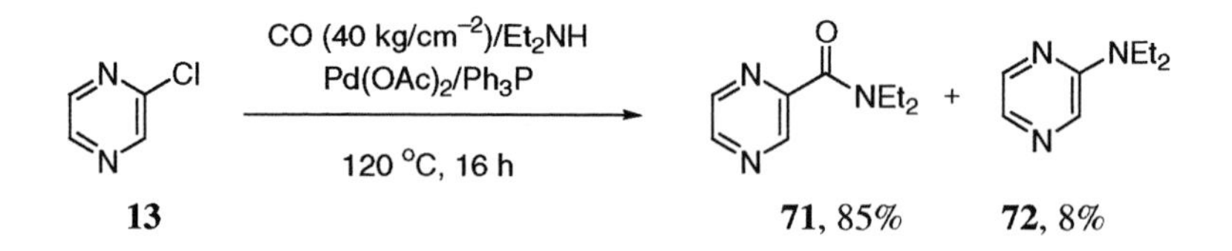


Pd-catalyzed carbonylation of 2-chloropyrazine 1-oxide failed, whereas that of 3-chloropyrazine 1-oxide (**40**) proceeded without deoxygenation of the *N*-oxide function to give 3-methoxycarbonylpyrazine 1-oxide (**73**). This observation was in accord with the failure of Stille reactions of 2-chloropyrazine 1-oxide [9, 18].

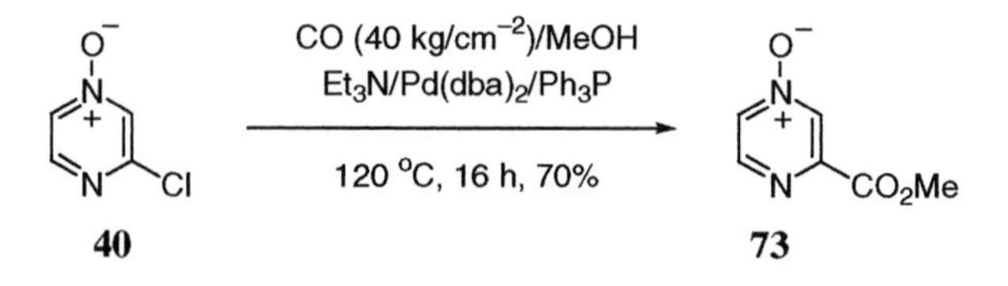


## 10.2 Quinoxalines

### 10.2.1 Coupling reactions with organometallic reagents

The Pd-catalyzed cross-coupling reactions on the benzene ring of quinoxalines with organometallic reagents are rather straightforward, and usually proceed in good yields. The Stille reaction has been utilized to synthesize aryl-substituted quinoxalines and related heteroarenes as novel herbicides [46]. To prepare the two regioisomeric aryldifluoromethoxy-quinoxalines **76** and **77**, a 2:5 mixture of bromoquinoxalines **74** and **75** was refluxed in toluene with *meta*-trifluoromethylphenylstannane in the presence of $Pd(Ph_3P)_4$. In parallel, two regioisomeric aryltrifluoromethylquinoxalines **80** and **81** were isolated from the Stille reaction of bromoquinoxalines **78** and **79** (**78/79** = 3:1) under the same conditions.

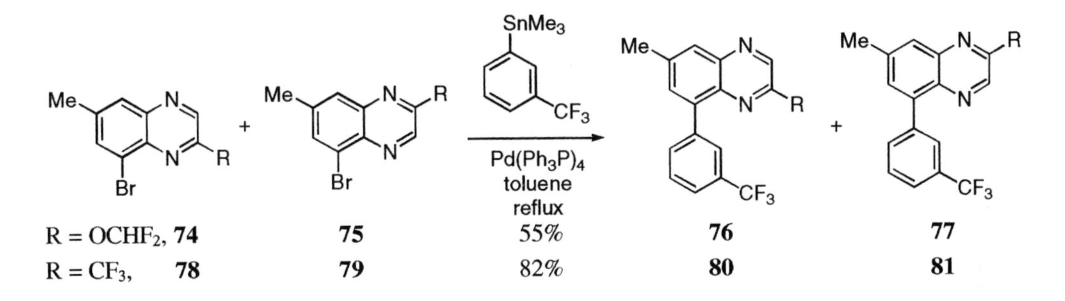


The Stille coupling of bromoquinoxaline-*N*-oxide **82** and *meta*-trifluoromethylphenylstannane produced aryldifluoromethoxyquinoxalines **83**. Later, **83** was deoxygenated by catalytic hydrogenation to give pure **76**. [46].

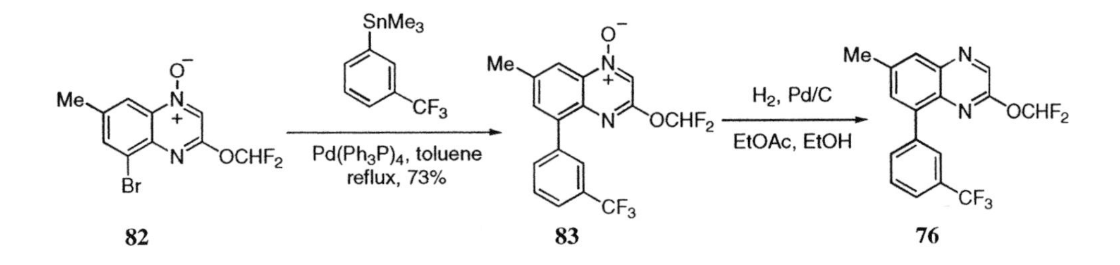


The Stille reaction featuring bromoquinoxaline **84** and vinylstannane delivered vinylquinoxaline **85**. In addition, **85** was further manipulated to a 5-aminomethylquinoxaline-2,3-dione **86** as an AMPA receptor antagonist [47]. Pd-catalyzed nucleophilic substitution on the benzene ring has also been described [48]. Thus, transformation of 5,8-diiodoquinoxalines to quinoxaline-5,8-dimalononitriles with sodium malononitrile was promoted by $PdCl_2$•$(Ph_3P)_2$.

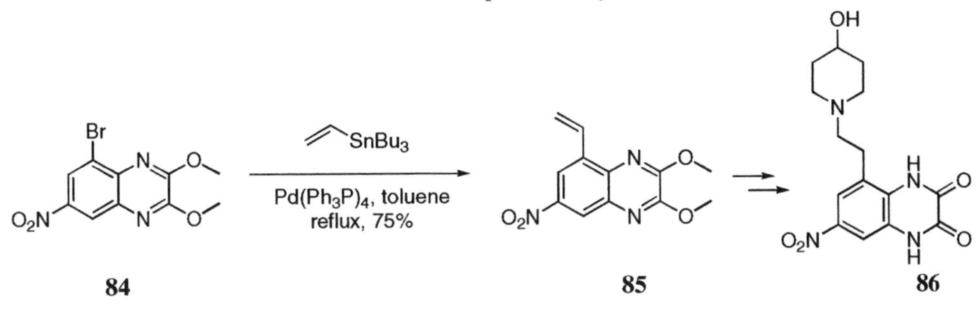


The Pd-mediated cross-coupling reactions on the pyrazine ring of quinoxaline with organometallic reagents are more cumbersome. In one example, the Stille reaction between 2-chloroquinoxaline **87** and benzylstannane **88**, attainable from the $S_N2$ displacement of corresponding biphenylbromide with tri-*n*-butyllithiostannane, produced only 32% of the coupling product **89** [49]. In another case, the Stille reaction of the simple 2-iodoquinoxaline and 1,3-dithiole-2-thione stannanes gave the adduct in only 0–10% yields although the use of copper thiophene-2-carboxylate (CuTC) gave the adduct in 54% yield [50].

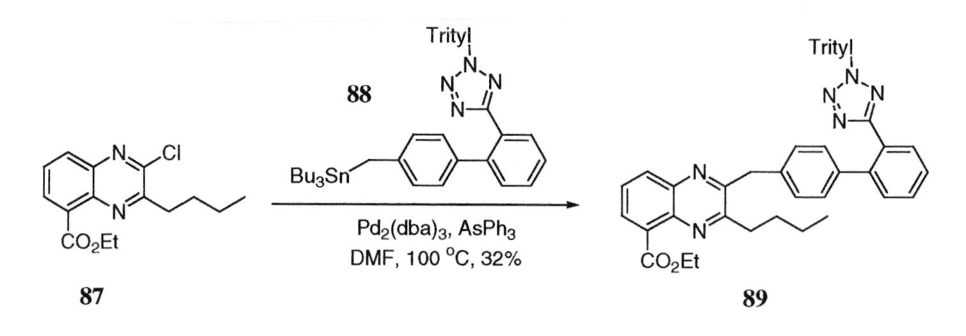


Recently, Li successfully conducted the Suzuki and Stille reactions on the pyrazine ring of 3-bromoquinoxalin-2-ylamines [51]. Indolylquinoxaline **92** was obtained from the union of 3-bromoquinoxalin-2-ylamine **90** and 1-phenylsulfonyl-2-tri-*n*-butylstannyl-1*H*-indole (**91**). A wide variety of heterocyclic stannanes bearing various functional groups underwent Stille coupling with **90** under the same conditions to give the corresponding adducts in 72–98% yields.

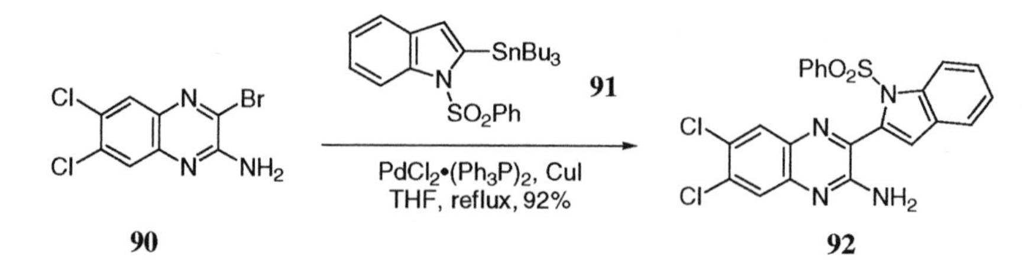


### 10.2.2 Sonogashira reaction

The Sonogashira reaction is of considerable value in heterocyclic synthesis. It has been conducted on the pyrazine ring of quinoxaline and the resulting alkynyl- and dialkynyl-quinoxalines were subsequently utilized to synthesize condensed quinoxalines [52–55]. Ames *et al.* prepared unsymmetrical diynes from 2,3-dichloroquinoxalines. Thus, condensation of 2-chloroquinoxaline (**93**) with an excess of phenylacetylene furnished 2-phenylethynylquinoxaline (**94**). Displacement of the chloride with the amine also occurred when the condensation was carried out in the presence of diethylamine. Treatment of **94** with a large excess of aqueous dimethylamine led to ketone **95** that exists predominantly in the intramolecularly hydrogen-bonded enol form **96**.

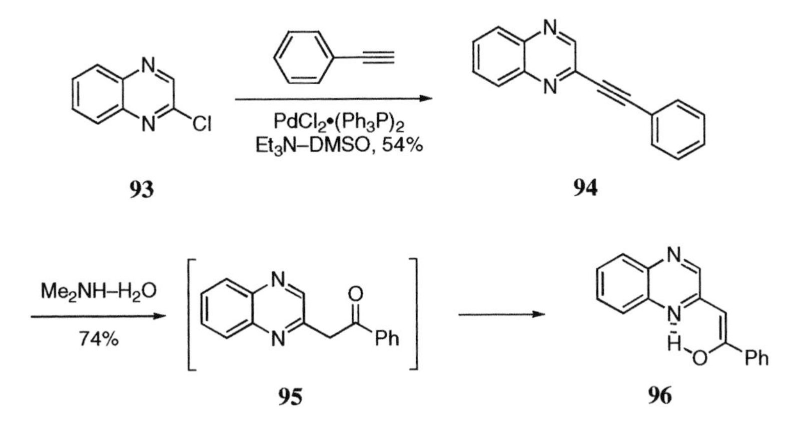


Condensed quinoxalines have been synthesized using 2-chloro-3-alkynylquinoxalines as common intermediates. 2-Chloro-3-phenylethynylquinoxaline **98** was prepared from 2,3-dichloroquinoxaline (**97**). Action of methylamine on **98** gave 1,2-disubstituted pyrrolo[2,3-*b*]quinoxaline **99**. Similarly, 2-phenylthieno[2,3-*b*]quinoxaline **100** was synthesized by treating **98** with ethanolic sodium sulfide.
Likewise, when **98** was refluxed with aqueous KOH in dioxane, the corresponding 2-phenylfurono[2,3-*b*]quinoxaline was produced in 67% yield [52]. The Sonogashira reactions of chloro- and dichloroquinoxalines and trimethylsilylacetylene [56, 57] or but-3-yn-2-ol [58] have also been documented.

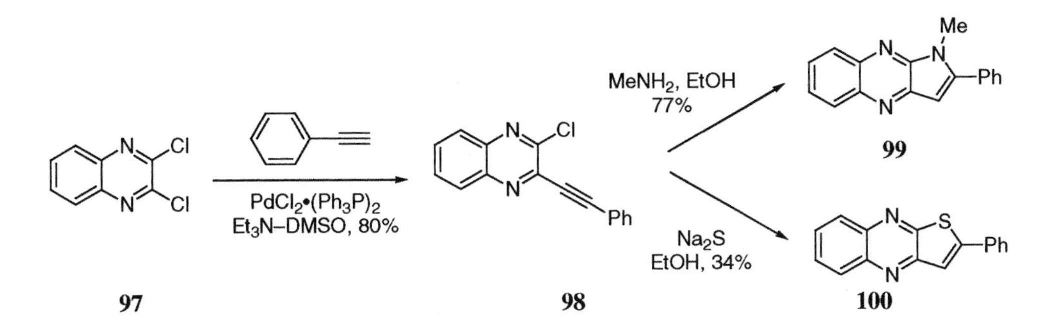


Pd-catalyzed C—P bond formation on the benzene ring of quinoxaline has been reported. Phosphoric acid ester **102** was prepared from 7-bromoquinoxaline **101** and diethylphosphite *via* a Heck-type reaction [59].

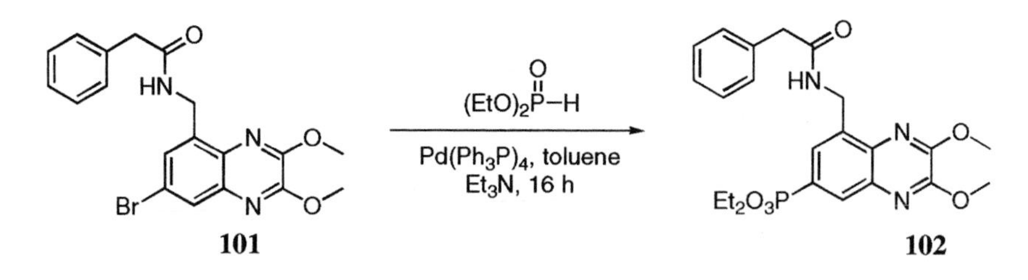


### 10.2.3 Intramolecular Heck reaction

Quinoxalines with their unique 1,4-diazine moiety are not ideal substrates for the Heck reaction because they are not only more labile than the corresponding naphthalenes, but also are strong chelating agents and behave as either monodentate or bidentate ligands. Aminoquinoxalines are especially strong chelating agents, and catalytic efficiency is difficult to achieve for such substrates. Li synthesized a variety of 3-substituted pyrrolo[2,3-*b*]quinoxalines *via* an intramolecular Heck reaction of allyl-3-haloquinoxalin-2-ylamines under Jeffery's "ligand-free" conditions. As an example, 3-methylpyrrolo[2,3-*b*]quinoxaline (**104**) was prepared from allyl-3-chloroquinoxalin-2-ylamine **103**. Apparently, the external double bond from the initial cyclization process underwent a facile rearrangement to deliver the thermodynamically more stable pyrrole ring. The enhanced reactivity and yield are presumably due to the coordination and thereby solvation of the palladium intermediates by chloride ions present in the reaction mixture. Once the "locked" palladium catalyst was released from the substrates, the catalytic cycle continues smoothly.

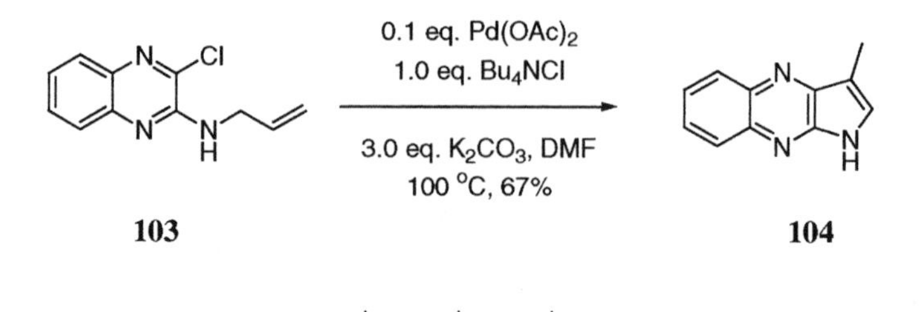


* * *

To summarize, both chloropyrazines and chloroquinoxalines are sufficiently activated to serve as viable substrates for palladium chemistry under standard conditions. In contrast to chlorobenzene, the inductive effect of the two nitrogen atoms polarizes the C—N bonds. Therefore, oxidative additions of both chloropyrazines and chloroquinoxalines to Pd(0) occur readily. One exception is 2-chloropyrazine *N*-oxide, which does not behave as a simple chloropyrazine. All Pd-catalyzed reactions with 2-chloropyrazine *N*-oxide failed, presumably because the nitrogen atom no longer possesses the electronegativity required for activation.

Although the boranes of pyrazines and quinoxalines are yet to be made, their halides are legitimate electrophilic coupling partners. Halopyrazines and haloquinoxalines undergo Kumada, Negishi, Suzuki, Sonogashira and Heck reactions. The heteroaryl Heck reactions of chloropyrazines are very well studied, whereas those of haloquinoxalines are yet to be seen. The heteroaryl Heck reactions are expected to provide good methods to prepare heterobiaryls without the neccessity of making heteroaryl organometallic reagents.

## 10.3 References


1. Noorduin, L.; Swen, S. *Cryst. Struct. Commun.* **1976**, *5*, 153–5.
2. Andre, D.; Fourme, R.; Renaud, M. *Acta Crystallogr., Sect. B.* **1971**, *27*, 2371–80.
3. Reetz, M. T.; Lohmer, G.; Schwickardi, R. *Angew. Chem., Int. Ed.* **1998**, *37*, 481–3.
4. (a) Littke, A. F.; Fu, G. C. *Angew. Chem., Int. Ed.* **1998**, *37*, 3387–8. (b) Littke, A. F.; Fu, G. C. *J. Org. Chem.* **1999**, *64*, 10–1. (c) Buchwald, S. L.; Wolfe, J. P. *Angew. Chem., Int. Ed.* **1999**, *38*, 2413–6.
5. Akita, Y.; Shimazaki, M.; Ohta, A. *Synthesis* **1981**, 974–5.
6. Sato, N.; Suzuki, M. *J. Heterocycl. Chem.* **1987**, *24*, 1371–2.
7. Ohta, A.; Inoue, A.; Watanabe, T. *Heterocycles* **1984**, *22*, 2317–21.
8. Ohta, A.; Inoue, A.; Ohtsuka, K.; Watanabe, T. *Heterocycles* **1985**, *23*, 133–7.
9. Watanabe, T.; Hayashi, K.; Sakurada, J.; Ohki, M.; Takamatsu, N.; Hirohata, H.; Takeuchi, K.; Yuasa, K.; Ohta, A. *Heterocycles* **1989**, *29*, 123–31.
10. Ohta, A.; Ohta, M.; Watanabe, T. *Heterocycles* **1986**, *24*, 785–92.
11. Nakamura, H.; Takeuchi, D. Murai, A. *Synlett* **1995**, 1227–8.
12. (a) Zoltewicz, J. A.; Cruskie, M. P., Jr. *J. Org. Chem.* **1995**, *60*, 3487–93; (b) Zoltewicz, J. A.; Maier, N. M.; Fabian, W. M. F. *J. Org. Chem.* **1997**, *62*, 3215–9.
13. Street, L. J.; Baker, R.; Book, T.; Reeve, A. J.; Saunders, J.; Willson, T.; Marwood, R. S.; Patel, S.; Freedman, S. B. *J. Med. Chem.* **1992**, *35*, 295–305.
14. Thompson, W. J.; Jones, J. H.; Lyle, P. A.; Thies, J. E. *J. Org. Chem.* **1988**, *53*, 2052–5.
15. Jones, K.; Keenan, M.; Hibbert, F. *Synlett* **1996**, 509–10.
16. Mitchell, M. B.; Wallbank, P. J. *Tetrahedron Lett.* **1991**, *32*, 2273–6.
17. Ali, N. M.; McKillop, A.; Mitchell, M. B.; Rebelo, R. A.; Wallbank, P. J. *Tetrahedron* **1992**, *48*, 8117–26.
18. Yang, Y.; Martin, A. R. *Heterocycles* **1992**, *34*, 1395–8.
19. Ohta, A.; Ohta, M.; Igarashi, Y.; Saeki, K.; Yuasa, K.; Mori, T. *Heterocycles* **1987**, *26*, 2449–54.
20. Ohta, A.; Itoh, R.; Kaneko, Y.; Koike, H.; Yuasa, K. *Heterocycles* **1989**, *29*, 939–45.
21. Helquist, P. *Tetrahedron Lett.* **1978**, 1913–4.
22. Akita, Y.; Ohta, A. *Heterocycles* **1981**, *16*, 1325–8.

23. Akita, Y.; Inoue, A.; Mori, Y.; Ohta, A. *Heterocycles* **1986**, *24*, 2093–7.
24. Akita, Y.; Ohta, A. *Heterocycles* **1982**, *19*, 329–31.
25. Akita, Y.; Inoue, A.; Ohta, A. *Chem. Pharm. Bull.* **1986**, *34*, 1447–58.
26. Akita, Y.; Kanekawa, H.; Kawasaki, T.; Shiratori, I.; Ohta, A. *J. Heterocycl. Chem.* **1988**, *25*, 975–7.
27. Sakamoto, T.; Shiraiwa, M.; Kondo, Y.; Yamanaka, H. *Synthesis* **1983**, 312–4.
28. Yamanaka, H.; Mizugaki, M.; Sakamoto, T.; Sagi, M.; Nakagawa, Y.; Takamoto, H.; Ishibashi, M.; Miyazaki, H. *Chem. Pharm. Bull.* **1983**, *31*, 4549–53.
29. Taylor, E. C.; Ray, P. S. *J. Org. Chem.* **1987**, *52*, 3997–4000.
30. Taylor, E. C.; Dötzer, R. *J. Org. Chem.* **1991**, *56*, 1816–22.
31. Nakamura, H.; Wu, C.; Takeuchi, D.; Murai, A. *Tetrahedron Lett.* **1998**, *39*, 301–4.
32. Taylor, E. C.; Ray, P. S. *J. Org. Chem.* **1988**, *53*, 35–8.
33. Sato, N.; Hayakawa, A.; Takeuchi, R. *J. Heterocycl. Chem.* **1990**, *27*, 503–6.
34. Chapdelaine, M. J.; Warwick, P. J.; Shaw, A. *J. Org. Chem.* **1989**, *54,* 1218–21.
35. Turck, A.; Trohay, D.; Mojovic, L.; Plé, N.; Quéguiner, G. *J. Organomet. Chem.* **1991**, *412*, 301–10.
36. Turck, A.; Plé, N.; Dognon, D.; Harmoy, C.; Quéguiner, G. *J. Heterocycl. Chem.* **1994**, *31*, 1449–53.
37. Akita, Y.; Noguchi, T.; Sugimoto, M.; Ohta, A. *J. Heterocycl. Chem.* **1986**, *23*, 1481–5.
38. Chen, J. J.; Walker, J. A., II; Liu, W.; Wise, D. S.; Townsend, L. B. *Tetrahedron Lett.* **1995**, *36*, 8363–6.
39. Walker, J. A., II; Chen, J. J.; Hinkley, J. M.; Wise, D. S.; Townsend, L. B. *Nucleosides Nucleotides* **1997**, *16*, 1999–2012.
40. Akita, Y.; Inoue, A.; Yamamoto, K.; Ohta, A.; Kurihara, T.; Shimizu, M. *Heterocycles* **1985**, *23*, 2327–33.
41. Akita, Y.; Itagaki, Y.; Tikazawa, S.; Ohta, A. *Chem. Pharm. Bull.* **1989**, *37*, 1477–80.
42. Ohta, A.; Akita, Y.; Ohkuwa, T.; Chiba, M.; Fukunaka, R.; Miyafuji, A.; Nakata, T.; Tani, N. Aoyagi, Y. *Heterocycles* **1990**, *31*, 1951–7.
43. Aoyagi, Y.; Inoue, A.; Koizumi, I.; Hashimoto, R.; Tokunaga, K.; Gohma, K.; Komatsu, J.; Sekine, K.; Miyafuji, A.; Konoh, J. Honma, R. Akita, Y.; Ohta, A. *Heterocycles* **1992**, *33*, 257–72.
44. Takeuchi, R.; Suzuki, K.; Sato, N. *Synthesis* **1990**, 923–4.
45. Takeuchi, R.; Suzuki, K.; Sato, N. *J. Mol. Cat.* **1991**, *66*, 277–8.
46. Selby, T. P.; Denes, R.; Kilama, J. J.; Smith, B. K. *ACS Symp. Ser.* **1995**, *584* (*Synthesis and Chemistry of Agrochemicals IV, Chapter 16*), 171–85.
47. Auberson, Y. P.; Bischoff, S.; Moretti, R.; Schmutz, M.; Veenstra, S. J. *Biorg. Med. Chem. Lett.* **1998**, *8*, 65–70.
48. Tsubata, Y.; Suzuki, T.; Miyashi, T.; Yamashita, Y. *J. Org. Chem.* **1992**, *57,* 6749–55.
49. Kim, K. S.; Qian, L.; Dickinson, K. E. J.; Delaney, C. L.; Bird, J. E.; Waldron, T. L.;

Moreland, S. *Biorg. Med. Chem. Lett.* **1993**, *3*, 2667–70.

50. Dinsmore, A.; Garner, C. D.; Joule, J. A. *Tetrahedron* **1998**, *54*, 3291–302.
51. Li, J. J.; Yue, W. S. *Tetrahedron Lett.* **1999**, *40*, 4507–10.
52. Ames, D. E.; Brohi, M. I. *J. Chem. Soc., Perkin Trans. 1* **1980**, *7*, 1384–9.
53. Ames, D. E.; Bull D.; Takundwa, C. *Synthesis* **1981**, 364–5.
54. Ames, D. E.; Mitchell, J. C.; Takundwa, C. C. *J. Chem. Res., (S)* **1985**, 144–5.
55. Ames, D. E.; Mitchell, J. C.; Takundwa, C. C. *J. Chem. Res., (M)* **1985**, 1683–96.
56. Dinsmore, A.; Birks, J. H.; Garner, C. D.; Joule, J. A. *J. Chem. Soc., Perkin Trans. 1* **1997**, 801–7.
57. Kim, C.-S.; Russell, K. C. *J. Org. Chem.* **198**, *63,* 8229–34.
58. Bradshaw, B.; Dinsmore, A.; Garner, C. D.; Joule, J. A. *J. Chem. Soc., Chem. Commun.* **1998**, 417–8.
59. Acklin, P.; Allgeier, H.; Auberson, Y. P.; Bischoff, S.; Ofner, S.; Sauer, D.; Schmutz, M. *Biorg. Med. Chem. Lett.* **1998**, *8*, 493–8.
60. Li, J. J. *J. Org. Chem.* **1999**, *64*, 8425–7.

# CHAPTER 11

## Pyrimidines

Pyrimidine-containing molecules are of paramount importance in nucleic acid chemistry. Their derivatives including uracil, thymine, cytosine, adenine and guanine are fundamental building blocks for deoxyribonucleic acid (DNA) and ribonucleic acid (RNA). Vitamin $B_1$ (thiamine) is another well-known example of naturally occurring pyrimidines encountered in our daily lives. Synthetic pyrimidine-containing compounds occupy a prominent place in the pharmaceutical arena. Pyrimethamine and Trimethoprim are two representative pyrimidine-containing chemotherapeutics. Pyrimethamine is a dihydrofolate reductase inhibitor; effective for toxoplasmosis in combination with a sulfonamide; whereas Trimethoprim is an antimalarial drug, widely used as a general systemic antibacterial agent in combination with sulfamethoxazole.

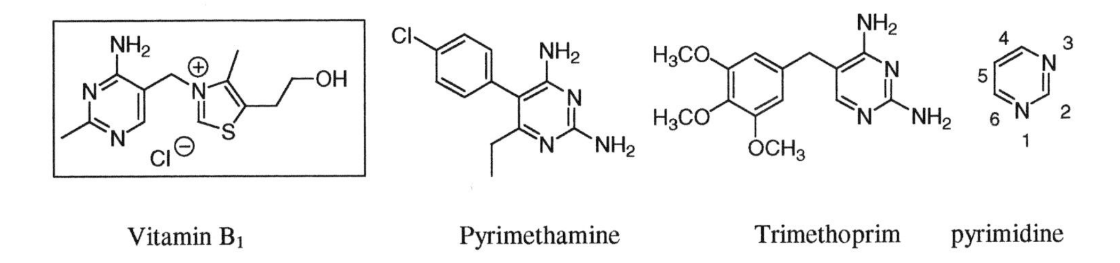


Vitamin $B_1$ Pyrimethamine Trimethoprim pyrimidine

Due to the electronegativity of the two nitrogen atoms, pyrimidine is a deactivated, π-electron-deficient heterocycle. Its chemical behavior is comparable to that of 1,3-dinitrobenzene or 3-nitropyridine. One or more electron-donating substituents on the pyrimidine ring is required for electrophilic substitution to occur. In contrast, nucleophilic displacement takes place on pyrimidine more readily than pyridine. The trend also translates to palladium chemistry: 4-chloropyrimidine oxidatively adds to Pd(0) more readily than does 2-chloropyridine.

Remarkable differences in reactivity for each position on pyrimidinyl halides and triflates have been observed. The C(4) and C(6) positions of a halopyrimidine are more prone to $S_NAr$ processes than the C(2) position. The order of $S_NAr$ displacement for halopyrimidines is:

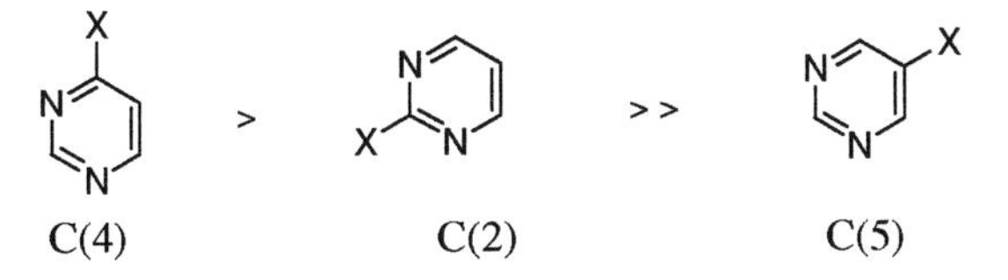

This trend is also observed in palladium chemistry where the general order for oxidative addition often correlates with that of nucleophilic substitution. Not only are 2-, 4- and 6-chloropyrimidines viable substrates for Pd-catalyzed reactions, but 4- and 6-chloropyrimidines react more readily than 2-chloropyrimidines.

Undheim and Benneche reviewed the Pd-catalyzed reactions of pyrimidines, among other $\pi$-deficient azaheterocycles including pyridines, quinolines and pyrazines, in 1990 [1] and 1995 [2]. A review by Kalinin also contains some early examples in which C—C formation on the pyrimidine ring is accomplished using Pd-catalyzed reactions [3]. In this chapter, we will systematically survey the palladium chemistry involving pyrimidines.

## 11.1 Synthesis of pyrimidinyl halides and triflates

Pyrimidinyl halides are not only precursors for Pd-catalyzed reactions, but also important pharmaceuticals in their own right. One of the most frequently employed approaches for halopyrimidine synthesis is direct halogenation. When pyrimidinium hydrochloride and 2-aminopyrimidine were treated with bromine, 5-bromopyrimidine and 2-amino-5-bromopyrimidine were obtained, respectively, *via* an addition–elimination process instead of an aromatic electrophilic substitution [4, 5]. Analogously, 2-chloro-5-bromopyrimidine (**1**) was generated from direct halogenation of 2-hydroxypyrimidine [6]. Treating **1** with HI then gave to 2-iodo-5-bromopyrimidine (**2**). In the preparation of 5-bromo-4,6-dimethoxypyrimidine (**4**), *N*-bromosuccinimide was found to be superior to bromine for the bromination of 4,6-dimethoxypyrimidine (**3**) [7].

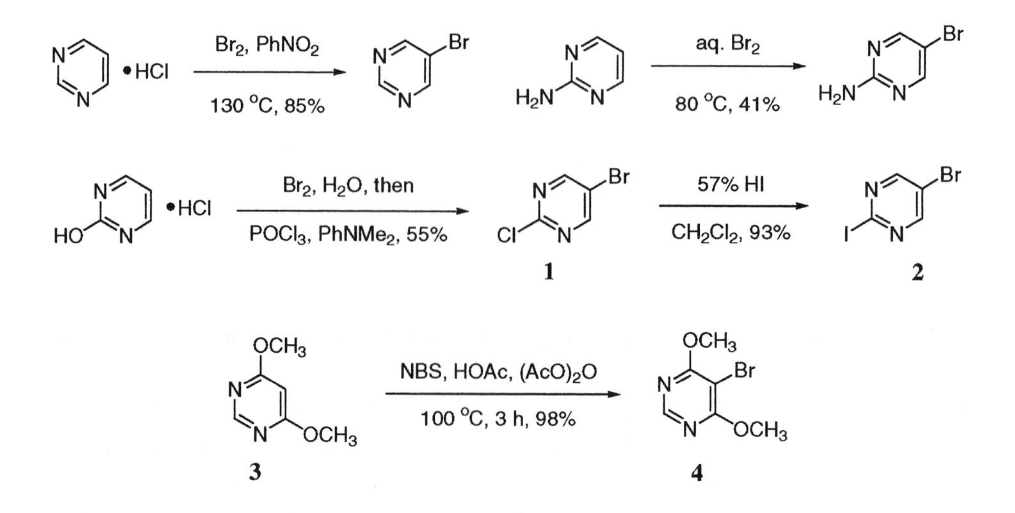

5-Iodopyrimidine **7** was prepared by iodination of 2,4-diaminopyrimidine **6**, which was derived from commercially available 2-amino-4-chloro-6-methylpyrimidine (**5**) *via* an $S_NAr$ reaction with ammonia [8]. Similarly, iodination of 6-chloro-2,4-dimethoxypyrimidine (**8**) with *N*-iodosuccinimide in trifluoroacetic acid led to dihalopyrimidine **9** [9].

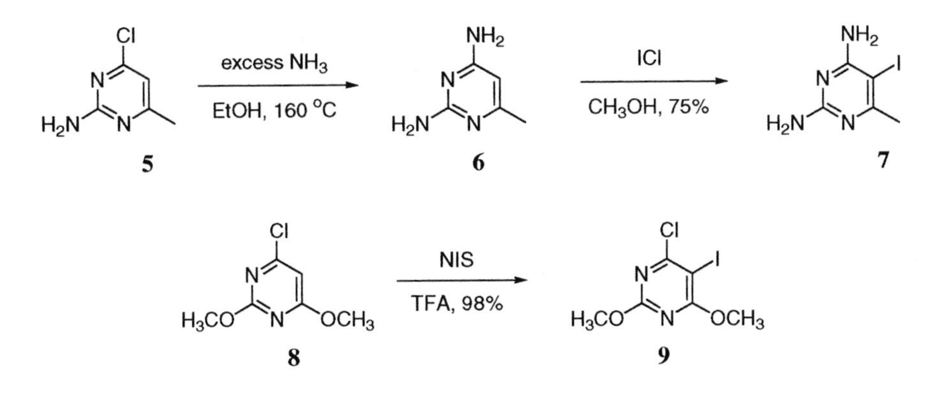


Another reliable method of halopyrimidine synthesis is "dehydroxy-halogenation". Refluxing pyrimidinones **10** and **13** with phosphorus oxychloride was followed by treating the resulting chloropyrimidines with hydroiodic acid to afford iodopyrimidine **11** and **14**, respectively [10, 11]. 4-Chloropyrimidinone **13**, on the other hand, was prepared by direct halogenation of pyrimidone **12**.

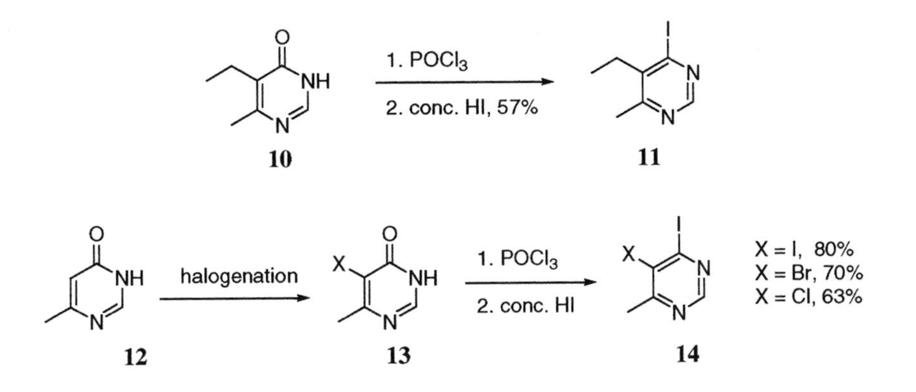


Although dehydroxy-halogenation of pyrimidones works well using phosphorus oxychloride and phosphorus oxybromide, both reagents are moisture-sensitive and phosphorus oxybromide is relatively expensive. Extending the well-known halogenation of alcohols using triphenylphosphine and *N*-halosuccinimide, Sugimoto *et al.* halogenated several π-deficient hydroxyheterocycles including hydroxypyridine, hydroxyquinoline, hydroxyquinoxaline and hydroxypyrimidine [12]. Thus, treating hydroxypyrimidine **15** with triphenylphosphine and *N*-halosuccinimide resulted in halopyrimidine **16** under mild conditions.

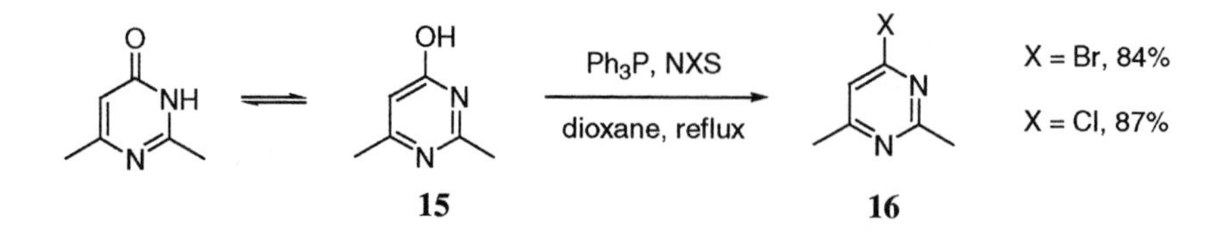


The preparation of 6-membered ring heterocyclic triflates including pyrimidinyl triflate is a challenging task [13]. Simple 2-pyrimidinyl triflates and 2-methylthio-4-pyrimidinyl triflates with hydrogen at the electron-deficient 4/6-positions are unstable. However, when there is an additional substituent on the pyrimidine ring, the corresponding pyrimidinyl triflate becomes sufficiently stable, allowing isolation *via* flash chromatography. Therefore, treating 6-methyluracil (**17**) with NaH followed by triflic anhydride gave rise to bis-triflate **18**. 2-Methylthio-4-pyrimidinyl triflate (**19**) and 2-pyrimidinyl triflate (**20**) were also prepared using triflic anhydride from respective pyrimidones [14]. In these cases, triflic anhydride was found to be more convenient than *N*-phenylmethanesulfonimide ($PhNTf_2$) due to comparative ease of purification of the resulting triflates.

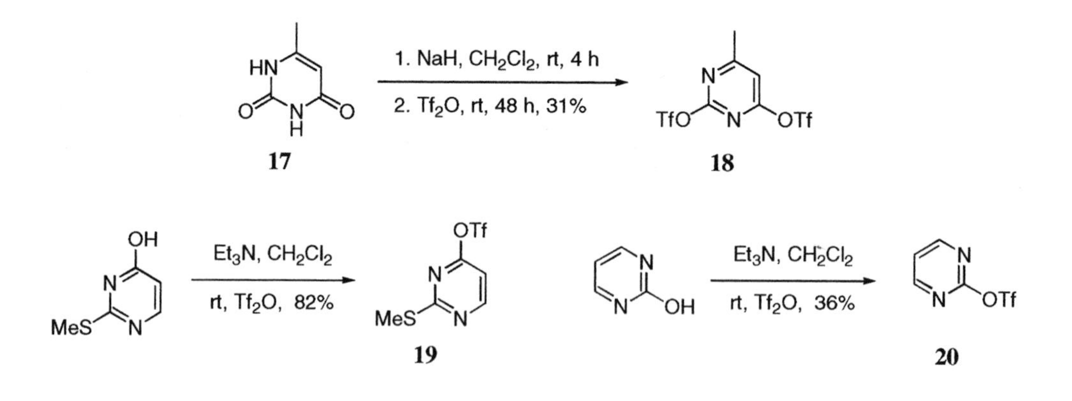


## 11.2 Coupling reactions with organometallic reagents

### 11.2.1 Negishi coupling

Pyrimidinylzinc chloride **22** was generated *in situ* by halogen-metal exchange of 5-bromo-4,6-dimethoxypyrimidine (**21**) with *n*-BuLi followed by treatment with $ZnCl_2$ [15]. The subsequent Negishi coupling of **22** with 3,4-dinitrobromobenzene gave phenylpyrimidine **23**.

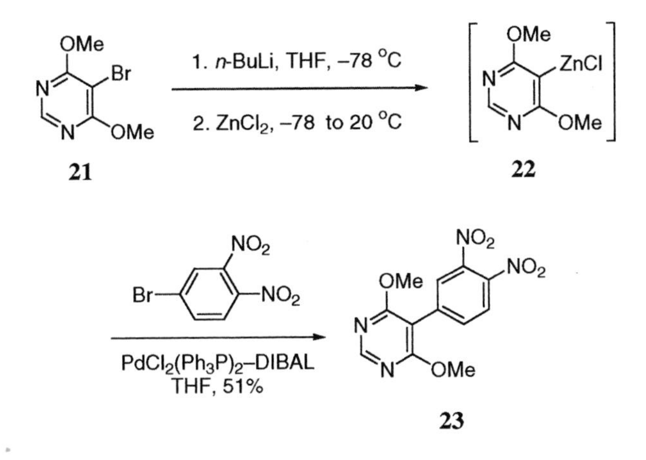


The Reformatsky reagent, a classic organozinc reagent, has proven useful for the Negishi reaction with organohalides including halopyrimidines. Reformatsky reagent **25** was prepared from metalation of ethyl bromoacetate with fresh zinc metal. While 2-iodopyrimidine **24** readily coupled with Reformatsky reagent **25** to afford ethyl 4,6-dimethyl-2-pyrimidine acetate (**26**), the reaction of 2-bromopyrimidine with **25** was much less efficient and 2-chloropyrimidine was virtually inert under such reaction conditions [16]. Ethyl 2,6-dimethyl-4-pyrimidineacetate (**28**) was synthesized in a similar fashion from 4-iodo-2,6-dimethylpyrimidine (**27**) and **25**. However, the desired coupling product was not observed for the same reaction of 5-iodo-2,6-dimethylpyrimidine.

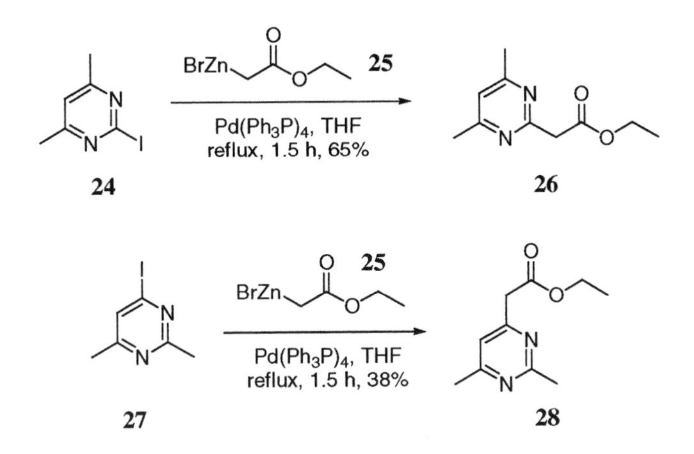


The marked increase of stability of organozinc reagents compared to lithium or magnesium organometallics allows the Negishi reactions to be carried out at high temperatures. Organozinc reagent **30**, derived from pyridazine **29** by *ortho*-lithiation and treatment with $ZnCl_2$, was coupled with 5-bromopyrimidine to form pyrimidinylpyradazine **31** at 65 °C [17]. Sonication was found to shorten the reaction time significantly and improve the yield. In a like fashion, *ortho*-lithiation of 2-methylthio-4-chloropyrimidine (**32**) using LTMP and subsequent treatment with $ZnCl_2$ led to organozinc reagent **33**, which was then cross-coupled with iodobenzene to furnish 5-arylpyrimidine **34**.

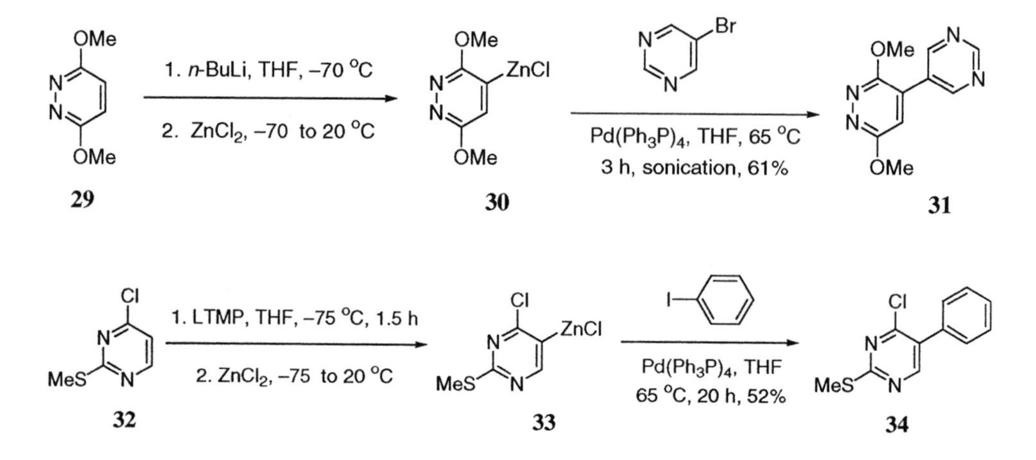


Pyrimidine triflates are also suitable substrates for Negishi reactions. The coupling of 2,4-pyrimidinyl triflate (**35**) with 2.5 equivalents of *p*-anisylzinc bromide gave 2,4-dianisylpyrimidine **36** [17]. The Negishi reaction of 2,4-pyrimidinyl bis-triflate **37** and *p*-anisylzinc bromide occurred predominantly at C(4). Interestingly, since *p*-anisylzinc bromide was generated by treating *p*-bromoanisole with *n*-BuLi followed by reaction with $ZnBr_2$, some butylzinc bromide was present in the reaction mixture. As a consequence, the second Negishi reaction with butylzinc bromide proceeded without detectable β-hydride elimination to produce 2-butylpyrimidine **38** [18].

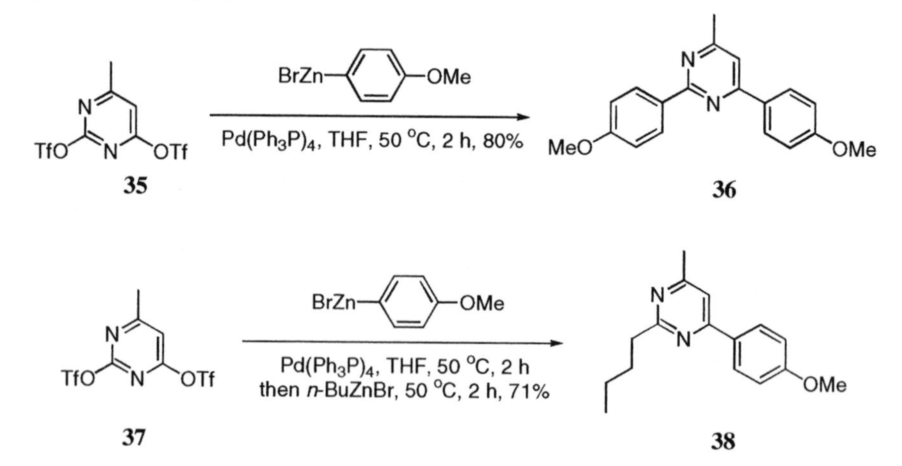

Although alkyl halides are prone to β-hydride elimination, alkylzinc or alkylboron reagents can take part in Negishi coupling or Suzuki coupling to install alkyl substituents onto pyrimidine rings. Indeed, alkylpyrimidine **40** was synthesized from pyrimidinyl triflate **39** using *n*-butylzinc chloride. Furthermore, vinylpyrimidine **41** and thienylpyrimidine **42** were prepared also from the corresponding triflates, respectively [19].

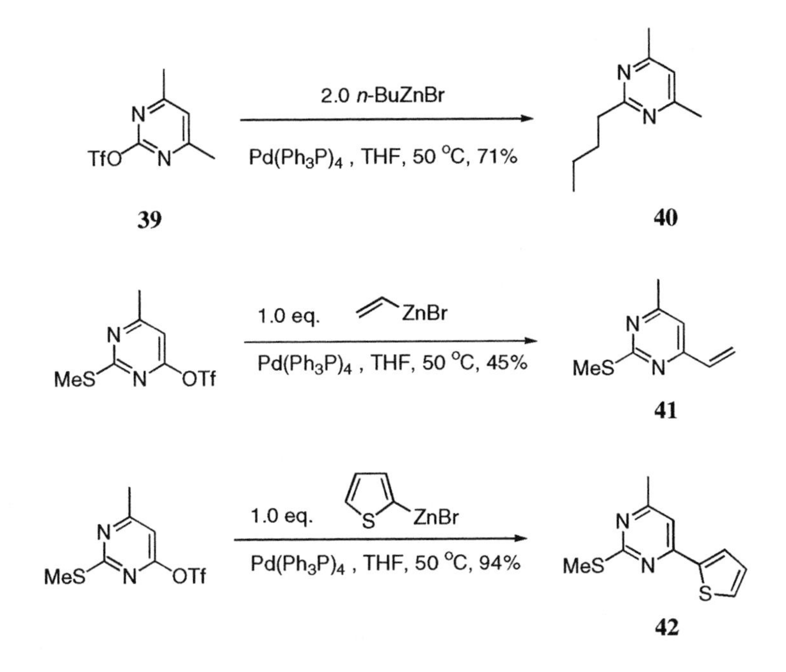


### 11.2.2 Suzuki coupling

Like simple carbocyclic arylboronic acids, pyrimidineboronic acids couple with organohalides and organotriflates under palladium catalysis. However, simple 5-pyrimidineboronic acid is not trivial to make due to competing nucleophilic addition of the anionic intermediate to the azomethine bond. It is preferable to reverse the coupling partners — using 5-halopyrimidine to couple with other easily accessible boronic acids to prepare the same Suzuki adduct. Nonetheless, the Gronowitz group successfully prepared 5-pyrimidineboronic acid from 5-bromopyrimidine *via* a halogen-metal exchange using *n*-BuLi at an extremely low temperature followed by treatment with tributylborate and basic hydrolysis [20].

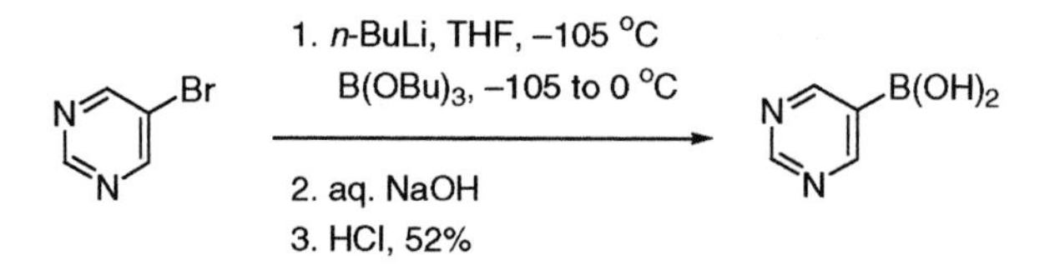


For the halogen-metal exchange reaction of bulkier halopyrimidines, steric hindrance retards the nucleophilic attack at the azomethine bond. As a consequence, halogen-metal exchange of 5-bromo-2,4-di-*t*-butoxypyrimidine (**43**) with *n*-BuLi could be carried out at –75 °C [20]. The resulting lithiated pyrimidine was then treated with *n*-butylborate followed by basic hydrolysis and acidification to provide 2,4-di-*t*-butoxy-5-pyrimidineboronic acid (**44**). 5-Bromopyrimidine **43** was prepared from 5-bromouracil in two steps consisting of a dehydroxy-halogenation with phosphorus oxychloride and an $S_NAr$ displacement with sodium *t*-butoxide.

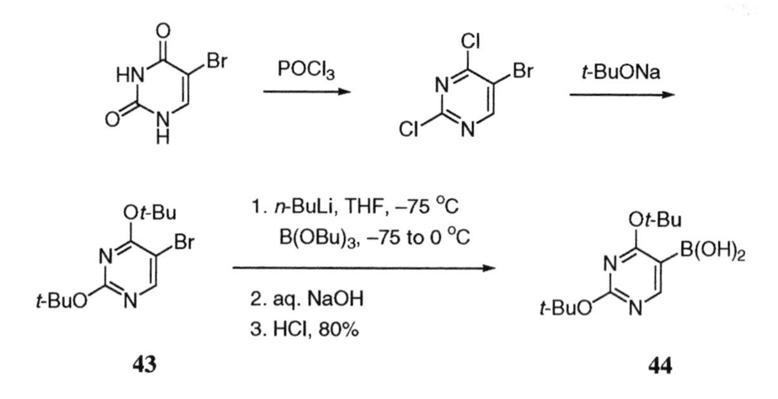


The Suzuki coupling of **44** was utilized to prepare 5-substituted uracils as potential antiviral agents [21, 22]. Adduct **45**, derived from **44** and 2-bromo-3-methylthiophene, was transformed to the corresponding uracil **46** *via* acidic hydrolysis. Conveniently, reversal of the coupling partners also resulted in formation of adduct **45**, assembled from the Suzuki coupling of 5-bromo-2,4-di-*t*-butoxypyrimidine (**43**) and 3-methyl-2-thiopheneboronic acid.

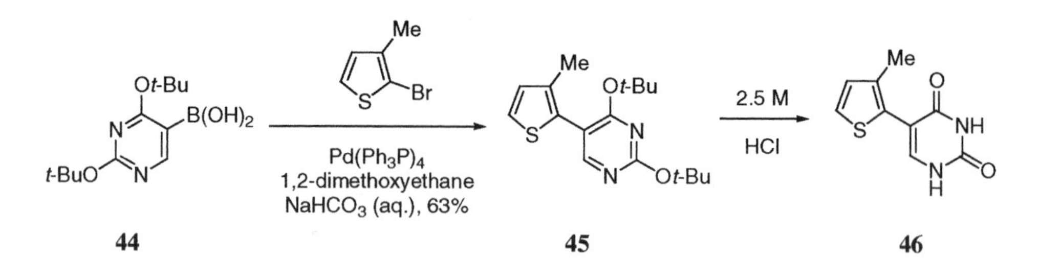

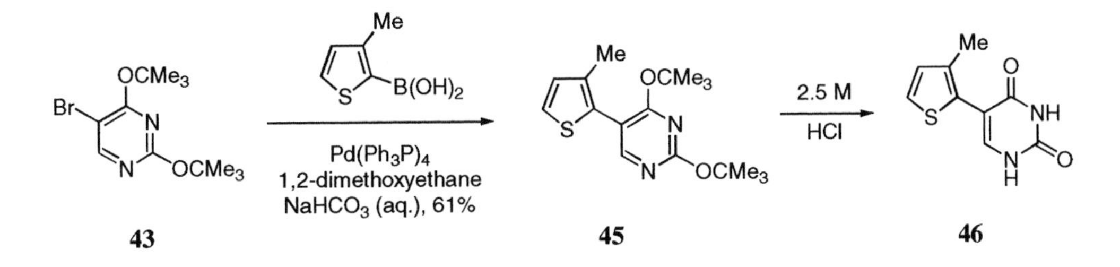


2-Chloropyrimidine was coupled with diethyl (3-pyridyl)borane in the presence of $Pd(Ph_3P)_4$, $Bu_4NBr$, and KOH to afford 3-(2'-pyrimidinyl)pyridine [23]. Likewise, the Suzuki coupling of 2-bromopyrimidine with diethyl (4-pyridyl)borane (**47**) led to 4-(2'-pyrimidinyl)pyridine (**48**) in 50% yield, whereas 2-chloropyrimidine produced **48** in only 20% yield under the same conditions [24]. Diethyl (4-pyridyl)borane (**47**), on the other hand, was readily accessible from sequential treatment of 4-bromopyridine with *n*-BuLi and diethylmethoxyborane.

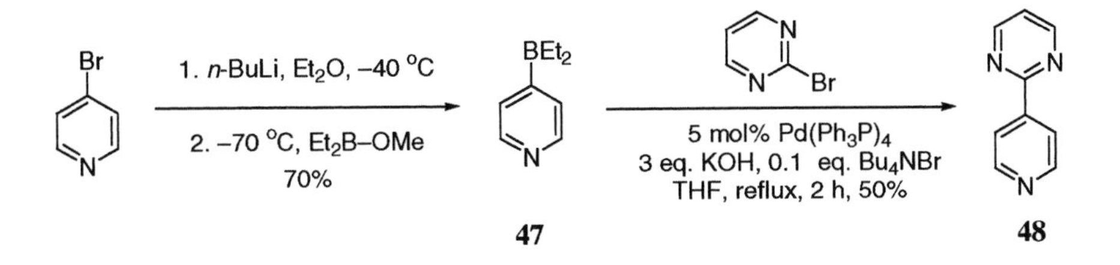


The coupling of 2- and 5-pyrimidinylbromides with 5-indolylboronic acid afforded 5-pyrimidinylindoles [25]. 5-Indolylboronic acid was readily prepared from commercially available 5-bromoindole by a one-pot process involving treating 5-lithio-1-potassioindole with tributylborate followed by acidic hydrolysis. Meanwhile, it was also discovered that $Pd(dppb)Cl_2$ possessed greater effectiveness than $Pd(Ph_3P)_4$ for the coupling of 2-chloropyrimidine **49** and arylboronic acid **50** to afford 2-arylpyrimidine **51** [26, 27].

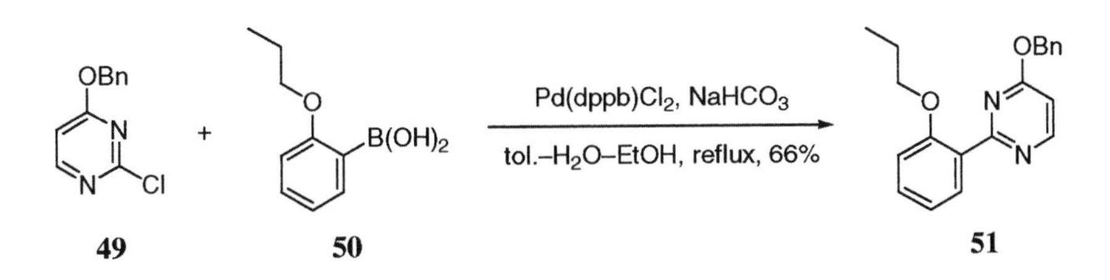


For the Suzuki coupling of 5-bromopyrimidine and protected *p*-boronophenylalanine **52** to assemble **53**, $Pd(dppf)Cl_2$ was found to be an effective catalyst [28], whereas using 2-bromopyrimidine as the substrate gave the corresponding adduct in only 20% yield.

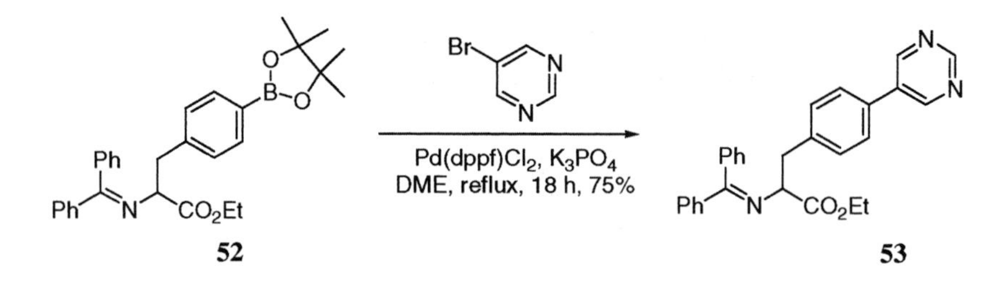


## 11.2.3 Stille coupling

### *11.2.3.a The pyrimidine motif as a nucleophile*

In 1989, Undheim and colleagues prepared 5-tributylstannylpyrimidine **55** *via* lithiation of 2-methylthio-5-bromopyrimidine and subsequent reaction with tributyltin chloride [29]. The same technique has been utilized to synthesize other 5-stannylpyrimidine derivatives such as 2-*tert*-butyldimethoxy-5-stannylpyrimidine [30]. The best yield for preparing 2-methylthio-4-tributylstannylpyrimidine (**57**) was obtained using 2-methylthio-4-iodopyrimidine (**56**) as the precursor. However, treating **56** with tributylstannyl lithium gave **57** in only 52% yield [31]. In addition, 4-stannylated pyrimidines were also prepared from the corresponding pyrimidinyl halides *via* a halogen-metal exchange followed by quenching with trialkyltin chloride.

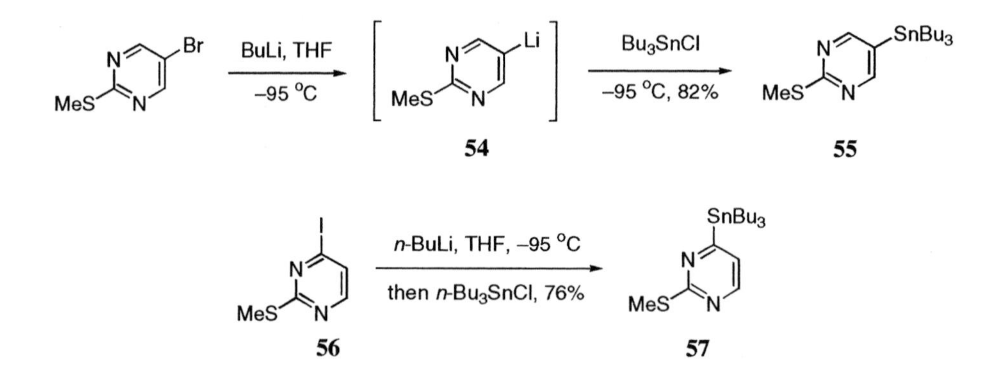


Another important approach for stannylpyrimidine synthesis is the $S_NAr$ reaction of a pyrimidinyl halide with a stannyl anion [31]. Treating 2-chloro-5-bromopyrimidine with $Bu_3SnLi$ led to 2-chloro-5-tributylstannyl-pyrimidine (**58**) in 53% yield. Similarly, treatment of 2-chloropyrimidine with either $Bu_3SnLi$ or $Me_3SnLi$ gave 2-tributylstannylpyrimidine or 2-trimethylstannylpyrimidine in 84% or 46% yield, respectively. Furthermore, a regioselective substitution was achieved for the $S_NAr$ reaction of 2,4-dibromopyrimidine with $Me_3SnNa$ to afford 4-bromo-2-trimethylstannylpyrimidine (**59**) in 78% yield, whereas using $Me_3SnLi$ resulted in, surprisingly, extensive polymerization. An attempt to conduct the stannylation of 2,5-dichloropyrimidine with $Bu_3SnLi$ was also unsuccessful.

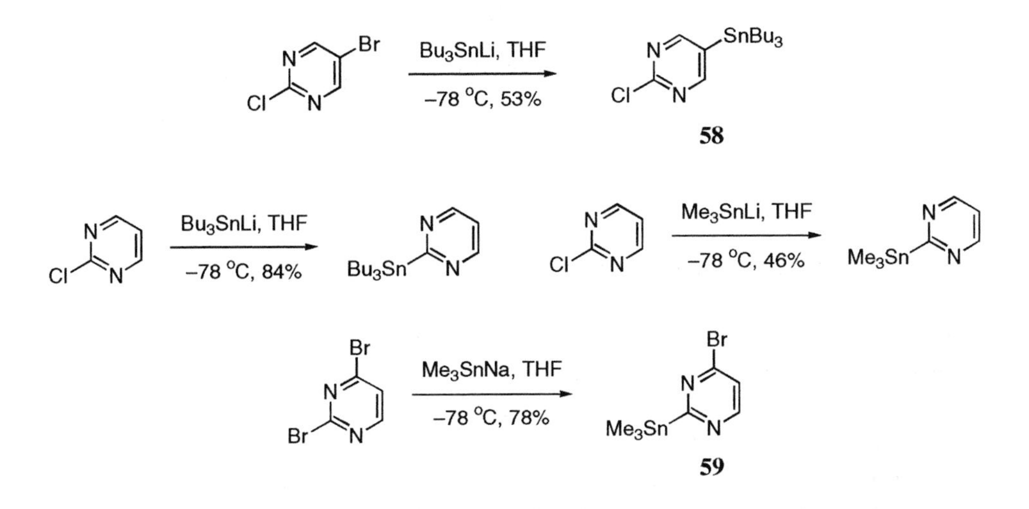


Both of the two aforementioned methods must be carried out at low temperatures. On the other hand, the Pd-catalyzed coupling reaction between 2-methylthio-5-bromopyrimidine and hexamethyldistannane in the presence of fluoride ion can be run at ambient temperature to prepare 2-methylthio-5-tributylstannylpyrimidine (**55**) [32]. The Stille reaction of **55** and 5-bromofurfural then afforded adduct **60**. Although Sn—Sn bonds are known to be thermally stable, the weakening of the Sn—Sn bond can be achieved through the formation of complexes such as $[(Bu_3SnSn_3X)^-Bu_4N^+]$ (X = F, or Cl). During the formation of **55**, it was found that bis(π-allylpalladium chloride) was the best catalyst in terms of reaction rate and yield. Furthermore, bromides of thiophene, pyridine and thiazole were coupled with **55**, and electron-withdrawing groups on the heteroaryl halides appeared to activate these electrophiles.

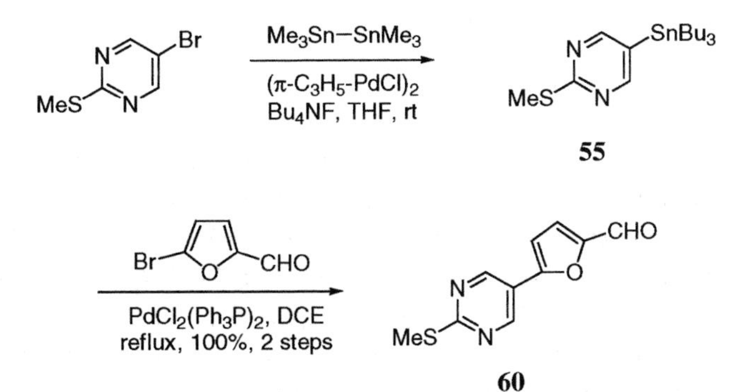

Thermal decarboxylation of pyrimidylcarboxylic organotin esters is another means to prepare the corresponding stannylpyrimidines [33]. This method obviates the intermediacy of lithiated pyrimidine species that would undergo undesired reactions at higher temperatures. The decarboxylation occurs at the activated positions. Therefore, thermal decarboxylation of tributyltin carboxylate **62**, derived from refluxing carboxylic acid **61** with bis(tributyltin) oxide, provided 4-stannylpyrimidine **63**. Addition of certain Pd(II) complexes such as bis(acetonitrile)palladium(II) dichloride improved the yields, whereas AIBN and illumination failed to significantly affect the yield.

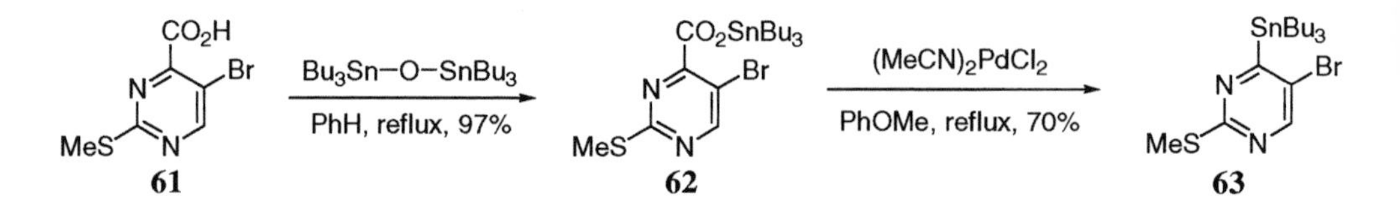


In the literature, most Stille couplings involving pyrimidines were conducted using pyrimidinyl halides and other organostannanes. There are limited applications of stannylpyrimidines in the Stille coupling. However, if the coupling partner has delicate functional groups, choosing a stannylpyrimidine as the nucleophile does have its merits as showcased by the coupling of **55** with 5-bromofurfural to make **60**, as well as the union of 5-tributylstannylpyrimidine with 4-bromo-2-nitrobenzylphosphonate (**64**) to provide heterobiaryl phosphonate **65** [34].

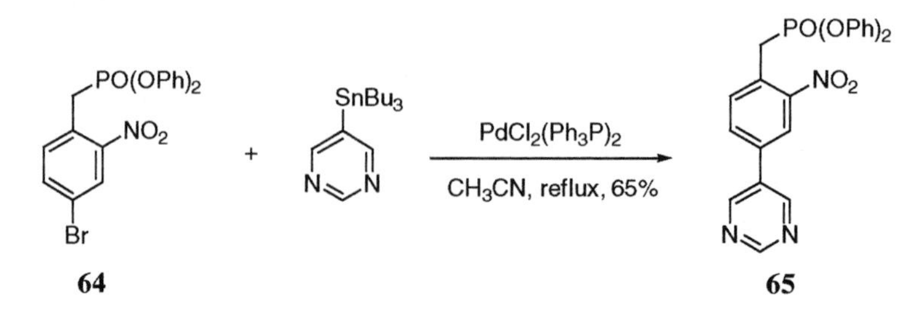


### *11.2.3.b The pyrimidine motif as an electrophile*

In the more prolific aspect of the Stille couplings involving a pyrimidine fragment, pyrimidinyl halides or triflates have been coupled with a variety of stannanes. When there is only one reactive halide on the pyrimidine ring, the reaction outcome is straightforward with no regiochemical concern. The simpler stannanes are vinyl stannanes [35–37]. More complicated variants include stannylquinones [38] and 1-(trialkylsilyloxy)vinyltin [39] as illustrated by the synthesis of **66**.

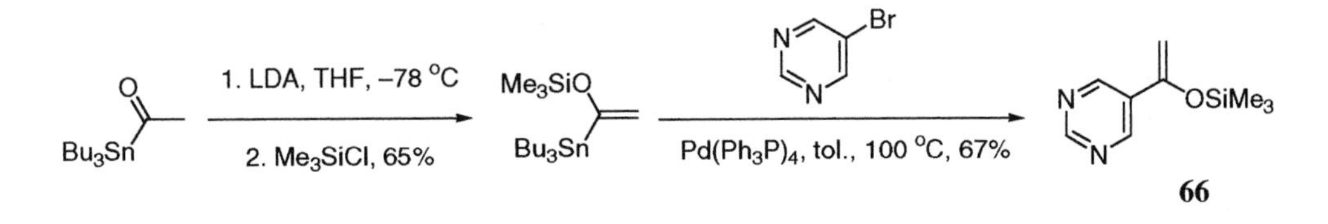


When there is more than one halide on the pyrimidine ring, multiple regioisomeric products can be obtained. The 4(6)-position in pyrimidine is more reactive than the 2-position and regiospecific coupling can be achieved. The reaction of 2,4-dichloropyrimidine and styrylstannane first proceeded regiospecifically at C(4), giving rise to **67**, which was subsequently coupled with phenylstannane under more forcing conditions to afford disubstituted pyrimidine **68** [36].

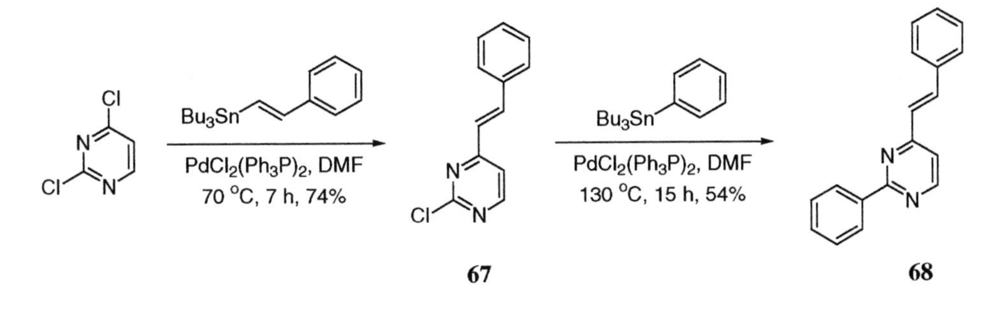


Only bromo- and iodopyrimidine provide sufficient activation for the Stille coupling of a 5-halopyrimidines to take place. Undheim *et al.* discovered that sequential substitution of 5-bromo-2,4-dichloropyrimidine occurred according to the order of reactivity, 4-Cl > 5-Br > 2-Cl. The regio- and chemoselectivity were elegantly demonstrated by stepwise introduction of three different substituents onto 5-bromo-2,4-dichloropyrimidine [33].

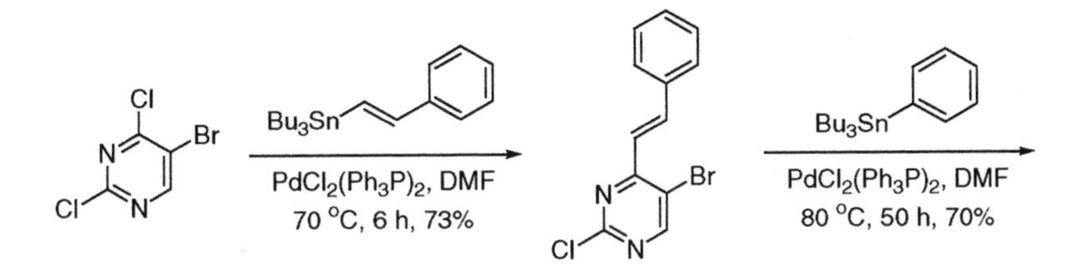

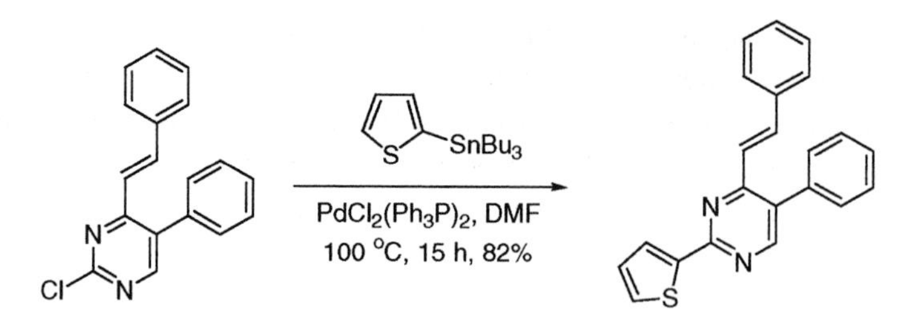


In addition, some Stille adducts have been further manipulated to form condensed heteroaromatic ring systems. The coupling of 4-acetylamino-5-bromopyrimidine **69** and (*E*)-1-ethoxy-2-(tributylstannyl)ethene resulted in (*E*)-4-acetylamino-5-(2-ethoxyethenyl)pyrimidine **70**, which then cyclized under acidic conditions to furnish pyrrolo[2,3-*d*]pyrimidine **71**. Pyrrolo[3,2-*d*]pyrimidines were also synthesized in a similar fashion by using 5-acetylamino-4-bromopyrimidine [40].

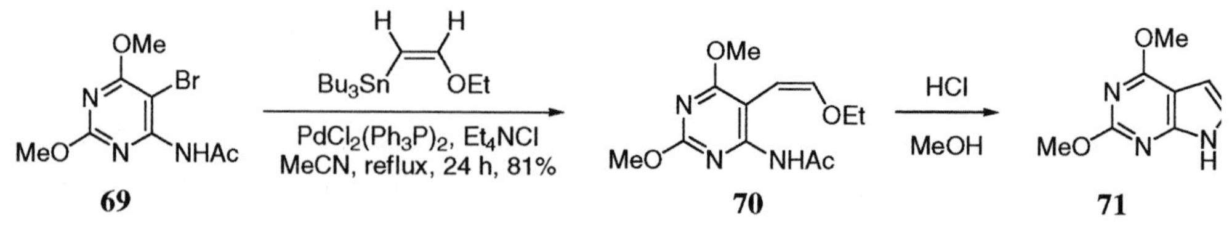


Halopyrimidines also couple with stannanes of heterocycles such as furans [41], azaindoles [42], pyridines [43–46], thiazoles, pyrroles [46] and thiophenes [47]. A representative example is the coupling of 3-tributylstannyl-7-azaindole **72** with 5-bromopyrimidine to furnish heterobiaryl **73** after acidic hydrolysis [42]. Moreover, a selective substitution at the 5-position was achieved when 4-chloro-5-iodopyrimidine **74** was allowed to react with 2-thienylstannane to provide thienylpyrimidine **75** [47].

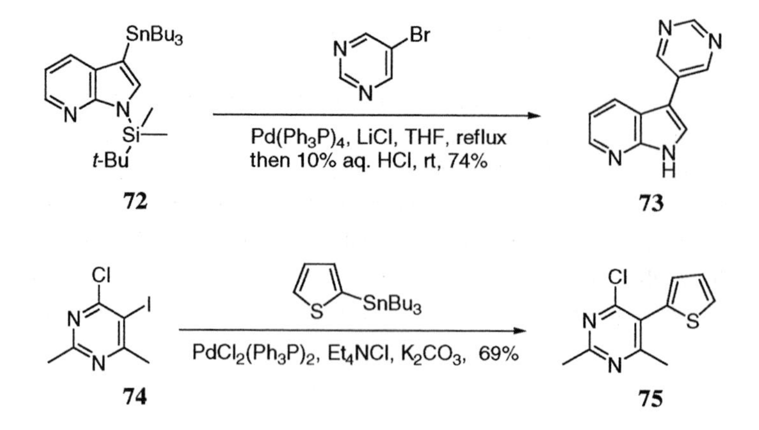

The synthesis of pyrimidinyl thioethers using palladium catalysis poses a synthetic challenge due to the potential multiple coordination of either pyrimidinyl halides or the product to the catalyst, resulting in inhibition of catalysis. Nonetheless, the Pd-catalyzed coupling of organotin sulfides with halopyrimidines was successfully carried out using the Stille coupling [48]. Both 5-bromopyrimidine and activated 2-chloropyrimidine were coupled with phenyltributyltin sulfide to give the expected phenylsufide derivatives.

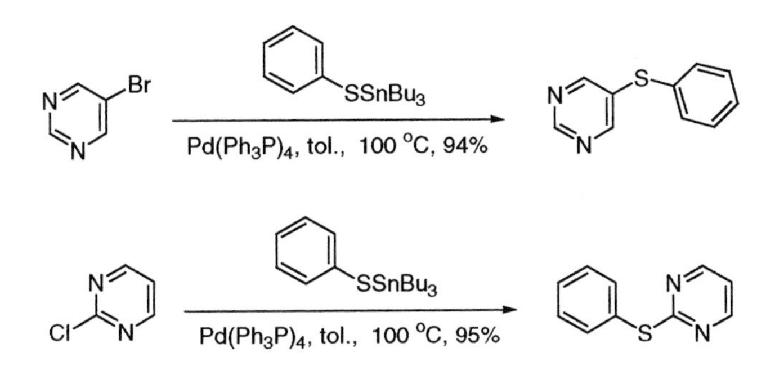


Pyrimidine thioethers may also be synthesized *via* direct Pd-catalyzed C—S bond formation between halopyrimidines and thiolate anions. For very unreactive thiol nucleophiles such as 2-thiopyrimidine, both a strong base and a palladium catalyst are essential. Without a palladium catalyst or replacing *t*-BuONa with $K_2CO_3$, the reaction failed to furnish the desired pyrimidine thioether [49].

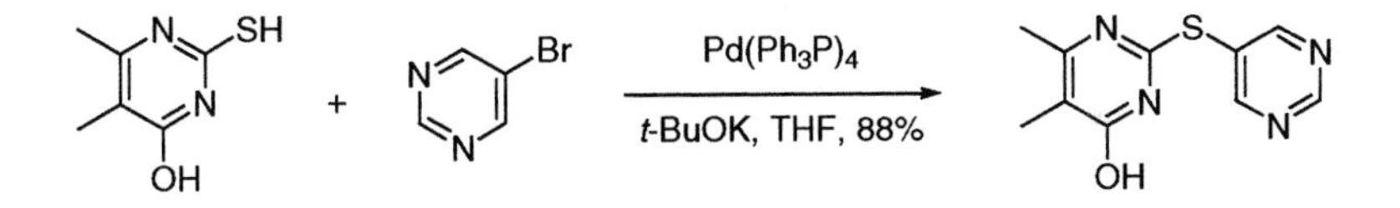


### 11.2.4 Organozirconium and organoaluminum reagents

Many examples exist for Pd-catalyzed cross-couplings of alkenylzirconocenes with simple carbocyclic aryl or alkenyl halides, whereas few precedents are seen for the coupling of alkenylzirconocenes with heteroaryl halides. Undheim and coworkers reported a Pd-catalyzed cross-coupling of 2,4-dichloropyrimidine with alkenylzirconocene [50]. Hydrozirconation of hexyne readily took place at room temperature with zirconocene chloride hydride in benzene. The resulting hexenylzirconocene chloride (**76**) was then coupled with 2,4-dichloropyrimidine at the more electrophilic 4 position, giving rise to 2-chloro-4-[(*E*)-1-hexenyl]pyrimidine (**77**).

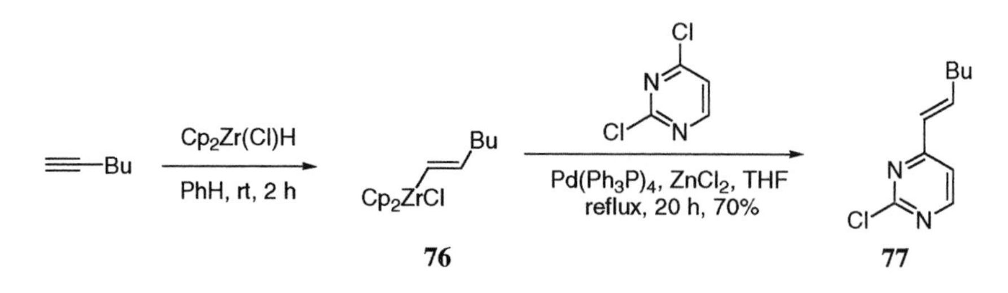


Pd-catalyzed alkylations are generally rare because of the ease with which β-hydride elimination occurs. However, alkylation of 2,4-dichloropyrimidine with trimethylaluminum in the presence of $Pd(Ph_3P)_4$ was achieved without detectable β-hydride elimination. The preference for coupling at the 4-position was maintained for the formation of 2-chloro-4-methylpyrimidine (**78**), which then underwent an additional cross-coupling with triisobutylaluminum to afford 2,4-dialkylpyrimidine **79** [51].

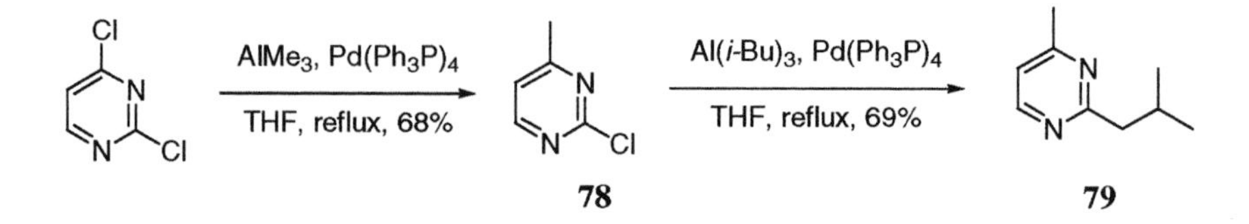


## 11.3 Sonogashira reaction

Activated and deactivated positions in halopyridines exhibit marked difference in reactivity in palladium chemistry, whereas little difference in reactivity was observed among 2-, 4- and 5-positions of halopyrimidines for their Sonogashira reactions [52]. While 2-iodo-4,6-dimethylpyrimidine was the most suitable substrate for preparing internal alkyne **80**, the reaction of either the corresponding bromide or chloride was less efficient [53]. Good to excellent yields were obtained for the preparation of alkynylpyrimidines from most terminal alkynes with the exception of propargyl alcohols. Later reports showed that at elevated temperature (100 °C), the Sonogashira reaction of both 2- and 4-chloropyrimidines with trimethylsilylacetylene proceeded to give, after basic hydrolysis, the corresponding ethynylpyrimidines in 64% and 61% yields, respectively [54].

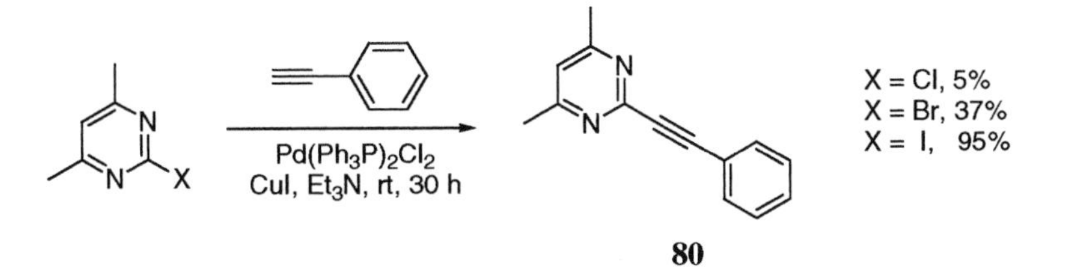


Due to its mild reaction conditions and tolerance of many functional groups, the Sonogashira reaction has been utilized extensively in the coupling of halopyrimidines with a variety of terminal alkynes. Halopyrimidine substrates including 2-iodo [55], 4-iodo- [56], 5-bromo- [57] and 5-iodopyrimidines [58] have been successfully coupled with terminal alkynes.

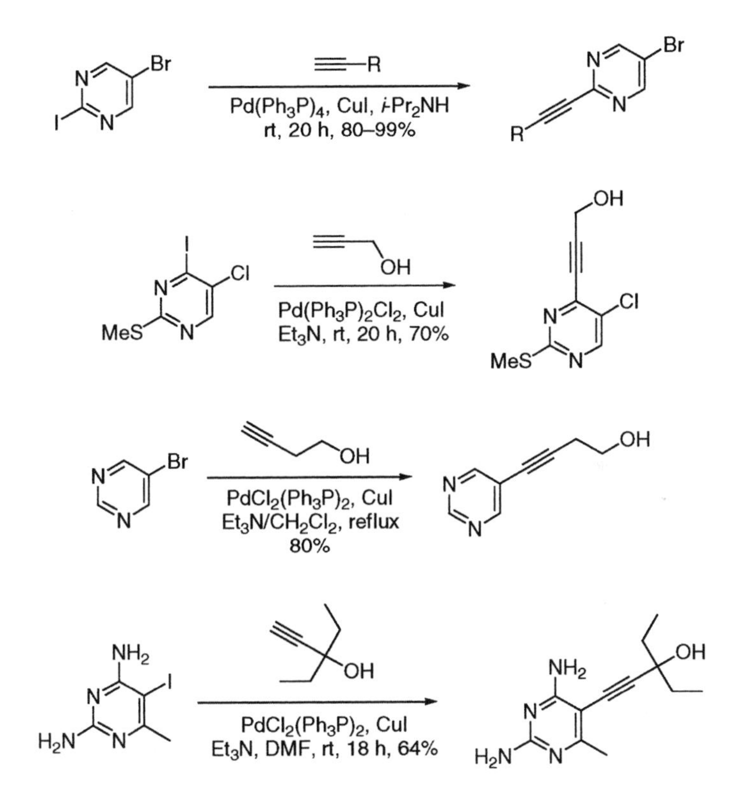


The Sonogashira reaction of 5-bromopyrimidine with *N,N*-dimethylpropargylamine gave aminoalkyne **81** [59], whereas the union of 6-methyl-2-phenyl-4-iodopyrimidine and 3,3,3-triethoxy-1-propyne afforded ester **82** after acidic hydrolysis [60].

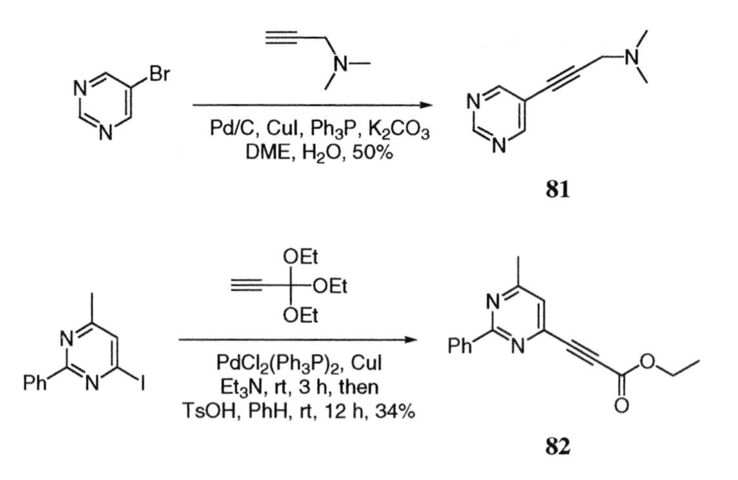


Kim and Russell synthesized 5,6-diethynyl-2,4-dimethoxypyrimidine (**85**) starting from iodination of 5-chloro-2,4-dimethoxypyrimidine [61]. Very careful experimentation resulted in optimal conditions for the Sonogashira reaction of dihalopyrimidine **83** with trimethylsilylacetylene to provide bis-alkyne **84**. The temperature appeared to be crucial. Only mono-substitution for the iodine was observed at lower temperature, whereas Bergman cyclization seemed to occur at temperatures higher than 120 °C. Subsequent desilylation of **84** then delivered diethynylpyrimidine **85**.

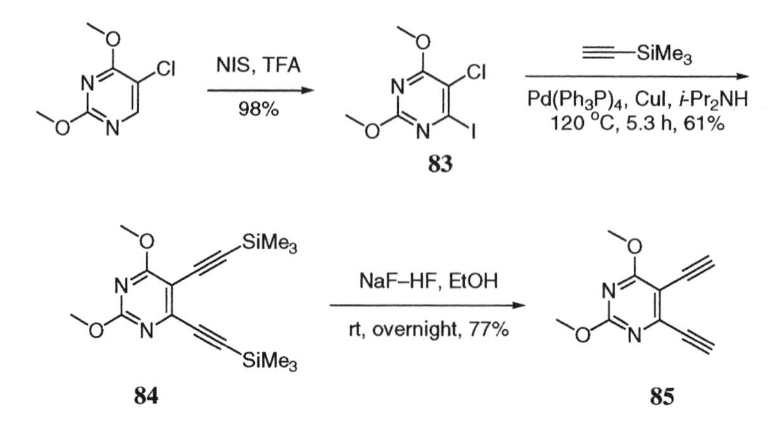


The Sonogashira adducts of halopyrimidines with pendant functional groups are good precursors for synthesizing condensed heteroaromatic compounds. Yamanaka's group prepared four different condensed heteroaromatics from further manipulations of such Sonogashira adducts. Those four condensed heteroaromatics are 5-oxo-7-pyrido[4,3-*d*]pyrimidine **88** [62], furo[2,3-

*d*]pyrimidine **91** [63, 64], thieno[2,3-*d*]pyrimidine **94** [63, 64] and pyrrolo[2,3-*d*]pyrimidine **98** [65–68].

First, 4-chloropyrimidine **86** was treated with phenylacetylene to give alkynylpyrimidine **87** [62]. The fact that the Sonogashira reaction proceeded readily at room temperature may be ascribed to the electron-withdrawing effect of the neighboring ethoxycarbonyl group on **86**. When heated with ammonia in EtOH, alkynylpyrimidine ester **87** cyclized efficiently to produce pyridyl lactam **88**.

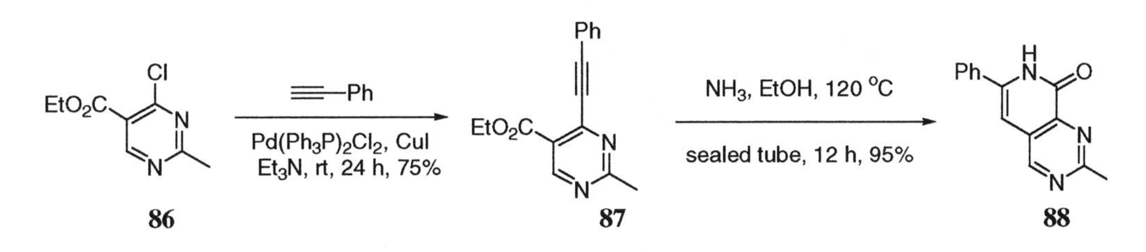


In addition, 2,6-dimethyl-5-iodo-4(3*H*)-pyrimidone (**89**) was allowed to react with phenylacetylene to give internal alkyne **90**, which underwent a spontaneous cyclization to furnish 2,4-dimethyl-6-phenylfuro[2,3-*d*]pyrimidine (**91**) [63, 64].

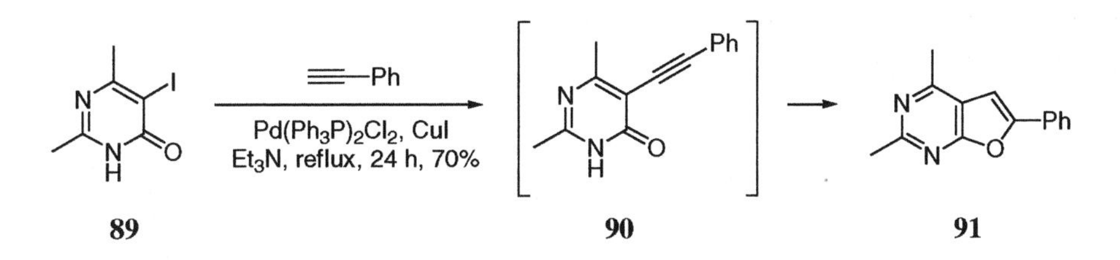


Dehydroxy-halogenation of **89** using $POCl_3$ led to dihalopyrimidine **92**, which was subsequently coupled with phenylacetylene to give 4-chloro-5-alkynylpyrimidine **93** [63, 64]. Subsequent treatment of **93** with sodium hydrosulfide in refluxing ethanol gave 2,4-dimethyl-6-phenylthieno[2,3-*d*]pyrimidine (**94**).

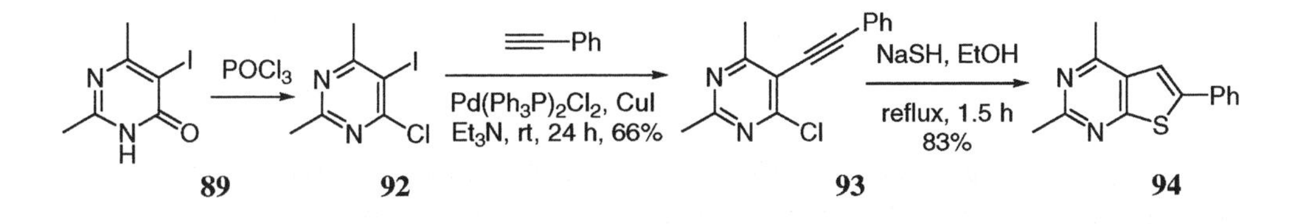


Yamanaka and associates developed a method for the synthesis of 2-butylindole from the Sonogashira adduct of ethyl 2-bromophenylcarbamate and 1-hexyne [65, 66]. Extension of that method to pyridines led to the synthesis of pyrrolopyridines [67]. However, the method was not applicable to the synthesis of pyrrolo[2,3-*d*]pyrimidines. They then developed an alternative route involving an initial $S_NAr$ displacement at the 4-position of 4,5-dihalopyrimidine followed by a Sonogashira coupling at the 5-position [68]. Thus, 5-iodopyrimidine **96** was obtained from an $S_NAr$ displacement at the 4-position of a 4-chloro-5-iodo-2-methylthiopyrimidine (**95**). The subsequent Sonogashira reaction of **96** with trimethylacetylene at 80 °C resulted in adduct **97**, which spontaneously cyclized to pyrrolo[2,3-*d*]pyrimidine **98**.

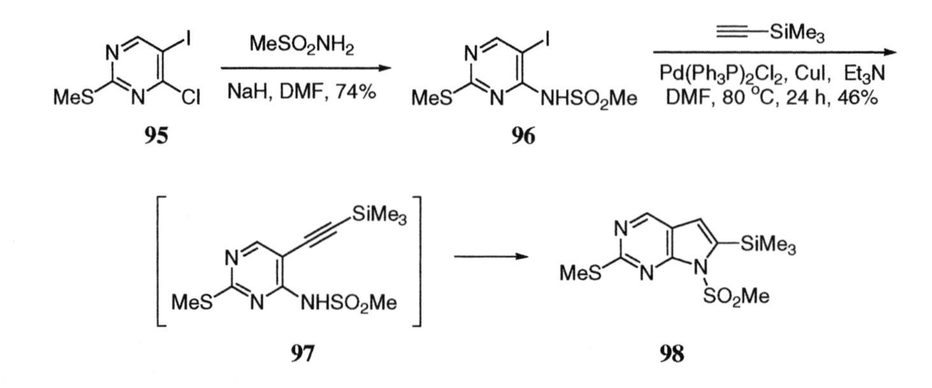


Several Sonogashira adducts of heteroaromatics including some pyridines (see Section 4.3) and pyrimidines underwent an unexpected isomerization [69]. This observed isomerization appeared to be idiosyncratic, and substrate-dependent. The normal Sonogashira adduct **100** was obtained when 2-methylthio-5-iodo-6-methylpyrimidine (**99**) was reacted with but-3-yn-ol, whereas chalcone **101**, derived from isomerization of the normal Sonogashira adduct, was the major product when the reaction was carried out with 1-phenylprop-2-yn-1-ol.

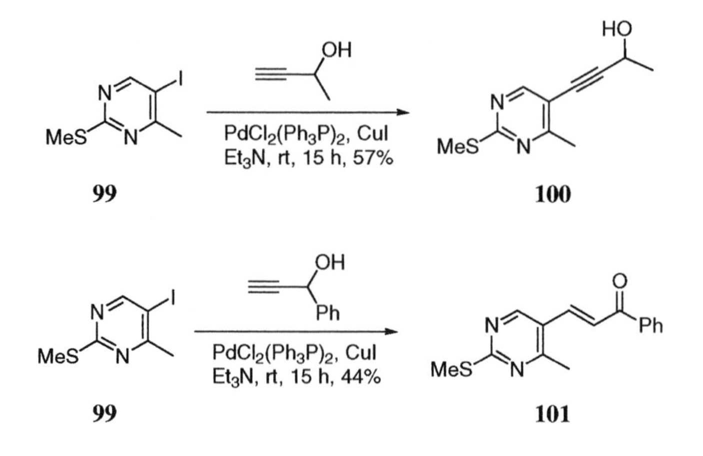

## 11.4 Heck reaction

Analogous to simple carbocyclic aryl halides, 5-halopyrimidines readily take part in Pd-catalyzed olefinations under standard Heck conditions. In a simple case, Yamanaka *et al.* synthesized ethyl 2,4-dimethyl-5-pyrimidineacrylate (**102**) *via* the Heck reaction of 5-iodo-2,4-dimethylpyrimidine and ethyl acrylate [70].

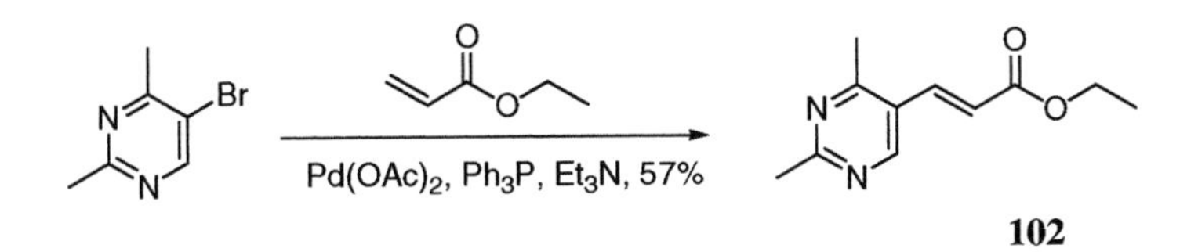


The Heck reaction of 2- or 4(6)-halopyrimidines was less straightforward. Initial attempts with the Heck reaction of 4-iodopyrimidines without a substituent at the 5-position were plagued by homocoupling, giving rise to bis-pyrimidine **103** as the major product [70]. Yamanaka and colleagues later discovered that the homocoupling could be eliminated if the reaction was carried out *in the absence of $Ph_3P$* [71–73]. The same was true for the elimination of $Ph_3P$ from the coupling reaction of 2-iodopyrimidine. Many palladium catalysts such as Pd/C, were found to be effective for these olefinations. If there was a substituent such as I, Br, Cl, EtO, Et on the 5-position of the 4-iodopyrimidines, $Ph_3P$ was still effective as the ligand [74]. For instance, selective olefination of 2-isopropyl-4-iodo-5-bromo-6-methylpyrimidine (**104**) with styrene was achieved to furnish styrylpyrimidine **105**.

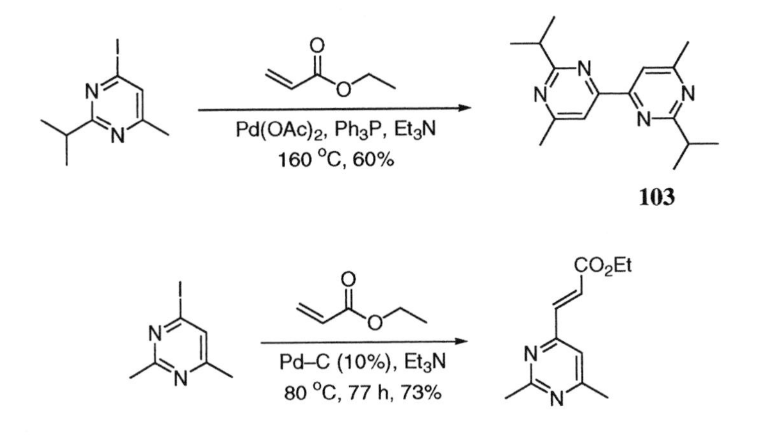

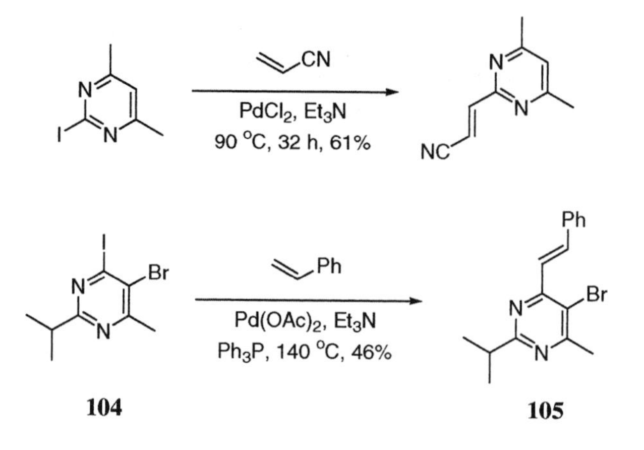


Studies of the Heck reaction of substituted halopyrimidines with methyl vinyl ketone revealed an interesting observation [75–77]: although the reactions of bromopyrimidines gave the usual olefinic substituted product, the reaction of 5-iodopyrimidine **106** with methyl vinyl ketone led to the *addition* of pyrimidine to the double bond, giving rise to substituted 5-(3-oxobutyl)-pyrimidine **107**. Two plausible pathways involve either a radical or an anion intermediate. Both possible mechanisms start with insertion of the initially formed σ-complex **108** to methyl vinyl ketone, leading to intermediate **109**. In pathway a, **109** undergoes a homolytic fission to afford **110**, which abstracts a hydrogen from either triethylamine or excess methyl vinyl ketone, giving rise to **107**. In pathway b, a heterolytic fission occurs, leading to carbanion **111**, which is then protonated by the tertiary ammonium salt generated in the catalytic cycle to give **107**.

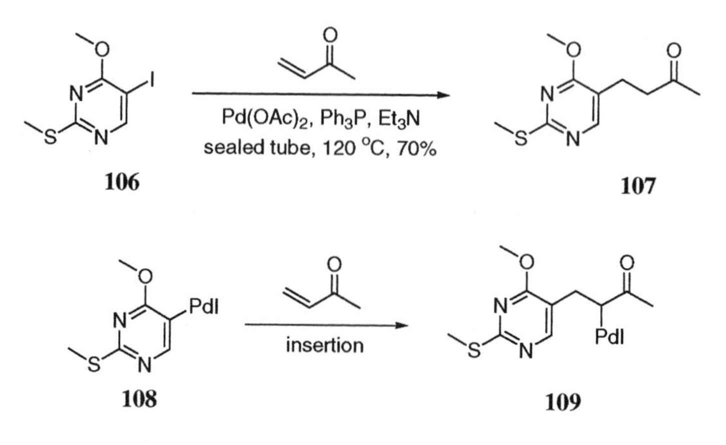

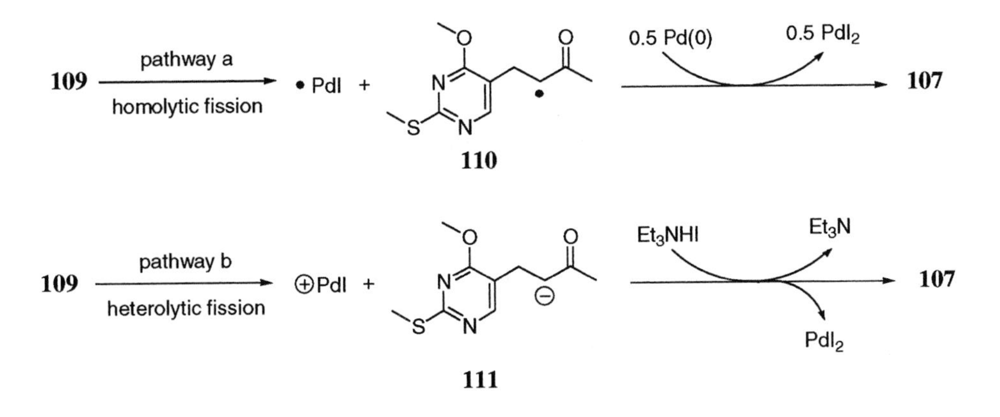


Walker and associates described a heteroaryl Heck reaction of 2,4-dimethoxy-5-iodopyrimidine with thiophene [78]. They found that it was advantageous to carry out the thienylation in the presence of water as opposed to anhydrous conditions. Thienylation of less reactive 2,4-dimethoxy-5-bromopyrimidine gave the product in a lower yield (38%).

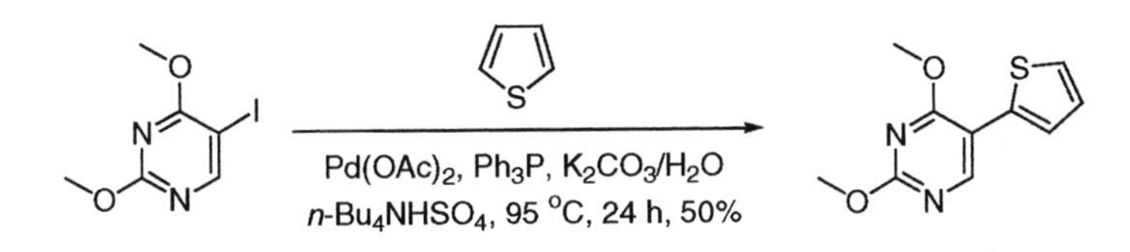


## 11.5 The Carbonylation reaction

Pd-catalyzed alkoxycarbonylation enables synthesis of a variety of heterocyclic esters that are otherwise not easily prepared. 5-Bromopyrimidine was transformed into 5-ethoxycarbonylpyrimidine in quantitative yield employing the Pd-catalyzed alkoxycarbonylation. The alkoxycarbonylation of 2-chloro-4,6-dimethoxypyrimidine, in turn, led to benzyl 4,6-dimethoxypyrimidine-2-carboxylate (**112**), whereas alkoxycarbonylation of 2-(chloromethyl)-4,6-dimethoxypyrimidine provided pyrimidinyl-2-acetate **113** [79]. 4,6-Dimethoxypyrimidines **112** and **113** are both important intermediates for the preparation of antihypertensive and antithrombotic drugs.

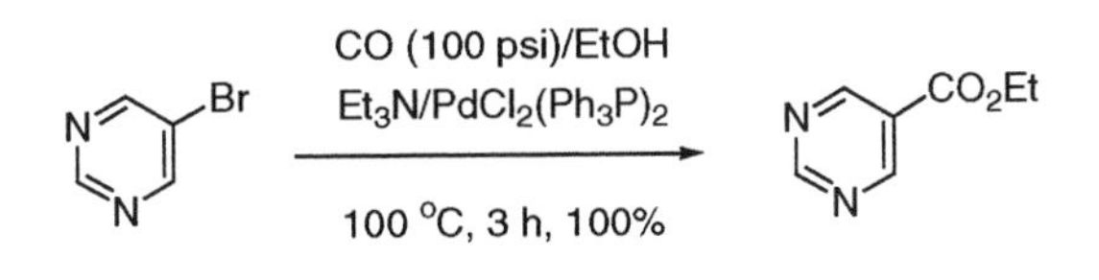

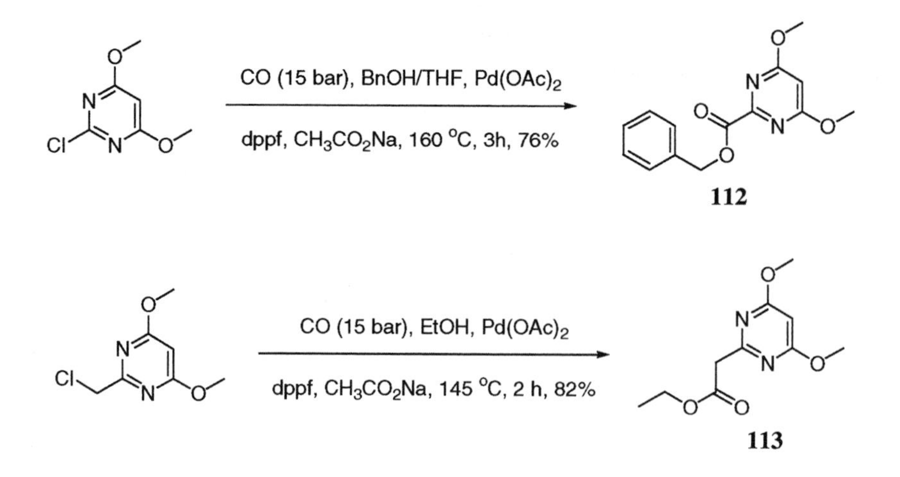


## 11.6 Heteroannulation

To make tryptophan analogs, Gronowitz and coworkers conducted a pyrrole annulation from an aminoiodopyrimidine utilizing the Larock indole synthesis conditions (see Section 1.10.) [80]. They prepared heterocondensed pyrrole **115** by treating 4-amino-5-iodopyrimidine **114** with trimethylsilyl propargyl alcohol under the influence of a palladium catalyst. The regiochemical outcome was governed by steric effects.

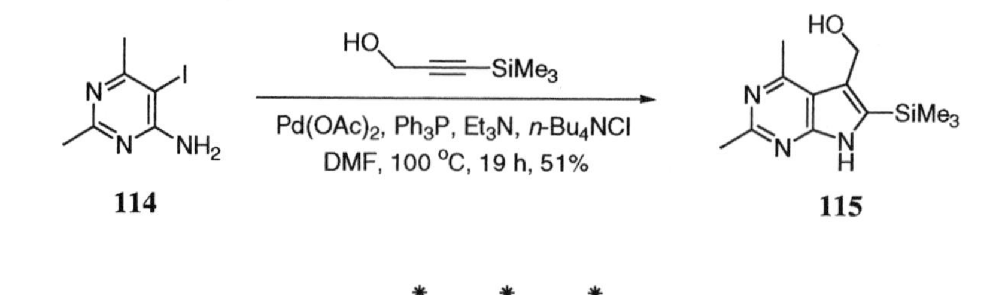


* * *

In conclusion, the Pd chemistry of pyrimidines has its own characteristics when compared to carbocyclic arenes and other nitrogen-containing heterocycles such as pyridine and imidazole. One salient feature of halopyrimidines is that the C(4) and C(6) positions are more activated than C(2). As a result, 2-, 4- and 6-chloropyrimidines are viable substrates for Pd-catalyzed reactions and 4- and 6-chloropyrimidines react more readily than 2-chloropyrimidines. For the Sonogashira reaction, though, there is little difference in the reactivity among 2-, 4- and 5-positions of substituted halopyrimidines. Not only is the Sonogashira reaction a reliable method to make alkynylpyrimidines, the Sonogashira adducts are also good substrates for synthesizing condensed heteroaromatics, as is the Pd-catalyzed heteroannulation strategy.

In choosing a particular cross-coupling reaction, the Stille coupling is the method of choice for two reasons. First, the Stille coupling conditions tolerate a wide variety of functional groups. Second, many pyrimidinylstannanes can be made using sodium or lithium stannanes without going through the intermediacy of an organolithium or organomagnesium species. On the other hand, although both Negishi and Suzuki reactions are less common than the Stille reaction, alkyl groups can be transferred from either alkylzinc or alkylboron reagents under somewhat forcing conditions, giving rise to alkylpyrimidines.

With regard to the Heck reaction, 5-halopyrimidines are the best substrates. The Heck reactions of both 4-halopyrimidines and 2-halopyrimidines are difficult unless a 5-substituent is present. Nonetheless, the Pd-catalyzed alkenylation of both 2- and 4-halopyrimidines proceeds smoothly in the absence of $Ph_3P$.

## 11.7 References


1. Undheim, K.; Benneche, T. *Heterocycles* **1990**, *30*, 1155–93.
2. Undheim, K.; Benneche, T. in *Adv. Heterocycl. Chem.* **1995**, *62*, 305–418.
3. Kalinin, K. N. *Synthesis* **1991**, 413–32.
4. Pews, R. G. *Heterocycles* **1990**, *31,* 109–14.
5. Sato, N.; Takeuchi, R. *Synthesis* **1990**, 659–60.
6. Falck-Pedersoen, M. L.; Benneche, T.; Undheim, K. *Acta Chem. Scand., Sec. B* **1989**, *B43*, 251–8.
7. Caton, P. L.; Grant, M. S.; Pain, D. L.; Slack, R. *J. Chem. Soc.* **1965**, 5467–73.
8. Jones, M. L.; Baccanari, D. P.; Tansik, R. L.; Boytos, C. M.; Rudolph, S. K.; Kuyper, L. F. *J. Heterocycl. Chem.* **1999**, *36*, 145–8.
9. Bhatt, R. S.; Kundu, N. G.; Chwang, T. L.; Heidelberger, C. *J. Heterocycl. Chem.* **1981**, *18*, 771–4.
10. Edo, K.; Sakamoto, T.; Yamanaka, H. *Heterocycles* **1979**, *12*, 383–6.
11. Solberg, J.; Undheim, K. *Acta Chem. Scand., Sec. B* **1986**, *B40*, 381–6.
12. Sugimoto, O.; Mori, M.; Tanji, K.-i. *Tetrahedron Lett.* **1999**, *40*, 7477–8.
13. Sandosham, J.; Undheim, K.; Rise, F. *Heterocycles* **1993**, *35*, 235–44.
14. Sandosham, J.; Undheim, K. *Heterocycles* **1994**, *37*, 501–14.
15. D'Alarcao, M.; Bakthavachalam, V.; Leonard, N. J. *J. Org. Chem.* **1985**, *50*, 2456–61.
16. Yamanaka, H.; An-Naka, M.; Kondo, Y.; Sakamoto, T. *Chem. Pharm. Bull.* **1985**, *33*, 4309–13.
17. Turck, A.; Plé, N.; Leprétre-Gaquére, A.; Quéguiner, G. *Heterocycles* **1998**, *49*, 205–14.
18. Sandosham, J.; Undheim, K. *Heterocycles* **1993**, *35*, 235–44.

19. Sandosham, J.; Undheim, K. *Heterocycles* **1994**, *37*, 501–14.
20. Gronowitz, S.; Hörnfeldt, A.-B.; Musil, T. *Chem. Scr.* **1986**, *26*, 305–9.
21. Peters, D.; Hörnfeldt, A.-B.; Gronowitz, S. *J. Heterocycl. Chem.* **1990**, *27*, 2165–73.
22. Wellmar, U.; Hörnfeldt, A.-B.; Gronowitz, S. *J. Heterocycl. Chem.* **1995**, *32*, 1159–63.
23. Ishikura, M.; Kamada, M.; Terashima, M. *Synthesis* **1984**, 936–8.
24. Ishikura, M.; Ohta, T.; Terashima, M. *Chem. Pharm. Bull.* **1985**, *33*, 4755–63.
25. Yang, Y.; Martin, A. R. *Heterocycles* **1992**, *34*, 1395–8.
26. Mitchell, M. B.; Walbank, P. J. *Tetrahedron Lett.* **1991**, *32*, 2273–6.
27. Ali, N. M.; McKillop, A.; Mitchell, M. B.; Rebelo, R. A.; Walbank, P. J. *Tetrahedron* **1992**, *48*, 8117–26.
28. Satoh, Y.; Gude, C.; Chan, K.; Firooznia, F. *Tetrahedron Lett.* **1997**, *38*, 7645–8.
29. Sandosham, J.; Benneche, T.; Moeller, B. S.; Undheim, K. *Acta Chem. Scand., Ser. B* **1988**, *B42*, 455–61.
30. Arukwe, J.; Benneche, T.; Undheim, K. *J. Chem. Soc., Perkin Trans. 1* **1989**, 255–9.
31. Sandosham, J.; Undheim, K. *Tetrahedron* **1994**, *50*, 275–84.
32. Sandosham, J.; Undheim, K. *Acta Chem. Scand.* **1989**, *43*, 684–9.
33. Majeed, A. J.; Antonsen, O.; Benneche, T.; Undheim, K. *Tetrahedron* **1989**, *45*, 993–1006.
34. Kennedy, G.; Perboni, A. D. *Tetrahedron Lett.* **1996**, *37*, 7611–4.
35. Solberg, J.; Undheim, K. *Acta Chem. Scand.* **1987**, *B41*, 712–6.
36. Sandosham, J.; Undheim, K. *Acta Chem. Scand.* **1989**, *43*, 62–8.
37. Benneche, T. *Acta Chem. Scand.* **1990**, *44*, 927–31.
38. Liebeskind, L. S.; Riesinger, S. W. *J. Org. Chem.* **1993**, *58*, 408–13.
39. Verlhac, J.-B.; Pereyre, M.; Shin, H. *Organometallics*, **1991**, *10*, 3007–9.
40. Sakamoto, T.; Satoh, C.; Kondo, Y.; Yamanaka, H. *Chem. Pharm. Bull.* **1993**, *41*, 81–6.
41. Sandosham, J.; Undheim, K. *Acta Chem. Scand.* **1994**, *48*, 279–82.
42. Alvarez, M.; Fernández, D.; Joule, J. A. *Synthesis* **1999**, 615–20.
43. Gros, P.; Fort, Y. *Synthesis* **1999**, 754–756.
44. Schubert, U. S.; Eschbaumer, C. *Org. Lett.* **1999**, *1*, 1027–9.
45. Gronowitz, S.; Hörnfeldt, A.-B.; Musil, T. *Chem. Scr.* **1986**, *26*, 305–9.
46. Peters, D.; Hörnfeldt, A.-B.; Gronowitz, S. *J. Heterocycl. Chem.* **1990**, *27*, 2165–73.
47. Kondo, Y.; Watanabe, R.; Sakamoto, T.; Yamanaka, H. *Chem. Pharm. Bull.* **1989**, *37*, 2933–6.
48. Chen, J.; Crisp, G. T. *Synth. Commun.* **1992**, *22*, 683–6.
49. Harr, M. S.; Presley, A. L.; Thoraresen, A. *Synlett* **1999**, 1579–81.
50. Mangalagiu, I.; Benneche, T.; Undheim, K. *Acta Chem. Scand.* **1996**, *50*, 914–7.
51. Lu, Q.; Mangalagiu, I.; Benneche, T.; Undheim, K. *Acta Chem. Scand.* **1997**, *51*, 302–6.
52. Edo, K.; Yamanaka, H.; Sakamoto, T. *Heterocycles* **1978**, *9*, 271–4.
53. Edo, K.; Sakamoto, T.; Yamanaka, H. *Chem. Pharm. Bull.* **1978**, *26*, 3843–50.

54. Sakamoto, T.; Shirawa, M.; Kondo, Y.; Yamanaka, H. *Synthesis* **1983**, 312–4.
55. Shibata, T.; Yonekubo, S.; Soai, K. *Angew. Chem. Int. Ed.* **1999**, *38*, 659–61.
56. Solberg, J.; Undheim, K. *Acta Chem. Scand.* **1986**, *B40*, 381–6.
57. Tilley, J. W.; Levitan, P.; Lind, J.; Welton, A. F.; Crowley, H. J.; Tobias, L. D.; O'Donnell, M. *J. Med. Chem.* **1987**, *30*, 185–93.
58. Jones, M. L.; Baccanari, D. P.; Tansik, R. L.; Boytos, C. M.; Rudolph, S. K.; Kuyper, L. F. *J. Heterocycl. Chem.* **1999**, *36*, 145–8.
59. Bleicher, L. S.; Cosford, N. D. P.; Herbaut, A.; McCallum, J. S.; McDonald, I. A. *J. Org. Chem.* **1998**, *63*, 1109–18.
60. Sakamoto, T.; Shiga, F.; Yasuhara, A.; Uchiyama, D.; Kondo, Y.; Yamanaka, H. *Synthesis* **1992**, 746–8.
61. Kim, C.-S.; Russell, K. C. *J. Org. Chem.* **1998**, *63*, 8229–34.
62. Sakamoto, T.; Kondo, Y.; Yamanaka, H. *Chem. Pharm. Bull.* **1982**, *30*, 2410–6.
63. Sakamoto, T.; Kondo, Y.; Yamanaka, H. *Chem. Pharm. Bull.* **1982**, *30*, 2417–20.
64. Sakamoto, T.; Kondo, Y.; Watanabe, R.; Yamanaka, H. *Chem. Pharm. Bull.* **1986**, *34*, 2719–24.
65. Sakamoto, T.; Kondo, Y.; Yamanaka, H. *Heterocycles* **1986**, *24*, 31–2.
66. Sakamoto, T.; Kondo, Y.; Iwashita, S.; Yamanaka, H. *Chem. Pharm. Bull.* **1987**, *35*, 1823–8.
67. Sakamoto, T.; Kondo, Y.; Iwashita, S.; Nagano, T.; Yamanaka, H. *Chem. Pharm. Bull.* **1988**, *36*, 1305–8.
68. Kondo, Y.; Watanabe, R.; Sakamoto, T.; Yamanaka, H. *Chem. Pharm. Bull.* **1989**, *37*, 2933–6.
69. Minn, K. *Synlett* **1991**, 115–6.
70. Edo, K.; Sakamoto, T.; Yamanaka, H. *Chem. Pharm. Bull.* **1979**, *27*, 193–7.
71. Sakamoto, T.; Arakida, H.; Edo, K.; Yamanaka, H. *Heterocycles* **1981**, *16*, 965–8.
72. Sakamoto, T.; Kondo, Y.; Yamanaka, H. *Chem. Pharm. Bull.* **1982**, *30*, 2417–20.
73. Sakamoto, T.; Arakida, H.; Edo, K.; Yamanaka, H. *Chem. Pharm. Bull.* **1982**, *30*, 3647–56.
74. Wada, A.; Yamamoto, J.; Hase, T.; Nagai, S.; Kanatomo, S. *Synthesis* **1986**, 555–6.
75. Wada, A.; Yasuda, H.; Kanatomo, S. *Synthesis* **1988**, 771–5.
76. Wada, A.; Ohki, K.; Nagai, S.; Kanatomo, S. *J. Heterocycl. Chem.* **1991**, *28*, 509–12.
77. Basnak, I.; Takatori, S.; Walker, R. T. *Tetrahedron Lett.* **1997**, *38*, 4869–72.
78. Head, R. A.; Ibbotson, A. *Tetrahedron Lett.* **1984**, *25*, 5939–42.
79. Bessard, Y.; Crettaz, R. *Tetrahedron* **1999**, *55*, 405–12.
80. Wensbo, D.; Eriksson, A.; Jeschke, T.; Annby, U.; Gronowitz, S. *Tetrahedron Lett.* **1993**, *34*, 2823–6.

# Subject Index

**Q**



**R**

**W**



**Y**



**Z**